Some Problems on Nonlinear Hyperbolic Equations and Applications

Series in Contemporary Applied Mathematics CAM

---

1 Jiongmin Yong, Rama Cont, eds., *Mathematical Finance: Theory and Practice* (2000)
2 F. Dubois, Huamo Wu, eds., *New Advances in Computational Fluid Dynamics: Theory, Methods and Applications* (2001)
3 Hanji Shang, Alain Tosseti, eds., *Actuarial Science: Theory and Practice* (2002)
4 Alexandre Ern, Weiping Liu, eds., *Mathematical Problems in Environmental Science and Engineering* (2002)
5 Haïm Brezis, Tatsien Li, eds., *Ginzburg-Landau Vortices* (2005)
6 Tatsien Li, Pingwen Zhang, eds., *Frontiers and Prospects of Contemporary Applied Mathematics* (2006)
7 Deguan Wang, Christian Duquennoi, Alexandre Ern, eds., *Mathematical Methods for Surface and Subsurface Hydrosystems* (2007)
8 Rolf Jeltsch, Tatsien Li, Ian H. Sloan, eds., *Some Topics in Industrial and Applied Mathematics* (2007)
9 Philippe G. Ciarlet, Tatsien Li, eds., *Differential Geometry: Theory and Applications* (2008)
10 Tatsien Li, Pingwen Zhang, eds., *Industrial and Applied Mathematics in China* (2009)
11 Zhien Ma, Yicang Zhou, Jianhong Wu, eds., *Modeling and Dynamics of Infectious Diseases* (2009)
12 Tomas Y. Hou, Chun Liu, Jianguo Liu, eds., *Multi-Scale Phenomena in Complex Fluids: Modeling, Analysis and Numerical Simulation* (2010)
13 Guiqiang Chen, Tatsien Li, Chun Liu, eds., *Nonlinear Conservation Laws, Fluid Systems and Related Topics* (2010)
14 Alain Damlamian, Stéphane Jaffard, eds., *Wavelet Methods in Mathematical Analysis and Engineering* (2010)
15 Tatsien Li, Yuejun Peng, Bopeng Rao, eds., *Some Problems on Nonlinear Hyperbolic Equations and Applications* (2010)

Series in Contemporary Applied Mathematics CAM 15

Some Problems on Nonlinear Hyperbolic Equations and Applications

editors

Tatsien Li

Fudan University, China

Yuejun Peng

Université Blaise Pascal, France

Bopeng Rao

Université de Strasbourg, France

Higher Education Press

World Scientific

NEW JERSEY • LONDON • SINGAPORE • BEIJING • SHANGHAI • HONG KONG • TAIPEI • CHENNAI

Tatsien Li
School of Mathematical Sciences
Fudan University
Shanghai 200433, China
Email: dqli@fudan.edu.cn

Yuejun Peng
Laboratoire de Mathématiques
Université Blaise Pascal
63177 Aubière Cedex, France
Email: peng@math.univ-bpclermont.fr

Bopeng Rao
Institut de Recherche Mathématique Avancée
Université de Strasbourg
67084 Strasbourg, France
Email: rao@math.u-strasbg.fr

Editorial Assistant: Chunlian Zhou

Higher Education Press
4 Dewai Dajie, Beijing 100120, P. R. China and
World Scientific Publishing Co Pte Ltd
5 Toh Tuch Link, Singapore 596224

ISBN 13: 978-981-4322-88-1
ISBN 10: 981-4322-88-1

Printed in P. R. China

Preface

This book is a collection of manuscripts from lectures given in the French-Chinese Summer Institute on Applied Mathematics, which was held at the School of Mathematical Sciences of Fudan University from September 1 to September 21, 2008. This Institute was mainly sponsored by the Centre National de Recherche Scientifique (CNRS) and the National Natural Science Foundation of China (NSFC). The activities were organized by the Institut Sino-Français de Mathématiques Appliquées (ISFMA). There were more than 70 participants, including graduate students, postdoctors and junior faculty members from universities and research institutions in China and France.

This volume is entitled *Some Problems on Nonlinear Hyperbolic Equations and Applications*. The volume is composed of two parts: Mathematical and Numerical Analysis for Strongly Nonlinear Plasma Models and Exact Controllability and Observability for Quasilinear Hyperbolic Systems and Applications, which represent two subjects of the Institute. These topics are important not only for industrial applications but also for the theory of partial differential equations itself.

The main propose of the Institute was to present recent progress and results obtained in the domains related to both subjects and to organize discussions for studying important problems by sustainable collaborations. We hope that this experience will be useful for the activities of the French-Chinese collaboration in the future.

During the activities of the Institute, more than 30 lectures of 50 minutes each were delivered. The speakers gave their presentation without attaching much importance to the details of proofs but rather to difficulties encountered, to open problems and possible ways to be exploited. Each lecture was followed by a free discussion of 30 minutes, so that the participants were able to clarify the situation of each problem and to find interesting subjects to be cooperated in the future. Three mini-courses of 3×1.5 hours each were given by Jean-Michel Coron (Université Paris 6, France), Vilmos Komornik (Université Louis Pasteur de Strasbourg, France) on the control theory and by Thierry Goudon (INRIA Lille-Nord Europe, France) on the mathematical theory for plasmas. The mini-course notes were prepared for all the students before the activities of the Institute. Moreover, in the middle and before the end of the Institute, we organized two sessions of general discussion on the open problems for future investigations by collaboration.

The editors would like to express their sincere thanks to all the authors in this volume for their contributions and to all the participants in the Summer Institute. Liqiang Lu, Zhiqiang Wang and Chunlian Zhou deserve our special thanks for their prompt and effective assistance to make the Institute run smoothly. The editors are grateful to the Centre National de Recherche Scientifique (CNRS), the Consulate General of France in Shanghai, the French Embassy in Beijing, the Institut Sino-Français de Mathématiques Appliquées (ISFMA), the National Natural Science Foundation of China (NSFC) and the School of Mathematical Sciences of Fudan University for their help and support. Finally, the editors wish to thank Tianfu Zhao (Senior Editor, Higher Education Press) and Chunlian Zhou for their patience and professional assistance.

Tatsien Li, Yuejun Peng, Bopeng Rao

April 2010

Contents

Part I

Mathematical and Numerical Analyses of Strongly Nonlinear Plasma Models

Open Boundary Conditions and Computational Schemes for Schrödinger Equations with General Potentials and Nonlinearities

Xavier Antoine, Pauline Klein
Institut Elie Cartan Nancy, Nancy-Université, CNRS UMR 7502
INRIA CORIDA Team
Boulevard des Aiguillettes B.P. 239
F-54506 Vandoeuvre-lès-Nancy, France
Email: Xavier.Antoine@iecn.u-nancy.fr

Christophe Besse
Project-Team SIMPAF, INRIA Lille Nord Europe Research Centre
Laboratoire Paul Painlevé, CNRS UMR 8524
Université des Sciences et Technologies de Lille
Cité Scientifique, 59655 Villeneuve d'Ascq Cedex, France
Email: Christophe.Besse@math.univ-lille1.fr

Abstract

This paper addresses the construction of absorbing boundary conditions for the one-dimensional Schrödinger equation with a general variable repulsive potential or with a cubic nonlinearity. Semi-discrete time schemes, based on Crank-Nicolson approximations, are built for the associated initial boundary value problems. Finally, some numerical simulations give a comparison of the various absorbing boundary conditions to analyse their accuracy and efficiency.

1 Introduction

We consider in this paper two kinds of initial value problems. The first one consists in a time-dependent Schrödinger equation with potential V set in an unbounded domain

$$\begin{cases} i\partial_t u + \partial_x^2 u + V\, u = 0, & (x,t) \in \mathbb{R} \times [0;T], \\ u(x,0) = u_0(x), & x \in \mathbb{R}, \end{cases} \tag{1.1}$$

where u_0 presents the initial data. The maximal time of computation is denoted by T. We assume in this article that V is a real-valued potential such that $V \in \mathcal{C}^\infty(\mathbb{R} \times \mathbb{R}^+, \mathbb{R})$. This kind of potential then creates acceleration of the field compared to the free-potential equation [10, 17].

Our second interest concerns the one-dimensional cubic nonlinear Schrödinger equation

$$\begin{cases} i\partial_t u + \partial_x^2 u + q\,|u|^2\, u = 0, & (x,t) \in \mathbb{R} \times [0;T], \\ u(x,0) = u_0(x), & x \in \mathbb{R}, \end{cases} \tag{1.2}$$

where the real parameter q corresponds to a focusing ($q > 0$) or defocusing ($q < 0$) effect of the cubic nonlinearity. This equation has the property to possess special solutions which propagate without dispersion, the so-called solitons.

For obvious reasons linked to the numerical solution of such problems, it is usual to truncate the spatial computational domain with a fictitious boundary $\Sigma := \partial\Omega = \{x_l, x_r\}$, where x_l and x_r respectively designate the left and right endpoints introduced to have a bounded domain of computation $\Omega =]x_l; x_r[$. Let us define the time domains $\Omega_T = \Omega \times [0;T]$ and $\Sigma_T = \Sigma \times [0;T]$. Considering the fictitious boundary Σ, we are now led to solve the problem

$$\begin{cases} i\partial_t u + \partial_x^2 u + \mathscr{V}\, u = 0, & (x,t) \in \Omega_T, \\ u(x,0) = u_0(x), & x \in \Omega, \end{cases} \tag{1.3}$$

where $\mathscr{V}$ denotes either the real potential $V(x,t)$ or the cubic nonlinearity $q\,|u|^2\,(x,t)$. In the sequel of the paper, we assume that the initial datum u_0 is compactly supported in the computational domain Ω.

Of course, a boundary condition set on Σ_T must be added to systems (1.3). An ideal exact boundary condition tackling the problem is the so-called Transparent Boundary Condition (TBC) which leads to a solution of (1.3) equal to the restriction of the solution of (1.1) or (1.2) on Ω_T. A first well-known case considers $\mathscr{V} = 0$. This situation has been treated by many authors [2]. In this case, according to what is precisely described in Section 2.2, we are able to build the following TBC in terms of the Dirichlet-to-Neumann (DtN) operator

$$\partial_{\mathbf{n}} u + e^{-i\pi/4}\partial_t^{1/2} u = 0, \qquad \text{on } \Sigma_T, \tag{1.4}$$

where $\mathbf{n}$ is the outwardly directed unit normal vector to Σ. The operator $\partial_t^{1/2}$ is known as the half-order derivative operator (see Eq. (2.7) for its definition). Its nonlocal character related to its convolutional structure has led to many developments concerning its accurate and efficient evaluation in the background of TBCs [2].

A second situation which is related to the above case is when the potential is only time varying: $\mathscr{V} = V(x,t) = V(t)$. In this case, the change of unknown

$$v(x,t) = e^{-i\mathcal{V}(t)}u(x,t), \tag{1.5}$$

with

$$\mathcal{V}(t) = \int_0^t V(s)\,ds \tag{1.6}$$

reduces the initial Schrödinger equation with potential to the free-potential Schrödinger equation [4]. Then, the TBC (1.4) can be used for v and the resulting DtN TBC for u is

$$\partial_{\mathbf{n}}u(x,t) + e^{-i\pi/4}e^{i\mathcal{V}(t)}\,\partial_t^{1/2}\left(e^{-i\mathcal{V}(t)}u(x,t)\right) = 0, \qquad \text{on } \Sigma_T. \tag{1.7}$$

This change of variables is fundamental and, coupled to a factorization theorem, and allowed to derive accurate approximations of the TBC, which are usually called artificial or Absorbing Boundary Conditions (ABCs), when $\mathscr{V} = V(x,t)$ [5] and $\mathscr{V} = q\,|u|^2$ [4]. Families of ABCs can be computed and are classified following their degree of accuracy. Typically, for a general function $\mathscr{V}$, the first ABC would be exactly (1.7), where $\mathcal{V}(t)$ has to be replaced by $\mathcal{V}(x,t) = \int_0^t \mathscr{V}(x,s)ds$. The ABC gives quite satisfactory accurate results but its evaluation remains costly since it involves the nonlocal time operator $\partial_t^{1/2}$. In [5], another kind of ABCs was introduced, with their numerical treatments being based on Padé approximants. It therefore gives rise to a local approximation scheme which is very competitive.

The aim of the present paper is to present precisely the link between the two different types of ABCs set up in [5] and [4] and to extend the local ABC derived for $\mathscr{V} = V(x,t)$ to the cubic nonlinear Schrödinger equation. Moreover, associated unconditionally stable schemes are given and numerical results are reported.

For completeness, we must mention that recent attempts have been directed towards the derivation of TBCs for special potentials. In [15], the case of a linear potential is considered in the background of parabolic equations in electromagnetism. Using the Airy functions, the TBC can still be written and its accuracy is tested. In [27], Zheng derives the TBC in the special case of a sinusoidal potential using Floquet's theory. All these solutions take care of the very special form of the potential. Let us remark that other solutions based on PML techniques have also been applied (see [26]). Concerning the nonlinear case, using paradifferential operators techniques, Szeftel [24] presented other kinds of ABCs. Moreover, a recent paper [6] gives a comprehensive review of current developments related to the derivation of artificial boundary conditions for nonlinear partial differential equations following various approaches.

The present paper is organized as follows. In Section 2, we recall the derivation of open boundary conditions for linear Schrödinger equations. Subsection 2.1 concerns the derivation of the TBC, and Subsection 2.2 gives some possible extensions and their interpretations in the context of pseudodifferential calculus. This tool is the essential ingredient used in Section 3 where two possible approaches for building ABCs for the one-dimensional Schrödinger equation with a variable repulsive potential are given. Section 4 is devoted to their numerical discretization and the underlying properties of the proposed schemes. Section 5 is concerned with the nonlinear case in which we explain the links between the different approaches and propose a new family of ABCs for the cubic nonlinear Schrödinger equation. Numerical schemes are also analysed. Section 6 presents some numerical computations. These simulations show the high accuracy and efficiency of the proposed ABCs. Moreover, comparisons are made between the different approaches. Finally, a conclusion is given in Section 7.

2 Open boundary conditions for linear Schrödinger equations

2.1 The constant coefficients case: derivation of the TBC

We recall in this Section the standard derivation of the Transparent Boundary Condition (TBC) in the context of the following 1D Schrödinger equation

$$\begin{aligned} & i\partial_t u + \partial_x^2 u + V(x,t)u = 0, \quad (x,t) \in \Omega_T, \\ & \lim_{|x|\to\infty} u(x,t) = 0, \\ & u(x,0) = u_0(x), \quad x \in \Omega, \end{aligned} \tag{2.1}$$

where the initial datum u_0 is compactly supported in Ω and the given real potential V is zero outside Ω. It is well known that the previous equation (2.1) is well posed in $L^2(\mathbb{R})$ (see e.g. [22, 23]) and that the "density" is time preserved, i.e., $\|u(t)\|_{L^2(\mathbb{R})} = \|u_0\|_{L^2(\mathbb{R})}$, $\forall t \geq 0$. The TBC for the Schrödinger equation (2.1) was independently derived by several authors from various application fields [20, 21, 8, 11, 13]. Such a TBC is nonlocal according to the time variable t and connects the Neumann datum $\partial_x v(x_{l,r}, t)$ to the Dirichlet one $v(x_{l,r}, t)$. As a Dirichlet-to-Neumann (DtN) map, it reads

$$\partial_{\mathbf{n}} v(x,t) = -\frac{e^{-i\pi/4}}{\sqrt{\pi}} \frac{d}{dt} \int_0^t \frac{v(x,\tau)}{\sqrt{t-\tau}} \, d\tau \quad \text{on } \Sigma_T, \tag{2.2}$$

where $\partial_{\mathbf{n}}$ is the outwardly directed unit normal derivative to Ω.

The derivation of the TBC (2.2) is performed from Equation (2.1) and is based on the decomposition of the Hilbert space $L^2(\mathbb{R})$ as $L^2(\Omega) \oplus L^2(\Omega_r \cup \Omega_l)$ where $\Omega =]x_l, x_r[$, $\Omega_l =]-\infty, x_l]$, and $\Omega_r = [x_r, \infty[$. Equation (2.1) is equivalent to the coupled system of equations

$$\begin{cases} (i\partial_t + \partial_x^2)v = -V(x,t)v, \ (x,t) \in \Omega_T, \\ \partial_x v(x,t) = \partial_x w(x,t), \ (x,t) \in \Sigma_T \\ v(x,0) = u_0(x), \ x \in \Omega, \end{cases} \tag{2.3}$$

$$\begin{cases} (i\partial_t + \partial_x^2)w = 0, \ x \in \Omega_l \cup \Omega_r, \ t > 0, \\ w(x,t) = v(x,t), \ (x,t) \in \Sigma_T, \\ \lim_{|x|\to\infty} w(x,t) = 0, \ t > 0, \\ w(x,0) = 0, \ x \in \Omega_l \cup \Omega_r. \end{cases} \tag{2.4}$$

This splitting of the spatial domain $\mathbb{R}$ into interior and exterior problems is explained in Fig. 2.1. It shows the basic idea for constructing the TBC. The Transparent Boundary Condition is obtained by applying the Laplace transformation $\mathcal{L}$ with respect to the time t to the exterior problems (2.4). The Laplace transform is defined through the relation $\hat{w}(s) := \mathcal{L}(w)(s) := \int_{\mathbb{R}^+} w(t)e^{-st}dt$, where $s = \sigma + i\tau$ is the time covariable with $\sigma > 0$.

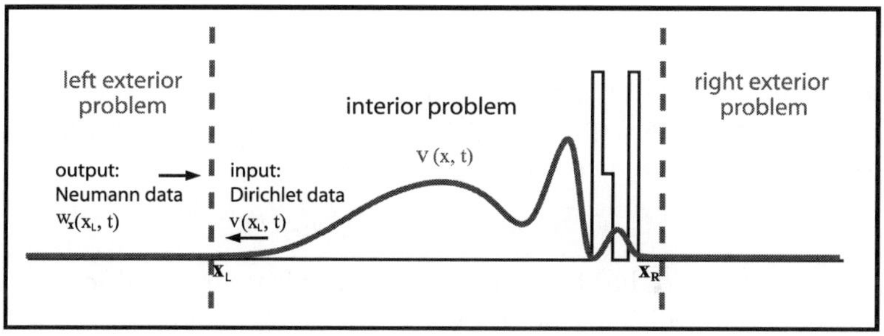

Figure 2.1 Domain decomposition for the construction of the TBC.

In the following, we focus on the derivation of the TBC at the right endpoint x_r. The Laplace transformation of (2.4) (on Ω_r) reads $is\widehat{w} + \partial_x^2\widehat{w} = 0$, $x \in \Omega_r$. The solution to this second-order ode with constant coefficients can be computed as $\hat{w}(x,s) = A^+(s)e^{\sqrt[+]{-is}\,x} + A^-(s)e^{-\sqrt[+]{-is}\,x}$, $x > x_r$, where the branch-cut of the square root $\sqrt[+]{\ }$ is taken such that the real part is positive. However, since the solution is an element of $L^2(\Omega_r)$, the coefficient A^+ must vanish. Using the Dirichlet data at the artificial boundary yields $\hat{w}(x,s) = e^{-\sqrt[+]{-is}\,(x-x_r)}\,\hat{w}(x,s)|_{x=x_r}$. Deriving $\hat{w}(x,s)$ with respect to x gives

$$\partial_x \hat{w}(x,s)|_{x=x_r} = -\sqrt[+]{-is}\,\hat{w}(x,s)|_{x=x_r}. \tag{2.5}$$

The analogous condition at the left boundary is $-\partial_x \hat{w}(x,s)|_{x=x_l} = -\sqrt[+]{-is}\,\hat{w}(x,s)|_{x=x_l}$. Applying an inverse Laplace transformation $\mathcal{L}^{-1}$ is able to obtain an expression of the Neumann datum $\partial_x w(x_{l,r},t)$ as a function of the Dirichlet one. Since we have continuity of the traces on Σ_T, the boundary condition of Eq. (2.3) is into

$$\partial_{\mathbf{n}} v(x,t) = \mathcal{L}^{-1}(-\sqrt[+]{-i\cdot}\,\hat{v}(x,\cdot))(t) = \int_0^t f(t-\tau)v(x,\tau)\,d\tau, \quad \text{on} \quad \Sigma_T, \tag{2.6}$$

where $\mathcal{L}(f)(s) = -\sqrt[+]{-is}$. By construction we have that u coincides with v on Ω, meaning that we have an exact or a Transparent Boundary Condition (TBC) given by the second equation of (2.6).

All this analysis could also be performed using the time Fourier transform $\mathscr{F}_t$

$$\mathscr{F}_t(u)(x,\tau) = \frac{1}{2\pi}\int_{\mathbb{R}} u(x,t)e^{-it\tau}\mathrm{d}t,$$

which roughly speaking corresponds to letting $\sigma \to 0$ in the expression of the Laplace transform and induces the following definition of the square root $\sqrt{\tau} = \sqrt{\tau}$ if $\tau \geq 0$ and $\sqrt{\tau} = -i\sqrt{-\tau}$ if $\tau < 0$. The condition (2.5) is thus replaced by

$$\partial_x \mathscr{F}_t w(x,\tau)|_{x=x_r} = i\sqrt{-\tau}\,\mathscr{F}_t w(x,\tau)|_{x=x_r}.$$

We recover the TBC on Σ_T with $\partial_{\mathbf{n}} v(x,t) = \mathscr{F}_t^{-1}(i\sqrt{-\cdot}\,\mathscr{F}_t v(x,\cdot))(t)$. This expression or its Laplace version $\partial_{\mathbf{n}} v(x,t) = \mathcal{L}^{-1}(-\sqrt[+]{-i\cdot}\,\hat{v}(x,\cdot))(t)$ can be simply written at points $x = x_{l,r}$ as follows:

$$\partial_{\mathbf{n}} v(x,t) = -e^{-i\pi/4}\partial_t^{1/2} v(x,t).$$

The term $\partial_t^{1/2} = \sqrt{\partial_t}$ has to be interpreted as a fractional half-order time derivative. We recall that the derivative $\partial_t^{k-\alpha} f(t)$ of order $k-\alpha > 0$ of a function f, with $k \in \mathbb{N}$ and $0 < \alpha \leq 1$, is defined by

$$\partial_t^{k-\alpha} f(t) = \frac{1}{\Gamma(\alpha)}\frac{d^k}{dt^k}\int_0^t (t-\tau)^{\alpha-1} f(\tau)d\tau, \tag{2.7}$$

where $\Gamma(z) = \int_0^{+\infty} e^{-t}t^{z-1}dz$ denotes the Gamma function. In the same spirit, one can also define the integration of real order $p > 0$ of a function f, denoted by $I_t^p f(t)$, by

$$I_t^p f(t) = \frac{1}{\Gamma(p)}\int_0^t (t-\tau)^{p-1} f(\tau)d\tau. \tag{2.8}$$

At this point, an interesting remark is that the Schrödinger equation can formally be factorized into left and right traveling waves (cf. [8]):

$$\left(\partial_x - e^{-i\frac{\pi}{4}}\partial_t^{1/2}\right)\left(\partial_x + e^{-i\frac{\pi}{4}}\partial_t^{1/2}\right)u = 0, \quad x > x_r. \tag{2.9}$$

This remark is crucial since it gives the idea to use a Nirenberg-like theorem in Section 3.2 for general variable coefficients equations (including potentials for instance).

2.2 Extensions and interpretations in the context of pseudodifferential operator calculus: introduction to the derivation of ABCs

The first possible extension is to consider a given real potential V which is constant in space outside Ω, *i.e.*, $V(x,t) = V_l(t)$ for $x < x_l$, $V(x,t) = V_r(t)$ for $x > x_r$. An easy computation, which consists in applying the following *gauge change* in (2.1), reduces this case to the zero exterior potential [3] for the new unknown

$$\psi_{l,r} = e^{-i\mathcal{V}_{l,r}(t)} u_{l,r}, \quad \text{with} \quad \mathcal{V}_{l,r}(t) = \int_0^t V_{l,r}(s)ds, \quad \forall t > 0. \tag{2.10}$$

The resulting TBC is then given by

$$\partial_{\mathbf{n}} u + e^{-i\pi/4} e^{i\mathcal{V}_{l,r}(t)} \partial_t^{1/2} (e^{-i\mathcal{V}_{l,r}(t))} u) = 0, \quad \text{on } \Sigma_T. \tag{2.11}$$

The analysis based on Laplace or Fourier transforms and performed in the previous subsection can also be done if the potential is constant outside Ω. This would lead to

$$\partial_{\mathbf{n}} u(x,t) = \int_0^t f(t-\tau) u(x,\tau)\, d\tau, \text{on } \Sigma_T, \tag{2.12}$$

where $\mathcal{L}(f)(s) = -\sqrt[+]{-is - V_{l,r}}$. Therefore, the Schrödinger equation can formally and exactly be factorized into left and right traveling waves (cf. [8]):

$$(\partial_x - e^{-i\frac{\pi}{4}} \sqrt[+]{\partial_t - iV_r})(\partial_x + e^{-i\frac{\pi}{4}} \sqrt[+]{\partial_t - iV_r}) u = 0, \quad x > x_r.$$

To understand and to make clearer the link between expressions (2.11) and (2.12), we have to introduce the notion of pseudodifferential operator. A pseudodifferential operator $P(x,t,\partial_t)$ is given by its symbol $p(x,t,\tau)$ in the Fourier space

$$\begin{aligned} P(x,t,\partial_t)u(x,t) &= \mathscr{F}_t^{-1}\Big(p(x,t,\tau)\hat{u}(x,\tau)\Big) \\ &= \int_{\mathbb{R}} p(x,t,\tau)\, \mathscr{F}_t(u)(x,\tau)\, e^{it\tau}\, \mathrm{d}\tau. \end{aligned} \tag{2.13}$$

The inhomogeneous pseudodifferential operator calculus used in the paper was first introduced in [14]. For self-conciseness reasons, we only present the useful notions required here. Let α be a real number and Ξ an open subset of $\mathbb{R}$. Then (see [19]), the symbol class $S^\alpha(\Xi \times \Xi)$ denotes the linear space of $\mathcal{C}^\infty$ functions $a(\cdot,\cdot,\cdot)$ in $\Xi \times \Xi \times \mathbb{R}$ such that for each $K \subseteq \Xi \times \Xi$ and for all indices β, δ, γ, there exists a constant $C_{\beta,\delta,\gamma}(K)$ such that $|\partial_\tau^\beta \partial_t^\delta \partial_x^\gamma a(x,t,\tau)| \leq C_{\beta,\delta,\gamma}(K)(1+|\tau|^2)^{\alpha-\beta}$, for all $(x,t) \in K$ and $\tau \in \mathbb{R}$. A function f is said to be inhomogeneous of degree m if: $f(x,t,\mu^2\tau) = \mu^m f(x,t,\tau)$, for any $\mu > 0$. Then, a pseudodifferential operator $P = P(x,t,\partial_t)$ is inhomogeneous and classical of order M, $M \in \mathbb{Z}/2$, if its total symbol, designated by $p = \sigma(P)$, has an asymptotic expansion in inhomogeneous symbols $\{p_{M-j/2}\}_{j=0}^{+\infty}$ as

$$p(x,t,\tau) \sim \sum_{j=0}^{+\infty} p_{M-j/2}(x,t,\tau),$$

where each function $p_{M-j/2}$ is inhomogeneous of degree $2M - j$, for $j \in \mathbb{N}$. The meaning of $\sim$ is that

$$\forall \widetilde{m} \in \mathbb{N}, \quad p - \sum_{j=0}^{\widetilde{m}} p_{M-j/2} \in S^{M-(\widetilde{m}+1)/2}.$$

A symbol p satisfying the above property is denoted by $p \in S_S^M$ and the associated operator $P = Op(p)$ by inverse Fourier transform (according to (2.13)) by $P \in \mathrm{OPS}_S^M$. Finally, let us remark that smoothness of the potential V is required for applying pseudodifferential operators theory. However, this is crucial for the complementary set of Ω but a much weaker regularity assumption could be expected for the interior problem set in Ω allowing therefore a wide class of potentials.

Let us come back to the comparison of relations (2.11) and (2.12) in the case of a constant potential outside Ω. With the previous definitions, Eqs. (2.11) and (2.12) respectively read

$$\partial_{\mathbf{n}} u(x,t) + ie^{iV_{l,r}t} Op\left(-\sqrt{-\tau}\right)(e^{-iV_{l,r}t}u)(x,t) = 0, \quad \text{on } \Sigma_T, \tag{2.14}$$

and

$$\partial_{\mathbf{n}} u(x,t) + iOp\left(-\sqrt{-\tau + V_{l,r}}\right)(u)(x,t), \quad \text{on } \Sigma_T. \tag{2.15}$$

Actually, these two formulations are equivalent thanks to the following Lemma (see [5] for a proof).

Lemma 2.1. *If a is a t-independent symbol of S^m and $V(x,t) = V(x)$, then the following identity holds*

$$Op\left(a(\tau - V(x))\right) u = e^{itV(x)} Op\left(a(\tau)\right)\left(e^{-itV(x)}u(x,t)\right). \tag{2.16}$$

In our case, since V is also x-independent, one gets

$$iOp\left(-\sqrt{-\tau + V_{l,r}}\right)(u)(x,t) = ie^{iV_{l,r}t} Op\left(-\sqrt{-\tau}\right)(e^{-iV_{l,r}t)}u)(x,t),$$

which explains the close link between (2.11) and (2.12).

Lemma 2.1 has other applications when the potential V depends on the spatial variable x. To emphasize this point, let us develop some approximations of the TBC for the case of a linear potential $V(x,t) = x$. Applying a Fourier transform in time, the Schrödinger equation: $i\partial_t u + \partial_x^2 u + xu = 0$ sets on Ω_T becomes the Airy equation $\partial_x^2 \mathscr{F}_t u + (-\tau + x)\mathscr{F}_t u = 0$. The solution to this equation which is outgoing is given by $\mathscr{F}_t u(x,\tau) = \mathrm{Ai}\left((x-\tau)e^{-i\pi/3}\right)$, where Ai stands for the Airy function [1]. Deriving this expression according to x, we obtain the exact relation expressing the corresponding DtN map in the Fourier space

$$\partial_{\mathbf{n}} \mathscr{F}_t u(x,\tau) = e^{-i\pi/3} \frac{\mathrm{Ai}'\left((x-\tau)e^{-i\pi/3}\right)}{\mathrm{Ai}\left((x-\tau)e^{-i\pi/3}\right)} \mathscr{F}_t u(x,\tau), \tag{2.17}$$

giving therefore the total symbol. The numerical approximation of the corresponding TBC is difficult to handle and approximations are needed. For sufficiently large values of $|\tau|$, one has the following approximation

$$e^{2i\pi/3} \frac{\mathrm{Ai}'\left((x-\tau)e^{-i\pi/3}\right)}{\mathrm{Ai}\left((x-\tau)e^{-i\pi/3}\right)} \approx -e^{-i\pi/6}\sqrt{-\tau + x}.$$

If we replace the total (left) symbol by its approximation, we obtain what is usually called an artificial or Absorbing Boundary Condition (ABC)

$$\partial_{\mathbf{n}} u + iOp\left(-\sqrt{-\tau + x}\right)(u) = 0, \quad \text{on } \Sigma_T. \tag{2.18}$$

Thanks to Lemma 2.1 and since $V(x,t) = x$, this ABC is strictly equivalent to

$$\partial_{\mathbf{n}} u + e^{-i\pi/4} e^{itx_{l,r}} \partial_t^{1/2}(e^{-itx_{l,r}} u) = 0, \quad \text{on } \Sigma_T. \tag{2.19}$$

Let us remark that, in the specific case of a linear potential, a change of unknown is allowed to transform the Schrödinger equation with linear potential into another Schrödinger equation without potential [10]. Indeed, if v is solution to $i\partial_t v + \partial_x^2 v = 0$, then $u(x,t) = e^{-i(-\alpha t x + \frac{t^3}{3}|\alpha|^2)} v(x - t^2\alpha, t)$ is solution to $i\partial_t u + \partial_x^2 u + \alpha x u = 0$.

At this point, some partial conclusions can be drawn:

- Formally, the operator $i\partial_t + \partial_x^2 + V$ can be (exactly or approximately) factorized as

$$i\partial_t + \partial_x^2 + V = \left(\partial_x + i\sqrt{i\partial_t + V}\right)\left(\partial_x - i\sqrt{i\partial_t + V}\right),$$

  according to the (x,t)-dependence of the potential. On the above right hand side, the second term characterizes the DtN map involved in the TBC or ABC.

- Transparent Boundary Conditions or Absorbing Boundary Conditions are written through a DtN operator either

$$\partial_{\mathbf{n}} u + iOp\left(-\sqrt{-\tau}\right)(u) = 0, \quad \text{on } \Sigma_T,$$

or

$$\partial_{\mathbf{n}} u + iOp\left(-\sqrt{-\tau + V}\right)(u) = 0, \quad \text{on } \Sigma_T.$$

- If $V(x,t) = V(t)$, the change of unknown $v(x,t) = e^{-i\mathcal{V}(t)}u(x,t)$, with $\mathcal{V}(t) = \int_0^t V(s)ds$, reduces the Schrödinger equation with potential to a free-Schrödinger equation and the TBC is then

$$\partial_{\mathbf{n}} u(x,t) + e^{-i\pi/4} e^{i\mathcal{V}(t)} \partial_t^{1/2}\left(e^{-i\mathcal{V}(t)}\right)(x,t) = 0, \quad \text{on } \Sigma_T.$$

3 ABCs for the linear Schrödinger equation with a general variable potential

3.1 Two possible strategies

It is clear from the above analysis that we cannot expect to get a TBC for real general potentials. We then need to derive some approximations and most specifically to compute ABCs using the previously introduced pseudodifferential operator calculus which extends the Laplace transform-based approach to variable coefficients operators. Furthermore, it enables to manipulate symbols of operators at the algebraic level instead of operators at the functional level. The partial conclusions given at the end of the previous section let one think that two possible strategies to build ABCs can be considered.

The first natural approach would consist in building an approximate boundary condition based on Eq. (1.1) with unknown u. However, even if this approach seems direct, it is quite intricate and for this reason it will be designated as strategy 2 in the sequel.

A second possibility, called strategy 1, is the following. Let us consider now that u is the solution to Eq. (1.1) and let us define $\mathcal{V}$ as a primitive in time of the potential V

$$\mathcal{V}(x,t) = \int_0^t V(x,s)\,\mathrm{d}s. \tag{3.1}$$

Following the Gauge change (2.10), let us introduce v as the new unknown defined by

$$v(x,t) = e^{-i\mathcal{V}(x,t)}u(x,t). \tag{3.2}$$

We obviously have $v_0(x) = u_0(x)$. Moreover, plugging u given by (3.1)–(3.2) into the Schrödinger equation with potential shows that v is solution to the variable coefficients Schrödinger equation

$$i\partial_t v + \partial_x^2 v + f\,\partial_x v + g\,v = 0, \quad \text{in } \Omega_T, \tag{3.3}$$

setting $f = 2i\partial_x \mathcal{V}$ and $g = i\partial_x^2 \mathcal{V} - (\partial_x \mathcal{V})^2$. The fundamental reason why this change of unknown is considered crucial is that this first step leads to the TBC (2.11) applied to v and associated with (3.3) for a time-dependent potential (since then $f = g = 0$).

We will see later that these two approaches lead to different absorbing boundary conditions which however coincide in some situations.

3.2 Practical computation of the asymptotic expansion of the DtN operator

We explain here how to compute the asymptotic expansion of the DtN operator for a given model Schrödinger equation with smooth variable coefficients A and B

$$L(x,t,\partial_x,\partial_t)w = i\partial_t w + \partial_x^2 w + A\partial_x w + Bw = 0. \tag{3.4}$$

Since we are trying to build an approximation of the DtN operator at the boundary, we must be able to write the normal derivative trace operator ∂_x (focusing on the right point x_r) as a function of the trace operator through an operator Λ^+ which involves some (fractional) time derivatives/integrals of w as well as the effects of the potential V and its (x,t) variations. This can be done in an approximate way thanks to the factorization of L given by relation (3.4)

$$L(x,t,\partial_x,\partial_t) = (\partial_x + i\Lambda^-)(\partial_x + i\Lambda^+) + R, \tag{3.5}$$

where $R \in \mathrm{OPS}^{-\infty}$ is a smoothing pseudodifferential operator. This relation corresponds to the formal factorization presented at the end of Section 2.2. The operators $\Lambda^\pm$ are pseudodifferential operators of order 1/2 (in time) and order zero in x. Computing the operators $\Lambda^\pm$ in an exact way through their respective total symbols $\sigma(\Lambda^\pm)$ cannot be expected in general (which would therefore provide a TBC). A more realistic approach consists in seeking an asymptotic form of the total symbol $\sigma(\Lambda^\pm)$ as

$$\sigma(\Lambda^\pm) = \lambda^\pm \sim \sum_{j=0}^{+\infty} \lambda^\pm_{1/2-j/2}, \tag{3.6}$$

where $\lambda^\pm_{1/2-j/2}$ are symbols corresponding to operators of order $1/2 - j/2$.

Now, expanding the factorization (3.5), identifying the terms in L in front of the operators ∂_x with the ones from the expanded factorization and finally using a few symbolic manipulations to yield the system of equations

$$\begin{cases} i(\lambda^- + \lambda^+) = a \\ i\partial_x \lambda^+ - \displaystyle\sum_{\alpha=0}^{\infty} \frac{(-i)^\alpha}{\alpha!} \partial_\tau^\alpha \lambda^- \partial_t^\alpha \lambda^+ = -\tau + b, \end{cases} \tag{3.7}$$

with $a(x,t) = \sigma(A) = A$, $b(x,t) = \sigma(B) = B$, since A and B are two functions of (x,t).

Looking at the first equation of system (3.7), we see that we must have: $\lambda_{1/2}^- = -\lambda_{1/2}^+$. Now, if we identify the highest order symbol in the second equation of system (3.7), then we get four possibilities

$$\lambda_{1/2}^+(\tau) = \mp\sqrt{-\tau} \tag{3.8}$$

and

$$\lambda_{1/2}^+(x,t,\tau) = \mp\sqrt{-\tau + b(x,t)}. \tag{3.9}$$

The first choice can be viewed as considering a principal classical symbol while the second possibility is rather referred to as a semi-classical symbol (see e.g. in [10]).

Let us now consider the strategy 1 based on the gauge change leading to computing v solution to (3.3) for $A = f$ and $B = g$. Following the derivation of the TBCs made in Section 2.2, the principal symbol with negative imaginary part characterizes the outgoing part of the solution u. A study of the sign of (3.8) (for a real-valued potential V) shows that the negative sign leads to the correct choice. Since g is a complex-valued potential with no controlled sign, we cannot determine the outgoing wave for (3.9). The only possible choice is then

$$\lambda_{1/2}^+ = -\sqrt{-\tau}. \tag{3.10}$$

Let us now consider the second strategy which consists in working on Eq. (1.1) for u setting $A = 0$ and $B = V$. The study of the signs of (3.8) and (3.9) for a real-valued potential V is possible in both cases and as for the first strategy, the negative sign provides the suitable solution. Therefore, we obtain the two possible symbols $\lambda_{1/2}^+ = -\sqrt{-\tau}$ and $\lambda_{1/2}^+ = -\sqrt{-\tau + V}$. However, considering $\lambda_{1/2}^+ = -\sqrt{-\tau}$ would give some symbols which are approximations of $\lambda_{1/2}^+ = -\sqrt{-\tau + V}$ by using a truncated Taylor expansion when $|\tau| \to +\infty$. Since this case leads to less accurate ABCs, we will only consider next the case

$$\lambda_{1/2}^+ = -\sqrt{-\tau + V}. \tag{3.11}$$

Choosing the principal symbol is a crucial point since it is directly related to the accuracy of the ABC. Moreover, for a given choice of the principal symbol, the corrective asymptotic terms $\{\lambda^+_{1/2-j/2}\}_{j\geq 1}$ are different since they are computed in cascade developing the infinite sum in the second equation of (3.7) as seen in the following Proposition.

Proposition 3.1. *Let us fix* $\lambda^+_{1/2}$ *by the expression* (3.10). *Then, the solution to system* (3.7) *is given by*

$$\lambda^+_0 = \frac{1}{2\lambda^+_{1/2}}\left(-i(\partial_x + a)\lambda^+_{1/2}\right), \tag{3.12}$$

and, for $j \in \mathbb{N}$, $j \geq 1$, *by*

$$\lambda^+_{-j/2} = \frac{1}{2\lambda^+_{1/2}}\left(b\,\delta_{j,1} - i(\partial_x + a)\lambda^+_{-j/2+1/2} + - \sum_{k=1}^{j} \lambda^+_{-j/2+k/2}\lambda^+_{1/2-k/2} \right.$$
$$\left. - \sum_{\alpha=1}^{(j+1)/2} \frac{(-i)^\alpha}{\alpha!} \sum_{k=0}^{j+1-2\alpha} \partial^\alpha_\tau \lambda^+_{-j/2+k/2+\alpha}\partial^\alpha_t \lambda^+_{1/2-k/2}\right) \tag{3.13}$$

where $\delta_{j,1} = 0$ *if* $j \neq 1$ *and* $\delta_{1,1} = 1$.

Applying the above proposition to our situation, one obtains the following corollary.

Corollary 3.2. *In strategy 1 (*$a = f$ *and* $b = g$*), if we fix the principal symbol* $\lambda^+_{1/2} = -\sqrt{-\tau}$ *in* (3.7)*, then the next three terms of the asymptotic symbolic expansion are given by using (3.12) as*

$$\lambda^+_0 = \partial_x V, \quad \lambda^+_{-1/2} = 0 \quad \textit{and} \quad \lambda^+_{-1} = \frac{i\partial_x V}{4\tau}. \tag{3.14}$$

In the case of the second strategy ($a = 0$ and $b = V$) and for $\lambda^+_{1/2} = -\sqrt{-\tau + V}$, we cannot obtain a general formula as for Proposition 3.1. However, the first terms can still be computed as

$$\lambda^+_0 = 0, \quad \lambda^+_{-1/2} = 0, \quad \text{and} \quad \lambda^+_{-1} = \frac{-i}{4}\frac{\partial_x V}{-\tau + V}. \tag{3.15}$$

In the case of a linear potential $V = x$, we saw that the total symbol is

$$\lambda^+ = e^{2i\pi/3}\frac{\mathrm{Ai}'\left((x-\tau)e^{-i\pi/3}\right)}{\mathrm{Ai}\left((x-\tau)e^{-i\pi/3}\right)}. \tag{3.16}$$

The application of Corollary 3.2 in the context of strategy 2 gives the first- and second-order approximate symbols $\sigma_1 = i\lambda^+_{1/2} = -i\sqrt{-\tau + x}$

and $\sigma_2 = i\lambda^+_{1/2} + i\lambda^+_0 = \sigma_1 + \frac{1}{4}\frac{1}{-\tau+x}$, setting $V = x$. These two relations give good approximations of $\lambda^+(x,\tau)$ for sufficiently large values of $|x-\tau|$ (see Fig. 3.1 for $x = 10$), corresponding to a high frequency spectrum approximation. This test case shows the validity of our approach in this situation.

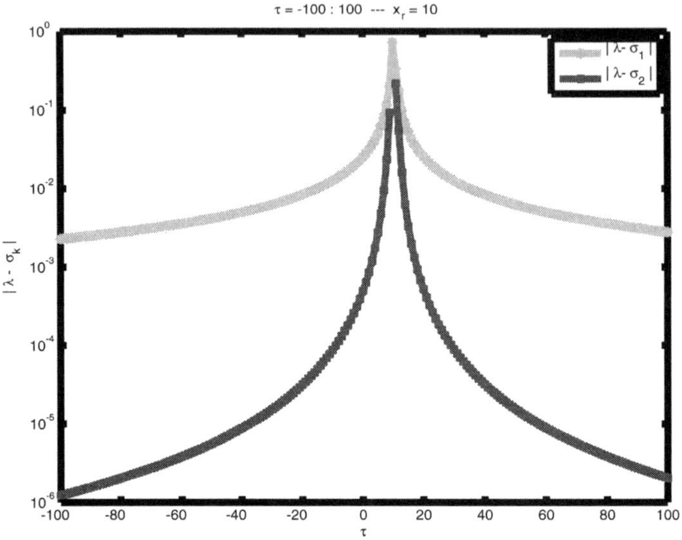

Figure 3.1 Logarithm of the absolute error $|\lambda^+ - \sigma_1|$ and $|\lambda^+ - \sigma_2|$ with respect to τ. A singularity is observed at $|\tau - x_r| = 0$, with $x_r = 10$.

3.3 Choosing the ABC in the context of strategy 1

If we assume that V is a real-valued smooth function, then the $L^2(\mathbb{R})$-norm of the solution u to system (1.1) is conserved. If we truncate the domain by introducing a fictitious boundary, then one can expect that the $L^2(\Omega)$-norm of the solution is bounded by $||u_0||_{L^2(\Omega)}$. This is for example proved in [3] in the case of the free-potential. In the case of a general potential, the expression of the artificial boundary condition is essential in the proof of a similar result. Particularly, by adapting the proof given in [7] and using the Plancherel theorem for Laplace transform, the following Lemma is the keypoint for proving a well-posedness result.

Lemma 3.3. *Let $\varphi \in H^{1/4}(0,T)$ be a function extended by zero for any time $s > T$. Then, we have the properties* $\Re\left(e^{i\pi/4}\int_0^\infty \overline{\varphi}\, \partial_t^{1/2}\varphi\, \mathrm{d}t\right) \geq 0$ *and* $\Re\left(\int_0^{+\infty} \overline{\varphi}\, I_t\, \varphi\, \mathrm{d}t\right) = 0$.

This Lemma emphasizes the fact that the absorbing boundary condition must have a symmetrical form. Since our approach gives the principal symbol of an operator, an infinite choice of corresponding operators with this principal symbol is possible. For symmetrization reasons, we propose to fix the choice of the artificial boundary condition based on the principal symbol $\lambda^+_{1/2} = -\sqrt{-\tau}$ and (3.14) as follows:

Cancelling the outgoing wave corresponding to $\lambda^+_{1/2}$ for v writes down

$$\partial_{\mathbf{n}} v + i\Lambda^+ v = 0, \qquad \text{on } \Sigma_T. \tag{3.17}$$

Retaining the M first symbols $\{\lambda^+_{1/2-j/2}\}_{M-1\geq j\geq 0}$, we consider the associated ABC

$$\partial_{\mathbf{n}} u_M - i(\partial_x \mathcal{V})\, u_M + ie^{i\mathcal{V}} \sum_{j=0}^{M-1} Op\left(\lambda^+_{1/2-j/2}\right)\left(e^{-i\mathcal{V}} u_M\right) = 0, \quad \text{on } \Sigma_T, \tag{3.18}$$

after replacing v in (3.17) by its expression (3.2). In Eq. (3.18), u_M designates an approximation of u since we do not have a TBC. However, u_M will be denoted by u in the sequel for conciseness. We adopt the following compact notation of (3.18)

$$\partial_{\mathbf{n}} u + \Lambda^M_\ell(x,t,\partial_t)\, u = 0, \quad \text{on } \Sigma_T, \tag{3.19}$$

where $M \geq 1$ corresponds to the order of the boundary condition and is equal to the total number of terms $\lambda^+_{j/2}$ retained in the sum. The subscript $\ell = 1$ (respectively $\ell = 2$) refers to the choice (3.10) (respectively (3.11)) of the principal symbol $\lambda^+_{1/2}$, and therefore to the two different strategies.

Let us begin by considering $\ell = 1$ and $M = 2$. Then one directly obtains

$$\Lambda^2_1(x,t,\partial_t)\, u = e^{-i\pi/4} e^{i\mathcal{V}(x,t)} \partial_t^{1/2}\left(e^{-i\mathcal{V}(x,t)} u\right) \tag{3.20}$$

which is a symmetrical operator. The case $M = 4$ is more ambiguous. Indeed, we only have access to the principal symbol $\lambda^+_{-1} = i\partial_x V/(4\tau)$ of an operator. In order to conserve a symmetrical operator for the definition of the ABC, our choice of operator is

$$Op\left(\lambda^+_{-1}\right) v = \mathrm{sg}(\partial_{\mathbf{n}} V) \frac{\sqrt{|\partial_{\mathbf{n}} V|}}{2} I_t \left(\frac{\sqrt{|\partial_{\mathbf{n}} V|}}{2} v\right) \quad \mathrm{mod}(\mathrm{OPS}_S^{-3/2}). \tag{3.21}$$

In the above equation, $\mathrm{sg}(\cdot)$ designates the sign function.

We finally obtain the following Proposition.

Proposition 3.4. *For $\ell = 1$, the ABC of order M is given by*

$$\partial_{\mathbf{n}} u + \Lambda^M_1 u = 0, \quad \textit{on } \Sigma_T, \tag{3.22}$$

with

$$\Lambda_1^2(x,t,\partial_t)\,u = e^{-i\pi/4}e^{i\mathcal{V}(x,t)}\partial_t^{1/2}\left(e^{-i\mathcal{V}(x,t)}u\right) \tag{3.23}$$

and

$$\begin{aligned}\Lambda_1^4(x,t,\partial_t)\,u &= \Lambda_1^2(x,t,\partial_t)\,u \\ &+ i\,\mathrm{sg}(\partial_{\mathbf{n}}V)\frac{\sqrt{|\partial_{\mathbf{n}}V|}}{2}e^{i\mathcal{V}(x,t)}I_t\left(\frac{\sqrt{|\partial_{\mathbf{n}}V|}}{2}e^{-i\mathcal{V}(x,t)}u\right).\end{aligned} \tag{3.24}$$

The boundary condition (3.22) *is referred to as* ABC_1^M *in the sequel.*

Considering the ABCs (3.22) of Proposition 3.4, we get the following well-posedness result (see proof in [5]).

Proposition 3.5. *Let* $u_0 \in L^2(\Omega)$ *be a compactly supported initial datum such that* $\mathrm{Supp}(u_0) \subset \Omega$. *Let* $V \in \mathcal{C}^\infty(\mathbb{R}\times\mathbb{R}^+,\mathbb{R})$ *be a real-valued potential. Let us denote by* u *a solution of the initial boundary value problem*

$$\begin{cases} i\partial_t u + \partial_x^2 u + Vu = 0, & \text{in } \Omega_T, \\ \partial_{\mathbf{n}} u + \Lambda_1^M u = 0, & \text{on } \Sigma_T, \\ u(x,0) = u_0(x), & \forall x \in \Omega, \end{cases} \tag{3.25}$$

where the operators Λ_1^M, *with* $M = 2, 4$, *are defined in Proposition 3.4. Then,* u *fulfils the following energy bound*

$$\forall t > 0, \quad ||u(t)||_{L^2(\Omega)} \le ||u_0||_{L^2(\Omega)}, \tag{3.26}$$

for $M = 2$. *Moreover, if* $\mathrm{sg}(\partial_{\mathbf{n}}V)$ *is constant on* Σ_T, *then the inequality* (3.26) *holds for* $M = 4$. *Particularly, this implies that we have the uniqueness of the solution* u *of the initial boundary value problem* (3.25).

3.4 Choosing the ABC in the context of strategy 2

Let us now consider the second strategy for building the absorbing boundary conditions ABC_2^M, for $M = 2$ and $M = 4$.

Proposition 3.6. *For* $\ell = 2$, *the ABC of order* M *based on the second strategy for symbols* (3.15) *is given by*

$$\partial_{\mathbf{n}} u + \Lambda_2^M u = 0, \quad \text{on } \Sigma_T, \tag{3.27}$$

with

$$\Lambda_2^2(x,t,\partial_t)\,u = Op\left(-i\sqrt{-\tau+V}\right)u \tag{3.28}$$

and

$$\Lambda_2^4(x,t,\partial_t)\,u = \Lambda_2^2(x,t,\partial_t)\,u + \frac{1}{4}Op\left(\frac{\partial_x V}{-\tau+V}\right)u. \tag{3.29}$$

The boundary condition (3.27) *is referred to as* ABC_2^M *in the sequel of the paper.*

Studying the well-posedness of the initial boundary value problem related to the boundary condition ABC_2^M (3.27)–(3.29) is more difficult than ABC_1^M except for the case $V(x,t) = V(x)$. Indeed, the well-posedness result is trivial since ABC_2^M is strictly equivalent to ABC_1^M. A direct application of Lemma 2.1 gives the following Corollary.

Corollary 3.7. *If the potential V is time independent, then ABC_1^M is equivalent ABC_2^M, for a fixed value of $M = 2, 4$, with $\mathcal{V}(x,t) = tV(x)$. Particularly, the well-posedness of the associated bounded initial value problem is immediate from Proposition 3.5.*

4 Semi-discretization schemes and their properties

The aim of this section is to proceed to the semi-discretization in time of the initial value problem

$$\begin{cases} i\partial_t u + \partial_x^2 u + V\, u = 0, & \text{in } \Omega_T, \\ \partial_{\mathbf{n}} u + \Lambda_{1,2}^M u = 0, & \text{on } \Sigma_T, \text{ for } M = 2 \text{ or } 4, \\ u(\cdot, 0) = u_0, & \text{in } \Omega, \end{cases} \tag{4.1}$$

for a maximal time of computation T.

We consider an interior Crank-Nicolson scheme for the time discretization of system (4.1). The interval $[0;T]$ is uniformly discretized using N intervals. Let $\Delta t = T/N$ be the time step and let us set $t_n = n\Delta t$. Furthermore, u^n stands for an approximation of $u(t_n)$ and $V^n = V(x, t_n)$. If $V = V(x)$ is a time-independent potential, then the Crank-Nicolson discretization of the Schrödinger equation is given by $i\,(u^{n+1} - u^n)/\Delta t + \partial_x^2 (u^{n+1} + u^n)/2 + V\,(u^{n+1} + u^n)/2 = 0$, for $n = 0, \ldots, N-1$. If $V = V(x,t)$, for matters of symmetry, we choose the following time-discretization of the interior equation

$$i\,\frac{u^{n+1} - u^n}{\Delta t} + \partial_x^2 \frac{u^{n+1} + u^n}{2} + \frac{V^{n+1} + V^n}{2}\frac{u^{n+1} + u^n}{2} = 0. \tag{4.2}$$

Let us remark that for implementation issues, it is often useful to set $v^{n+1} = (u^{n+1} + u^n)/2 = u^{n+1/2}$, with $u^{-1} = 0$ and $u^0 = u_0$. Similarly, we define $W^{n+1} = (V^{n+1} + V^n)/2 = V^{n+1/2}$. Then, the time scheme (4.2) reads

$$2i\,\frac{v^{n+1}}{\Delta t} + \partial_x^2 v^{n+1} + W^{n+1} v^{n+1} = 2i\,\frac{u^n}{\Delta t}. \tag{4.3}$$

We propose here one approximation for each kind of ABC. The approach for strategy 1 is based on semi-discrete convolutions for the fractional operators involved in (4.4)–(4.5), which leads to an unconditionally stable semi-discrete scheme. Considering strategy 2, we propose a scheme based on the approximation of the fractional operators through the solution of auxiliary differential equations which can be solved explicitly. The evaluation is then extremely efficient but at the same time, no stability proof is at hand.

4.1 Discrete convolutions-based discretizations for ABC_1^M

We first consider the boundary conditions ABC_1^M. According to Proposition 3.4, we have

$$\mathrm{ABC}_1^2: \quad \partial_{\mathbf{n}} u + e^{-i\pi/4} e^{i\mathcal{V}} \partial_t^{1/2}\left(e^{-i\mathcal{V}} u\right) = 0, \tag{4.4}$$

and

$$\begin{aligned}\mathrm{ABC}_1^4: \quad & \partial_{\mathbf{n}} u + e^{-i\pi/4} e^{i\mathcal{V}} \partial_t^{1/2}\left(e^{-i\mathcal{V}} u\right) \\ & + i\,\mathrm{sg}(\partial_{\mathbf{n}} V) \frac{\sqrt{|\partial_{\mathbf{n}} V|}}{2} e^{i\mathcal{V}} I_t\left(\frac{\sqrt{|\partial_{\mathbf{n}} V|}}{2} e^{-i\mathcal{V}} u\right) = 0.\end{aligned} \tag{4.5}$$

We use the symmetrical form of ABC_1^4, which is a keypoint in the case $V = V(x,t)$. The associated initial boundary value problem is then

$$\begin{cases} i\partial_t u + \partial_x^2 u + V u = 0, & \text{in } \Omega_T, \\ \partial_{\mathbf{n}} u + \Lambda_1^M u = 0, & \text{on } \Sigma_T, \text{ for } M = 2 \text{ or } 4, \\ u(\cdot, 0) = u_0, & \text{in } \Omega. \end{cases} \tag{4.6}$$

We will use in the sequel the following discrete convolutions approximating the continuous convolution operators.

Proposition 4.1. *If $\{f^n\}_{n\in\mathbb{N}}$ is a sequence of complex numbers approximating $\{f(t_n)\}_{n\in\mathbb{N}}$, then the approximations of $\partial_t^{1/2} f(t_n)$, $I_t^{1/2} f(t_n)$ and $I_t f(t_n)$ with respect to the Crank-Nicolson scheme for a time step Δt are given by the numerical quadrature formulas $\partial_t^{1/2} f(t_n) \approx \sqrt{\frac{2}{\Delta t}} \sum_{k=0}^{n} \beta_{n-k} f^k$, $I_t^{1/2} f(t_n) \approx \sqrt{\frac{\Delta t}{2}} \sum_{k=0}^{n} \alpha_{n-k} f^k$, $I_t f(t_n) \approx \frac{\Delta t}{2} \sum_{k=0}^{n} \gamma_{n-k} f^k$, where the sequences $(\alpha_n)_{n\in\mathbb{N}}$, $(\beta_n)_{n\in\mathbb{N}}$ and $(\gamma_n)_{n\in\mathbb{N}}$ are*

$$\begin{cases} (\alpha_0, \alpha_1, \alpha_2, \alpha_3, \alpha_4, \alpha_5, \dots) = \left(1, 1, \frac{1}{2}, \frac{1}{2}, \frac{3}{8}, \frac{3}{8}, \dots\right), \\ \beta_k = (-1)^k \alpha_k, \quad \forall k \geq 0, \\ (\gamma_0, \gamma_1, \gamma_2, \gamma_3, \dots) = (1, 2, 2, \dots). \end{cases}$$

The weak formulation of (4.2) writes, for any test-function ψ in $H^1(\Omega)$,

$$\frac{2i}{\Delta t}\int_{x_l}^{x_r}(v^{n+1}-u^n)\psi \mathrm{d}x + \left[\partial_x v^{n+1}\psi\right]_{x_l}^{x_r} - \int_{x_l}^{x_r}\partial_x v^{n+1}\partial_x\psi \mathrm{d}x + \int_{x_l}^{x_r} W^{n+1}v^{n+1}\psi \mathrm{d}x = 0. \tag{4.7}$$

According to the interior scheme (4.2), the semi-discretization of ABC_1^2 for v at time t_{n+1} is given by

$$\partial_{\mathbf{n}} v^{n+1}(x_{l,r}) = -\, e^{-i\pi/4}e^{i\mathscr{W}^{n+1}}\sqrt{\frac{2}{\Delta t}}\sum_{k=0}^{n+1}\beta_{n+1-k}e^{-i\mathscr{W}^k}v^k(x_{l,r}),$$

and, for ABC_1^4, by

$$\begin{aligned}\partial_{\mathbf{n}} v^{n+1}(x_{l,r}) = &-e^{-i\pi/4}e^{i\mathscr{W}^{n+1}}\sqrt{\frac{2}{\Delta t}}\sum_{k=0}^{n+1}\beta_{n+1-k}e^{-i\mathscr{W}^k}v^k(x_{l,r})\\ &-i\,\mathrm{sg}(\partial_{\mathbf{n}}W^{n+1})\frac{\sqrt{|\partial_{\mathbf{n}}W^{n+1}|}}{2}e^{i\mathscr{W}^{n+1}}\frac{\Delta t}{2}\\ &\sum_{k=0}^{n+1}\gamma_{n+1-k}\frac{\sqrt{|\partial_{\mathbf{n}}W^k|}}{2}e^{-i\mathscr{W}^k}v^k(x_{l,r}),\end{aligned} \tag{4.8}$$

with the notation $\mathscr{W}^{n+1} = (\mathcal{V}^{n+1}+\mathcal{V}^n)/2$. Then, the following Proposition can be proved (see [5]).

Proposition 4.2. *The semi-discrete Crank-Nicolson scheme for the initial boundary value problem* (4.6) *is given by*

$$\begin{cases} 2i\,\dfrac{v^{n+1}-u^n}{\Delta t} + \partial_x^2 v^{n+1} + W^{n+1}\,v^{n+1} = 0, \quad \text{in } \Omega,\\ \partial_{\mathbf{n}} v^{n+1} + \Lambda_1^{M,n+1}v^{n+1} = 0, \quad \text{on } \Sigma, \quad \text{for } M = 2 \text{ or } 4,\\ u^0 = u_0, \quad \text{in } \Omega,\end{cases} \tag{4.9}$$

for $n = 0,\ldots,N-1$, *setting* $v^{n+1} = (u^{n+1}+u^n)/2$, $W^{n+1} = (V^{n+1}+V^n)/2$, *where the semi-discrete operators* $\Lambda_1^{2,n+1}$ *and* $\Lambda_1^{4,n+1}$ *are defined by*

$$\Lambda_1^{2,n+1}v^{n+1} = e^{-i\pi/4}e^{i\mathscr{W}^{n+1}}\sqrt{\frac{2}{\Delta t}}\sum_{k=0}^{n+1}\beta_{n+1-k}e^{-i\mathscr{W}^k}v^k, \tag{4.10}$$

$$\Lambda_1^{4,n+1}v^{n+1} = \Lambda_1^{2,n+1}v^{n+1} \tag{4.11}$$

$$+\,i\,\mathrm{sg}(\partial_{\mathbf{n}}W^{n+1})\frac{\sqrt{|\partial_{\mathbf{n}}W^{n+1}|}}{2}e^{i\mathscr{W}^{n+1}}\frac{\Delta t}{2}\sum_{k=0}^{n+1}\gamma_{n+1-k}\frac{\sqrt{|\partial_{\mathbf{n}}W^k|}}{2}e^{-i\mathscr{W}^k}v^k. \tag{4.12}$$

Here, $\mathscr{W}^{n+1}$ is defined by $\mathscr{W}^{n+1} = (\mathcal{V}^{n+1} + \mathcal{V}^n)/2$, $\mathcal{V}^n(x)$ being the approximation of $\mathcal{V}(x, t_n)$ using the trapezoidal rule ($\mathcal{V}$ is given by (3.1)). Moreover, for $M = 2$, one has the following energy inequality

$$\forall n \in \{0, \ldots, N\}, \quad \|u^n\|_{L^2(\Omega)} \leq \|u^0\|_{L^2(\Omega)}, \tag{4.13}$$

and if $\mathrm{sg}(\partial_{\mathbf{n}} W^k)$ is constant, then (4.13) also holds for $M = 4$. This proves the $L^2(\Omega)$ stability of the scheme. Inequality (4.13) is the semi-discrete version of (3.26) under the corresponding assumptions.

4.2 Auxiliary functions-based discretizations for ABC_2^M

While the previous strategy based on discrete convolution operators is accurate and provides a stability result, it may lead to significantly long computational time. For ABC_2^M, the discretizations of the resulting pseudodifferential operators involved are not easy to obtain. Particularly, the operators with square-root symbols cannot be expressed in terms of fractional time operators since Lemma 2.1 cannot be applied. Let us consider the following additional approximations which will provide a more suitable way to discretize the ABCs.

Lemma 4.3. *We have the approximations $Op\left(\sqrt{-\tau + V}\right) = \sqrt{i\partial_t + V}$, mod $(OPS_S^{-3/2})$ and $Op\left(\frac{\partial_x V}{4}\frac{1}{-\tau+V}\right) = \mathrm{sg}(\partial_{\mathbf{n}} V)\frac{\sqrt{|\partial_{\mathbf{n}} V|}}{2}(i\partial_t + V)^{-1}$ $\frac{\sqrt{|\partial_{\mathbf{n}} V|}}{2}$, mod (OPS_S^{-3}).*

Thanks to Lemma 4.3, we now define the new approximations of ABC_2^M (see Proposition 3.6)

$$\widetilde{\mathrm{ABC}_2^2}: \quad \partial_{\mathbf{n}} u - i\sqrt{i\partial_t + V}u = 0, \tag{4.14}$$

and

$$\widetilde{\mathrm{ABC}_2^4}: \quad \partial_{\mathbf{n}} u - i\sqrt{i\partial_t + V}u + \mathrm{sg}(\partial_{\mathbf{n}} V)\frac{\sqrt{|\partial_{\mathbf{n}} V|}}{2}(i\partial_t + V)^{-1}\left(\frac{\sqrt{|\partial_{\mathbf{n}} V|}}{2}u\right) = 0. \tag{4.15}$$

Let us begin by the second-order condition (4.14). An alternative approach to discrete convolutions (which cannot be applied here) consists in approximating the square-root operator $\sqrt{i\partial_t + V}$ by using rational functions. More specifically here, we consider the m-th order Padé approximants [18]

$$\sqrt{z} \approx R_m(z) = a_0^m + \sum_{k=1}^{m} \frac{a_k^m z}{z + d_k^m} = \sum_{k=0}^{m} a_k^m - \sum_{k=1}^{m} \frac{a_k^m d_k^m}{z + d_k^m}, \tag{4.16}$$

where the coefficients $(a_k^m)_{0\leq k\leq m}$ and $(d_k^m)_{1\leq k\leq m}$ are given by

$$a_0^m = 0 \quad , \quad a_k^m = \frac{1}{m\cos^2\left(\frac{(2k+1)\pi}{4m}\right)} \quad , \quad d_k^m = \tan^2\left(\frac{(2k+1)\pi}{4m}\right).$$

Formally, $\sqrt{i\partial_t + V}$ is approximated by

$$R_m(i\partial_t + V) = \sum_{k=0}^{m} a_k^m - \sum_{k=1}^{m} a_k^m\, d_k^m (i\partial_t + V + d_k^m)^{-1}. \tag{4.17}$$

Applying this process to Eq. (4.14), we have the new relation

$$\partial_{\mathbf{n}} u - i\sum_{k=0}^{m} a_k^m\, u + i\sum_{k=1}^{m} a_k^m\, d_k^m (i\partial_t + V + d_k^m)^{-1}\, u = 0, \tag{4.18}$$

defining then a second-order artificial boundary condition referred to as $\mathrm{ABC}_{2,m}^2$ in the sequel. To write a suitable form of the equation in view of an efficient numerical treatment, we classically introduce m auxiliary functions φ_k, for $1 \leq k \leq m$, (see Lindmann [16]) as follows

$$\varphi_k = (i\partial_t + V + d_k^m)^{-1}\, u, \tag{4.19}$$

leading to the following equation

$$i\partial_t\varphi_k + (V + d_k^m)\,\varphi_k = u, \qquad \text{for } 1 \leq k \leq m, \text{ at } x = x_{l,r}, \tag{4.20}$$

with the initial condition $\varphi_k(x,0) = 0$. The corresponding complete local artificial boundary condition is written as a system

$$\begin{cases} \partial_{\mathbf{n}} u - i\displaystyle\sum_{k=0}^{m} a_k^m u + i\sum_{k=1}^{m} a_k^m\, d_k^m\, \varphi_k = 0, \\ i\partial_t\varphi_k + (V + d_k^m)\,\varphi_k = u, \qquad \text{for } 1 \leq k \leq m,\ x = x_{l,r}, \\ \varphi_k(x,0) = 0. \end{cases} \tag{4.21}$$

The semi-discretization of the interior scheme remains the same as before (4.2), and consequently, (4.21) becomes in terms of v_k^n functions

$$\begin{cases} \partial_{\mathbf{n}} v^{n+1} - i\displaystyle\sum_{k=0}^{m} a_k^m v^{n+1} + i\sum_{k=1}^{m} a_k^m d_k^m \varphi_k^{n+1/2} = 0, \\ i\,\dfrac{\varphi_k^{n+1} - \varphi_k^n}{\Delta t} + (W^{n+1} + d_k^m)\,\varphi_k^{n+1/2} = v^{n+1}, \\ \varphi_k^0 = 0. \end{cases} \tag{4.22}$$

for $1 \leq k \leq m$ and $x = x_{l,r}$.

Now, let us consider the fourth-order condition $\widetilde{\mathrm{ABC}_2^4}$ given by (4.15)

$$\partial_{\mathbf{n}} u - i\sqrt{i\partial_t + V}\, u + \mathrm{sg}(\partial_{\mathbf{n}} V)\frac{\sqrt{|\partial_{\mathbf{n}} V|}}{2}(i\partial_t + V)^{-1}\left(\frac{\sqrt{|\partial_{\mathbf{n}} V|}}{2}u\right) = 0, \quad \text{on } \Sigma \times \mathbb{R}. \tag{4.23}$$

Then, one has to introduce one more additional auxiliary function ψ such that $(i\partial_t + V)\ \psi = \frac{\sqrt{|\partial_{\mathbf{n}} V|}}{2}\ u$.

We call $\mathrm{ABC}_{2,m}^4$ the resulting approximation of $\widetilde{\mathrm{ABC}_2^4}$. Using again a Crank-Nicolson discretization of ψ, one gets the following approximate representation of $\mathrm{ABC}_{1,m}^4$

$$\begin{cases} \partial_{\mathbf{n}} v^{n+1} - i\displaystyle\sum_{k=0}^{m} a_k^m v^{n+1} + i\sum_{k=1}^{m} a_k^m d_k^m \varphi_k^{n+1/2} \\ \qquad\qquad + \mathrm{sg}(\partial_{\mathbf{n}} W^{n+1})\dfrac{\sqrt{|\partial_{\mathbf{n}} W^{n+1}|}}{2}\psi^{n+1/2} = 0, \\ i\,\dfrac{\varphi_k^{n+1} - \varphi_k^n}{\Delta t} + (W^{n+1} + d_k^m)\,\varphi_k^{n+1/2} = v^{n+1}, \\ i\,\dfrac{\psi^{n+1} - \psi^n}{\Delta t} + W^{n+1}\,\psi^{n+1/2} = \dfrac{\sqrt{|\partial_{\mathbf{n}} W^{n+1}|}}{2}v^{n+1}, \\ \varphi_k^0 = \psi^0 = 0, \end{cases} \tag{4.24}$$

for $1 \le k \le m$ and $x = x_{l,r}$.

5 Extensions to nonlinear problems

Following the developments in [4] for the cubic nonlinear Schrödinger (NLS), one can extend the derivation performed in Section 3.2 to cases in which the potential is formally replaced by a nonlinearity. To be more precise, we consider the following cubic (NLS) equation

$$\begin{cases} i\partial_t u + \partial_x^2 u + q|u|^2 u = 0, \ (x,t) \in \Omega_T, \\ u(x,0) = u_0(x), \qquad\qquad x \in \Omega. \end{cases} \tag{5.1}$$

The role of the potential $V(x,t)$ is now replaced by the cubic nonlinear term $q|u|^2(x,t)$. If $q > 0$ (resp. $q < 0$), the (NLS) equation is said to be focusing (resp. defocusing). This equation is well posed and has special solutions when dispersion and nonlinearity compensate, namely the soliton solution, which exhibits the specific behavior to propagate without modification of its amplitude. The cubic NLS equation is extremely interesting since it is the prototype of more general nonlinear dispersive equations and therefore it has received much attention these years. In the context of TBCs and ABCs, contributions can be found in the papers [4, 24, 26, 6].

If we formally replace the potential by the nonlinearity $q|u|^2$, the two strategies developped in the previous sections lead respectively to two different ABCs of order M that will be denoted by NLABC_1^M for strategy 1 and $\widetilde{\text{NLABC}}_2^M$-$\text{NLABC}_{2,m}^M$ for strategy 2. For strategy 1, the ABCs of order M are given by

$$\partial_{\mathbf{n}} u + \Upsilon_1^M u = 0, \quad \text{on } \Sigma_T, \tag{5.2}$$

with (NLABC_1^2)

$$\Upsilon_1^2 u = e^{-i\pi/4} e^{i\mathbb{V}(x,t)} \partial_t^{1/2} \left(e^{-i\mathbb{V}(x,t)} u \right)$$

and (NLABC_1^4)

$$\Upsilon_1^4 u = \Upsilon_1^2 u + i \,\text{sg}(\partial_{\mathbf{n}} q|u|^2) \frac{\sqrt{|\partial_{\mathbf{n}} q|u|^2|}}{2} e^{i\mathbb{V}(x,t)} I_t \left(\frac{\sqrt{|\partial_{\mathbf{n}} q|u|^2|}}{2} e^{-i\mathbb{V}(x,t)} u \right),$$

setting $\mathbb{V}(x,t) = \int_0^t q|u|^2(x,s)\, ds$.

For strategy 2, one gets $\widetilde{\text{NLABC}}_2^2$: $\partial_{\mathbf{n}} u - i\sqrt{i\partial_t + q|u|^2} u = 0$ and $\widetilde{\text{NLABC}}_2^4$:

$$\begin{aligned} &\partial_{\mathbf{n}} u - i\sqrt{i\partial_t + q|u|^2} u \\ &\quad + \text{sg}(\partial_{\mathbf{n}} q|u|^2) \frac{\sqrt{|\partial_{\mathbf{n}} q|u|^2|}}{2} (i\partial_t + q|u|^2)^{-1} \left(\frac{\sqrt{|\partial_{\mathbf{n}} q|u|^2|}}{2} u \right) = 0. \end{aligned}$$

The numerical treatment is slightly different from the linear Schrödinger equation with potential. Indeed, the semi-discrete approximation of the nonlinear term $q|u|^2 u$ is done following the Durán and Sanz-Serna scheme [12]. More precisely, we use the midpoint approximation $q|(u^{n+1} + u^n)/2|^2 (u^{n+1} + u^n)/2$. This differs from $q(|u^{n+1}|^2 + |u^n|^2)(u^{n+1} + u^n)/4$ which is the classical Crank-Nicolson approximation and corresponds to Eq. (4.2). Therefore, the semi-discrete time scheme reads

$$i\frac{u^{n+1} - u^n}{\Delta t} + \partial_x^2 \frac{u^{n+1} + u^n}{2} + q\left| \frac{u^{n+1} + u^n}{2} \right|^2 \frac{u^{n+1} + u^n}{2} = 0$$

which can be recast as follows

$$2i\frac{v^{n+1}}{\Delta t} + \partial_x^2 v^{n+1} + q|v^{n+1}|^2 v^{n+1} = 2i\frac{u^n}{\Delta t}, \tag{5.3}$$

where v^{n+1} denotes the midpoint term $(u^{n+1} + u^n)/2$. Since this scheme is now nonlinear, we solve it by a fixed-point procedure with error tolerance ε. The algorithm is described below:

```
let ζ^0 = v^n, s = 0
repeat
        solve the linear elliptic problem
        2i ζ^{s+1}/Δt + ∂_x^2 ζ^{s+1} = 2i u^n/Δt − q|ζ^s|^2 ζ^s
        s = s + 1
until ||ζ^{s+1} − ζ^s||_{L^2(Ω)} ≤ ε
v^{n+1} = ζ^{s+1}, u^{n+1} = 2v^{n+1} − u^n
```

The rule is to replace v^{n+1} by ζ^{s+1} if the corresponding term is linear and by ζ^s if one deals with a nonlinear one. We do not detail this step further and this principle is also applied to the numerical treatment of other nonlinear ABCs.

The numerical approximation of NLABC_1^4 is

$$\partial_{\mathbf{n}}\zeta^{s+1} + e^{-i\frac{\pi}{4}}\sqrt{\frac{2}{\Delta t}}\zeta^{s+1} = g^s \text{ on } \Sigma_T, \tag{5.4}$$

with

$$\begin{aligned} g^s = -e^{-i\frac{\pi}{4}}\sqrt{\frac{2}{\Delta t}}&\left(\widetilde{\mathbb{E}^n}\exp\left(iq\Delta t\frac{|\zeta^s|^2}{2}\right)\sum_{k=1}^{n}\beta_{n+1-k}\overline{\mathbb{E}^k}v^k\right) \\ -i\frac{q}{4}\partial_{\mathbf{n}}(|\zeta^s|^2)&\left(\frac{\Delta t}{2}\zeta^s + \Delta t\widetilde{\mathbb{E}^n}\exp\left(iq\Delta t\frac{|\zeta^s|^2}{2}\right)\sum_{k=1}^{n}\overline{\mathbb{E}^k}v^k\right). \end{aligned}$$

The notations $\mathbb{E}^p$ and $\widetilde{\mathbb{E}^{p-1}}$ are the quantities defined by

$$\begin{aligned} \mathbb{E}^p = \exp(i\mathbb{U}^p) &= \exp\left(iq\Delta t\sum_{l=1}^{p-1}\left|u^l\right|^2\right)\exp\left(iq\frac{\Delta t}{2}\left|u^p\right|^2\right) \\ &= \widetilde{\mathbb{E}^{p-1}}\exp\left(iq\frac{\Delta t}{2}\left|u^p\right|^2\right), \end{aligned} \tag{5.5}$$

setting $\mathbb{E}^0 = 1$ and $\mathbb{E}^1 = \exp(i\mathbb{U}^1)$.

The Crank-Nicolson scheme (5.3) coupled to (5.4) remains nonlocal in time since we have to deal with convolution terms. In this direction, $\mathrm{NLABC}_{2,m}^M$ are computationally more efficient since they are based on the Padé approximants and are therefore local in time. For example, $\mathrm{NLABC}_{2,m}^4$ reads

$$\begin{aligned} \partial_{\mathbf{n}}\zeta^{s+1} = i\sum_{k=0}^{m}a_k^m\zeta^{s+1} &- i\sum_{k=1}^{m}a_k^m d_k^m\left(\frac{\phi_k^s + \varphi_k^n}{2}\right) \\ &- \mathrm{sg}(\partial_{\mathbf{n}}q|\zeta^s|^2)\frac{\sqrt{|\partial_{\mathbf{n}}q|\zeta^s|^2|}}{2}\left(\frac{\chi^s + \psi^n}{2}\right), \end{aligned} \tag{5.6}$$

with

$$\phi_k^s = \left(\frac{i}{\Delta t} + q\frac{|\zeta^s|^2}{2} + \frac{d_k^m}{2}\right)^{-1} \left(\zeta^s + \varphi_k^n \left(\frac{i}{\Delta t} - q\frac{|\zeta^s|^2}{2} - \frac{d_k^m}{2}\right)\right),$$

where $\varphi_k^0 = 0$, $\forall k$, $\phi_k^0 = \varphi_k^n$ and

$$\chi^s = \left(\frac{i}{\Delta t} + q\frac{|\zeta^s|^2}{2}\right)^{-1} \left(\frac{\sqrt{|\partial_{\mathbf{n}} q|\zeta^s|^2|}}{2}\zeta^s + \left(\frac{i}{\Delta t} - q\frac{|\zeta^s|^2}{2}\right)\psi^n\right)$$

where $\psi^0 = 0$ and $\chi^0 = \psi^n$. When the convergence assumption

$$\|\zeta^{s+1} - \zeta^s\|_{L^2} \leq \varepsilon$$

is reached, one affects $\varphi_k^{n+1} = \phi_k^s$ and $\psi^{n+1} = \chi^s$.

6 Numerical examples

The aim of this section is to provide some test cases to validate our approach. We perform some experiments for Schrödinger equations with both variable potentials and nonlinearities. For each situation, we use a variational formulation of the semi-discrete time problem with n_h linear finite elements (with spatial size h) and integrate the ABCs in the corresponding scheme as a Fourier-Robin boundary condition. This leads to a tridiagonal banded matrix. The solution to the associated linear system is then simple and is realized by a direct LU solver.

6.1 Linear Schrödinger equation with variable potential

We consider the initial gaussian datum $u_0(x) = e^{ik_0x-x^2}$, where k_0 designates the wave number fixed to $k_0 = 10$ in our simulations. This choice, like the nonlinear Schrödinger equation, is related to the fact that our ABCs are derived under a high frequency hypothesis. We present here one kind of potential: $V(x,t) = 5xt$ (more examples are available in [5]). The computational domain is $\Omega =]-5;10[$. The final time of computation is $T = 2.5$. The spatial step size is $h = 2.5 \times 10^{-3}$ for the linear finite element method and the time step is $\Delta t = 10^{-4}$. We present in Figure 6.1 the quantity $\log_{10}(|u(x,t)|)$ in the domain Ω_T. We begin by reporting the reference solution (top left) computed on a larger domain to avoid any effect related to spurious reflection at the boundary. Next, we present (top) the solutions using ABC_1^2 and ABC_1^4 which show that increasing the order of the boundary conditions yields smaller undesired back reflections. Finally, we compare the effect of the localization based on the Padé approximation of order m for the second-order ABC and strategy 2. We choose $m = 20$ ($\text{ABC}^2_{2,20}$) and $m = 50$ ($\text{ABC}^2_{2,50}$) terms. To give an equivalent precision to ABC^2_1, $m = 50$ is required. However, we note here that this leads to a negligible additional cost compared to $m = 20$. We also see on the right bottom picture that the precision of ABC^4_2 is conserved for $\text{ABC}^4_{2,50}$. All these simulations show that the proposed ABCs have increasing accuracy according to the order M, with similar accuracy for the same order when a localization process is applied.

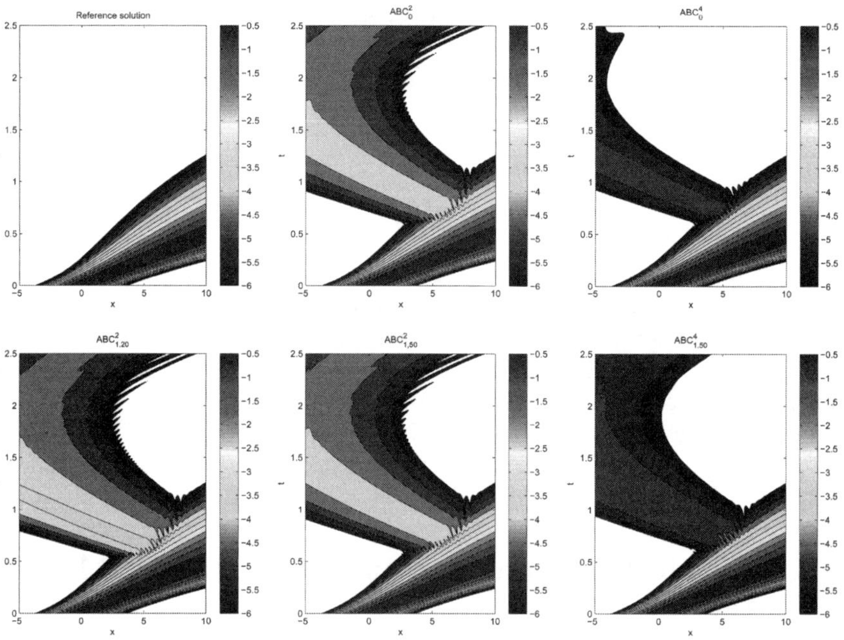

Figure 6.1 Log_{10} representation of the amplitude of the computed solutions for $V(x, t) = 5xt$. From left to right, top: reference solution, ABC^2_1, ABC^4_1; bottom: $\text{ABC}^2_{2,20}$, $\text{ABC}^2_{2,50}$, $\text{ABC}^4_{2,50}$.

6.2 Nonlinear Schröinger equation

The one-dimensional cubic nonlinear Schrödinger equation is integrable according to the inverse scattering theory [25]. This approach yields the so-called exact *soliton* solution given by

$$u_{\text{ex}}(x,t) = \sqrt{\frac{2a}{q}} \text{sech}\left(\sqrt{a}(x-ct)\right) \exp\left(i\frac{c}{2}(x-ct)\right) \exp\left(i\left(a+\frac{c^2}{4}\right)t\right).$$

From now on, we fix the focusing parameter q to 1. The real parameter a, equaling 2 here, characterizes the amplitude of the wavefield. Finally, c is the velocity of the soliton. Like what is described in the previous subsection, since the derivation of the nonlinear artificial boundary conditions has been constructed under a high-frequency assumption ($|\tau|$ large), we take $c = 15$. All along the computations, we consider $\varepsilon = 10^{-6}$ in the fixed-point algorithm. The numerical parameters are $\Delta t = 10^{-3}$ for a final time $T = 2$. The finite computational spatial domain is $\Omega =]-10, 10[$ discretized with $n_h = 4000$ equally spaced points ($h = 0.5 \times 10^{-2}$). Concerning the Padé approximation, we choose $m = 50$ since this is an optimal choice for the potential test cases.

To focus on the spurious reflections link to the different methods, we plot the contour of $\log_{10}(|u|)$ in Figs. 6.2–6.6. We see in Fig. 6.2 that the maximal reflection is approximately equal to 10^{-2} for an initial amplitude of 2 and the linear TBC (2.2). For Figs. 6.3–6.6, the reflection attains a maximal value around 5×10^{-3}. The reflection occurring at the right boundary decreases according to the order M of the different conditions NLABC_1^M or $\text{NLABC}_{2,m}^M$. Moreover, the most accurate results are obtained for the condition NLABC_4^2 with a minimal region of maximal reflection. Unlike the linear TBC, the reflection at the left boundary has an amplitude smaller than 10^{-4}.

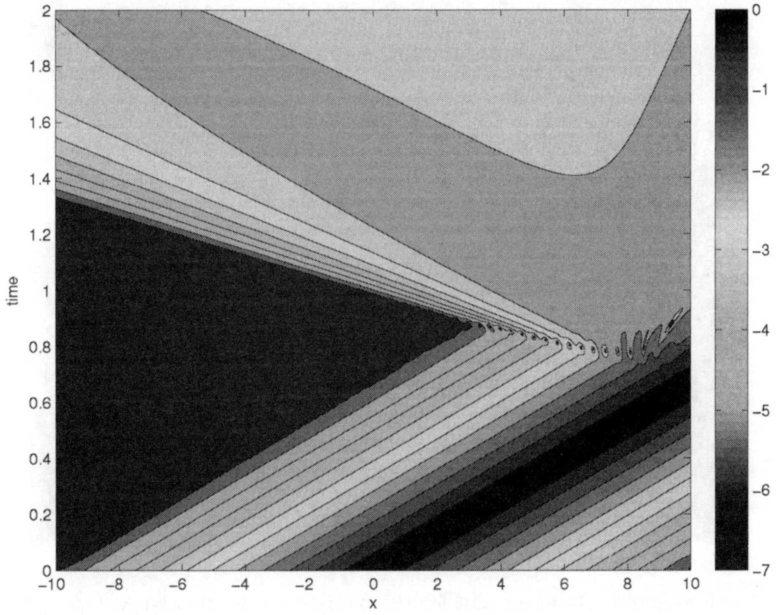

Figure 6.2 Contour plot of $\log_{10}(|u|)$ for the linear TBC (2.2).

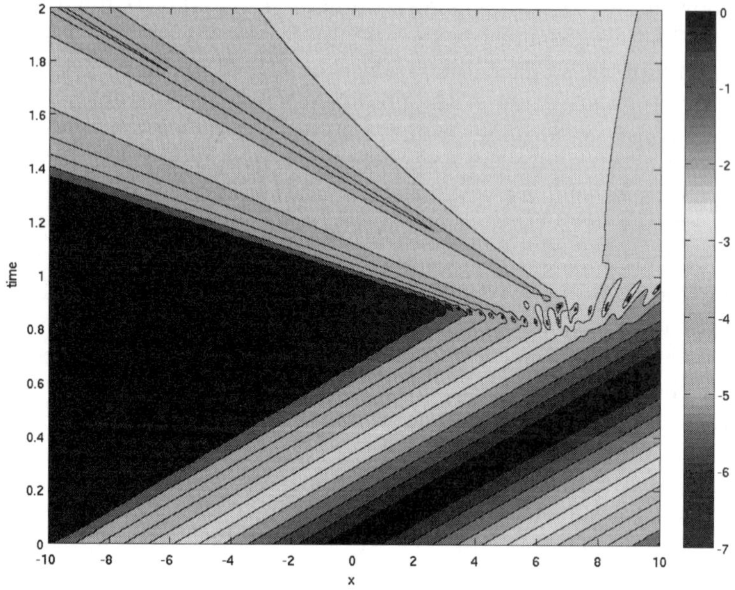

Figure 6.3 Contour plot for the boundary condition NLABC_1^2.

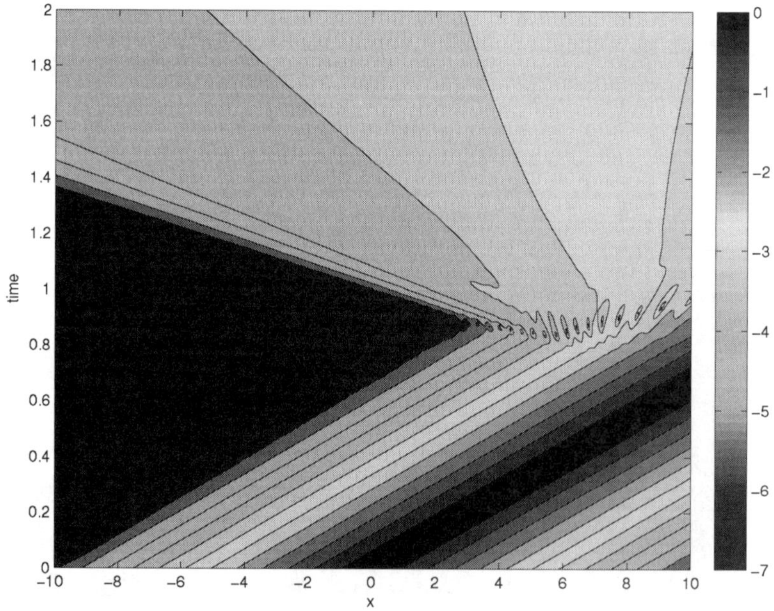

Figure 6.4 Contour plot for the boundary condition NLABC_1^4.

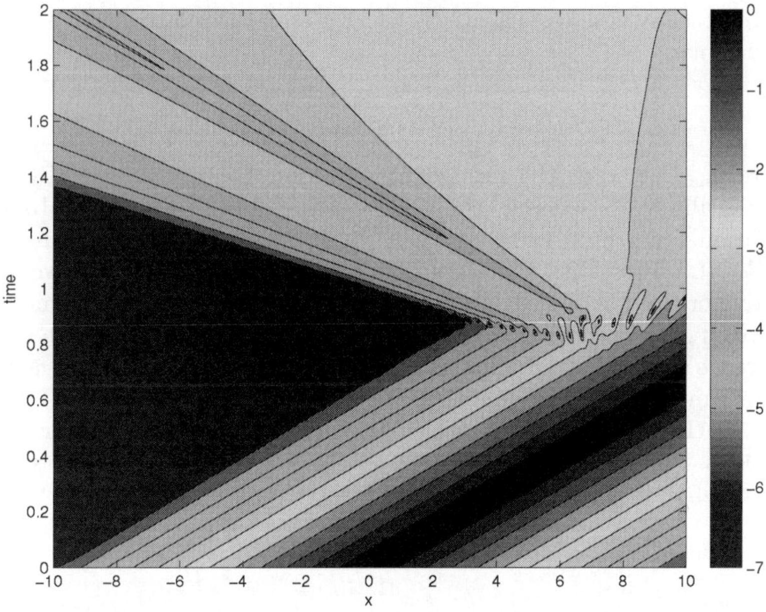

Figure 6.5 Contour plot for the boundary condition $\mathrm{NLABC}_{2,50}^{2}$.

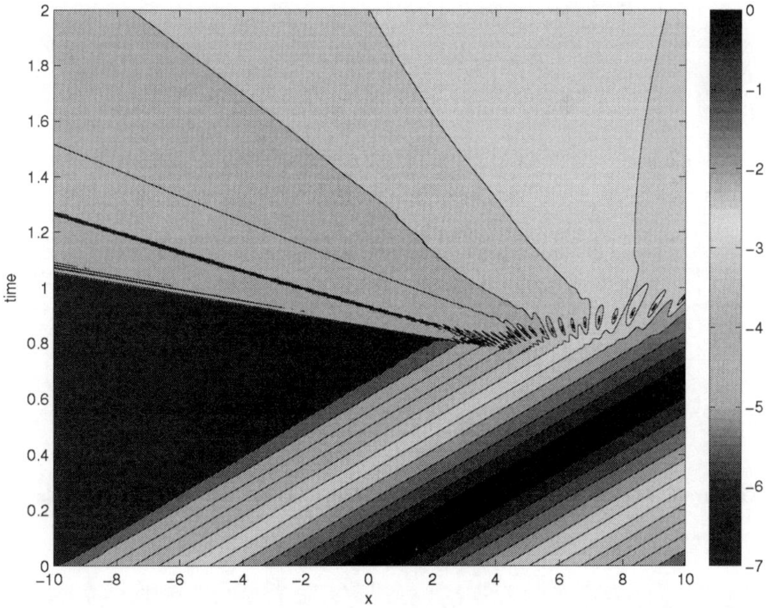

Figure 6.6 Contour plot for the boundary condition $\mathrm{NLABC}_{2,50}^{4}$.

To precise these results, we plot in Fig. 6.7 the relative errors for the $L^2(\Omega)$-norm

$$\frac{\|u_{\text{ex}} - u_{\text{num}}\|_{L^2(\Omega)}}{\|u_{\text{num}}\|_{L^2(\Omega)}},$$

where u_{num} denotes the numerical solution. For the linear TBC, the error is about 2% whereas the best result is obtained for the NLABC_1^4 condition for a final error of 0.2%. It is interesting to note that the ABCs NLABC_2^M with Padé approximations are very competitive. The relative errors for NLABC_2^2 and $\text{NLABC}_{2,50}^2$ are exactly the same, and $\text{NLABC}_{2,50}^4$ is between NLABC_2^2 and NLABC_2^4 methods, with the main difference being that methods based on Padé approximations are local in time and easy to implement. However, the fact that NLABC_2^4 and $\text{NLABC}_{2,m}^4$ are not numerically equivalent requires further investigations. Indeed, for the variable potential cases, we obtained similar results while it is no longer the case here.

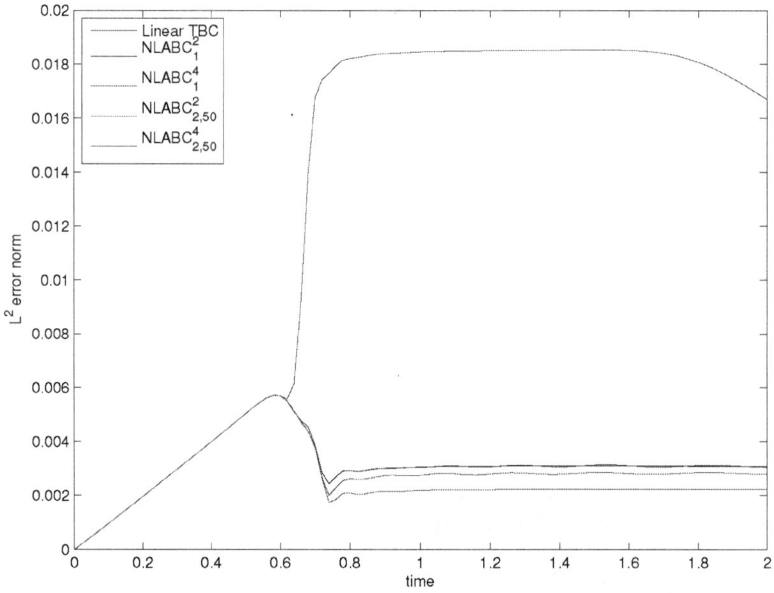

Figure 6.7 Relative error for the different linear and nonlinear ABCs.

7 Conclusion

We have introduced various constructions of Absorbing Boundary Conditions (ABCs) for the one-dimensional Schrödinger equation with time-

and space-variable repulsive potentials and for the one-dimensional nonlinear cubic Schrödinger equation. They are derived with the help of general pseudodifferential techniques and applied to variable potentials and nonlinear equations. New accurate and efficient Absorbing Boundary Conditions for the nonlinear cubic Schrödinger equation are proposed. Numerical examples compare the different ABCs of various orders, showing that fourth-order ABCs yield accurate computations, and that Padé-based approximations are accurate while they are also efficient. Further studies will include other nonlinearities as well as extensions to higher dimensions.

References

[1] M. Abramowitz and I. A. Stegun. *Handbook of Mathematical Functions with Formulas, Graphs, and Mathematical Tables, 10th Printing*, volume 55 of *National Bureau of Standards Applied Mathematics Series.* For sale by the Superintendent of Documents, U.S. Government Printing Office, Washington, D.C., 1964.

[2] X. Antoine, A. Arnold, C. Besse, M. Ehrhardt and A. Schädle. A review of transparent and artificial boundary conditions techniques for linear and nonlinear Schrödinger equations. *Commun. Comput. Phys.*, **4**(4): 729–796, 2008.

[3] X. Antoine and C. Besse. Unconditionally stable discretization schemes of non-reflecting boundary conditions for the one-dimensional Schrödinger equation. *J. Comput. Phys.*, **188**(1): 157–175, 2003.

[4] X. Antoine, C. Besse and S. Descombes. Artificial boundary conditions for one-dimensional cubic nonlinear Schrödinger equations. *SIAM J. Numer. Anal.*, **43**(6): 2272–2293 (electronic), 2006.

[5] X. Antoine, C. Besse and P. Klein. Absorbing boundary conditions for the one-dimensional Schrödinger equation with an exterior repulsive potential. *J. Comput. Phys.*, **228**(2): 312–335, 2009.

[6] X. Antoine, C. Besse and J. Szeftel. Towards accurate artificial boundary conditions for nonlinear pdes through examples. *Cubo, A Mathematical Journal*, 2009.

[7] A. Arnold. Numerically absorbing boundary conditions for quantum evolution equations. *VLSI Design*, **6**(1–4): 313–319, 1998.

[8] V. Baskakov and A. Popov. Implementation of transparent boundaries for numerical solution of the Schrödinger equation. *Wave Motion*, **14**: 123–128, 1991.

[9] R. Carles. Linear vs. nonlinear effects for nonlinear Schrödinger equations with potential. *Commun. Contemp. Math.*, **7**(4): 483–508, 2005.

[10] R. Carles. *Semi-classical analysis for nonlinear Schrödinger equations.* World Scientific, 2008.

[11] R. Dalrymple and P. Martin. Perfect boundary conditions for parabolic water-wave models. *Proc. R. Soc. Lond. A*, **437**: 41–54, 1992.

[12] A. Durán and J. M. Sanz-Serna. The numerical integration of relative equilibrium solutions, the nonlinear Schrödinger equation. *IMA J. Numer. Anal.*, **20**(2): 235–261, 2000.

[13] J. Hellums and W. Frensley. Non-markovian open-system boundary conditions for the time-dependent Schrödinger equation. *Phys. Rev. B*, **49**: 2904–2906, 1994.

[14] R. Lascar. Propagation des singularités des solutions d'équations pseudodifférentielles quasi-homogènes. *Ann. Inst. Fourier (Grenoble)*, **27**(2): 79–123, 1977.

[15] M. Levy. *Parabolic equation methods for electromagnetic wave propagation*, volume 45 of *IEE Electromagnetic Waves Series*. Institution of Electrical Engineers (IEE), London, 2000.

[16] E. Lindmann. Free-space boundary conditions for the time dependent wave equation. *J. Comput. Phys.*, **18**: 16–78, 1985.

[17] E. Lorin, A. Bandrauk and S. Chelkowski. Numerical Maxwell-Schrödinger model for laser-molecule interaction and propagation. *Comput. Phys. Comm.*, **177**(12): 908–932, 2007.

[18] F. Milinazzo, C. Zala and G. Brooke. Rational square-root approximations for parabolic equation algorithms. *J. Acoust. Soc. Amer.*, **101**(2): 760–766, 1997.

[19] L. Nirenberg. *Pseudodifferential operators and some applications*, volume 17 of *Regional Conf. Ser. in Math. AMS 17, Lectures on Linear Partial Differential Equations.* AMS, 1973.

[20] J. Papadakis. Impedance formulation of the bottom boundary condition for the parabolic equation model in underwater acoustics. Technical Report 143, NORDA Tech. Note, 1982. NORDA Parabolic Equation Workshop.

[21] J. Papadakis. Exact nonreflecting boundary conditions for parabolic-type approximations in underwater acoustics. *J. Comput. Acoust.*, **2**: 83–98, 1994.

[22] A. Pazy. *Semigroups of Linear Operators and Applications to Partial Differential Equations.* Springer, New York, 1983.

[23] M. Reed and B. Simon. *Methods of Modern Mathematical Physics II: Fourier Analysis, Self-adjointness.* Academic Press, San Diego, 1975.

[24] J. Szeftel. Absorbing boundary conditions for nonlinear Schrödinger equations. *Numer. Math.*, **104**: 103–127, 2006.

[25] V. Zakharov. *Theory of Solitons, The Inverse Scattering Method.* Contemp. Soviet Math., 1968.

[26] C. Zheng. A perfectly matched layer approach to the nonlinear Schrödinger wave equations. *J. Comput. Phys*, **227**: 537–556, 2007.

[27] C. Zheng. An exact absorbing boundary condition for the Schrödinger equation with sinusoidal potentials at infinity. *Comm. Comput. Phys.*, **3**(3): 641–658, 2008.

On Hydrodynamic Models for LEO Spacecraft Charging*

Christophe Besse[†], Saja Borghol[†], Jean-Paul Dudon[‡], Thierry Goudon[†], Ingrid Lacroix-Violet[†]

[†]*Project Team SIMPAF*
INRIA Lille Nord Europe Research Centre
Park Plazza, 40 avenue Halley
59650 Villeneuve d'Ascq cedex, France
& Laboratoire Paul Painlevé UMR 8524 CNRS-USTLille

[‡]*Thales Alenia Space*
100, bd. du Midi, 06156 Cannes La Bocca Cedex, France

Email: christophe.besse@math.univ-lille1.fr
jean-paul.dudon@thalesaleniaspace.com
saja.borghol@math.univ-lille1.fr
thierry.goudon@inria.fr
ingrid.violet@math.univ-lille1.fr

Abstract

This paper is devoted to hydrodynamic models intending to describe charging phenomena and the spacecraft evolving in Low Earth Orbits (LEO) are dealt with. The models we are interested in couple the stationary Euler equations to the Poisson equation which defines the electric potential. Furthermore, the charging dynamics is embodied into the boundary conditions where the time derivative of the potential appears. We point out the main mathematical difficulties by restricting to a 1D caricature model for which we present rigorous existence results and numerical simulations.

*This research is partly supported by a contract with Thalesalenia Space.

1 Introduction

A spacecraft evolves in the space plasma and interacts with it. These complex interactions, due to the different dielectric properties of the materials on the surface of the spacecraft, can induce the apparition of severe potential differences which, in turn, produce electric arcing. These phenomena are sources of in-orbit failures since the arcing can lead to irreversible damage on the in-board devices or on the solar arrays. Therefore, the prevention of the apparition of excessive electric charges has motivated an intense research in space engineering in order to design efficient procedures of numerical simulations (see [21], [6], [22]). This effort requires an important preliminary step on modeling issues. A basis model is clearly based on the Vlasov-Maxwell-Boltzmann (or Fokker-Planck) equations for describing both the motion of the charged particles and the variations of the electro-magnetic fields. The nonlinear system of PDEs is completed by suitable boundary conditions on the surfaces of the satellite and equilibrium conditions at infinity. The charging phenomenon is precisely driven by the boundary conditions on the spacecraft surface for the electromagnetic field and the densities. We shall see that their expression, which involves the dielectric properties of the different materials on the surface, makes the problem highly non-standard. Moreover, taking account of the specific features of the plasma environment can help to reduce the complexity of the model, and we can actually decline a hierarchy of possible models. In the next section we describe some aspects of the derivation of the models, emphasizing the specificities of GEostationary Orbits (GEO) and Low Earth Orbits (LEO) environments. In Section 3 we derive a simpler one-dimensional model which helps point out several interesting features of the problem. This is completed by theoretical results in Section 4 and numerical simulations in Section 5.

2 Kinetic, hydrodynamics models and potential boundary conditions

2.1 Generalities

We suppose that the plasma consists in two charged particles species: ions H^+ and electrons. We denote by f_i and f_e respectively the distribution functions of these species: $f_{i/e}(t,x,v)\,\mathrm{d}v\,\mathrm{d}x$ stands for the number of ions (respectively electrons) in the domain centered at the point (x,v) of the phase space with infinitesimal volume $\mathrm{d}v\,\mathrm{d}x$ at time $t \geq 0$. Let $q_i = -q_e = q > 0$ be the elementary charge, let m_i and m_e be the ion mass and the electron mass, respectively. The evolution of the charged particles obeys the following PDEs

$$\partial_t f_{i/e} + v \cdot \nabla_x f_{i/e} + \frac{q_{i/e}}{m_{i/e}}(E + v \wedge B) \cdot \nabla_v f_{i/e} = C_{i/e}(f_i, f_e), \quad (2.1)$$

which is coupled to the Maxwell equations for the electro-magnetic field (E, B):

$$\epsilon_0\big(- \partial_t E + c^2 \text{curl}_x B\big) = J_i + J_e, \quad (2.2)$$

$$\text{div}_x(\epsilon_0 E) = q(n_i - n_e), \quad (2.3)$$

$$c^2 \partial_t B + \text{curl}_x E = 0, \quad (2.4)$$

$$\text{div}_x B = 0, \quad (2.5)$$

where ε_0 and c stand for the vacuum permittivity and the light speed respectively and we denote

$$n_{i/e} = \int_{\mathbb{R}^3} f_{i/e}\, \mathrm{d}v, \qquad J_{i/e} = q_{i/e} \int_{\mathbb{R}^3} v\ f_{i/e}\, \mathrm{d}v.$$

In (2.1) the right hand side contains the collision dynamics between the particles (electron/electron, ion/ion and electron/ion), and the operator $C_{i/e}$ being of Boltzmann or Fokker-Planck type (see [10, 13]). However, except for very specific flights (e. g. in Polar Earth Orbits), the magnetic effects can be neglected so that the Maxwell equations (2.2)–(2.5) can be replaced by a mere Poisson equation for the electric potential. Indeed, let us introduce the electric potential $\Phi(t, x)$: the electric field is defined by $E = -\nabla_x \Phi$. Then, (2.1) reduces to

$$\partial_t f_{i/e} + v \cdot \nabla_x f_{i/e} - \frac{q_{i/e}}{m_{i/e}} \nabla_x \Phi \cdot \nabla_v f_{i/e} = C_{i/e}(f_i, f_e), \quad (2.6)$$

where (2.3) leads to the Poisson equation

$$-\text{div}_x(\epsilon_0 \nabla_x \Phi) = q(n_i - n_e). \quad (2.7)$$

Equations (2.1)–(2.5) hold for $t \geq 0$, $x \in \Omega$, $v \in \mathbb{R}^3$, where $\Omega \subset \mathbb{R}^3$ represents the exterior of the satellite. Therefore the problem should be tackled with boundary conditions for the potential and the distribution functions. First of all, far from the spacecraft the plasma is supposed to be in an equilibrium state, thus, at infinity, we assume that

$$\begin{aligned} &\Phi(t, x) \xrightarrow[|x| \to \infty]{} 0, \\ &f_{i/e}(t, x, v) \xrightarrow[|x| \to \infty]{} \frac{n^\infty_{i/e}}{(2\pi \Theta^\infty_{i/e})^{3/2}} \exp\Big(- \frac{v^2}{2\Theta^\infty_{i/e}}\Big) \end{aligned} \quad (2.8)$$

holds with $n^\infty_{i/e} > 0$ and $\Theta^\infty_{i/e} > 0$ giving densities and temperatures for the ions and the electrons.

Second of all, on the spacecraft the particles distributions obey

$$\gamma_{inc} f_{i/e} = \mathcal{R}\big(\gamma_{out} f_{i/e}\big) + S \quad \text{for } v \cdot \nu(x) < 0 \tag{2.9}$$

where $\nu(x)$ stands for the outward unit vector at point $x \in \partial\Omega$, γ_{inc} denotes the trace operator on the incoming set $\big\{(x, v) \in \partial\Omega \times \mathbb{R}^3 \text{ s. t. } v \cdot \nu(x) < 0\big\}$, and γ_{out} denotes the trace operator on the outgoing set $\big\{(x, v) \in \partial\Omega \times \mathbb{R}^3 \text{ s. t. } v \cdot \nu(x) > 0\big\}$. The linear operator $\mathcal{R}$ describes how impinging particles are reflected by the walls; for instance we can use the simple specular reflection law

$$\mathcal{R}f(x, v) = \alpha\ f\big(x, v - 2(v \cdot \nu(x))\nu(x)\big)$$

with $\alpha \in (0, 1)$ being an accommodation coefficient. Varying the value of α can be regarded as a model of the photo-emission. When flying in darkness the spacecraft surfaces are absorbing ($\alpha = 0$) whereas exposition to light causes emission of particles ($\alpha > 0$). Finally, S is a source term accounting for possible emission of charged particles by the surface. Let us now describe, according to [5], the boundary condition for the potential which is the most original part of the model.

The spacecraft can be considered a perfect conductor, partially covered by an assembly of dielectric materials. We denote by $\mathcal{O}_0$ the conductor, and $\mathcal{O}_k$, $k \in \{1, \ldots, N_d\}$ the dielectrics which are characterized by their permittivity $\varepsilon_k > 0$ and conductivity $\sigma_k > 0$. The height of the kth dielectric layer is denoted by d_k. The plasma fills the domain $\Omega = \mathbb{R}^3 \setminus \overline{\bigcup_{k=0}^{N_d} \mathcal{O}_k}$. We set $\Gamma = \bigcup_{k=0}^{N_d} \partial\mathcal{O}_k$ and for a given point $x \in \Gamma$, $\nu(x)$ stands for the normal vector at the surface Γ (pointing outward the considered domain). We consider the following interfaces (see Fig. 2.1):

- $\Gamma_{c/v} = \Gamma \setminus \bigcap_{k=0}^{N_d} \partial\mathcal{O}_k$, the interface between the conductor and the vacuum,
- $\Gamma_{c/d} = \partial\mathcal{O}_0 \setminus \Gamma_{c/v}$, the interface between the conductor and the dielectrics,
- $\Gamma_{d/v} = \partial\Omega \setminus \Gamma_{c/v}$, the interface between the dielectrics and the vacuum,
- $\Gamma_{d/d} = \Gamma \setminus (\partial\mathcal{O}_0 \cup \partial\Omega)$, the interface between neighbors dielectrics.

The boundary conditions for Φ can be deduced from the Maxwell equations considered in the whole space $\mathbb{R}^3$ and bearing in mind that the different parts of the spacecraft have different electric behavior. At any place of the conductor, the electric potential remains at a constant value: we denote by $\phi_{abs}(t)$, the so-called "absolute potential", the value of the potential at time t. Particularly, we have

$$\Phi(t, x) = \phi_{abs}(t) \text{ on } \Gamma_{c/v}. \tag{2.10}$$

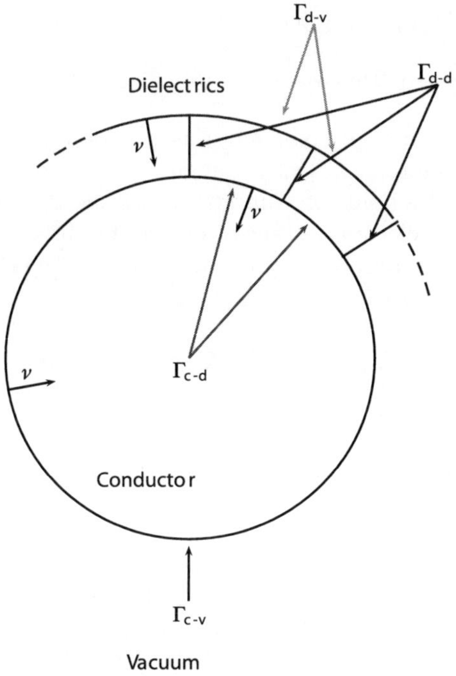

Figure 2.1 Domain and notations of interfaces.

In the dielectrics, there exists a runaway current, proportional to the electric field $J_k = -\sigma_k \nabla_x \Phi_{diel}$. Then, we consider the jump relations associated with the Ampère law (2.2) recasts as $\partial_t \mathrm{div}_x(\varepsilon \nabla_x \Phi) = \mathrm{div}_x(J)$. Denoting $J_{ext} = J_i + J_e$, we get

$$\partial_t(\epsilon_k \partial_\nu \Phi_{diel} - \epsilon_0 \partial_\nu \Phi) + J_{ext} \cdot \nu + \sigma_k \partial_\nu \Phi_{diel} = 0 \quad \text{on } \Gamma_{d/v} \tag{2.11}$$

together with the relation

$$\int_{\Gamma_{c/v}} \Big[\partial_t(-\epsilon_0 \partial_\nu \Phi) + J_{ext} \cdot \nu\Big] \, d\gamma + \int_{\Gamma_{c/d}} \Big[\partial_t(-\epsilon_k \partial_\nu \Phi_{diel}) - \sigma_k \partial_\nu \Phi_{diel}\Big] \, d\gamma = 0. \tag{2.12}$$

Since the dielectric layer is very thin, which means that the d_k's are small compared to the characteristic lengths of the spacecraft, and the normal derivative of the dielectric potential on $\Gamma_{c/d}$ and $\Gamma_{d/v}$ is approached by

$$\partial_\nu \Phi_{diel}(t, x) \simeq \frac{\phi_{abs}(t) - \Phi(t, x)}{d_k}. \tag{2.13}$$

Finally (2.10), (2.11), (2.12) and (2.13) define the boundary conditions for the potential.

2.2 From GEO to LEO

The most studied environment relies on the geostationary orbits (GEO) which yields further simplifications, based on asymptotic considerations. There, the plasma can be considered as collisionless, that is, $C_{i/e} = 0$ in (2.6). Furthermore, the Debye length is large and the evolution of the charged particles holds on a larger time scale than on the time scale of evolution of the electric potential on the boundary. Eventually, the GEO charging of a spacecraft is thus described by the stationary Vlasov-Poisson equations

$$\begin{cases} v \cdot \nabla_x f_{i/e} - \dfrac{q_{i/e}}{m_{i/e}} \nabla_x \Phi \cdot \nabla_v f_{i/e} = 0, \\ \Delta_x \Phi = 0, \end{cases}$$

with the boundary conditions (2.8), (2.9) and (2.10)–(2.12). Note that the problem remains time-dependent due to the time derivative in (2.11) which governs the evolution of the charging phenomena. We refer to [5] for an introduction to this model, particularly for the discussion of the potential boundary conditions. The model is currently used in GEO codes (see [3, 6, 4, 1]). In this paper we are rather interested in Low Earth Orbits (it means orbits with an altitude between 100 and 2000 km whereas GEO is around 36.000 km). Since the plasma is more dense with a smaller mean free path, the use of hydrodynamic models becomes reasonable, at least in the first approximation. This is interesting for numerical purposes since by getting rid of the velocity variable, it is allowed to reduce the size of the unknowns. The model can be derived as follows: Bearing in mind the standard collision operators in plasma physics, electron/electron and ion/ion collisions preserve mass, impulsion and energy and relax towards equilibrium state which are the Maxwellian functions. After integration of (2.6) we obtain

$$\partial_t \int_{\mathbb{R}^3} \begin{pmatrix} 1 \\ v \\ v^2 \end{pmatrix} f_{i/e}\, \mathrm{d}v + \nabla_x \int_{\mathbb{R}^3} v \begin{pmatrix} 1 \\ v \\ v^2 \end{pmatrix} f_{i/e}\, \mathrm{d}v \\ + \frac{q_{i/e}}{m_{i/e}} \nabla_x \Phi \cdot \int_{\mathbb{R}^3} \begin{pmatrix} 0 \\ 1 \\ 2v \end{pmatrix} f_{i/e}\, \mathrm{d}v = \int_{\mathbb{R}^3} \begin{pmatrix} 1 \\ v \\ v^2 \end{pmatrix} C_{i/e}(f_i, f_e)\, \mathrm{d}v. \tag{2.14}$$

Actually, the right hand side only retains the momentum and energy exchanges between the two species due to the electron/ion collisions. Of course, this set of moment equations is not closed since higher moments appear in the convection terms. However, dealing with collision-dominated flows, the distribution functions relax to Maxwellians and replacing $f_{i/e}$ by the corresponding $\frac{n_{i/e}}{(2\pi\Theta_{i/e})^{3/2}} \exp\left(-\frac{|v-u_{i/e}|^2}{2\Theta_{i/e}}\right)$ we are led to the Euler equations satisfied by the density $n_{i/e}$, velocity $u_{i/e}$ and temperature $\Theta_{i/e}$

$$\begin{cases} \partial_t n_{i/e} + \mathrm{div}_x(n_{i/e}u_{i/e}) = 0, \\ m_{i/e}\big(\partial_t(n_{i/e}u_{i/e}) + \mathrm{Div}_x((n_{i/e}u_{i/e}) \otimes u_{i/e})\big) + \nabla_x p_{i/e} \\ \qquad = -q_{i/e}n_{i/e}\nabla\Phi - kq_{i/e}n_e n_i(u_i - u_e), \\ \partial_t w_{i/e} + \mathrm{div}_x(w_{i/e}u_{i/e} + p_{i/e}u_{i/e}) \\ \qquad = -q_{i/e}n_{i/e}\nabla_x\Phi \cdot u_{i/e} - kq_{i/e}n_e n_i(u_i - u_e)\cdot u_{i/e} \\ \qquad\qquad -\kappa q_{i/e}n_e n_i(\Theta_i - \Theta_e) \end{cases} \tag{2.15}$$

with $w_{i/e} = \frac{m_{i/e}}{2}n_{i/e}|u_{i/e}|^2 + \frac{3}{2}n_{i/e}\Theta_{i/e}$. Here we denote by div the standard divergence of a vector and by Div the divergence of a matrix.

On the right hand side, the term $kq_{i/e}n_e n_i(u_i - u_e)$ is a drag force associated with the momentum exchanges between the two species, due to the ion-electron collisions. Similarly $\kappa q_{i/e}n_e n_i(\Theta_i - \Theta_e)$ represents the energy exchanges due to the ion-electron collisions. We refer on this derivation to classical textbooks in plasmas physics [2, 10, 13]. The equation is completed by the perfect gas law

$$p_{i/e} = n_{i/e}\Theta_{i/e}.$$

The force field is still given by the Poisson equation

$$-\varepsilon_0\Delta_x\Phi = q(n_i - n_e) \tag{2.16}$$

endowed with the boundary conditions

$$\begin{gathered} \int_{\Gamma_{c/v}} \big[\partial_t(-\epsilon_0\partial_\nu\Phi) + J_{ext}\cdot\nu\big]\,\mathrm{d}\gamma \\ \qquad + \int_{\Gamma_{c/d}} \big[\partial_t(-\epsilon_k\frac{\phi_{abs}-\Phi}{d_k}) - \sigma_k\frac{\phi_{abs}-\Phi}{d_k}\big]\,\mathrm{d}\gamma = 0, \\ \partial_t\big(\epsilon_k\frac{\phi_{abs}-\Phi}{d_k} - \epsilon_0\partial_\nu\Phi\big) + \sigma_k\frac{\phi_{abs}-\Phi}{d_k} + J_{ext}\cdot\nu = 0 \quad \text{on } \Gamma_{d/v}, \\ \Phi(t,x) = \phi_{abs}(t) \text{ on } \Gamma_{c/v}, \\ \lim_{\|x\|\to+\infty} \Phi(t,x) = 0, \end{gathered} \tag{2.17}$$

with

$$J_{ext} = q(n_i u_i - n_e u_e) + J_S,$$

where J_S describes the possible emission current of particles from the boundary.

The derivation of relevant boundary conditions for the macroscopic quantities $(n_{i/e}, u_{i/e}, \Theta_{i/e})$ is an issue. The difficulty is two-fold:

– On the one hand, we deal with a hyperbolic system so that we should prescribe only the incoming fields. We refer to [12] for a deep discussion in this aspect.

– On the other hand, the Maxwellian state is usually not compatible with the kinetic boundary condition (2.9). Hence a kinetic boundary layer, the so-called Knudsen layer, should be taken into account (see [17, 25]), for a more practical viewpoint [11]. Remark that a conservative boundary condition that $J_{ext} \cdot \nu = 0$, for instance, with full reflection $\alpha = 1$ and no source $S = 0$ in (2.9), has no interest for the charging phenomena; we refer to [1] for similar remarks.

This aspect of the problem is particularly relevant, but it is beyond the scope of the present paper.

Next, asymptotic considerations are allowed to derive a hierarchy of possible models. Indeed, for LEO regimes the following reasoning can be applied:

- The charging time can still be considered small compared to the typical time scale of the fluid evolution. This leads to replacing the evolution equation in (2.15) by their stationary version:

$$\begin{cases} \mathrm{div}_x(n_{i/e}u_{i/e}) = 0, \\ m_{i/e}\mathrm{Div}_x((n_{i/e}u_{i/e}) \otimes u_{i/e}) + \nabla p_{i/e} \\ \qquad\qquad = -q_{i/e}n_{i/e}\nabla_x\Phi - kq_{i/e}n_e n_i(u_i - u_e), \\ \mathrm{div}_x(w_{i/e}u_{i/e} + p_{i/e}u_{i/e}) \\ \qquad\qquad = -q_{i/e}n_{i/e}\nabla_x\Phi \cdot u_{i/e} - kq_{i/e}n_e n_i(u_i - u_e) \cdot u_{i/e} \\ \qquad\qquad\qquad -\kappa q_{i/e}n_e n_i(\Theta_i - \Theta_e) \end{cases} \tag{2.18}$$

  coupled to (2.16). Time appears as a parameter in these equations and the problem remains subject to time evolution through the boundary conditions (2.17).

- A further approximation comes by assuming that the ions/electrons temperatures depend only on the densities

$$\Theta_{i/e} = \Theta^0_{i/e}\, n_{i/e}^{\gamma_{i/e}-1}, \qquad \gamma_{i/e} \geq 1, \quad \Theta^0_{i/e} > 0,$$

  which leads to isentropic ($\gamma_{i/e} > 1$) or isothermal ($\gamma_{i/e} = 1$) models.

- Then the classical asymptotics $m_e/m_i \ll 1$ and the quasi-neutral regime where the Debye length is small compared to the characteristic length of the spacecraft make sense for this application. The situation differs completely from the GEO case: in GEO the Debye length is of order 10–100m, but it is of order of a few centimeters in LEO. A rigorous justification of these asymptotics is a very tough piece of analysis; we mention for instance [18, 24, 27] for the treatment of some specific situations, including a complete description of the boundary layers, and further references to these topics.

3 A simple 1D model

In this section we consider a one-dimensional caricature of the LEO charging problem. Despite its simplicity, this model is interesting since it is permitted to bring out certain mathematical difficulties and to evaluate easily the efficiency of numerical schemes. In this model the spacecraft is treated as a scatterer occupying the domain $\mathcal{O} = (-h_d, h_c)$ where $\mathcal{O}_1 = (-h_d, 0)$ is occupied by a dielectric material whereas $\mathcal{O}_0 = (0, h_c)$ is the conductor domain. The plasma fills the domain $\Omega = (-L-h_d, -h_d) \cup (h_c, L+h_c)$. Bearing in mind numerical purposes, we consider a bounded domain, characterized by $0 < L < \infty$, but L is thought of as a "large" quantity, far from the scatterer. We consider only the population of positive particles, described by the density $n \geq 0$ and current J. They obey the following stationary Euler equations:

$$\partial_x J = 0, \tag{3.1}$$

$$\partial_x \left(\frac{J^2}{n} + p(n) \right) = -\frac{q}{m_i} n \partial_x \Phi \tag{3.2}$$

for $x \in \Omega$, with the pressure function

$$p(n) = n^\gamma \qquad \gamma > 1.$$

We assume the following Dirichlet boundary conditions for the density

$$n(t, -h_d) = n_0^l > 0, \qquad n(t, h_c) = n_0^r > 0, \tag{3.3}$$

$$n(t, L + h_c) = n(t, -L - h_d) = n_\infty > 0. \tag{3.4}$$

The potential Φ is required to satisfy the Poisson equation

$$-\varepsilon_0 \partial^2_{xx} \Phi = q(n - C), \tag{3.5}$$

for $x \in \Omega$ where $C(x)$ is a given positive function describing the electrons background. The neutrality far from the spacecraft is guaranteed by $C(L + h_c) = C(-L - h_d) = n_\infty$. The potential verifies

$$\Phi(t, -L - h_d) = \Phi(t, L + h_c) = 0. \tag{3.6}$$

This set of equations can be roughly obtained from (2.18) by assuming $k = 0$, $\kappa = 0$ (no impulsion nor energy exchanges), $n_e = C = n_\infty$ is constant, $u_e = 0$ (hence there is no electron current) and $\nabla_x \Theta_e = q\nabla_x \Phi$ with the isentropic approximation for the ions. It remains to write the boundary conditions for the potential on $-h_d$ and h_c. For the sake of completeness, we give the main hints of the derivation, following [1]. The basis of the derivation consists in keeping in mind that the potential is actually defined on the whole domain $(-L - h_d, L + h_c)$ and that electrodynamics relations should be used in the scatterer. We introduce a reference potential Φ_{ref} defined by

$$\begin{cases} -\partial^2_{xx}\Phi_{ref} = 0, \\ \Phi_{ref}(-h_d) = \Phi_{ref}(h_c) = 1, \;\; \Phi_{ref}(L + h_c) = \Phi_{ref}(-L - h_d) = 0. \end{cases} \tag{3.7}$$

In the conductor domain, the potential is constant: $\Phi(t, x) = \phi_{abs}(t)$ for any $x \in (0, h_c)$ where the absolute potential ϕ_{abs} is a function of time to be determined. We regard J_{cond} the current in the conductor. We split $\Phi(t, x) = \phi_{abs}(t)\Phi_{ref}(x) + \Phi'(t, x)$, so that the differential potential Φ' verifies

$$\begin{cases} -\partial^2_{xx}\Phi'(t, x) = n - C \qquad \text{on } \Omega, \\ \Phi'(t, L + h_c) = \Phi'(t, -L - h_d) = 0, \\ \Phi'(t, h_c) = 0, \;\; \Phi'(t, -h_d) = \Phi(t, -h_d) - \phi_{abs}(t). \end{cases} \tag{3.8}$$

The boundary condition on the spacecraft will take the form of equations satisfied by $\phi_{abs}(t)$ and $\Phi'(t, -h_d)$. Note that for the spacecraft engineering application, the crucial quantity to be controlled is precisely the differential potential.

Since the dielectric layer is very thin, $h_d \ll h_c \ll L$, there is no volumic charge in the dielectric and the derivative of the potential in the dielectric can be approximated by the finite difference

$$\partial_x \Phi(t, -h_d) \simeq \frac{\phi_{abs}(t) - \Phi(t, -h_d)}{h_d}.$$

The runaway current in the dielectric domain is defined by

$$J_{diel} = -\sigma_d \frac{\phi_{abs}(t) - \Phi(t, -h_d)}{h_d},$$

with σ_d being the conductivity of the dielectric. Therefore, the Ampère law yields the following relations

- At $x = -h_d$

$$\partial_t \left(\epsilon_0 \partial_x \Phi(t, -h_d) - \epsilon_d \frac{\phi_{abs}(t) - \Phi(t, -h_d)}{h_d} \right) = J(t, -h_d) + \sigma_d \frac{\phi_{abs}(t) - \Phi(t, -h_d)}{h_d}. \tag{3.9}$$

- At $x = 0$

$$\partial_t \left(\epsilon_d \frac{\phi_{abs}(t) - \Phi(t, -h_d)}{h_d} \right) = -\sigma_d \frac{\phi_{abs}(t) - \Phi(t, -h_d)}{h_d} - J_{cond}. \tag{3.10}$$

- At $x = -h_c$

$$-\partial_t(\epsilon_0 \partial_x \Phi(t, h_c)) = J_{cond} - J(t, h_c). \tag{3.11}$$

Adding (3.10) and (3.11) leads to

$$\partial_t \left(\epsilon_d \frac{\phi_{abs}(t) - \Phi(t, -h_d)}{h_d} - \epsilon_0 \partial_x \Phi(t, h_c) \right) = -\sigma_d \frac{\phi_{abs}(t) - \Phi(t, -h_d)}{h_d} - J(t, h_c).$$

Combining with (3.9) yields

$$\epsilon_0 \partial_t \partial_x \big(\Phi(t, -h_d) - \Phi(t, h_c) \big) = J(t, -h_d) - J(t, h_c) \tag{3.12}$$

Eq. (3.9) can also be recast as

$$\epsilon_0 \partial_t \Big((\partial_x \Phi'(t, -h_d) + \phi_{abs}(t) \partial_x \Phi_{ref}(-h_d) \Big) + \frac{\epsilon_d}{h_d} \partial_t \Phi'(t, -h_d) + \frac{\sigma_d}{h_d} \Phi'(t, -h_d) = J(t, -h_d). \tag{3.13}$$

The quantities $\Phi'(t, -h_d)$ and $\phi_{abs}(t)$ are entirely defined by (3.12) and (3.13).

Taking account of the scaling of the dielectric thickness $0 < h_d/h_c \ll 1$ the boundary relations become

$$J(t, -h_d) = J(t, h_c), \tag{3.14}$$

$$C_d \partial_t \big(\Phi(t, -h_d) - \phi_{abs}(t) \big) + S_d \big(\Phi(t, -h_d) - \phi_{abs}(t) \big) = J_{ext}(t, -h_d) \tag{3.15}$$

where $C_d = \epsilon_d/\epsilon_0$ and S_d are the dimensionless capacity and conductance of the dielectric respectively. Eventually, we recap the charging equations, written here in dimensionless form, as follows:

$$\begin{cases} \partial_x J = 0, \ \partial_x (J^2/n + p(n)) = -n \partial_x \Phi \\ -\lambda^2 \partial_{xx}^2 \Phi = n - C, \end{cases} \tag{3.16}$$

hold on the domain $\Omega = \Omega^l \cup \Omega^r = (-L - h_d, -h_d) \cup (h_c, L + h_c)$, where λ is the ratio between the Debye length and the characteristic length, and with the boundary conditions

$$\begin{cases} n(t, -h_d) = n_0^l, \ n(t, h_c) = n_0^r \\ n(t, L + h_c) = n(t, -L - h_d) = n_\infty, \\ \Phi(t, L + h_c) = \Phi(t, -L - h_d) = 0, \end{cases} \tag{3.17}$$

together with (3.14) and(3.15). Hence we deduce that $J = J(t)$ is actually constant on the whole set Ω. Next, combining the momentum equation with the Poisson equation we get

$$\begin{aligned}-\partial^2_{xx}\Phi &= \frac{1}{\lambda^2}(n-C)\\ &= \partial_x\Big(\frac{1}{n}\partial_x(J^2/n+p(n))\Big) = \partial_x\big(F'_J(n)\partial_x n\big)\end{aligned}$$

with $F'_J(n) = -J^2/n^3 + p'(n)/n$. Therefore the density verifies the following second order equation

$$\begin{cases} -\partial^2_{xx}F_J(n) + \dfrac{1}{\lambda^2}\Big(n-C\Big) = 0 \qquad \text{on } \Omega \\ F_J(n) = \dfrac{J^2}{2n^2} + h(n) \\ h(n) = \displaystyle\int_1^n \frac{p'(y)}{y}dy = \frac{\gamma}{\gamma-1}\Big(n^{\gamma-1}-1\Big), \end{cases} \tag{3.18}$$

endowed with Dirichlet boundary conditions.

We can also show that J is solution of a simple ODE. Indeed, we have

$$-\partial_x\Phi = \frac{-J^2/n^2+p'(n)}{n}\partial_x n = \partial_x F_J(n). \tag{3.19}$$

Integrating this relation and using $\Phi(t,-L-h_d) = \Phi(t, L+h_c) = 0$, we obtain

$$\begin{cases} \phi_{abs}(t) = F_J(n_\infty) - F_J(n_0^r), \\ \Phi(t,-h_d) = F_J(n_\infty) - F_J(n_0^l). \end{cases}$$

Obviously if $n_0^r = n_0^l$ we get $\phi_{abs}(t) = \Phi(t,-h_d)$ for any $t \geq 0$ and (3.15) implies that there is no current at all: $J = 0$. From now on we suppose $n_0^r \neq n_0^l$. Hence the differential equation (3.15) becomes

$$\begin{aligned}&\partial_t\Big(\frac{\gamma}{\gamma-1}((n_0^r)^{\gamma-1}-(n_0^l)^{\gamma-1}) + \frac{J^2}{2}\Big(\frac{1}{(n_0^r)^2} - \frac{1}{(n_0^l)^2}\Big)\Big)\\ &+\frac{S_d}{C_d}\Big(\frac{\gamma}{\gamma-1}((n_0^r)^{\gamma-1}-(n_0^l)^{\gamma-1}) + \frac{J^2}{2}\Big(\frac{1}{(n_0^r)^2} - \frac{1}{(n_0^l)^2}\Big)\Big) = \frac{J}{C_d}\end{aligned}$$

which, as soon as $J(t) \neq 0$, can be recast as

$$J'(t) + \frac{S_d}{C_d}\frac{J(t)}{2} = \frac{s}{J(t)} + \beta, \tag{3.20}$$

with

$$s = \frac{S_d}{C_d}\frac{\gamma}{\gamma-1}\frac{(n_0^l)^{\gamma-1}-(n_0^r)^{\gamma-1}}{(n_0^l)^2-(n_0^r)^2}(n_0^r)^2(n_0^l)^2,$$

and

$$\beta = \frac{1}{C_d} \frac{(n_0^r)^2 (n_0^l)^2}{(n_0^l)^2 - (n_0^r)^2}.$$

We observe that the equation admits two stationary solutions

$$J_1 = \frac{C_d}{S_d}(\beta + \sqrt{\beta^2 + 2sS_d/C_d}) > 0, \; J_2 = \frac{C_d}{S_d}(\beta - \sqrt{\beta^2 + 2sS_d/C_d}) < 0$$

4 Analysis of the one-dimensional problem

According to the previous manipulations, the evolution of the current is decoupled from the density variations. In turn, there is no difficulty in analyzing the current equation and we obtain the following statement.

Proposition 4.1. *Let $n_0^r, n_0^l > 0$, $n_0^r \neq n_0^l$ and let J_{Init} be the initial current. Then, Eq.* (3.20) *has a unique global solution. Furthermore, the solution has the following behavior*

- *if $J_{Init} > J_1$ then $J(t)$ is a positive non increasing function which converges to J_1 as t goes to ∞,*
- *if $0 < J_{Init} < J_1$ then $J(t)$ is a positive non decreasing function which converges to J_1 as t goes to ∞,*
- *if $J_{Init} < J_2$ then $J(t)$ is a negative non decreasing function which converges to J_2 as t goes to ∞,*
- *if $J_2 < J_{Init} < 0$ then $J(t)$ is a negative non increasing function which converges to J_2 as t goes to ∞.*

Therefore, the density $n(t, x)$ is determined by (3.18), which is parametrized by the time variable, via the definition of the current $J(t)$ by (3.20). Nevertheless, while J is globally defined, this is not enough to ensure the well-posedness of (3.18) due to possible change of type of the equation.

Definition 4.2. When the pair (n, J) is $F_J'(n) > 0$, we say that the regime is subsonic; when the pair (n, J) is $F_J'(n) < 0$, we say that the regime is supersonic.

We are able to justify the existence of solutions, as far as the estimates guarantee that we remain in the subsonic region, so that (3.18) is a nonlinear elliptic equation.

Theorem 4.3. *(Existence, uniqueness and regularity of subsonic solutions) Let n_0^r, n_0^l and n_∞ be positive. We set $\underline{n} = \min(n_0^r, n_0^l, n_\infty, \min C)$ and $\overline{n} = \max(n_0^r, n_0^l, n_\infty, \max C)$. We set*

$$J_{crit} = \underline{n}\sqrt{\gamma \underline{n}^{\gamma-1}}. \tag{4.1}$$

Then for any $|J_{Init}| \leq J_{crit}$ *there exists a time* $T_\star$ *and a unique solution* (n, Φ) *of* (3.18), (3.17) *defined on* $[0, T_\star]$. *The solution lies in* $C^1([0, T_\star]; C^2(\overline{\Omega}))$ *and it verifies* $\underline{n} \leq n(t,x) \leq \overline{n}$. *If the data are* $0 < J_{Init} < J_1 \leq J_{crit}$ *or* $0 < J_1 \leq J_{Init} < J_{crit}$ *(resp.* $-J_{crit} < J_2 \leq J_{Init} < 0$ *or* $-J_{crit} \leq J_{Init} < J_2 < 0$*), then the solution is globally defined.*

We plot in Fig. 4.1 the phase portrait of the current J which summarizes the different situations described in Proposition 4.1 and Theorem 4.3. We depict the subsonic and supersonic regions respectively by white and grey colored areas. Values of J_1 and J_2 are 0.8035 and -4.0250, whereas $J_{crit} = 1.1832$. The trajectories converge very fast to J_1 for positive current, and the contrary is observed for J_2. This situation changes if we switch n_0^l and n_0^r.

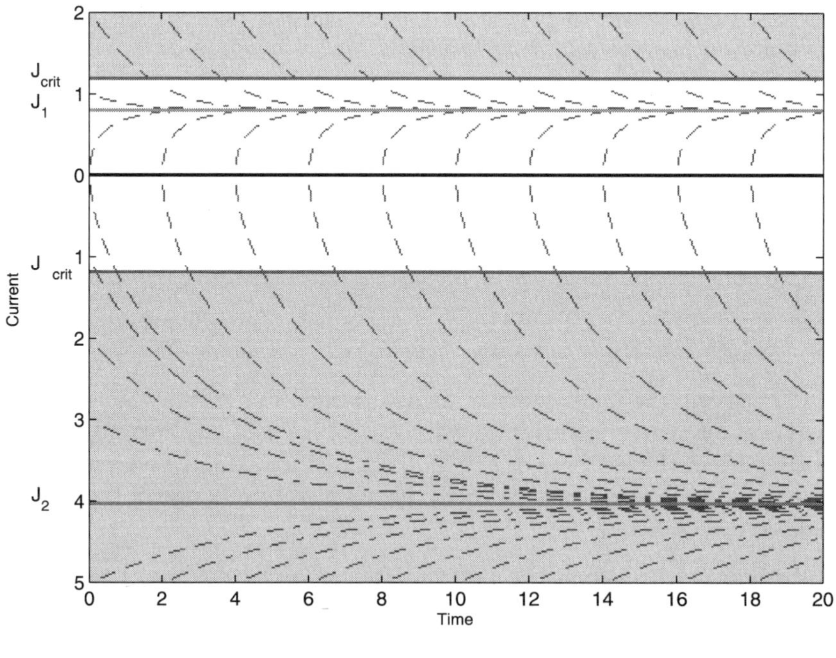

Figure 4.1 Phase portrait of current J for $n_0^l = 1.1$, $n_0^r = 1.9$, $n_\infty = 1$, $C = 1$, $\gamma = 1.4$, $S_d = 1.13$ and $C_d = 3$.

The proof of Theorem 4.3 follows the lines of [14] and it is based on a suitable fixed point method. Indeed, we show that the mapping $\mathcal{T} : n \mapsto \tilde{n}$ defined by

$$\tilde{n} - \lambda^2 \partial_x (F_J'(n) \partial_x \tilde{n}) = C$$

endowed with the Dirichlet boundary conditions (3.17) has a unique fixed point. The proof uses the regularizing effect of elliptic equations. Hence it works as soon as the regime becomes subsonic, which leads to the condition (4.1) on the current [14]. Going back to (3.20), we can exhibit conditions on the data (that is on n_0^l, n_0^r, n_∞) such that the current $J(t)$ remains in the interval $0 < J(t) < J_1 < J_{crit}$ for any time $t \geq 0$, and therefore the solution of the whole problem is globally defined.

According to [23], we guess that we can exhibit some $J^{crit} > J_{crit}$ such that if the initial current is large enough $|J_{Init}| \geq J^{crit}$ then, we remain in a supersonic case and we can also show the existence-uniqueness of a smooth solution. The proof is much more delicate since we do not have in the supersonic case a so nice elliptic structure and helpful estimates (like in particular the maximum principle) are not easily available. The analysis of the possible change of type and transonic regimes would be very interesting and challenging; we refer to [15, 16] for results in this direction.

5 Numerical simulation of the one-dimensional problem

We investigate numerically the following system

$$J'(t) + \frac{S_d}{C_d}\frac{J(t)}{2} = \frac{s}{J(t)} + \beta, \qquad t \in [0,T], \tag{5.1}$$

$$-\partial^2_{xx} F_J(n) + \frac{1}{\lambda^2}\Big(n - C\Big) = 0, \ (x,t) \in \Omega \times [0,T] \tag{5.2}$$

$$-\partial^2_{xx}\Phi = \frac{1}{\lambda^2}\Big(n - C\Big), \ (x,t) \in \Omega \times [0,T] \tag{5.3}$$

$$n(t,-h_d) = n_0^l, \quad n(t,h_c) = n_0^r, \qquad t \in [0,T], \tag{5.4}$$

$$n(t, L + h_c) = n(t, -L - h_d) = n_\infty, \qquad t \in [0,T], \tag{5.5}$$

$$\Phi(t,h_c) = \phi_{abs}(t) = F_J(n_\infty) - F_J(n_0^r), \qquad t \in [0,T], \tag{5.6}$$

$$\Phi(t,-h_d) = F_J(n_\infty) - F_J(n_0^l), \qquad t \in [0,T], \tag{5.7}$$

$$\Phi(t, L + h_c) = \Phi(t, -L - h_d) = 0, \qquad t \in [0,T], \tag{5.8}$$

with $F_J(n) = \dfrac{J^2}{2n^2} + h(n)$ and $h(n) = \displaystyle\int_1^n \frac{p'(y)}{y}dy = \frac{\gamma}{\gamma - 1}\Big(n^{\gamma-1} - 1\Big)$.

We solve the current Eq. (5.1) for the variable $y(t) = J(t)^2$, with the equation being transformed into

$$y'(t) + \frac{S_d}{C_d} y(t) = 2\Big(s \pm \beta\sqrt{y(t)}\Big). \tag{5.9}$$

The choice of the sign in the r.h.s of (5.9) is determined by the sign of the initial datum J_{Init} since the sign of J remains constant in time. Therefore, $J(t) = \pm\sqrt{y(t)}$. Equation (5.9) is solved once for all by a standard Runge-Kutta scheme.

Then, knowing the current J^k, approximation of $J(k\Delta t)$, we approach (5.2) with a basic finite difference scheme

$$n_j^k - \frac{\lambda^2}{\Delta x}\left(F'_{J^k}(n_{j+1}^k)\frac{n_{j+1}^k - n_j^k}{\Delta x} - F'_{J^k}(n_j^k)\frac{n_j^k - n_{j-1}^k}{\Delta x}\right) = C_j. \quad (5.10)$$

The nonlinear equation (5.10) is solved by a Newton algorithm. The elliptic Poisson equation (5.3) is also solved by classical finite difference scheme. Although those equations seem stationary, they depend on time by their boundary condition (5.4)–(5.8).

The simulation reveals the threshold effect in the choice of the initial current: for a small enough J_{Init} the scheme works well and reproduces a smooth density profile, as expected. But, starting with a larger initial current, singularity might appear characterized by the non invertibility of the linear systems involved in the resolution of (5.10). To emphasize this point we make the following experiment with $-h_d = h_c = 0$ and $L = 1$ for the domain Ω. We consider $\gamma = 1.4$, $S_d = 1.13$, $C_d = 3$, $n_\infty = 1$, $n_0^l = 1.1$, $n_0^r = 1.9$ and $C = 1$. In this case we recall that the critical current is $J_{crit} = 1.1832$. In Figs. 5.1, 5.2, 5.3 we take $J_{Init} = 1.15$ such that $J_{Init} < J_{crit}$ and we are in the subsonic case. Here the current is a smooth decreasing function of time. With the same values of parameters, taking $J_{Init} = 1.2$, singularities appear directly from the begining. If $J_{Init} = -0.5$, we also observe a problem when $J(t)$ crosses the value of $-J_{crit}$ and singularities appear.

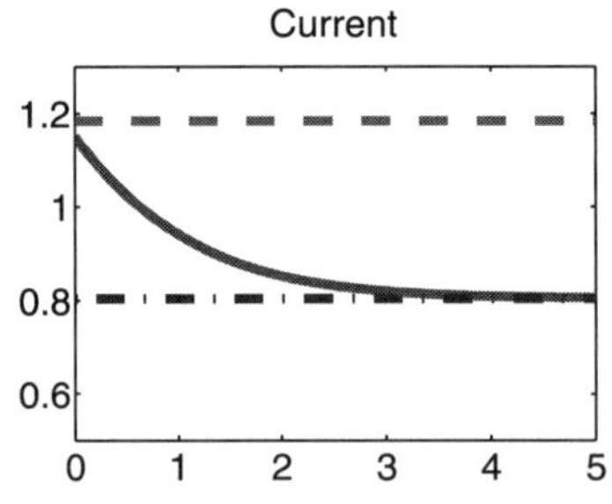

Figure 5.1 Evolution of current (line at top corresponds to the value of J_{crit}).

As a final comment, it is worth having in mind that in the rescaled problem (3.16)–(3.17) the Debye length might be small compared to the characteristic length scale. Hence, in LEO environment we usually have

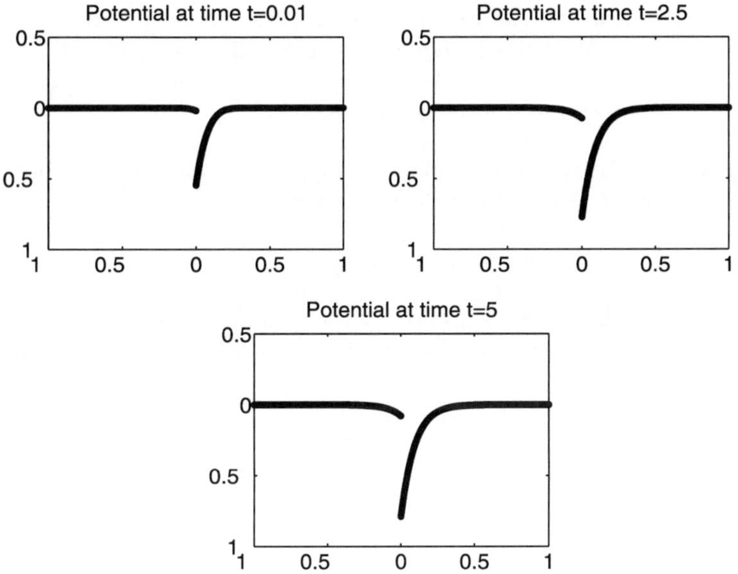

Figure 5.2 Evolution of potential.

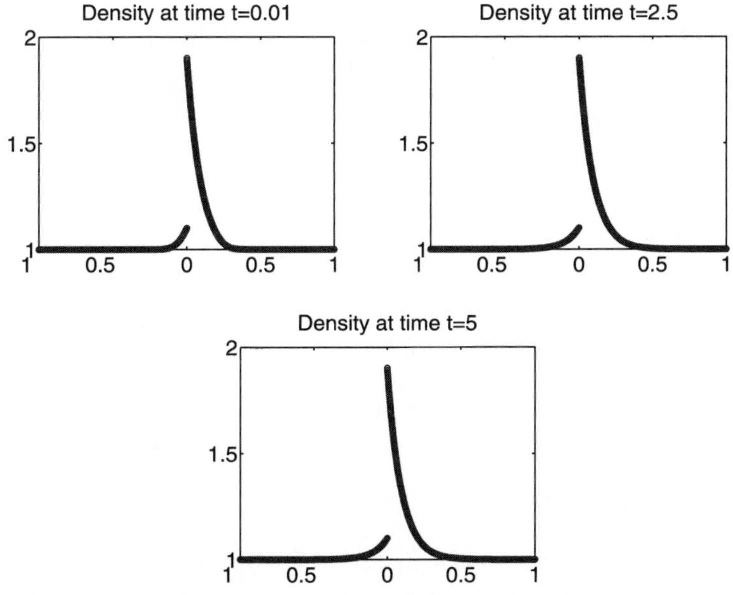

Figure 5.3 Evolution of density.

$0 < \lambda \ll 1$. It leads to the formation of boundary layers. Indeed, let us set $C = n_\infty = 1$. Writing the equation for $\lambda = 0$ we simply get

$$\partial_x j = 0, \quad F_j'(n)\partial_x n = \partial_x \Phi, \quad n = 1.$$

Taking account of the condition at infinity (or at the far end boundary $x = L + h_c$ or $-L - h_d$), the solution reads

$$\overline{n} = 1, \quad \overline{j} = j(t), \quad \overline{\Phi} = 0.$$

However, this solution does not verify the boundary condition at $x = h_c$ nor $x = -h_d$. Let us expand the solution of (3.16)–(3.17) as follows:

$$j = \overline{j} + \widetilde{j}(x/\lambda) + \lambda\widehat{j}, \quad \Phi = \overline{\Phi} + \widetilde{\Phi}(x/\lambda) + \lambda\widehat{\Phi}, \quad n = 1 + \widetilde{n}(x/\lambda) + \lambda\widehat{n}.$$

At leading order we obtain the following relations satisfied by the boundary correctors:

$$\begin{cases} \dfrac{1}{\lambda}\partial_y \widetilde{j}(x/\lambda) = 0, \\ \dfrac{\gamma - \overline{j}^2}{\lambda}\partial_y \widetilde{n} = \dfrac{1}{\lambda}\partial_y \widetilde{\Phi}, \\ -\partial^2_{yy}\widetilde{\Phi} = \widetilde{n}. \end{cases}$$

The equation is completed by the boundary condition matching the data to the solution corresponding to $\lambda = 0$, that is,

$$\begin{aligned} &\widetilde{\Phi}(y = 0) = \phi_{abs}(t) = F_{J(t)}(n_\infty) - F_{J(t)}(n_0^r) \\ &\qquad\qquad\qquad \text{or } \Phi(t, -h_d) = F_{J(t)}(n_\infty) - F_{J(t)}(n_0^l), \\ &\widetilde{\Phi}(y \to \infty) = 0, \\ &\widetilde{n}(y = 0) = n_0^r - 1 \text{ or } n_0^l - 1, \\ &\widetilde{n}(y \to \infty) = 0. \end{aligned}$$

The numerical treatment of this kind of asymptotic problem leads to severe stiff problems, which require a specific treatment. A deep understanding of the boundary layer formation and of the scale separation helps to design an efficient numerical scheme, as in [26].

6 Conclusions

Considering LEO environment instead of GEO, it can be tempted to describe spacecraft charge phenomena by using hydrodynamic models, at least as a first approximation. Such models are indeed less complicated than a full kinetic description of the plasma and can be treated for a reduced numerical cost. The underlying Euler equations are thus coupled to the Poisson equation for the electric potential, with complex and non standard boundary conditions. These boundary conditions for the potential, which consider that different places on the spacecraft surface can have a different electrical behavior, are at the origin of the charging phenomena. We point out several difficulties related to the hydrodynamic modeling:

- A crucial issue concerns the boundary condition to be satisfied by the hydrodynamic unknowns. A convincing derivation should certainly go back to the kinetic model and the hydrodynamic limit through a fine analysis of the kinetic boundary layer.
- Due to the time evolution through the boundary condition, change of type of the flow can occur. Such passage from subsonic to supersonic regimes makes the mathematical analysis difficult and might lead to breakdown of the numerical methods. This is illustrated on a simple one-dimensional caricature model.
- Eventually, a careful discussion of the various scales involved in the equations is necessary. The multiscale features of the problem definitely make it challenge the numerical simulations which require the design of refined and dedicated schemes.

Acknowledgements: We are gratefully indebted to Jean-Paul Dudon from Thalesalenia Space for introduction to the problem and many fruitful discussions.

References

[1] C. Besse, J.-P. Dudon, T. Goudon and N. Vauchelet. *Comparison of Vlasov solvers for spacecraft charging simulation.* Preprint INRIA 2009.

[2] S. I. Braginskii. Transport processes in a plasma. In *Reviews of plasma physics*, M. Leontovich ed., **1**: 205–311. Consultants Bureau, New York, 1965.

[3] M. Chane-Yook. *Etude d'une équation cinétique liée à l'effet Compton; modélisation et simulation 3D de la charge d'un satellite en environnement plasmique.* PhD thesis, Université de Provence, 2004.

[4] M. Chane-Yook, S. Clerc and S. Piperno. Space charge and potential distribution around a spacecraft in an isotropic plasma. *J. Geophys. Res.*, **111**(A4), A04211, 2006.

[5] O. Chanrion. *Simulation de l'influence de la propulsion plasmique sur la charge électrostatique d'un satellite en milieu magnétosphérique.* PhD thesis, Ecole Nationale des Ponts et Chaussées, 2001.

[6] S. Clerc, S. Brosse and M. Chane-Yook. Sparcs: an advanced software for spacecraft charging analysis. *Proceedings of the 8th Spacecraft Charging Tech. Conf.*, Huntsville, AL, October 20–24, 2003.

[7] F. Coron, F. Golse and C. Sulem. A classification of well-posed kinetic layer problems. *Commun. Pure Appl. Math.*, **41**: 409–435, 1988.

[8] P. Crispel. *Etude de la transition de la décharge primaire vers l'arc secondaire sur un générateur solaire de satellite.* PhD thesis, Université Toulouse 3, 2005.

[9] P. Crispel, P. Degond and M.-H. Vignal. Quasi-neutral fluid models for current-carrying plasmas. *J. Comput. Phys.*, **205**(2): 408–438, 2005.

[10] A. Decoster. *Modelling of collisions.* Series in Applied Mathematics, vol. 2. Elsevier, 1998.

[11] S. Dellacherie. Coupling of the Wang Chang-Uhlenbeck equations with the multispecies Euler system. *J. Comp. Phys.*, **189**: 239–276, 2003.

[12] F. Dubois and P. LeFloch. Boundary conditions for nonlinear hyperbolic systems of conservation laws. *J. Differential Equations*, **71**(1): 93–122, 1988.

[13] P. Degond. Macroscopic models of charged-particle transport derived from kinetic theory. ICIAM 95, Hamburg, 1995. *Math. Res.*, **87**: 39–53, Akademie Verlag, Berlin, 1996.

[14] P. Degond and P. A. Markowich. On a one-dimensional steady-state hydrodynamic model for semiconductors. *Appl. Math. Lett.*, **3**: 25–29, 1990.

[15] I. M. Gamba. Stationary transonic solutions of a one-dimensional hydrodynamicmodel for semiconductors. *Comm. Part. Diff. Eqs.*, **17**: 553–577, 1992.

[16] I. M. Gamba and C. S.Morawetz. A viscous approximation for a 2-D steadysemiconductor or transonic gas dynamic flow: existence theorem for potentialflow. *Comm. Pure Appl. Math.*, **XLIX**: 999–1049, 1996.

[17] F. Golse, B. Perthame and C. Sulem. On a boundary layer problem for the nonlinear Boltzmann equation. *Arch. Rational Mech. Anal.*, **103**(1): 81–96, 1988.

[18] A. Jüngel and Y. J. Peng. A hierarchy of hydrodynamic models for plasmas.Quasineutral limits in the drift-diffusion equations. *Asymptotic Analysis*, **28**: 49-73, 2001.

[19] R. J. LeVeque. *Numerical Methods for Conservation Laws.* Lectures in Mathematics, ETH-Zurich, Birkhauser-Verlag, Basel, 1990.

[20] L. Lévy. Charge des matériaux et systèmes en environnement spatial. In *Space Environment: prevention of risks related to spacecraft charging.* Cepadués-Editions, 1996.

[21] M. J. Mandell, V. A. Davies and L. G. Mikelides. NASCAP-2K Preliminary Documentation. Science Applications International Corp, San Diego, CA, *Scientific rept.* (2): A555024, 2002.

[22] J. F. Roussel, F. Rogier, M. Lemoine, D. Volpert, G. Rousseau, G. Sookahet, P. Sng and A. Hilgers. Design of a new modular spacecraft plasma interaction modeling software (SPIS). *Proceedings of the 8th Spacecraft Charging Tech. Conf.*, Huntsville, AL, October 20–24, 2003.

[23] Y. J. Peng and I. Violet. Example of supersonic solutions to a steady-state Euler-Poisson system. *Appl. Math. Lett.*, **19**: 1335–1340, 2006.

[24] Y. J. Peng and I.Violet. Asymptotic expansions in a steady state Euler-Poisson system andconvergence to the incompressible Euler equations. *M3AS*, **15**(5): 717–736, 2005.

[25] S. Ukai, T. Yang and S.-H. Yu. Nonlinear Boundary Layers of the Boltzmann Equation: I. Existence. *Commun. Math. Phys.*, **236**: 373–393, 2003.

[26] M.-H. Vignal, A boundary layer problem for an asymptotic preserving scheme in the quasi-neutral limit for the Euler-Poisson system. Preprint 2008.

[27] I. Violet. High-order expansions in quasi-neutral limit of the Euler-Poisson system for apotential flow. *Proc. Royal Soc. Edinburgh*, **137A**: 1101–1118, 2007.

Asymptotic Regimes for Plasma Physics with Strong Magnetic Fields

Mihai Bostan
Laboratoire de Mathématiques Université de Besançon
16 route de Gray, 25030 Besançon Cedex, France
Email: mbostan@univ-fcomte.fr

Abstract

The subject matter of this paper concerns the asymptotic regimes for transport equations. Such models arise in the magnetic confinement context, where charged particles move under the action of strong magnetic fields. The main difficulty comes from the multi-scale character of the problem. According to the different possible orderings between the typical physical scales (Larmor radius, Debye length, cyclotronic frequency, plasma frequency) we distinguish several regimes. The main purpose is to derive the limit models: we justify rigorously the convergence towards these limit models and investigate the well-posedness of them.

1 Introduction

Motivated by the magnetic confinement fusion, the study of strong magnetic field effect is now of crucial importance. We are concerned with the dynamics of a population of charged particles interacting through electro-magnetic fields. We consider a population of non relativistic electrons whose density is denoted by f. This particle density satisfies the Vlasov equation

$$\partial_t f + \frac{p}{m_e} \cdot \nabla_x f - e\left(E(t,x) + \frac{p}{m_e} \wedge B(t,x)\right) \cdot \nabla_p f = 0$$

where $-e < 0$ is the electron charge and $m_e > 0$ is the electron mass. The self-consistent electro-magnetic field (E, B) verifies the Maxwell equations

$$\partial_t E - c_0^2 \,\mathrm{curl}_x B = \frac{e}{\varepsilon_0} \int_{\mathbb{R}^3} \frac{p}{m_e} f \,\mathrm{d}p, \quad \partial_t B + \mathrm{curl}_x E = 0$$

$$\mathrm{div}_x E = \frac{e}{\varepsilon_0}\left(n - \int_{\mathbb{R}^3} f \, \mathrm{d}p\right), \quad \mathrm{div}_x B = 0.$$

Here ε_0 is the vacuum permittivity, c_0 is the light speed in the vacuum and n is the concentration of the background ion distribution. One of the asymptotic regimes we wish to address here is the gyro-kinetic model with finite Larmor radius. Let us denote by ω_{p} the plasma frequency

$$\omega_{\mathrm{p}}^2 = \frac{e^2 n}{m_{\mathrm{e}}\varepsilon_0}$$

and by ω_{c} the cyclotronic frequency

$$\omega_{\mathrm{c}} = \frac{eB}{m_{\mathrm{e}}}.$$

Assuming that the cyclotronic frequency is much higher than the plasma frequency we deduce that the typical magnetic field magnitude satisfies

$$B = \frac{m_{\mathrm{e}}\omega_{\mathrm{p}}}{e} \cdot \frac{\omega_{\mathrm{c}}}{\omega_{\mathrm{p}}} = \frac{m_{\mathrm{e}}\omega_{\mathrm{p}}}{e} \cdot \frac{1}{\varepsilon}$$

where $\omega_{\mathrm{c}}/\omega_{\mathrm{p}} = 1/\varepsilon$, $0 < \varepsilon \ll 1$. We assume also that the electron momentum in the plane orthogonal to the magnetic field is much larger than the thermal momentum p_{th} given by

$$\frac{p_{\mathrm{th}}^2}{m_{\mathrm{e}}} = K_B T_{\mathrm{th}}$$

where K_{B} is the Boltzmann constant and T_{th} is the temperature. Note that in this case the Larmor radius corresponding to the cyclotronic frequency ω_{c} and the typical momentum $\frac{p_{\mathrm{th}}}{\varepsilon}$ remains of order of the Debye length

$$\rho_{\mathrm{L}} = \frac{p_{\mathrm{th}}}{\varepsilon m_{\mathrm{e}}\omega_{\mathrm{c}}} = \frac{p_{\mathrm{th}}}{m_{\mathrm{e}}\omega_{\mathrm{p}}} = \left(\frac{\varepsilon_0 K_{\mathrm{B}} T_{\mathrm{th}}}{e^2 n}\right)^{1/2} = \lambda_{\mathrm{D}}.$$

This model is called the finite Larmor radius regime. For example, in the two-dimensional setting and assuming that the magnetic field has a constant direction

$$f = f(t,x,p), \quad (E,B) = (E_1, E_2, 0, 0, 0, B_3)(t,x), \quad (t,x,p) \in \mathbb{R}_+ \times \mathbb{R}^2 \times \mathbb{R}^2$$

we are led, up to a multiplicative constant of order 1, to the following Vlasov equation (see [8], [2])

$$\partial_t f^\varepsilon + \frac{p}{\varepsilon} \cdot \nabla_x f^\varepsilon - \left(E^\varepsilon(t,x) + B_3^\varepsilon(t,x)\frac{{}^\perp p}{\varepsilon}\right) \cdot \nabla_p f^\varepsilon = 0 \qquad (1.1)$$

where the notation ${}^{\perp}p$ stands for $(p_2, -p_1)$ for any $p = (p_1, p_2) \in \mathbb{R}^2$. When the typical momentum is supposed to remain of order of the thermal momentum, the Larmor radius vanishes as the magnetic field becomes very large; we are dealing with the guiding-center approximation. The guiding-center approximation for the Vlasov-Maxwell system was studied in [4] by the modulated energy method (see [3], [6]) for other results obtained by this method. The analysis of the Vlasov or Vlasov-Poisson equations with large external magnetic field has been carried out in [9], [11], [10], [5].

For simplifying we assume that the self-consistent electric field in the Vlasov equation derives from a potential determined by solving the Poisson equation

$$E^\varepsilon = \nabla_x \phi^\varepsilon, \quad \Delta_x \phi^\varepsilon = 1 - \int_{\mathbb{R}^2} f^\varepsilon \, dp.$$

We suppose also that $B_3 = B_3(x)$ is a given stationary external magnetic field. The Vlasov equation leads naturally to problems like

$$\partial_t u^\varepsilon + a(t,y) \cdot \nabla_y u^\varepsilon + \frac{b(y)}{\varepsilon} \cdot \nabla_y u^\varepsilon = 0, \quad (t,y) \in \mathbb{R}_+ \times \mathbb{R}^m \tag{1.2}$$

with the initial condition

$$u^\varepsilon(0,y) = u_0^\varepsilon(y), \quad y \in \mathbb{R}^m.$$

For example, (1.1) can be recast in the form (1.2) by taking $m = 4, y = (x,p) \in \mathbb{R}^2 \times \mathbb{R}^2$, $u^\varepsilon(t,y) = f^\varepsilon(t,x,p)$, $a(t,y) = -(0,0,E(t,x))$, $b(y) = (p, -B_3(x)^{\perp}p)$.

In this work we focus on the linear transport equation (1.2) when a and b are given smooth fields. Formally, multiplying (1.2) by ε one gets $b(y) \cdot \nabla_y u^\varepsilon = \mathcal{O}(\varepsilon)$, saying that the variation of u^ε along the trajectories of b vanishes as ε goes to zero. Following this observation it may seem reasonable to interpret the asymptotic $\varepsilon \searrow 0$ in (1.2) as homogenization procedure with respect to the flow of b. More precisely we appeal here to the ergodic theory.

By Hilbert's method we have the formal expansion

$$u^\varepsilon = u + \varepsilon u_1 + \varepsilon^2 u_2 + \ldots \tag{1.3}$$

and thus, plugging the ansatz (1.3) in (1.2) yields the equations

$$\varepsilon^{-1} \ : \ b(y) \cdot \nabla_y u = 0 \tag{1.4}$$

$$\varepsilon^{0} \ : \ \partial_t u + a(t,y) \cdot \nabla_y u + b(y) \cdot \nabla_y u_1 = 0 \tag{1.5}$$

$$\varepsilon^{1} \ : \ \partial_t u_1 + a(t,y) \cdot \nabla_y u_1 + b(y) \cdot \nabla_y u_2 = 0 \tag{1.6}$$

$$\vdots$$

The operator $\mathcal{T} = b(y) \cdot \nabla_y$ will play a crucial role in our analysis: Eq. (1.4) says that at any time $t \in \mathbb{R}_+$ the leading order term in the expansion (1.3) belongs to the kernel of $\mathcal{T}$. Unfortunately this information (which will be interpreted later on as a constraint) is not sufficient for uniquely determining u. The use of (1.5) is mandatory, despite the coupling with the next term u_1 in the asymptotic expansion (1.3). Actually, at least in a first step, we do not need all the information in (1.5), but only some consequence of it, such that, supplemented by the constraint (1.4), it will allow us to determine u. Since we need to eliminate u_1 in (1.5), the idea is to project (1.5) at any time $t \in \mathbb{R}_+$ to the orthogonal complement of the image of $\mathcal{T}$, for example, in $L^2(\mathbb{R}^m)$. Indeed, we will see that this consequence of (1.5) together with the constraint (1.4) provide a well posed limit model for $u = \lim_{\varepsilon \searrow 0} u^\varepsilon$. And the same procedure applies to computing $u_1, u_2, \ldots$ For example, once we have determined u, by (1.5) we know the image by $\mathcal{T}$ of u_1

$$\mathcal{T} u_1 = -\partial_t u - a(t, y) \cdot \nabla_y u. \tag{1.7}$$

Projecting now (1.6) on the orthogonal complement of the image of $\mathcal{T}$ we eliminate u_2 and get another equation for u_1, which combined with (1.7) provides a well posed problem for u_1.

Our paper is organized as follows. In Section 2 we recall some notions of ergodic theory. We introduce the average over a flow associated with a smooth field and discuss the main properties of this operator. Section 3 is devoted to the study of the limit model. We prove existence, uniqueness and regularity results. The convergence towards the limit model is justified rigorously in Section 4. Based on the concept of prime integrals, an equivalent limit model is derived in Section 5. We end this paper with some examples.

2 Ergodic theory and average over a flow

We assume that $b : \mathbb{R}^m \to \mathbb{R}^m$ is a given field satisfying

$$b \in W^{1,\infty}_{\text{loc}}(\mathbb{R}^m) \tag{2.1}$$

$$\text{div}_y b = 0 \tag{2.2}$$

and the growth condition

$$\exists\, C > 0 \;:\; |b(y)| \leq C(1 + |y|), \;\; y \in \mathbb{R}^m. \tag{2.3}$$

Upon the above hypotheses the characteristic flow $Y = Y(s; y)$ is well defined

$$\frac{dY}{ds} = b(Y(s;y)), \quad (s,y) \in \mathbb{R} \times \mathbb{R}^m \tag{2.4}$$

$$Y(0;y) = y, \quad y \in \mathbb{R}^m, \tag{2.5}$$

and has the regularity $Y \in W^{1,\infty}_{\text{loc}}(\mathbb{R} \times \mathbb{R}^m)$. By (2.2) we deduce that for any $s \in \mathbb{R}$, the map $y \to Y(s;y)$ is measure preserving

$$\int_{\mathbb{R}^m} \theta(Y(s;y))\, \mathrm{d}y = \int_{\mathbb{R}^m} \theta(y)\, \mathrm{d}y, \quad \forall\, \theta \in L^1(\mathbb{R}^m).$$

We have the following standard result concerning the kernel of $u \to \mathcal{T}u = \mathrm{div}_y(b(y)u(y))$.

Proposition 2.1. *Let $u \in L^1_{\text{loc}}(\mathbb{R}^m)$. Then $\mathrm{div}_y(b(y)u(y)) = 0$ in $\mathcal{D}'(\mathbb{R}^m)$ iff for any $s \in \mathbb{R}$ we have $u(Y(s;y)) = u(y)$ for a.a. $y \in \mathbb{R}^m$.*

Remark 2.2. Sometimes we will write $u \in \ker \mathcal{T}$ meaning that u is constant along the characteristics, *i.e.*, $u(Y(s;y)) = u(y)$ for all $s \in \mathbb{R}$ and a.a. $y \in \mathbb{R}^m$.

For any $q \in [1, +\infty]$ we denote by $\mathcal{T}_q$ the linear operator defined by $\mathcal{T}_q u = \mathrm{div}_y(b(y)u(y))$ for any u in the domain

$$\mathrm{D}_q = \{u \in L^q(\mathbb{R}^m) \;:\; \mathrm{div}_y(b(y)u(y)) \in L^q(\mathbb{R}^m)\}.$$

Thanks to Proposition 2.1 we have for any $q \in [1, +\infty]$

$$\ker \mathcal{T}_q = \{u \in L^q(\mathbb{R}^m) \;:\; u(Y(s;y)) = u(y),\ s \in \mathbb{R},\ \text{a.e. } y \in \mathbb{R}^m\}.$$

For any continuous function $h \in C([a,b]; L^q(\mathbb{R}^m))$, with $q \in [1, +\infty]$, we denote by $\int_a^b h(t)\, \mathrm{d}t \in L^q(\mathbb{R}^m)$ the Riemann integral of the function $t \to h(t) \in L^q(\mathbb{R}^m)$ on the interval $[a,b]$. Consider now a function $u \in L^q(\mathbb{R}^m)$. Observing that for any $q \in [1, +\infty)$ the application $s \to u(Y(s;\cdot))$ belongs to $C(\mathbb{R}; L^q(\mathbb{R}^m))$, we deduce that for any $T > 0$ the function $\langle u \rangle_T := \frac{1}{T}\int_0^T u(Y(s;\cdot))\, \mathrm{d}s$ is well defined as an element of $L^q(\mathbb{R}^m)$ and $\|\langle u \rangle_T\|_{L^q(\mathbb{R}^m)} \le \|u\|_{L^q(\mathbb{R}^m)}$. Observe that for any function $h \in L^\infty([a,b]; L^\infty(\mathbb{R}^m))$, the map $\varphi \in L^1(\mathbb{R}^m) \to \int_a^b \int_{\mathbb{R}^m} h(t,y)\varphi(y)\, \mathrm{d}y\, \mathrm{d}t$ belongs to $(L^1(\mathbb{R}^m))' = L^\infty(\mathbb{R}^m)$. Therefore there is a unique function in $L^\infty(\mathbb{R}^m)$, denoted $\int_a^b h(t)\, \mathrm{d}t$, such that for any $\varphi \in L^1(\mathbb{R}^m)$ we have

$$\int_{\mathbb{R}^m} \left(\int_a^b h(t)\, \mathrm{d}t \right)(y)\varphi(y)\, \mathrm{d}y = \int_a^b \left(\int_{\mathbb{R}^m} h(t,y)\varphi(y)\, \mathrm{d}y \right) \mathrm{d}t.$$

Particularly, we have

$$\left\| \int_a^b h(t)\, \mathrm{d}t \right\|_{L^\infty(\mathbb{R}^m)} \le \int_a^b \|h(t)\|_{L^\infty(\mathbb{R}^m)}\, \mathrm{d}t$$

and

$$\left(\int_a^b h(t)\,\mathrm{d}t\right)(y) = \int_a^b h(t,y)\,\mathrm{d}t, \quad \text{a.e. } y \in \mathbb{R}^m.$$

Notice that for any function $u \in L^\infty(\mathbb{R}^m)$, the map $s \to u(Y(s;\cdot))$ belongs to $L^\infty(\mathbb{R}; L^\infty(\mathbb{R}^m))$ and thus we deduce that for any $T > 0$ the function $\langle u \rangle_T := \frac{1}{T}\int_0^T u(Y(s;\cdot))\,\mathrm{d}s$ is well defined as an element of $L^\infty(\mathbb{R}^m)$ and $\|\langle u\rangle_T\|_{L^\infty(\mathbb{R}^m)} \le \|u\|_{L^\infty(\mathbb{R}^m)}$.

Obviously, when u belongs to $\ker \mathcal{T}_q$ we have $\langle u\rangle_T = u$ for any $q \in [1, +\infty]$ and $T > 0$. Generally, when $q \in (1, +\infty)$ we prove the weak convergence of $\langle u\rangle_T$ as T goes to $+\infty$ towards some element in $\ker \mathcal{T}_q$. The arguments are standard [12].

Proposition 2.3. *Assume that $q \in (1, +\infty)$ and $u \in L^q(\mathbb{R}^m)$. Then there is a unique function $\langle u \rangle \in \ker \mathcal{T}_q$ such that for any $\varphi \in \ker \mathcal{T}_{q'}$ we have*

$$\int_{\mathbb{R}^m} (u(y) - \langle u\rangle(y))\varphi(y)\,\mathrm{d}y = 0. \tag{2.6}$$

Moreover we have the weak convergence in $L^q(\mathbb{R}^m)$

$$\langle u \rangle = \lim_{T\to+\infty} \frac{1}{T}\int_0^T u(Y(s;\cdot))\,\mathrm{d}s = \lim_{T\to+\infty} \frac{1}{T}\int_{-T}^0 u(Y(s;\cdot))\,\mathrm{d}s$$

and the inequality $\|\langle u\rangle\|_{L^q(\mathbb{R}^m)} \le \|u\|_{L^q(\mathbb{R}^m)}$. Particularly the application $u \in L^q(\mathbb{R}^m) \to \langle u\rangle \in L^q(\mathbb{R}^m)$ is linear, and continuous.

Corollary 2.4. *Assume that $q \in (1, +\infty)$ and $u \in L^q(\mathbb{R}^m)$. Let us denote by $\langle u\rangle \in L^q(\mathbb{R}^m)$ the function constructed in Proposition 2.3.*
a) If $u \ge -M$ for some constant $M \ge 0$ then $\langle u\rangle \ge -M$.
b) If $u \le M$ for some constant $M \ge 0$ then $\langle u \rangle \le M$.

Corollary 2.5. *Assume that $1 < q_1 < q_2 < +\infty$ and $u \in L^{q_1}(\mathbb{R}^m) \cap L^{q_2}(\mathbb{R}^m)$. We denote by $\langle u\rangle^{(q)}$ the function of $L^q(\mathbb{R}^m)$ constructed in Proposition 2.3 for $q \in \{q_1, q_2\}$. Then we have $\langle u\rangle^{(q_1)} = \langle u\rangle^{(q_2)} \in \ker \mathcal{T}_{q_1} \cap \ker \mathcal{T}_{q_2}$.*

It is possible to prove that the convergence in Proposition 2.3 is strong. This is the object of the next proposition. Actually the case $q = 2$ corresponds to the mean ergodic theorem, or von Neumann's ergodic theorem (see [12], p. 57).

Proposition 2.6. *Assume that $q \in (1, +\infty)$ and $u \in L^q(\mathbb{R}^m)$. Then*

$$\lim_{T\to+\infty} \frac{1}{T}\int_0^T u(Y(s;\cdot))\,\mathrm{d}s = \lim_{T\to+\infty} \frac{1}{T}\int_{-T}^0 u(Y(s;\cdot))\,\mathrm{d}s = \langle u\rangle$$

strongly in $L^q(\mathbb{R}^m)$.

It is also possible to define the operator $\langle\cdot\rangle$ for functions in $L^1(\mathbb{R}^m)$ and $L^\infty(\mathbb{R}^m)$. These constructions are a little bit more delicate and require some additional hypotheses on the flow. On these hypotheses we have, for $q \in \{1, +\infty\}$, similar results to those in Proposition 2.3.

We inquire now about the symmetry between the operators $\langle\cdot\rangle^{(q)}$, $\langle\cdot\rangle^{(q')}$ when q, q' are conjugate exponents. We have the natural duality result.

Proposition 2.7. *a) Assume that* $q, q' \in (1, +\infty)$, $1/q + 1/q' = 1$, $u \in L^q(\mathbb{R}^m)$, $\varphi \in L^{q'}(\mathbb{R}^m)$. *Then*

$$\int_{\mathbb{R}^m} u\, \langle\varphi\rangle^{(q')}\, dy = \int_{\mathbb{R}^m} \langle u\rangle^{(q)} \varphi\, dy.$$

b) Particularly $\langle\cdot\rangle^{(2)}$ *is symmetric on* $L^2(\mathbb{R}^m)$ *and coincides with the orthogonal projection on* $\ker \mathcal{T}_2$. *Moreover we have the orthogonal decomposition* $L^2(\mathbb{R}^m) = \ker \mathcal{T}_2 \oplus \ker\langle\cdot\rangle^{(2)}$.

Proof. a) The function $\langle\varphi\rangle^{(q')}$ belongs to $\ker \mathcal{T}_{q'}$ and therefore

$$\int_{\mathbb{R}^m} (u - \langle u\rangle^{(q)})\langle\varphi\rangle^{(q')}\, dy = 0. \tag{2.7}$$

Similarly $\langle u\rangle^{(q)}$ belongs to $\ker \mathcal{T}_q$ and thus

$$\int_{\mathbb{R}^m} (\varphi - \langle\varphi\rangle^{(q')})\langle u\rangle^{(q)}\, dy = 0. \tag{2.8}$$

Combining (2.7) with (2.8) yields

$$\int_{\mathbb{R}^m} u\, \langle\varphi\rangle^{(q')}\, dy = \int_{\mathbb{R}^m} \langle u\rangle^{(q)}\langle\varphi\rangle^{(q')}\, dy = \int_{\mathbb{R}^m} \langle u\rangle^{(q)} \varphi\, dy.$$

b) When $q = 2$ we obtain

$$\int_{\mathbb{R}^m} u\, \langle\varphi\rangle^{(2)}\, dy = \int_{\mathbb{R}^m} \langle u\rangle^{(2)} \varphi\, dy, \quad \forall\, u, \varphi \in L^2(\mathbb{R}^m).$$

By the characterization in Proposition 2.3 we deduce that $\langle\cdot\rangle^{(2)} = \mathrm{Proj}_{\ker \mathcal{T}_2}$. Since $\ker \mathcal{T}_2$ is closed we have the orthogonal decomposition

$$L^2(\mathbb{R}^m) = \ker \mathcal{T}_2 \oplus (\ker \mathcal{T}_2)^\perp = \ker \mathcal{T}_2 \oplus \ker\langle\cdot\rangle^{(2)}.$$

□

The following result is a straightforward consequence of the characterization for $\langle\cdot\rangle^{(r)}$ with $r \in [1, +\infty]$. The proof is left to readers.

Corollary 2.8. *Let $u \in L^p(\mathbb{R}^m)$, $v \in L^q(\mathbb{R}^m)$ and $1/r = 1/p + 1/q$ with $p, q, r \in [1, +\infty]$. Assume that u is constant along the flow. Then*

$$\langle uv \rangle^{(r)} = u \, \langle v \rangle^{(q)}.$$

By the orthogonal decompositions in Propositions 2.6 and 2.7 we deduce that $\ker\langle\cdot\rangle^{(2)} = \overline{\mathrm{range}\mathcal{T}_2}$. We have the general result.

Proposition 2.9. *Assume that $q \in (1, +\infty)$. Then* $\ker\langle\cdot\rangle^{(q)} = \overline{\mathrm{range}\mathcal{T}_q}$.

Proof. For any $v = \mathcal{T}_q u \in \mathrm{range}\mathcal{T}_q$ and $\varphi \in \ker \mathcal{T}_{q'}$ we have

$$\int_{\mathbb{R}^m} (v - 0)\varphi \,\mathrm{d}y = \int_{\mathbb{R}^m} \mathcal{T}_q u \,\varphi \,\mathrm{d}y = -\int_{\mathbb{R}^m} u\mathcal{T}_{q'}\varphi \,\mathrm{d}y = 0$$

saying that $\langle v \rangle^{(q)} = 0$. Therefore $\mathrm{range}\mathcal{T}_q \subset \ker\langle\cdot\rangle^{(q)}$ and also $\overline{\mathrm{range}\mathcal{T}_q} \subset \ker\langle\cdot\rangle^{(q)}$. Consider now a linear form h on $L^q(\mathbb{R}^m)$ vanishing on $\mathrm{range}\mathcal{T}_q$. There is $v \in L^{q'}(\mathbb{R}^m)$ such that $h(w) = \int_{\mathbb{R}^m} wv \,\mathrm{d}y$ for any $w \in L^q(\mathbb{R}^m)$. Particularly

$$\int_{\mathbb{R}^m} \mathcal{T}_q u \; v \,\mathrm{d}y = 0, \;\; \forall \, u \in \mathrm{D}_q$$

saying that $v \in \ker \mathcal{T}_{q'}$. For any $\varphi \in \ker\langle\cdot\rangle^{(q)}$ we can write by Proposition 2.7

$$h(\varphi) = \int_{\mathbb{R}^m} v\varphi \,\mathrm{d}y = \int_{\mathbb{R}^m} \langle v \rangle^{(q')}\varphi \,\mathrm{d}y = \int_{\mathbb{R}^m} v \, \langle \varphi \rangle^{(q)} \,\mathrm{d}y = 0$$

and thus h vanishes on $\ker\langle\cdot\rangle^{(q)}$. Consequently we have $\overline{\mathrm{range}\mathcal{T}_q} = \ker\langle\cdot\rangle^{(q)}$. □

Generally we have the following characterization for $\ker\langle\cdot\rangle^{(q)} = \overline{\mathrm{range}\mathcal{T}_q}$.

Proposition 2.10. *Let f be a function in $L^q(\mathbb{R}^m)$ for some $q \in (1, +\infty)$. For any $\mu > 0$ we denote by u_μ the unique solution of*

$$\mu u_\mu + \mathcal{T}_q u_\mu = f \tag{2.9}$$

which is given by

$$u_\mu = \int_{-\infty}^{0} \mathrm{e}^{\mu s} f(Y(s;\cdot)) \,\mathrm{d}s. \tag{2.10}$$

Then the following statements are equivalet
a) $\langle f \rangle^{(q)} = 0$.
b) $\lim_{\mu \searrow 0}(\mu u_\mu) = 0$ *in* $L^q(\mathbb{R}^m)$.

Proof. Assume that b) holds true. Applying the operator $\langle\cdot\rangle^{(q)}$ in (2.9) one gets

$$\langle f \rangle^{(q)} = \langle \mu u_\mu \rangle^{(q)} + \langle \mathcal{T}_q u_\mu \rangle^{(q)} = \langle \mu u_\mu \rangle^{(q)}, \;\; \forall \, \mu > 0$$

and therefore

$$\langle f\rangle^{(q)} = \lim_{\mu\searrow 0}\langle \mu u_\mu\rangle^{(q)} = \langle \lim_{\mu\searrow 0}(\mu u_\mu)\rangle^{(q)} = 0.$$

Conversely, suppose that a) holds true. Considering the function $G(s;y) = \int_s^0 f(Y(\tau;y))\,\mathrm{d}\tau$ we obtain by formula (2.10) (use the inequality $\|G(s;\cdot)\|_{L^q(\mathbb{R}^m)} \leq |s|\|f\|_{L^q(\mathbb{R}^m)}$ in order to justify the integration by parts)

$$\begin{aligned} u_\mu &= -\int_{-\infty}^0 \mathrm{e}^{\mu s}\frac{\partial G}{\partial s}(s;\cdot)\,\mathrm{d}s = \int_{-\infty}^0 \mu s\,\mathrm{e}^{\mu s}\frac{G(s;\cdot)}{s}\,\mathrm{d}s \\ &= \frac{1}{\mu}\int_{-\infty}^0 t\,\mathrm{e}^t \frac{G(t\mu^{-1};\cdot)}{t\mu^{-1}}\,\mathrm{d}t. \end{aligned}$$

We know that $\|G(t\mu^{-1})/(t\mu^{-1})\|_{L^q(\mathbb{R}^m)} \leq \|f\|_{L^q(\mathbb{R}^m)}$ and by Proposition 2.6 we have for any $t<0$

$$\begin{aligned} \lim_{\mu\searrow 0}\frac{G(t\mu^{-1};\cdot)}{t\mu^{-1}} &= \lim_{\mu\searrow 0}\frac{\int_{t/\mu}^0 f(Y(s;\cdot))\,\mathrm{d}s}{t/\mu} \\ &= -\langle f\rangle^{(q)} = 0, \quad \text{strongly in } L^q(\mathbb{R}^m). \end{aligned}$$

Consequently, by the dominated convergence theorem, one gets

$$\|\mu u_\mu\|_{L^q(\mathbb{R}^m)} \leq \int_{-\infty}^0 |t|\mathrm{e}^t \left\|\frac{G(t\mu^{-1};\cdot)}{t\mu^{-1}}\right\|_{L^q(\mathbb{R}^m)} dt \to 0 \text{ as } \mu\searrow 0.$$

□

Remark 2.11. With the above notations we have $\|\mu u_\mu\|_{L^q(\mathbb{R}^m)} \leq \|f\|_{L^q(\mathbb{R}^m)}$.

Up to this point we have investigated the properties of $\langle\cdot\rangle^{(q)}$ operating from $L^q(\mathbb{R}^m)$ to $L^q(\mathbb{R}^m)$ with $q\in[1,+\infty]$. In view of further regularity results for transport equations with singular coefficients we investigate now how $\langle\cdot\rangle^{(q)}$ acts on some particular subspaces of smooth functions. For this purpose we recall here the following basic results concerning the derivation operators along fields in $\mathbb{R}^m$. For any $\xi = (\xi_1(y),\ldots,\xi_m(y))$, where $y\in\mathbb{R}^m$, we denote by L_ξ the operator $\xi\cdot\nabla_y$. A direct computation shows that for any smooth fields ξ,η, the commutator between L_ξ and L_η is still a first order operator, given by

$$[L_\xi, L_\eta] := L_\xi L_\eta - L_\eta L_\xi = L_\chi$$

where χ is the Poisson bracket of ξ and η

$$\chi = [\xi,\eta],\ \ [\xi,\eta]_i = (\xi\cdot\nabla_y)\eta_i - (\eta\cdot\nabla_y)\xi_i = L_\xi(\eta_i) - L_\eta(\xi_i),\ \ i\in\{1,\ldots,m\}.$$

It is well known (see [1], p. 93) that L_ξ, L_η commute (or equivalently the Poisson bracket $[\xi, \eta]$ vanishes) iff the flows corresponding to ξ, η, let's say, Z_1, Z_2, commute

$$Z_1(s_1; Z_2(s_2; y)) = Z_2(s_2; Z_1(s_1; y)), \quad s_1, s_2 \in \mathbb{R}, \quad y \in \mathbb{R}^m.$$

Considering a smooth field c in involution with b and having bounded divergence, one gets

$$c \in W^{1,\infty}_{\text{loc}}(\mathbb{R}^m), \quad \operatorname{div}_y c \in L^\infty(\mathbb{R}^m), \quad [c, b] = 0$$

and let us denote by Z the flow associated with c (we assume that Z is well defined for any $(s, y) \in \mathbb{R} \times \mathbb{R}^m$). For any $h \in \mathbb{R}$ we denote by τ_h the map associated with a function u with its translation on a time h along the flow Z

$$(\tau_h u)(y) = u(Z(h; y)), \quad y \in \mathbb{R}^m \quad h \in \mathbb{R}.$$

We claim that for any $h \in \mathbb{R}$ the operators $\langle \cdot \rangle^{(q)}$ and τ_h commute. We use the following easy lemma.

Lemma 2.12. *Let c be a smooth field in involution with b. Then the divergence of c is invariant along the flow of b.*

Proposition 2.13. *Assume that c is a smooth field in involution with b, with bounded divergence and well defined flow. Then for any $q \in (1, +\infty)$ the operator $\langle \cdot \rangle^{(q)}$ commutes with the translations along the flow of c*

$$\langle u \circ Z(h; \cdot) \rangle^{(q)} = \langle u \rangle^{(q)} \circ Z(h; \cdot), \quad u \in L^q(\mathbb{R}^m), \quad h \in \mathbb{R}.$$

Proof. First of all observe that τ_h maps $L^q(\mathbb{R}^m)$ to $L^q(\mathbb{R}^m)$ (use Liouville's theorem and the hypothesis $\operatorname{div}_y c \in L^\infty(\mathbb{R}^m)$). Assume that $q \in (1, +\infty)$. By Proposition 2.3 we know that for any $\varphi \in \mathcal{T}_{q'}$ we have

$$\int_{\mathbb{R}^m} (u - \langle u \rangle^{(q)}) \varphi \, \mathrm{d}y = 0. \tag{2.11}$$

We denote by φ_{-h} the function

$$\varphi_{-h}(z) = \varphi(Z(-h; z)) \, \mathrm{e}^{-\int_0^h (\operatorname{div}_y c)(Z(-t; z)) \, \mathrm{d}t}.$$

Notice that $\varphi_{-h} \in \ker \mathcal{T}_{q'}$. Indeed, replacing z by $Y(s; y)$ and by considering that the flows Y and Z commute we obtain

$$\varphi(Z(-h; Y(s; y))) = \varphi(Y(s; Z(-h; y))) = \varphi(Z(-h; y)).$$

Thanks to Lemma 2.12 we have

$$(\operatorname{div}_y c)(Z(-t; Y(s; y))) = (\operatorname{div}_y c)(Y(s; Z(-t; y))) = (\operatorname{div}_y c)(Z(-t; y)).$$

Consequently one gets $\varphi_{-h}(Y(s;y)) = \varphi_{-h}(y)$ and it is easily seen that φ_{-h} belongs to $L^{q'}(\mathbb{R}^m)$. Applying (2.11) with the trial function φ_{-h} and using the variable change $z = Z(h;y)$ we deduce that

$$\int_{\mathbb{R}^m} (u(Z(h;y)) - \langle u\rangle^{(q)}(Z(h;y))\varphi(y)\ \mathrm{d}y = 0.$$

Observe also that $\langle u\rangle^{(q)}(Z(h;\cdot))$ belongs to $L^q(\mathbb{R}^m)$ and that it is invariant along the flow of b

$$\langle u\rangle^{(q)}(Z(h;Y(s;y))) = \langle u\rangle^{(q)}(Y(s;Z(h;y))) = \langle u\rangle^{(q)}(Z(h;y)).$$

Consequently, by Proposition 2.3 we deduce that $\langle u\circ Z(h;\cdot)\rangle^{(q)} = \langle u\rangle^{(q)}\circ Z(h;\cdot)$. □

Remark 2.14. Particularly we have $[b,b] = 0$ and therefore $\langle\cdot\rangle^{(q)}$ commutes with the translations along the flow of b. We have for any $h \in \mathbb{R}$, $u \in L^q(\mathbb{R}^m)$, $q \in (1,+\infty)$

$$\langle u(Y(h;\cdot))\rangle^{(q)} = \langle u\rangle^{(q)}(Y(h;\cdot)) = \langle u\rangle^{(q)}.$$

We shall show that for any smooth field c in involution with b, the operator $\langle\cdot\rangle^{(q)}$ commutes with $c\cdot\nabla_y$. We denote by $\mathcal{T}_q^c$ the operator given by

$$\mathrm{D}(\mathcal{T}_q^c) = \{u \in L^q(\mathbb{R}^m)\ :\ \mathrm{div}_y(cu) \in L^q(\mathbb{R}^m)\},$$

$$\mathcal{T}_q^c u = \mathrm{div}_y(cu) - (\mathrm{div}_y c)u,\quad u \in \mathrm{D}(\mathcal{T}_q^c).$$

We have the standard result (see [7], Proposition IX.3, p. 153 for similar results).

Lemma 2.15. *Assume that $q \in (1,+\infty)$ and let u be a function in $L^q(\mathbb{R}^m)$. Then the following statements are equivalent*
a) $u \in \mathrm{D}(\mathcal{T}_q^c)$.
b) $(h^{-1}(u(Z(h;\cdot)) - u))_h$ is bounded in $L^q(\mathbb{R}^m)$.
Moreover, for any $u \in \mathrm{D}(\mathcal{T}_q^c)$ we have the convergence

$$\lim_{h\to 0} \frac{u(Z(h;\cdot)) - u}{h} = \mathcal{T}_q^c u,\ \ \textit{strongly in}\ \ L^q(\mathbb{R}^m).$$

Thanks to Proposition 2.13 and Lemma 2.15 it is easily seen that

Proposition 2.16. *On the hypotheses of Proposition 2.13, assume that $u \in \mathrm{D}(\mathcal{T}_q^c)$ for some $q \in (1,+\infty)$. Then $\langle u\rangle^{(q)} \in \mathrm{D}(\mathcal{T}_q^c)$ and $\mathcal{T}_q^c\langle u\rangle^{(q)} = \langle \mathcal{T}_q^c u\rangle^{(q)}$.*

Remark 2.17. Particularly Proposition 2.16 applies to $c = b$. Actually, for any $u \in \mathrm{D}(\mathcal{T}_q)$, $q \in (1, +\infty)$ we have $\mathcal{T}_q \langle u \rangle^{(q)} = \langle \mathcal{T}_q u \rangle^{(q)} = 0$.

Remark 2.18. Upon the hypotheses of Proposition 2.13 we check immediately thanks to Lemma 2.15 that if $u \in \mathrm{D}(\mathcal{T}_q^c)$, then for any $s \in \mathbb{R}$, $u \circ Y(s; \cdot) \in \mathrm{D}(\mathcal{T}_q^c)$ and

$$\mathcal{T}_q^c (u \circ Y(s; \cdot)) = (\mathcal{T}_q^c u) \circ Y(s; \cdot).$$

Particularly if $u \in \ker \mathcal{T}_q \cap \mathrm{D}(\mathcal{T}_q^c)$ then $\mathcal{T}_q^c u \in \ker \mathcal{T}_q$.

The last result in this section states that $\langle \cdot \rangle^{(q)}$ commutes with the time derivation. The proof is standard and comes easily by observing that

$$\frac{\langle u(t+h) \rangle^{(q)} - \langle u(t) \rangle^{(q)}}{h} = \left\langle \frac{u(t+h) - u(t)}{h} \right\rangle^{(q)}$$

and by adapting the arguments in Lemma 2.15.

Proposition 2.19. *Assume that $u \in W^{1,p}([0,T]; L^q(\mathbb{R}^m))$ for some $p, q \in (1, +\infty)$. Then the application $(t, y) \to \langle u(t, \cdot) \rangle^{(q)}(y)$ belongs to $W^{1,p}([0,T]; L^q(\mathbb{R}^m))$ and we have $\partial_t \langle u \rangle^{(q)} = \langle \partial_t u \rangle^{(q)}$.*

3 Well-posedness of the limit model

This section is devoted to the study of the limit model, when ε goes to 0, for the transport equation

$$\begin{cases} \partial_t u^\varepsilon + a(t,y) \cdot \nabla_y u^\varepsilon + \dfrac{b(y)}{\varepsilon} \cdot \nabla_y u^\varepsilon = 0, & (t,y) \in (0,T) \times \mathbb{R}^m \\ u^\varepsilon(0,y) = u_0^\varepsilon(y), & y \in \mathbb{R}^m. \end{cases} \tag{3.1}$$

Recall that b is a given smooth field satisfying (2.1), (2.2), (2.3). We assume that a satisfies the conditions

$$a \in L^1([0,T]; W^{1,\infty}(\mathbb{R}^m)), \ \ \mathrm{div}_y a = 0. \tag{3.2}$$

Based on Hilbert's expansion method we have obtained (see (1.4), (1.5)) the formula $u^\varepsilon = u + \varepsilon u_1 + \varepsilon^2 \mathcal{O}(\varepsilon)$ where

$$b(y) \cdot \nabla_y u = 0, \ \ \partial_t u + a(t,y) \cdot \nabla_y u + b(y) \cdot \nabla_y u_1 = 0.$$

Projecting the second equation on the kernel of $\mathcal{T}$ leads to the model

$$\partial_t \langle u \rangle + \langle a(t) \cdot \nabla_y u(t) \rangle = 0, \ \ (t,y) \in (0,T) \times \mathbb{R}^m.$$

Notice that $\mathcal{T}u = 0$ and thus $\langle u \rangle = u$. Finally we obtain

$$\begin{cases} \partial_t u + \langle a(t) \cdot \nabla_y u(t) \rangle = 0, \;\; b(y) \cdot \nabla_y u = 0, & (t,y) \in (0,T) \times \mathbb{R}^m \\ u(0,y) = u_0(y), & y \in \mathbb{R}^m. \end{cases} \tag{3.3}$$

We work in the $L^q(\mathbb{R}^m)$ setting, with $q \in (1, +\infty)$. For any $\varphi \in \ker \mathcal{T}_{q'}$ we have

$$\int_{\mathbb{R}^m} (a(t,y) \cdot \nabla_y u - \langle a(t) \cdot \nabla_y u(t) \rangle^{(q)}) \varphi(y) \; \mathrm{d}y = 0$$

and we introduce the notion of weak solution for (3.3) as follows:

Definition 3.1. Assume that $u_0 \in \ker \mathcal{T}_q$, $f \in L^1([0,T]; \ker \mathcal{T}_q)$ (*i.e.*, $f \in L^1([0,T]; L^q(\mathbb{R}^m))$ and $f(t) \in \ker \mathcal{T}_q$, $t \in [0,T]$). We say that $u \in L^\infty([0,T]; \ker \mathcal{T}_q)$ is a weak solution for

$$\begin{cases} \partial_t u + \langle a(t) \cdot \nabla_y u(t) \rangle^{(q)} = f(t,y), \;\; \mathcal{T}_q u = 0, & (t,y) \in (0,T) \times \mathbb{R}^m \\ u(0,y) = u_0(y), & y \in \mathbb{R}^m \end{cases} \tag{3.4}$$

iff for any $\varphi \in C^1_c([0,T) \times \mathbb{R}^m)$ satisfying $\mathcal{T}\varphi = 0$ we have

$$\int_0^T \int_{\mathbb{R}^m} u(t,y)(\partial_t \varphi + \mathrm{div}_y(\varphi a)) \; \mathrm{d}y\mathrm{d}t + \int_{\mathbb{R}^m} u_0(y) \varphi(0,y) \; \mathrm{d}y + \int_0^T \int_{\mathbb{R}^m} f(t,y) \varphi(t,y) \; \mathrm{d}y\mathrm{d}t = 0. \tag{3.5}$$

We start by establishing existence and regularity results for the solution of (3.4).

Proposition 3.2. *Assume that $u_0 \in \ker \mathcal{T}_q$, $f \in L^1([0,T]; \ker \mathcal{T}_q)$ for some $q \in (1, +\infty)$. Then there is at least a weak solution $u \in L^\infty([0,T];$ $\ker \mathcal{T}_q)$ of (3.4) satisfying*

$$\|u(t)\|_{L^q(\mathbb{R}^m)} \leq \|u_0\|_{L^q(\mathbb{R}^m)} + \int_0^t \|f(s)\|_{L^q(\mathbb{R}^m)}, \;\; t \in [0,T].$$

Moreover, if $u_0 \geq 0$ and $f \geq 0$ then $u \geq 0$.

Proof. For any $\varepsilon > 0$ there is a unique weak solution u^ε of

$$\begin{cases} \partial_t u^\varepsilon + a(t,y) \cdot \nabla_y u^\varepsilon + \frac{b(y)}{\varepsilon} \cdot \nabla_y u^\varepsilon = f(t,y), & (t,y) \in (0,T) \times \mathbb{R}^m \\ u^\varepsilon(0,y) = u_0(y), & y \in \mathbb{R}^m. \end{cases} \tag{3.6}$$

The solution is given by

$$u^\varepsilon(t,y) = u_0(Z^\varepsilon(0;t,y)) + \int_0^t f(s, Z^\varepsilon(s;t,y)) \; \mathrm{d}s, \;\; (t,y) \in [0,T] \times \mathbb{R}^m$$

where Z^ε are the characteristics corresponding to the field $a + \varepsilon^{-1}b$. Multiplying by $u^\varepsilon(t,y)|u^\varepsilon(t,y)|^{q-2}$ and integrating with $y \in \mathbb{R}^m$, we obtain thanks to Hölder's inequality

$$\|u^\varepsilon\|_{L^q(\mathbb{R}^m)} \le \|u_0\|_{L^q(\mathbb{R}^m)} + \int_0^t \|f(s)\|_{L^q(\mathbb{R}^m)}, \quad t \in [0,T].$$

We extract a sequence $(\varepsilon_k)_k$ converging towards 0 such that $u^{\varepsilon_k} \rightharpoonup u$ weakly $\star$ in $L^\infty([0,T];L^q(\mathbb{R}^m))$ for some function $u \in L^\infty([0,T]; L^q(\mathbb{R}^m))$ satisfying

$$\|u\|_{L^\infty([0,T];L^q(\mathbb{R}^m))} \le \|u_0\|_{L^q(\mathbb{R}^m)} + \|f\|_{L^1([0,T];L^q(\mathbb{R}^m))}.$$

By the weak formulation of (3.6) with a function $\varphi \in C_c^1([0,T) \times \mathbb{R}^m)$ we deduce that

$$\int_0^T\!\!\int_{\mathbb{R}^m} u^{\varepsilon_k}\left(\partial_t\varphi + \left(a + \frac{b}{\varepsilon_k}\right)\cdot\nabla_y\varphi\right)\,\mathrm{d}y\mathrm{d}t + \int_{\mathbb{R}^m} u_0\varphi(0,y)\,\mathrm{d}y + \int_0^T\!\!\int_{\mathbb{R}^m} f\varphi\,\mathrm{d}y\mathrm{d}t = 0. \tag{3.7}$$

Multiplying by ε_k and passing to the limit as $k \to +\infty$ one gets easily by Proposition 2.1 that $u(t) \in \ker \mathcal{T}_q$, $t \in [0,T)$. If the test function in (3.7) verifies $\mathcal{T}\varphi = 0$ we obtain

$$\int_0^T\!\!\int_{\mathbb{R}^m} u^{\varepsilon_k}\left(\partial_t\varphi + a\cdot\nabla_y\varphi\right)\,\mathrm{d}y\mathrm{d}t + \int_{\mathbb{R}^m} u_0\varphi(0,y)\,\mathrm{d}y + \int_0^T\!\!\int_{\mathbb{R}^m} f\varphi\,\mathrm{d}y\mathrm{d}t = 0.$$

Passing to the limit for $k \to +\infty$ we deduce that the weak $\star$ limit u satisfies the weak formulation of (3.4). If $u_0 \ge 0, f \ge 0$ then $u^\varepsilon \ge 0$ for any $\varepsilon > 0$ and thus the solution constructed above is non negative. □

Whereas Proposition 3.2 yields a satisfactory theoretical result for solving the limit model (3.4), its numerical approximation remains a difficult problem. The main drawback of the weak formulation (3.5) is the particular form of the trial functions $\varphi \in \ker \mathcal{T} \cap C_c^1([0,T)\times\mathbb{R}^m)$. Generally, the choice of such test functions could be a difficult task. Accordingly, we are looking for a strong formulation of (3.4). Therefore we inquire about the smoothness of the solution. A complete regularity analysis can be carried out on the following hypothesis: we will assume that the field a is a linear combination of fields in involution with $b^0 := b$

$$a(t,y) = \sum_{i=0}^{r} \alpha_i(t,y)b^i(y), \quad b^i \in W^{1,\infty}(\mathbb{R}^m), \quad [b^i, b] = 0,\ i \in \{1,\dots,r\} \tag{3.8}$$

where $(\alpha_i)_i$ are smooth coefficients verifying

$$\alpha_i \in L^1([0,T]; L^\infty(\mathbb{R}^m)),\ b^j \cdot \nabla_y \alpha_i \in L^1([0,T]; L^\infty(\mathbb{R}^m)),\ i,j \in \{0,1,\dots,r\}. \tag{3.9}$$

For any $i \in \{1,\dots,r\}$ we denote by $\mathcal{T}_q^i : \mathrm{D}(\mathcal{T}_q^i) \subset L^q(\mathbb{R}^m) \to L^q(\mathbb{R}^m)$ the operator given by

$$\mathrm{D}(\mathcal{T}_q^i) = \{u \in L^q(\mathbb{R}^m) \ :\ \mathrm{div}_y(b^i u) \in L^q(\mathbb{R}^m)\},$$

$$\mathcal{T}_q^i u = \mathrm{div}_y(b^i u) - (\mathrm{div}_y b^i) u,\quad u \in \mathrm{D}(\mathcal{T}_q^i)$$

and by Y^i the flow associated with b^i. Since $[b^i, b] = 0$ then Y^i commutes with Y for any $i \in \{1,\dots,r\}$.

Proposition 3.3. *Assume that (3.8), (3.9) hold, $u_0 \in \ker \mathcal{T}_q \cap (\cap_{i=1}^r \mathrm{D}(\mathcal{T}_q^i))$, $f \in L^1([0,T]; \ker \mathcal{T}_q \cap (\cap_{i=1}^r \mathrm{D}(\mathcal{T}_q^i)))$ (i.e., $f \in L^1([0,T]; L^q(\mathbb{R}^m))$, $\mathcal{T}_q f = 0$ and $\mathcal{T}_q^i f \in L^1([0,T]; L^q(\mathbb{R}^m))$, $i \in \{1,\dots,r\}$) and let us denote by u the weak solution of (3.4) constructed in Proposition 3.2. Then we have $u(t) \in \ker \mathcal{T}_q \cap (\cap_{i=1}^r \mathrm{D}(\mathcal{T}_q^i))$, $t \in [0,T]$ and*

$$\|\partial_t u\|_{L^1([0,T];L^q(\mathbb{R}^m))} + \sum_{i=1}^r \|\mathcal{T}_q^i u\|_{L^\infty([0,T];L^q(\mathbb{R}^m))} \le C(\|f\|_{L^1([0,T];L^q(\mathbb{R}^m))} + \sum_{i=1}^r \|\mathcal{T}_q^i f\|_{L^1([0,T];L^q(\mathbb{R}^m))} + \sum_{i=1}^r \|\mathcal{T}_q^i u_0\|_{L^q(\mathbb{R}^m)})$$

for some constant depending on $\sum_{0\le i,j\le r} \|b^i \cdot \nabla_y \alpha_j\|_{L^1([0,T];L^\infty(\mathbb{R}^m))}$, $\sum_{i=0}^r \|\alpha_i\|_{L^1([0,T];L^\infty(\mathbb{R}^m))}$. Moreover, if $f \in L^\infty([0,T]; L^q(\mathbb{R}^m))$, $\alpha_i \in L^\infty([0,T]; L^\infty(\mathbb{R}^m))$ for any $i \in \{1,\dots,r\}$ then $\partial_t u \in L^\infty([0,T]; L^q(\mathbb{R}^m))$.

Proof. For any $\varepsilon > 0$ let u^ε be the solution of (3.6). We intend to estimate $\|\mathcal{T}_q u^\varepsilon\|_{L^\infty([0,T];L^q(\mathbb{R}^m))} + \sum_{i=1}^r \|\mathcal{T}_q^i u^\varepsilon\|_{L^\infty([0,T];L^q(\mathbb{R}^m))}$ and $\|\partial_t u^\varepsilon\|_{L^1([0,T];L^q(\mathbb{R}^m))}$ uniformly with respect to $\varepsilon > 0$. Consider the sequences of smooth functions $(u_{0n})_n$, $(f_n)_n$ such that

$$\lim_{n\to+\infty} u_{0n} = u_0,\ \lim_{n\to+\infty} \mathcal{T}_q^i u_{0n} = \mathcal{T}_q^i u_0,\ i \in \{0,1,\dots,r\} \text{ in } L^q(\mathbb{R}^m)$$

$$\lim_{n\to+\infty} f_n = f,\ \lim_{n\to+\infty} \mathcal{T}_q^i f_n = \mathcal{T}_q^i f,\ i \in \{0,1,\dots,r\} \text{ in } L^1([0,T]; L^q(\mathbb{R}^m))$$

and let us denote by $(u_n^\varepsilon)_n$ the solutions of (3.6) corresponding to the initial conditions $(u_{0n})_n$ and the source terms $(f_n)_n$. Actually $(u_n^\varepsilon)_n$ are strong solutions. It is easily seen that for any $t \in [0,T]$

$$\|u_n^\varepsilon(t) - u^\varepsilon(t)\|_{L^q(\mathbb{R}^m)} \le \|u_{0n} - u_0\|_{L^q(\mathbb{R}^m)} + \int_0^t \|f_n(s) - f(s)\|_{L^q(\mathbb{R}^m)}\,ds$$

and therefore $\lim_{n\to+\infty} u_n^\varepsilon = u^\varepsilon$ in $L^\infty([0,T];L^q(\mathbb{R}^m))$. Assume for the moment that ε, n are fixed and let us estimate $\sum_{i=0}^r \|\mathcal{T}_q^i u_n^\varepsilon\|_{L^\infty([0,T];L^q(\mathbb{R}^m))}$ and $\|\partial_t u_n^\varepsilon\|_{L^1([0,T];L^q(\mathbb{R}^m))}$. Take $h \in \mathbb{R}$, $i \in \{1,\dots,r\}$ and consider the functions

$$u_{nh}^\varepsilon(t,y) = u_n^\varepsilon(t, Y^i(h;y)),\quad a_h(t,y) = \frac{\partial Y^i}{\partial y}(-h; Y^i(h;y))a(t, Y^i(h;y))$$

$$b_h(y) = \frac{\partial Y^i}{\partial y}(-h; Y^i(h;y))b(Y^i(h;y)),\quad u_{0nh}(y) = u_{0n}(Y^i(h;y)),$$

$$f_{nh}(t,y) = f_n(t, Y^i(h;y)).$$

A direct computation shows that

$$\begin{cases} \partial_t u_{nh}^\varepsilon + a_h(t,y)\cdot\nabla_y u_{nh}^\varepsilon + \frac{b_h(y)}{\varepsilon}\cdot\nabla_y u_{nh}^\varepsilon = f_{nh}(t,y), & (t,y)\in(0,T)\times\mathbb{R}^m \\ u_{nh}^\varepsilon(0,y) = u_{0nh}(y), & y\in\mathbb{R}^m. \end{cases} \tag{3.10}$$

Combining with the formulation (3.6) of u_n^ε one gets

$$\begin{cases} \partial_t\left(\frac{u_{nh}^\varepsilon - u_n^\varepsilon}{h}\right) + \frac{a_h - a}{h}\cdot\nabla_y u_{nh}^\varepsilon + a(t,y)\cdot\nabla_y\left(\frac{u_{nh}^\varepsilon - u_n^\varepsilon}{h}\right) & \\ +\frac{b_h - b}{\varepsilon h}\cdot\nabla_y u_{nh}^\varepsilon + \frac{b(y)}{\varepsilon}\cdot\nabla_y\left(\frac{u_{nh}^\varepsilon - u_n^\varepsilon}{h}\right) = \frac{f_{nh} - f_n}{h}, & (t,y)\in(0,T)\times\mathbb{R}^m \\ \frac{u_{nh}^\varepsilon(0,y) - u_n^\varepsilon(0,y)}{h} = \frac{u_{0nh}(y) - u_{0n}(y)}{h}, & y\in\mathbb{R}^m. \end{cases} \tag{3.11}$$

Obviously we have

$$\lim_{h\to 0}\frac{u_{nh}^\varepsilon - u_n^\varepsilon}{h} = \lim_{h\to 0}\frac{u_n^\varepsilon(t, Y^i(h;y)) - u_n^\varepsilon(t,y)}{h} = b^i(y)\cdot\nabla_y u_n^\varepsilon(t,y) = \mathcal{T}_q^i u_n^\varepsilon$$

$$\lim_{h\to 0}\frac{f_{nh} - f_n}{h} = \lim_{h\to 0}\frac{f_n(t, Y^i(h;y)) - f_n(t,y)}{h} = b^i(y)\cdot\nabla_y f_n(t,y) = \mathcal{T}_q^i f_n$$

$$\lim_{h\to 0}\frac{u_{0nh} - u_{0n}}{h} = \lim_{h\to 0}\frac{u_{0n}(Y^i(h;y)) - u_{0n}(y)}{h} = b^i(y)\cdot\nabla_y u_{0n}(y) = \mathcal{T}_q^i u_{0n}.$$

Taking the derivatives with respect to y and then with respect to h in the equality $Y^i(-h; Y^i(h;y)) = y$, we deduce after some easy manipulations that

$$\lim_{h\to 0}\frac{1}{h}\left\{\frac{\partial Y^i}{\partial y}(-h; Y^i(h;y)) - I_m\right\} = -\frac{\partial b^i}{\partial y}(y).$$

By direct computations we obtain immediately

$$\lim_{h\to 0}\frac{a_h - a}{h} = (b^i\cdot\nabla_y)a - (a\cdot\nabla_y)b^i = [b^i, a]$$

$$\lim_{h\to 0}\frac{b_h - b}{h} = (b^i\cdot\nabla_y)b - (b\cdot\nabla_y)b^i = [b^i, b] = 0.$$

By passing to the limit for $h \to 0$ in (3.11) we deduce that $\mathcal{T}_q^i u_n^\varepsilon$ solves weakly the problem

$$\begin{cases} \partial_t(\mathcal{T}_q^i u_n^\varepsilon) + a \cdot \nabla_y(\mathcal{T}_q^i u_n^\varepsilon) + \dfrac{b}{\varepsilon} \cdot \nabla_y(\mathcal{T}_q^i u_n^\varepsilon) = \mathcal{T}_q^i f_n - [b^i, a] \cdot \nabla_y u_n^\varepsilon \\ \mathcal{T}_q^i u_n^\varepsilon(0, \cdot) = \mathcal{T}_q^i u_{0n}. \end{cases} \tag{3.12}$$

As shown in the proof of Proposition 3.2 we obtain for any $t \in [0, T]$ and $i \in \{1, \dots, r\}$

$$\|\mathcal{T}_q^i u_n^\varepsilon(t)\|_{L^q(\mathbb{R}^m)} \le \|\mathcal{T}_q^i u_{0n}\|_{L^q(\mathbb{R}^m)} + \int_0^t \|\mathcal{T}_q^i f_n(s) - [b^i, a(s)] \cdot \nabla_y u_n^\varepsilon(s)\|_{L^q(\mathbb{R}^m)} \,\mathrm{d}s. \tag{3.13}$$

Since $a = \sum_{k=0}^r \alpha_k b^k$ we obtain by direct computation, with the notation $\mathcal{T}_q^0 := \mathcal{T}_q$

$$[b^i, a] = \sum_{k=0}^r (\mathcal{T}_q^i \alpha_k) b^k$$

and therefore

$$[b^i, a] \cdot \nabla_y u_n^\varepsilon = \sum_{k=0}^r (\mathcal{T}_q^i \alpha_k)(\mathcal{T}_q^k u_n^\varepsilon).$$

Consequently (3.13) implies

$$\|\mathcal{T}_q^i u_n^\varepsilon(t)\|_{L^q(\mathbb{R}^m)} \le \|\mathcal{T}_q^i u_{0n}\|_{L^q(\mathbb{R}^m)} + \int_0^t \|\mathcal{T}_q^i f_n(s)\|_{L^q(\mathbb{R}^m)} \,\mathrm{d}s + \int_0^t \sum_{k=0}^r \|b^i \cdot \nabla_y \alpha_k(s)\|_{L^\infty(\mathbb{R}^m)} \|\mathcal{T}_q^k u_n^\varepsilon(s)\|_{L^q(\mathbb{R}^m)} \,\mathrm{d}s. \tag{3.14}$$

Actually (3.14) holds also for b^i replaced by $b^0 = b$ since $[b, b] = 0$

$$\|\mathcal{T}_q^0 u_n^\varepsilon(t)\|_{L^q(\mathbb{R}^m)} \le \|\mathcal{T}_q^0 u_{0n}\|_{L^q(\mathbb{R}^m)} + \int_0^t \|\mathcal{T}_q^0 f_n(s)\|_{L^q(\mathbb{R}^m)} \,\mathrm{d}s + \int_0^t \sum_{k=0}^r \|b^0 \cdot \nabla_y \alpha_k(s)\|_{L^\infty(\mathbb{R}^m)} \|\mathcal{T}_q^k u_n^\varepsilon(s)\|_{L^q(\mathbb{R}^m)} \,\mathrm{d}s. \tag{3.15}$$

Summing up the above inequalities one gets

$$\sum_{i=0}^r \|\mathcal{T}_q^i u_n^\varepsilon(t)\|_{L^q(\mathbb{R}^m)} \le \sum_{i=0}^r \|\mathcal{T}_q^i u_{0n}\|_{L^q(\mathbb{R}^m)} + \int_0^t \sum_{i=0}^r \|\mathcal{T}_q^i f_n(s)\|_{L^q(\mathbb{R}^m)} \,\mathrm{d}s + \sum_{i=0}^r \sum_{k=0}^r \int_0^t \|b^i \cdot \nabla_y \alpha_k(s)\|_{L^\infty(\mathbb{R}^m)} \|\mathcal{T}_q^k u_n^\varepsilon(s)\|_{L^q(\mathbb{R}^m)}. \tag{3.16}$$

By Gronwall's lemma we deduce that for any $t \in [0, T]$

$$\sum_{i=0}^{r} \|\mathcal{T}_q^i u_n^\varepsilon\|_{L^\infty([0,T];L^q(\mathbb{R}^m))} \leq C \sum_{i=0}^{r} \{\|\mathcal{T}_q^i u_{0n}\|_{L^q(\mathbb{R}^m)} + \|\mathcal{T}_q^i f_n\|_{L^1([0,T];L^q(\mathbb{R}^m))}\} \quad (3.17)$$

for some constant depending on $\sum_{0\leq i,j\leq r} \|b^i \cdot \nabla_y \alpha_j\|_{L^1([0,T];L^\infty(\mathbb{R}^m))}$. After extraction eventually we can assume that $(\mathcal{T}_q^i u_n^\varepsilon)_n$ converges weakly $\star$ in $L^\infty([0,T]; L^q(\mathbb{R}^m))$ towards some function $w^i \in L^\infty([0,T]; L^q(\mathbb{R}^m))$ for any $i \in \{0, 1, \ldots, r\}$. Since we know that $\lim_{n\to+\infty} u_n^\varepsilon = u^\varepsilon$ in $L^\infty([0,T]; L^q(\mathbb{R}^m))$ it is easily seen that

$$u^\varepsilon(t) \in \cap_{i=0}^r \mathrm{D}(\mathcal{T}_q^i), \quad \mathcal{T}_q^i u^\varepsilon(t) = w^i(t), \quad t \in [0, T].$$

Moreover, passing to the limit with respect to n in (3.17) and taking account of that $\lim_{n\to+\infty} \mathcal{T}_q u_{0n} = \mathcal{T}_q u_0 = 0$ in $L^q(\mathbb{R}^m)$ and $\lim_{n\to+\infty} \mathcal{T}_q f_n = \mathcal{T}_q f = 0$ in $L^1([0,T]; L^q(\mathbb{R}^m))$ we obtain

$$\sum_{i=1}^{r} \|\mathcal{T}_q^i u^\varepsilon\|_{L^\infty([0,T];L^q(\mathbb{R}^m))} \leq C \sum_{i=1}^{r} \{\|\mathcal{T}_q^i u_0\|_{L^q(\mathbb{R}^m)} + \|\mathcal{T}_q^i f\|_{L^1([0,T];L^q(\mathbb{R}^m))}\}. \quad (3.18)$$

Recall that the weak solution u constructed in Proposition 3.2 has been obtained by taking a weak $\star$ limit point of the family $(u^\varepsilon)_{\varepsilon>0}$ in $L^\infty([0,T]; L^q(\mathbb{R}^m))$. Therefore we deduce by passing to the limit for $\varepsilon \searrow 0$ in (3.18) that $u(t) \in \cap_{i=1}^r \mathrm{D}(\mathcal{T}_q^i)$, $t \in [0, T]$ and

$$\sum_{i=1}^{r} \|\mathcal{T}_q^i u\|_{L^\infty([0,T];L^q(\mathbb{R}^m))} \leq C \sum_{i=1}^{r} \{\|\mathcal{T}_q^i u_0\|_{L^q(\mathbb{R}^m)} + \|\mathcal{T}_q^i f\|_{L^1([0,T];L^q(\mathbb{R}^m))}\}. \quad (3.19)$$

Since $\mathcal{T}_q u = 0$, observe also that

$$\|a(t) \cdot \nabla_y u(t)\|_{L^q(\mathbb{R}^m)} = \|\sum_{i=1}^{r} \alpha_i(t) b^i \cdot \nabla_y u(t)\|_{L^q(\mathbb{R}^m)} \leq \sum_{i=1}^{r} \|\alpha_i(t)\|_{L^\infty(\mathbb{R}^m)} \|\mathcal{T}_q^i u(t)\|_{L^q(\mathbb{R}^m)}$$

and thus

$$\begin{aligned}\|\partial_t u\|_{L^1([0,T];L^q(\mathbb{R}^m))} &= \|f - \langle a\cdot\nabla_y u\rangle^{(q)}\|_{L^1([0,T];L^q(\mathbb{R}^m))}\\ &\le \|f\|_{L^1([0,T];L^q(\mathbb{R}^m))} + \sum_{i=1}^{r}\\ &\quad \|\mathcal{T}_q^i u\|_{L^\infty([0,T];L^q(\mathbb{R}^m))}\|\alpha_i\|_{L^1([0,T];L^\infty(\mathbb{R}^m))}\\ &\le \|f\|_{L^1([0,T];L^q(\mathbb{R}^m))} + C\sum_{i=1}^{r}\\ &\quad \{\|\mathcal{T}_q^i f\|_{L^1([0,T];L^q(\mathbb{R}^m))} + \|\mathcal{T}_q^i u_0\|_{L^q(\mathbb{R}^m)}\}.\end{aligned}$$

When f belongs to $L^\infty([0,T];L^q(\mathbb{R}^m))$ and $\alpha_i \in L^\infty([0,T];L^\infty(\mathbb{R}^m))$ for any $i \in \{1,\dots,r\}$ we obtain

$$\begin{aligned}\|\partial_t u\|_{L^\infty([0,T];L^q(\mathbb{R}^m))} &\le \|f\|_{L^\infty([0,T];L^q(\mathbb{R}^m))} + \sum_{i=1}^{r}\\ &\quad \|\alpha_i\|_{L^\infty([0,T];L^\infty(\mathbb{R}^m))}\|\mathcal{T}_q^i u\|_{L^\infty([0,T];L^q(\mathbb{R}^m))}\\ &\le \|f\|_{L^\infty([0,T];L^q(\mathbb{R}^m))} + C\sum_{i=1}^{r}\\ &\quad \{\|\mathcal{T}_q^i f\|_{L^1([0,T];L^q(\mathbb{R}^m))} + \|\mathcal{T}_q^i u_0\|_{L^q(\mathbb{R}^m)}\}.\end{aligned}$$

□

Thanks to the previous regularity result we are able to establish the existence of strong solution for (3.4).

Definition 3.4. Upon the hypotheses (3.8), (3.9) we say that u is a strong solution of (3.4) iff $u \in L^\infty([0,T];L^q(\mathbb{R}^m))$, $\partial_t u \in L^1([0,T];L^q(\mathbb{R}^m))$, $\mathcal{T}_q^i u \in L^\infty([0,T];L^q(\mathbb{R}^m))$ for any $i \in \{1,\dots,r\}$ and

$$\begin{cases} \partial_t u + \sum_{i=1}^r \langle\alpha_i(t)\rangle^{(\infty)}\mathcal{T}_q^i u(t) = f(t), & t\in(0,T)\\ u(0) = u_0. \end{cases} \tag{3.20}$$

Corollary 3.5. *Assume that (3.8), (3.9) hold. Then for any* $u_0 \in (\cap_{i=1}^r \mathrm{D}(\mathcal{T}_q^i)) \cap \ker \mathcal{T}_q$ *and* $f \in L^1([0,T];(\cap_{i=1}^r \mathrm{D}(\mathcal{T}_q^i))\cap\ker\mathcal{T}_q)$, *there is a strong solution* u *for (3.4) verifying*

$$\begin{aligned}\|\partial_t u\|_{L^1([0,T];L^q(\mathbb{R}^m))} + \sum_{i=1}^{r}\|\mathcal{T}_q^i u\|_{L^\infty([0,T];L^q(\mathbb{R}^m))} &\le C\|f\|_{L^1([0,T];L^q(\mathbb{R}^m))}\\ &+ C\sum_{i=1}^{r}\{\|\mathcal{T}_q^i f\|_{L^1([0,T];L^q(\mathbb{R}^m))} + \|\mathcal{T}_q^i u_0\|_{L^q}\}. \end{aligned}\tag{3.21}$$

As usual, the existence of strong solution for the adjoint problem implies the uniqueness of weak solution.

Proposition 3.6. *Assume that (3.8), (3.9) hold. Then for any* $u_0 \in \ker \mathcal{T}_q$ *and* $f \in L^1([0,T]; \ker \mathcal{T}_q)$, *with* $q \in (1, +\infty)$, *there is at most one weak solution of (3.4).*

Remark 3.7. The uniqueness of the weak solution for (3.4) guarantees the uniqueness of the strong solution in Corollary 3.5.

Corollary 3.8. *Assume that (3.8), (3.9) hold and that* $u_0 \in \ker \mathcal{T}_q$, $f \in L^1([0,T]; \ker \mathcal{T}_q)$ *for some* $q \in (1, +\infty)$. *Then the weak solution of (3.4) satisfies for any* $t \in [0,T]$

$$\frac{1}{q}\int_{\mathbb{R}^m} |u(t,y)|^q \, \mathrm{d}y = \frac{1}{q}\int_{\mathbb{R}^m} |u_0(y)|^q \, \mathrm{d}y + \int_0^t \int_{\mathbb{R}^m} f(s,y)|u(s,y)|^{q-2}u(s,y) \, \mathrm{d}y \, \mathrm{d}s.$$

Particularly, when $f = 0$ *the* L^q *norm is preserved.*

Naturally we can obtain more smoothness for the solution provided that the data are more regular. We present here a simplified version for the homogeneous problem. The proof is a direct consequence of Propositions 3.3, 2.16.

Proposition 3.9. *Assume that (3.8), (3.9) hold and let us denote by* u *the solution of (3.4) with* $f = 0$ *and the initial condition* u_0 *satisfying for some* $q \in (1, +\infty)$

$$u_0 \in (\cap_{i=1}^r \mathrm{D}(\mathcal{T}_q^i)) \cap \ker \mathcal{T}_q, \quad \mathcal{T}_q^j u_0 \in \cap_{i=1}^r \mathrm{D}(\mathcal{T}_q^i), \ \forall\, j \in \{1, \ldots, r\}.$$

Then we have

$$\sum_{i=1}^r \sum_{j=1}^r \|\mathcal{T}_q^i \mathcal{T}_q^j u\|_{L^\infty([0,T];L^q(\mathbb{R}^m))}$$
$$\leq C \left(\sum_{i=1}^r \sum_{j=1}^r \|\mathcal{T}_q^i \mathcal{T}_q^j u_0\|_{L^q(\mathbb{R}^m)} + \sum_{i=1}^r \|\mathcal{T}_q^i u_0\|_{L^q(\mathbb{R}^m)} \right)$$

with C *depending on* $\sum_{1 \leq i,j,k \leq r} \|\mathcal{T}_q^i \mathcal{T}_q^j \alpha_k\|_{L^1([0,T];L^\infty(\mathbb{R}^m))}$, $\sum_{1 \leq i,j \leq r} \|\mathcal{T}_q^i \alpha_j\|_{L^1([0,T];L^\infty(\mathbb{R}^m))}$ *and*

$$\|\partial_t^2 u\|_{L^1([0,T];L^q)} + \sum_{i=1}^r \|\partial_t \mathcal{T}_q^i u\|_{L^1([0,T];L^q)}$$
$$\leq C \left(\sum_{i=1}^r \sum_{j=1}^r \|\mathcal{T}_q^i \mathcal{T}_q^j u_0\|_{L^q} + \sum_{i=1}^r \|\mathcal{T}_q^i u_0\|_{L^q} \right)$$

with C *depending on* $\sum_{1\le i,j,k\le r} \|\mathcal{T}_q^i\mathcal{T}_q^j\alpha_k\|_{L^1([0,T];\ L^\infty(\mathbb{R}^m))}$, $\sum_{1\le i,j\le r} \|\mathcal{T}_q^i\alpha_j\|_{L^1([0,T];\ L^\infty(\mathbb{R}^m))}$, $\sum_{i=1}^r \|\alpha_i\|_{L^1([0,T];\ L^\infty(\mathbb{R}^m))}$ *and* $\sum_{i=1}^r \|\partial_t \alpha_i\|_{L^1([0,T];\ L^\infty(\mathbb{R}^m))}$.

4 Convergence towards the limit model

This section is devoted to the asymptotic behavior of the solutions $(u^\varepsilon)_{\varepsilon>0}$ of

$$\begin{cases} \partial_t u^\varepsilon + a(t,y)\cdot\nabla_y u^\varepsilon + \dfrac{b(y)}{\varepsilon}\cdot\nabla_y u^\varepsilon = 0, & (t,y)\in(0,T)\times\mathbb{R}^m \\ u^\varepsilon(0,y) = u_0^\varepsilon(y), & y\in\mathbb{R}^m. \end{cases} \tag{4.1}$$

We assume that b, a satisfy the hypotheses (2.1), (2.2), (2.3), (3.8) and we work in the $L^2(\mathbb{R}^m)$ setting ($q=2$). Motivated by Hilbert's expansion method, we intend to show the convergence of $(u^\varepsilon)_{\varepsilon>0}$ as ε goes to 0 towards the solution u of

$$\begin{cases} \partial_t u + \langle a(t)\cdot\nabla_y u(t)\rangle^{(2)} = 0, & (t,y)\in(0,T)\times\mathbb{R}^m \\ u(0,y) = u_0(y), & y\in\mathbb{R}^m. \end{cases} \tag{4.2}$$

Our main result is the following.

Theorem 4.1. *Assume that* $(\alpha_i)_{i\in\{1,\dots,r\}}$ *are smooth and satisfy*

$$\sum_{i=1}^r \|\alpha_i\|_{L^1([0,T];L^\infty(\mathbb{R}^m))} + \sum_{i=1}^r \|\partial_t\alpha_i\|_{L^1([0,T];L^\infty(\mathbb{R}^m))} < +\infty$$

$$\sum_{i=1}^r\sum_{j=1}^r \|\mathcal{T}_2^i\alpha_j\|_{L^1([0,T];L^\infty(\mathbb{R}^m))} + \sum_{i=1}^r\sum_{j=1}^r\sum_{k=1}^r \|\mathcal{T}_2^i\mathcal{T}_2^j\alpha_k\|_{L^1([0,T];L^\infty(\mathbb{R}^m))} < +\infty.$$

Suppose that

$$u_0 \in (\cap_{i=1}^r \mathrm{D}(\mathcal{T}_2^i)) \cap \ker\mathcal{T}_2,\quad \mathcal{T}_2^j u_0 \in \cap_{i=1}^r \mathrm{D}(\mathcal{T}_2^i),\ \forall\, j\in\{1,\dots,r\}$$

and that $(u_0^\varepsilon)_{\varepsilon>0}$ *are smooth initial conditions such that* $\lim_{\varepsilon\searrow 0} u_0^\varepsilon = u_0$ *in* $L^2(\mathbb{R}^m)$. *We denote by* u^ε, u *the solutions of (4.1), (4.2) respectively. Then we have* $\lim_{\varepsilon\searrow 0} u^\varepsilon = u$, *in* $L^\infty([0,T];L^2(\mathbb{R}^m))$.

Proof. According to Propositions 3.3, 3.6 and Corollary 3.8 there is a unique strong solution u for (4.2), satisfying $\|u(t)\|_{L^2(\mathbb{R}^m)} = \|u_0\|_{L^2(\mathbb{R}^m)}$ for any $t\in[0,T]$ and

$$\|\partial_t u\|_{L^\infty([0,T];L^2(\mathbb{R}^m))} + \sum_{i=1}^r \|\mathcal{T}_2^i u\|_{L^\infty([0,T];L^2(\mathbb{R}^m))} \le C\sum_{i=1}^r \|\mathcal{T}_2^i u_0\|_{L^2(\mathbb{R}^m)}.$$

Since $u(t) \in \ker \mathcal{T}_2$, $t \in [0,T]$, we have

$$\begin{aligned}\langle \partial_t u + a(t) \cdot \nabla_y u(t) \rangle^{(2)} &= \partial_t \langle u \rangle^{(2)} + \langle a(t) \cdot \nabla_y u(t) \rangle^{(2)} \\ &= \partial_t u + \langle a(t) \cdot \nabla_y u(t) \rangle^{(2)} = 0\end{aligned}$$

and thus by Proposition 2.10 there are $(v_\mu)_{\mu>0}$ such that

$$\begin{aligned}&\partial_t u + a(t,y) \cdot \nabla_y u + \mu v_\mu(t,y) + \mathcal{T}_2 v_\mu \\ &= 0, \quad \lim_{\mu \searrow 0} (\mu v_\mu(t)) = 0 \text{ in } L^2(\mathbb{R}^m), \ t \in [0,T]. \end{aligned} \tag{4.3}$$

Moreover, by Remark 2.11 we know that

$$\begin{aligned}\|\mu v_\mu\|_{L^\infty([0,T];L^2(\mathbb{R}^m))} &\le \|\partial_t u + a(t) \cdot \nabla_y u\|_{L^\infty([0,T];L^2(\mathbb{R}^m))} \\ &\le \|\partial_t u\|_{L^\infty([0,T];L^2(\mathbb{R}^m))} \\ &\quad + C \sum_{i=1}^{r} \|\alpha_i\|_{W^{1,1}([0,T];L^\infty(\mathbb{R}^m))} \|\mathcal{T}_2^i u\|_{L^\infty([0,T];L^2(\mathbb{R}^m))} \\ &\le C \sum_{i=1}^{r} \|\mathcal{T}_2^i u_0\|_{L^2(\mathbb{R}^m)}. \end{aligned} \tag{4.4}$$

Combining (4.1), (4.2) and the equation $\mathcal{T}_2 u = 0$ yields

$$\begin{aligned}&\left(\partial_t + a(t,y) \cdot \nabla_y + \frac{b(y)}{\varepsilon} \cdot \nabla_y\right)(u^\varepsilon - u - \varepsilon v_\mu) \\ &= \mu v_\mu - \varepsilon(\partial_t v_\mu + a(t,y) \cdot \nabla_y v_\mu). \end{aligned} \tag{4.5}$$

We investigate now the regularity of v_μ. By Remark 2.11 we have

$$\mu \|\partial_t v_\mu(t)\|_{L^2(\mathbb{R}^m)} \le \left\| \partial_t^2 u + \sum_{i=1}^{r} \partial_t \alpha_i \mathcal{T}_2^i u + \sum_{i=1}^{r} \alpha_i(t) \partial_t \mathcal{T}_2^i u \right\|_{L^2(\mathbb{R}^m)}$$

and thus Proposition 3.9 implies

$$\begin{aligned}&\mu \|\partial_t v_\mu\|_{L^1([0,T];L^2(\mathbb{R}^m))} \\ &\le C \left(\sum_{i=1}^{r} \sum_{j=1}^{r} \|\mathcal{T}_2^i \mathcal{T}_2^j u_0\|_{L^2(\mathbb{R}^m)} + \sum_{i=1}^{r} \|\mathcal{T}_2^i u_0\|_{L^2(\mathbb{R}^m)} \right). \end{aligned} \tag{4.6}$$

Applying now the operator $\mathcal{T}_2^i$, $i \in \{0, 1, \ldots, r\}$, in (4.3), yields

$$\partial_t \mathcal{T}_2^i u + \sum_{j=1}^{r} \{(\mathcal{T}_2^i \alpha_j)(\mathcal{T}_2^j u) + \alpha_j (\mathcal{T}_2^i \mathcal{T}_2^j u)\} + \mu \mathcal{T}_2^i v_\mu + \mathcal{T}_2 \mathcal{T}_2^i v_\mu = 0.$$

By Remark 2.11 and Proposition 3.9 we obtain as before

$$\mu\|\mathcal{T}_2^i v_\mu(t)\|_{L^2(\mathbb{R}^m)}$$
$$\leq \|\partial_t \mathcal{T}_2^i u(t) + \sum_{j=1}^{r}\{(\mathcal{T}_2^i \alpha_j(t))(\mathcal{T}_2^j u(t)) + \alpha_j(t)(\mathcal{T}_2^i \mathcal{T}_2^j u(t))\}\|_{L^2(\mathbb{R}^m)}$$

implying that

$$\mu \sum_{i=0}^{r} \|\mathcal{T}_2^i v_\mu\|_{L^1([0,T];L^2(\mathbb{R}^m))}$$
$$\leq C\left(\sum_{i=1}^{r}\sum_{j=1}^{r}\|\mathcal{T}_2^i \mathcal{T}_2^j u_0\|_{L^2(\mathbb{R}^m)} + \sum_{i=1}^{r}\|\mathcal{T}_2^i u_0\|_{L^2(\mathbb{R}^m)}\right). \quad (4.7)$$

Multiplying (4.5) by $u^\varepsilon - u - \varepsilon v_\mu$ and integrating over $\mathbb{R}^m$ yield

$$\frac{1}{2}\frac{d}{dt}\|(u^\varepsilon - u - \varepsilon v_\mu)(t)\|^2_{L^2(\mathbb{R}^m)} \leq \|\mu v_\mu(t)\|_{L^2(\mathbb{R}^m)}\|(u^\varepsilon - u - \varepsilon v_\mu)(t)\|_{L^2(\mathbb{R}^m)}$$
$$+\varepsilon \left\|\partial_t v_\mu(t) + \sum_{i=0}^{r}\alpha_i(t)\mathcal{T}_2^i v_\mu(t)\right\|_{L^2(\mathbb{R}^m)}$$
$$\times\|(u^\varepsilon - u - \varepsilon v_\mu)(t)\|_{L^2(\mathbb{R}^m)}$$

and we deduce that

$$\frac{d}{dt}\|(u^\varepsilon - u - \varepsilon v_\mu)(t)\|_{L^2(\mathbb{R}^m)}$$
$$\leq \|\mu v_\mu(t)\|_{L^2(\mathbb{R}^m)} + C\varepsilon(\|\partial_t v_\mu(t)\|_{L^2(\mathbb{R}^m)} + \sum_{i=0}^{r}\|\mathcal{T}_2^i v_\mu(t)\|_{L^2(\mathbb{R}^m)}).$$

Combining with (4.6), (4.7), we obtain for any $t \in [0,T]$

$$\|(u^\varepsilon - u - \varepsilon v_\mu)(t)\|_{L^2(\mathbb{R}^m)} \leq \|u_0^\varepsilon - u_0 - \varepsilon v_\mu(0)\|_{L^2(\mathbb{R}^m)}$$
$$+\int_0^T \|\mu v_\mu(s)\|_{L^2(\mathbb{R}^m)}\,\mathrm{d}s$$
$$+C\frac{\varepsilon}{\mu}(\|\mu\partial_t v_\mu\|_{L^1([0,T];L^2(\mathbb{R}^m))}$$
$$+\sum_{i=0}^{r}\|\mu \mathcal{T}_2^i v_\mu\|_{L^1([0,T];L^2(\mathbb{R}^m))})$$
$$\leq \|u_0^\varepsilon - u_0 - \varepsilon v_\mu(0)\|_{L^2(\mathbb{R}^m)}$$
$$+\int_0^T \|\mu v_\mu(s)\|_{L^2(\mathbb{R}^m)}\,\mathrm{d}s + C\frac{\varepsilon}{\mu}.$$

Consequently one gets by (4.4) for any $t \in [0, T]$

$$
\begin{aligned}
\|(u^\varepsilon - u)(t)\|_{L^2(\mathbb{R}^m)} &\leq \|u_0^\varepsilon - u_0\|_{L^2(\mathbb{R}^m)} \\
&\quad + \frac{\varepsilon}{\mu}(\|\mu v_\mu(t)\|_{L^2(\mathbb{R}^m)} + \|\mu v_\mu(0)\|_{L^2(\mathbb{R}^m)}) \\
&\quad + C\frac{\varepsilon}{\mu} + \|\mu v_\mu\|_{L^1([0,T];L^2(\mathbb{R}^m))} \\
&\leq \|u_0^\varepsilon - u_0\|_{L^2(\mathbb{R}^m)} + C\frac{\varepsilon}{\mu} + \|\mu v_\mu\|_{L^1([0,T];L^2(\mathbb{R}^m))}.
\end{aligned}
$$

Since the functions $t \to \|\mu v_\mu(t)\|_{L^2(\mathbb{R}^m)}$ converge pointwise to 0 as $\mu \searrow 0$ (cf. (4.3)) and they are uniformly bounded on $[0, T]$ (cf. (4.4)) we deduce by dominated convergence theorem that

$$\lim_{\mu \searrow 0} \|\mu v_\mu\|_{L^1([0,T];L^2(\mathbb{R}^m))} = 0.$$

Particularly, for $\mu = \varepsilon^\delta$, with $\delta \in (0, 1)$ we have, for $\varepsilon \searrow 0$

$$
\begin{aligned}
&\|u^\varepsilon - u\|_{L^\infty([0,T];L^2(\mathbb{R}^m))} \\
&\leq \|u_0^\varepsilon - u_0\|_{L^2(\mathbb{R}^m)} + C\varepsilon^{1-\delta} + \|\varepsilon^\delta v_{\varepsilon^\delta}\|_{L^1([0,T];L^2(\mathbb{R}^m))} \to 0.
\end{aligned}
$$

□

5 The limit model in terms of prime integrals

In the previous section we have derived a limit model for the transport equation (4.1) based on the computation of the fields $(b^i)_i$ in involution with b. We investigate now the same limit model from the view point of prime integral concept. Surely, this approach will provide an equivalent analysis. Nevertheless, in practical situations (see the examples in the next section) the computations are simplified when prime integrals are employed. We assume that there are $m-1$ prime integrals, independent of $\mathbb{R}^m$, and associated with the field b

$$b \cdot \nabla_y \psi^i = 0, \quad i \in \{1, \ldots, m-1\} \tag{5.1}$$

$$\text{rank}\left(\frac{\partial \psi^i}{\partial y_j}(y)\right)_{(m-1)\times m} = m-1, \quad y \in \mathbb{R}^m. \tag{5.2}$$

Let us recall that generally, around any non singular point y_0 of b (*i.e.*, $b(y_0) \neq 0$) there are $(m-1)$ independent prime integrals, defined only locally, in a small enough neighborhood of y_0 (see [1], p. 95). For any $y \in$

$\mathbb{R}^m$ we denote by $M(y)$ the matrix whose lines are $\nabla_y \psi^1, \ldots, \nabla_y \psi^{m-1}$ and b. The hypotheses (5.1), (5.2) imply that $\det M(y) \neq 0$ for any $y \in \mathbb{R}^m$. The idea is to search for fields $c = c(y)$ such that $c(y) \cdot \nabla_y u$ remains constant along the flow of b for any function u which is constant along the same flow. If u is constant on the characteristics of b, there is a function $v = v(z) : \mathbb{R}^{m-1} \to \mathbb{R}$ such that

$$u(y) = v(\psi^1(y), \ldots, \psi^{m-1}(y)), \;\; y \in \mathbb{R}^m.$$

Therefore one gets

$$\frac{\partial u}{\partial y_j} = \sum_{k=1}^{m-1} \frac{\partial v}{\partial z_k}(\psi^1(y), \ldots, \psi^{m-1}(y)) \frac{\partial \psi^k}{\partial y_j}$$

implying that

$$c \cdot \nabla_y u = \sum_{k=1}^{m-1} \frac{\partial v}{\partial z_k}(\psi^1(y), \ldots, \psi^{m-1}(y)) \sum_{j=1}^{m} \frac{\partial \psi^k}{\partial y_j} c_j = (\nabla_z v)(\psi(y)) \cdot \frac{\partial \psi}{\partial y} c(y).$$

Particularly, if $\frac{\partial \psi}{\partial y} c(y)$ do not depend on y, the directional derivative $c \cdot \nabla_y$ remains constant along the trajectories of b. Actually, the following more general result holds.

Lemma 5.1. *Assume that (5.1), (5.2) hold and let c be a smooth field such that $y \to \frac{\partial \psi}{\partial y}(y)c(y)$ and $y \to b(y) \cdot c(y)$ are constant along the flow of b. Then we have*

$$[c, b](y) = \frac{c(y) \cdot \left(\nabla_y \frac{|b|^2}{2} + \mathcal{T}b\right)}{|b(y)|^2} \, b(y), \;\; y \in \mathbb{R}^m.$$

For any $i \in \{1, \ldots, m-1\}$ let us denote by $c^i(y)$ the unique solution of the linear system

$$M(y)c^i(y) = e^i := (\delta_{ij})_{1 \le j \le m}$$

where δ_{ij} are the Kronecker's symbols. Notice that $M(y)\frac{b(y)}{|b(y)|^2} = e^m$ and thus $c^1(y), \ldots, c^{m-1}(y), b(y)$ are linearly independent at any $y \in \mathbb{R}^m$. According to Lemma 5.1 we have for any $i \in \{1, \ldots, m-1\}$

$$(c^i \cdot \nabla_y)(b \cdot \nabla_y) - (b \cdot \nabla_y)(c^i \cdot \nabla_y) = \frac{c^i(y) \cdot \left(\nabla_y \frac{|b|^2}{2} + (b(y) \cdot \nabla_y) b\right)}{|b(y)|^2} (b \cdot \nabla_y).$$

Particularly, for any function u constant along the flow of b, the directional derivative $c^i \cdot \nabla_y u$ remains constant along the same flow for any

$i \in \{1, \ldots, m-1\}$. We denote by $\beta_0, \beta_1, \ldots, \beta_{m-1}$ the coordinates of a with respect to $b, c^1, \ldots, c^{m-1}$ and we assume that $(\beta_i)_i$ are smooth and bounded

$$a(t,y) = \beta_0(t,y)b(y) + \sum_{i=1}^{m-1} \beta_i(t,y)c^i(y), \quad (t,y) \in [0,T] \times \mathbb{R}^m. \tag{5.3}$$

Thanks to Corollary 2.8, one gets for any function $u \in (\cap_{i=1}^{m-1} \mathrm{D}(\mathcal{T}_q^{c^i})) \cap \ker \mathcal{T}_q$

$$\begin{aligned}\langle a(t) \cdot \nabla_y u(t) \rangle^{(q)} &= \left\langle \sum_{i=1}^{m-1} \beta_i(t) c^i(y) \cdot \nabla_y u(t) \right\rangle^{(q)} \\ &= \sum_{i=1}^{m-1} \langle \beta_i(t) \rangle^{(\infty)} c^i(y) \cdot \nabla_y u(t).\end{aligned}$$

It remains computing $(\beta_i)_i$. Multiplying (5.3) by $M(y)$ yields

$$M(y)a(t,y) = \beta_0(t,y)|b(y)|^2 e^m + \sum_{i=1}^{m-1} \beta_i(t,y)e^i$$

implying that

$$\begin{aligned}&\beta_i(t,y) = M(y)a(t,y) \cdot e^i, \quad i \in \{1, \ldots, m-1\}, \\ &\beta_0(t,y)|b(y)|^2 = M(y)a(t,y) \cdot e^m\end{aligned}$$

or equivalently to

$$\beta_i(t,y) = a(t,y) \cdot \nabla_y \psi^i, \quad i \in \{1, \ldots, m-1\}, \quad \beta_0(t,y) = \frac{a(t,y) \cdot b(y)}{|b(y)|^2}.$$

Finally one gets the following form of the limit model

$$\partial_t u + \sum_{i=1}^{m-1} \langle a(t) \cdot \nabla_y \psi^i \rangle^{(\infty)} M^{-1}(y) e^i \cdot \nabla_y u = 0 \tag{5.4}$$

supplemented by the constraint $\mathcal{T}_q u = 0$. Actually we check that this constraint is a consequence of Eq. (5.4), provided that the initial condition satisfies $\mathcal{T}_q u_0 = 0$. Indeed, by Lemma 5.1 it is easily seen that for any $i \in \{1, \ldots, m-1\}$ we have

$$\begin{aligned}\mathcal{T}_q\left(\langle a(t) \cdot \nabla_y \psi^i \rangle^{(\infty)} c^i \cdot \nabla_y u \right) &= \langle a(t) \cdot \nabla_y \psi^i \rangle^{(\infty)} \mathcal{T}_q (c^i \cdot \nabla_y u) \\ &= \langle a(t) \cdot \nabla_y \psi^i \rangle^{(\infty)} (c^i \cdot \nabla_y) \mathcal{T}_q u \\ &\quad - \langle a(t) \cdot \nabla_y \psi^i \rangle^{(\infty)} \cdot \\ &\quad \frac{c^i(y) \cdot \left(\nabla_y \frac{|b|^2}{2} + (b(y) \cdot \nabla_y) b \right)}{|b(y)|^2} \mathcal{T}_q u.\end{aligned}$$

Therefore, by applying $\mathcal{T}_q$ to (5.4) we obtain

$$\partial_t \mathcal{T}_q u + \sum_{i=1}^{m-1} \langle a(t) \cdot \nabla_y \psi^i \rangle^{(\infty)} (c^i \cdot \nabla_y) \mathcal{T}_q u$$
$$- \left(\sum_{i=1}^{m-1} \langle a(t) \cdot \nabla_y \psi^i \rangle^{(\infty)} \frac{c^i(y) \cdot \left(\nabla_y \frac{|b|^2}{2} + (b(y) \cdot \nabla_y) b \right)}{|b(y)|^2} \right) \mathcal{T}_q u = 0$$

and thus it is clear that if $\mathcal{T}_q u_0 = 0$, then $\mathcal{T}_q u(t) = 0$, $t \in [0, T]$.

5.1 Examples

We apply the previous theoretical results to the finite Larmor radius regime, in the particular case of a constant magnetic field $B_3 \neq 0$. We have $m = 4$, $y = (x, p) \in \mathbb{R}^2 \times \mathbb{R}^2$, $\tilde{a}(t, y) = (0, 0, -E_1(t, x), -E_2(t, x))$, $\tilde{b}(y) = (p_1, p_2, -B_3 p_2, B_3 p_1) = (p, -B_3 {}^\perp p)$. The characteristic flow $Y = (X, P)$ associated with $\tilde{b}$ satisfies

$$\frac{dX}{ds} = P(s; x, p), \quad \frac{dP}{ds} = -B_3 {}^\perp P(s; x, p).$$

It is easily seen that a set of independent prime integrals is given by

$$\tilde{\psi}^1(x, p) = B_3 x_2 + p_1, \quad \tilde{\psi}^2(x, p) = -B_3 x_1 + p_2, \quad \tilde{\psi}^3(p) = \frac{1}{2}|p|^2$$

and thus we need to invert the matrix

$$\tilde{M}(p) = \begin{pmatrix} 0 & B_3 & 1 & 0 \\ -B_3 & 0 & 0 & 1 \\ 0 & 0 & p_1 & p_2 \\ p_1 & p_2 & -B_3 p_2 & B_3 p_1 \end{pmatrix}.$$

In order to simplify our computations it is very convenient to introduce the new variable $z = x - \frac{{}^\perp p}{B_3} = (-\tilde{\psi}^2, \tilde{\psi}^1)/B_3$ and the new unknown $g^\varepsilon(t, z, p) = f^\varepsilon(t, x, p)$. The equation for g^ε becomes

$$\partial_t g^\varepsilon + \frac{1}{B_3} {}^\perp E\left(t, z + \frac{{}^\perp p}{B_3}\right) \cdot \nabla_z g^\varepsilon - E\left(t, z + \frac{{}^\perp p}{B_3}\right) \cdot \nabla_p g^\varepsilon - \frac{1}{\varepsilon} B_3 {}^\perp p \cdot \nabla_p g^\varepsilon = 0$$

and thus the fields to analyze in this case are

$$a(t, z, p) = \left(\frac{1}{B_3} {}^\perp E\left(t, z + \frac{{}^\perp p}{B_3}\right), -E\left(t, z + \frac{{}^\perp p}{B_3}\right) \right), \quad b(p) = (0, 0, -B_3 {}^\perp p).$$

A set of independent prime integrals is given by

$$\psi^1 = z_1, \quad \psi^2 = z_2, \quad \psi^3 = \frac{1}{2}|p|^2.$$

The matrix to be inverted is

$$M(p) = \begin{pmatrix} 1 & 0 & 0 & 0 \\ 0 & 1 & 0 & 0 \\ 0 & 0 & p_1 & p_2 \\ 0 & 0 & -B_3 p_2 & B_3 p_1 \end{pmatrix}.$$

It is easily seen that M^{-1} is given by

$$M^{-1}(p) = \begin{pmatrix} 1 & 0 & 0 & 0 \\ 0 & 1 & 0 & 0 \\ 0 & 0 & \frac{p_1}{|p|^2} & -\frac{p_2}{B_3|p|^2} \\ 0 & 0 & \frac{p_2}{|p|^2} & \frac{p_1}{B_3|p|^2} \end{pmatrix}.$$

In view of (5.4) we need to compute $\langle a(t) \cdot \nabla_{(z,p)} \psi^i \rangle^{(\infty)}$, $i \in \{1, 2, 3\}$. A direct computation shows that the flow $(Z, P)(s; z, p)$ associated with b is given by

$$Z(s; z, p) = z, \quad P(s; z, p) = R(sB_3)p, \quad R(\theta) = \begin{pmatrix} \cos\theta & -\sin\theta \\ \sin\theta & \cos\theta \end{pmatrix}.$$

Consequently the constant functions along the flow are the functions with radial symmetry with respect to p. Since all the trajectories are $2\pi/B_3$ periodic, we have

$$\langle u \rangle^{(\infty)}(z, p) = \frac{B_3}{2\pi} \int_0^{\frac{2\pi}{B_3}} u(z, R(sB_3)p)\, \mathrm{d}s = \frac{1}{2\pi} \int_0^{2\pi} u(z, R(\theta)p)\, \mathrm{d}\theta$$

for any bounded function $u \in L^\infty(\mathbb{R}^4)$. We have

$$\begin{aligned} \langle a(t) \cdot \nabla_{(z,p)} \psi^1 \rangle^{(\infty)} &= \left\langle \frac{1}{B_3} E_2 \left(t, z + \frac{{}^\perp p}{B_3} \right) \right\rangle^{(\infty)} \\ &= \frac{1}{2\pi B_3} \int_0^{2\pi} E_2 \left(t, z + \frac{{}^\perp (R(\theta)p)}{B_3} \right) \mathrm{d}\theta \end{aligned}$$

$$\begin{aligned} \langle a(t) \cdot \nabla_{(z,p)} \psi^2 \rangle^{(\infty)} &= -\left\langle \frac{1}{B_3} E_1 \left(t, z + \frac{{}^\perp p}{B_3} \right) \right\rangle^{(\infty)} \\ &= -\frac{1}{2\pi B_3} \int_0^{2\pi} E_1 \left(t, z + \frac{{}^\perp (R(\theta)p)}{B_3} \right) \mathrm{d}\theta. \end{aligned}$$

We claim that the coefficient $\langle a(t)\cdot\nabla_{(z,p)}\psi^3\rangle^{(\infty)}$ vanishes. Indeed

$$\langle a(t)\cdot\nabla_{(z,p)}\psi^3\rangle^{(\infty)} = -\frac{B_3}{2\pi}\int_0^{\frac{2\pi}{B_3}} E\left(t, z+\frac{{}^\perp P(s;z,p)}{B_3}\right)\cdot P(s;z,p)\,\mathrm{d}s.$$

Taking account of that $E(t)$ derives from a potential $\phi(t)$ and that

$$\frac{d}{ds}\phi\left(t, z+\frac{{}^\perp P(s;z,p)}{B_3}\right) = E\left(t, z+\frac{{}^\perp P(s;z,p)}{B_3}\right)\cdot P(s;z,p)$$

we deduce that

$$\langle a(t)\cdot\nabla_{(z,p)}\psi^3\rangle^{(\infty)} = -\frac{B_3}{2\pi}\int_0^{\frac{2\pi}{B_3}} \frac{d}{ds}\phi\left(t, z+\frac{{}^\perp P(s;z,p)}{B_3}\right)\,\mathrm{d}s = 0.$$

Plugging into (5.4) all these computations yield the limit model

$$\partial_t g + \frac{1}{2\pi B_3}\int_0^{2\pi} {}^\perp E\left(t, z+\frac{{}^\perp(R(\theta)p)}{B_3}\right)\,\mathrm{d}\theta\cdot\nabla_z g = 0$$

which is equivalent to

$$\partial_t f + \frac{1}{2\pi B_3}\int_0^{2\pi} {}^\perp E\left(t, x-\frac{{}^\perp p}{B_3}+\frac{{}^\perp(R(\theta)p)}{B_3}\right)\,\mathrm{d}\theta\cdot\nabla_x f = 0.$$

Therefore the finite Larmor radius regime leads to a transport equation for the particle density, whose advection field is given by a gyro-average type operator. For more details, the reader can refer to [2] where a complete analysis of the coupled Vlasov-Poisson equations (with finite Larmor radius) was performed.

References

[1] V. I. Arnold, Ecuaţii diferenţiale ordinare, Editura Ştiinţifica şi Enciclopedica, Bucureşti, 1978.

[2] M. Bostan, The Vlasov-Poisson system with strong external magnetic field. Finite Larmor radius regime, to appear in *Asymptot. Anal.*

[3] M. Bostan, T. Goudon, Electric high-field limit for the Vlasov-Maxwell-Fokker-Planck system, to appear in *Ann. Inst. H. Poincaé, Anal. non linéaire.*

[4] M. Bostan, The Vlasov-Maxwell system with strong initial magnetic field. Guiding-center approximation, *SIAM J. Multiscale Model. Simul.*, **6**: 1026–1058, 2007.

[5] Y. Brenier, Convergence of the Vlasov-Poisson system to the incompressible Euler equations, *Comm. Partial Differential Equations*, **25**: 737–754, 2000.

[6] Y. Brenier, N. Mauser, M. Puel, Incompressible Euler and e-MHD as scaling limits of the Vlasov-Maxwell system, *Commun. Math. Sci.* **1**: 437–447, 2003.

[7] H. Brezis, *Analyse fonctionnelle. Théorie et applications*, Masson 1983.

[8] E. Frénod, P.-A. Raviart, E. Sonnendrücker, Two-scale expansion of a singularly perturbed convection equation, *J. Math. Pures Appl.*, **80**: 815–843, 2001.

[9] E. Frénod, E. Sonnendrücker, Homogenization of the Vlasov equation and of the Vlasov-Poisson system with strong external magnetic field, *Asymptotic Anal.*, **18**: 193–213, 1998.

[10] E. Frénod, E. Sonnendrücker, Long time behavior of the Vlasov equation with a strong external magnetic field, *Math. Models Methods Appl. Sci.*, **10**: 539–553, 2000.

[11] F. Golse, L. Saint-Raymond, The Vlasov-Poisson system with strong magnetic field, *J. Math. Pures Appl.*, **78**: 791–817, 1999.

[12] M. Reed, B. Simon, *Methods of Modern Mathematical Physics, Vol. I, Functional Analysis*, Academic Press 1980.

The Zero-Electron-Mass Limit in the Hydrodynamic Model (Euler-Poisson System)*

Li Chen
Department of Mathematical Sciences
Tsinghua University, Beijing, 100084, China
Email: lchen@math.tsinghua.edu.cn

Abstract

The Euler-Poisson system consists of the conservation laws for the electron and ion densities and their current densities coupled to the Poisson equation for the electrostatic potential. We report in this paper that the limit of vanishing electron mass of the system for given ion density (unipolar case) with both well and ill prepared initial data is proved. The limit is related to the low Mach number limit in Euler system, which was proved by Klainerman and Majda for well prepared initial data, and by Ukai and Schochet for ill prepared initial data. More precisely, in the zero mass limit, the limit velocity satisfies the incompressible Euler equations with damping. The difference between the zero mass limit and the low Mach number limit comes from the singular coupling of the electrostatic potential. This additional singular coupling gives the most difficult part in the proof. We prove the limit for both well and ill prepared initial data. By a reformulation of the equations in terms of the enthalpy, we are able to show the uniform higher-order energy estimates with a careful use of the Poisson equation. With similar idea we can show the uniform estimate for time derivative for well prepared initial data. For ill prepared initial data, a careful analysis on the structure of the linear perturbation has been done to show that the convergence occurs away from time $t = 0$.

*The results reported in this Chinese French Summer Institute are joint work with Giuseppe Alì, Ansgar Jüngel and Yue-Jun Peng. The work is partially supported by the National Natural Science Foundation of China (NSFC), grant numbers 10571101 and 10871112.

1 Introduction

This is a review on zero mass limit of the Euler-Poisson system. In plasma physics and semiconductor simulation, the Euler-Poisson system is derived to study the time evolution of charged fluids. These models can be obtained from Boltzmann equation for electrons and ions (or holes in semiconductor) (see [23, 14]). The system consists of the conservation laws for the electron(ion) density and current density for electron(ion), coupled to the Poisson equation for the electrostatic potential.

1.1 Quasi-hydrodynamic models (Euler-Poisson system)

The following are the scaled hydrodynamic equations for the electron density n_e with charge $q_e = -1$, the density n_i of the positively charged ions with charge $q_i = +1$, the respective velocities v_e, v_i, and the electrostatic potential ϕ,

$$\begin{aligned} &\partial_t n_\alpha + \nabla \cdot (n_\alpha v_\alpha) = 0, \quad \alpha = e, i, \\ &m_\alpha \partial_t (n_\alpha v_\alpha) + m_\alpha \nabla \cdot (n_\alpha v_\alpha \otimes v_\alpha) + \nabla p_\alpha (n_\alpha) = -q_\alpha n_\alpha \nabla \phi - m_\alpha \frac{n_\alpha v_\alpha}{\tau_\alpha}, \\ &-\lambda^2 \Delta \phi = n_i - n_e - C(x) \quad \text{for } x \in \Omega,\ t > 0, \end{aligned} \tag{1}$$

where $d \geq 1$, Ω is $\mathbb{R}^d$ or $\mathbb{T}^d$. The initial conditions are given by

$$n_\alpha(\cdot, 0) = n_{I,\alpha}, \quad v_\alpha(\cdot, 0) = v_{I,\alpha} \quad \text{in } \Omega,\ \alpha = e, i.$$

p_α are the pressure functions, usually given by $p_\alpha(x) = c_\alpha x^{\gamma_\alpha}$, $x \geq 0$, where $c_\alpha > 0$ and $\gamma_\alpha \geq 1$. In the following discussions, we only assume that p_α are strictly monotone and smooth. The function $C(x)$ models fixed charged background ions (doping profile). The (scaled) physical parameters are the particle mass m_α, the relaxation time τ_α, and the Debye length λ.

1.2 Some known results on Euler-Poisson system

There are a lot of mathematical works, on both well-posedness and different kinds of singular limit problems, for Euler-Poisson system in the literature. We only list here a few of them.

In the stationary case, Degond-Markowich [6] discussed the existence and uniqueness of the steady state solution in the subsonic case, while Gamba [7] studied the same problem in the transonic case.

In the time evolution case, Zhang [32] and Marcati-Natalini [22] got the global existence of weak solutions of the initial boundary value problem and the Cauchy problem respectively by using compensated compactness.

Singular limit problems in this system include

- Relaxation-time limit: $\tau_\alpha \to 0$, $\alpha = i$ or e;
- Quasineutral limit: $\lambda \to 0$;
- The zero-electron-mass limit: $\dfrac{m_e}{m_i} \to 0$.

We also list some of the known results on singular limit.

- The relaxation limit ($\tau_{i,e} \to 0$) of Euler-Poisson system to the drift-diffusion equations for the Cauchy and initial boundary value problem was studied separately by Marcati-Natalini [22] and Hsiao-Zhang [13]. In the consideration of smooth solution, Luo-Natalini-Xin [17], Hsiao-Yang [12] and Li-Markowich-Mei [16] investigated the asymptotic behavior of solutions to the Cauchy and initial boundary value problem respectively.
- The quasineutral limit ($\lambda \to 0$) in the Euler-Poisson system has been analyzed for transient smooth solutions by Cordier and Grenier [5] in the one-dimensional case and independently in [26, 30] in the multi-dimensional case.
- For the zero mass limit ($m_e/m_i \to 0$), there has been very limited work on zero electron mass limit up to now, under restrictive assumptions (see [10]). We got some results recently for unipolar case in [1] and [2] with $\Omega = \mathbb{T}^d$ for well prepared initial data and $\Omega = \mathbb{R}^d$ for both well and ill prepared initial data. We will report these results in this review paper. Recently, L. Chen, X. Chen and C. Zhang also got a result on well prepared initial data in [4] for bipolar case.

1.3 Zero-electron-mass limit for given ion density

In this review paper we restrict ourselves to a situation in which the ion density is given and $\Omega = \mathbb{R}^d$ or $\mathbb{T}^d$ (the case for $\Omega = \mathbb{T}^d$ with well prepared initial data is much easier), i.e., we wish to perform rigorously the limit $m_e \to 0$ in the system

$$\partial_t n_e + \nabla \cdot (n_e v_e) = 0, \tag{2}$$

$$m_e \partial_t (n_e v_e) + m_e \nabla \cdot (n_e v_e \otimes v_e) + \nabla p_e(n_e) = n_e E - m_e \frac{n_e v_e}{\tau_e}, \tag{3}$$

$$\lambda^2 E = \nabla(\Gamma * (n_e - N)) \quad \text{for } x \in \mathbb{R}^d,\ t > 0, \tag{4}$$

where $N = n_i - C$ is given (In fact, we need that N is a constant.), Γ is the fundamental solution of Poisson equation.

Remark 1. *The reason we use the electric field E instead of $\nabla\phi$ here is that $\phi = \Gamma * (n_e - N)$ might not be well defined for not good enough $n_e - N$ in the whole space $\mathbb{R}^d$ case, while E could be defined for $n_e - N \in H^s(\mathbb{R}^d)$. If $\Omega = \mathbb{T}^d$, we will still use ϕ with $\int_{\mathbb{T}^d} \phi dx$ fixed.*

The parameter m_e is essentially the ratio of the electron mass to the ion mass (see [15] for details on the scaling). We assume that the ion mass is much larger than the electron mass such that the limit $m_e \to 0$ makes sense. The limit has the goal to achieve simpler models containing the essential physical phenomena. We notice that in plasma physics, zero-electron-mass assumptions are widely used (see [11, 18]).

To present the main ideas, it is convenient to write the main part of the systems (2)–(3) in symmetric hyperbolic form. Setting $n = n_e$, $v = v_e$, $p(n) = p_e(n_e)$, and $\varepsilon^2 = m_e$ and introducing the *enthalpy* $h = h(n_e)$, defined by $h'(n) = p'(n)/n$ and $h(1) = 0$, the systems (2)–(4) can be rewritten as

$$
\begin{aligned}
\partial_t n + \nabla \cdot (nv) &= 0, \\
\varepsilon^2(\partial_t v + v \cdot \nabla v) + \nabla h(n) &= E - \varepsilon^2 v, \\
E &= \nabla(\Gamma * (n - N)), \quad x \in \mathbb{R}^d, \ t > 0,
\end{aligned}
\tag{5}
$$

with initial conditions

$$
n(\cdot, 0) = n_I^\varepsilon, \quad v(\cdot, 0) = v_I^\varepsilon \quad \text{in } \mathbb{R}^d. \tag{6}
$$

Here, we have set $\tau_e = \lambda = 1$ in order to simplify the notation. Clearly, for smooth solutions, this system is equivalent to the following system with symmetric hyperbolic structure

$$
\begin{aligned}
(\partial_t + v \cdot \nabla)h + p'(n)\nabla \cdot v &= 0, \\
\varepsilon^2(\partial_t + v \cdot \nabla)v + \nabla h &= E - \varepsilon^2 v, \\
E &= \nabla(\Gamma * (n(h) - N)), \quad x \in \mathbb{R}^d, \ t > 0,
\end{aligned}
\tag{7}
$$

with initial conditions

$$
h(\cdot, 0) = h_I^\varepsilon = h(n_I), \quad v(\cdot, 0) = v_I^\varepsilon \quad \text{in } \mathbb{R}^d. \tag{8}
$$

As we suppose that the pressure function is invertible, so is $h(n)$ and we denote its inverse by $n(h)$. Sometimes we will analyse this system instead of the original one (5)–(6).

Now the objective is to perform the limit $\varepsilon \to 0$ in (5).

1.4 Formal asymptotic analysis

To express the idea more clearly, in this part, we will take $\Omega = \mathbb{T}^d$. In this case, we can write the system into

$$\begin{aligned}(\partial_t + v\cdot\nabla)h + p'(n)\nabla\cdot v = 0,\\ \varepsilon^2(\partial_t + v\cdot\nabla)v + \nabla h = \nabla\phi - \varepsilon^2 v,\\ \Delta\phi = n(h) - N, \quad x\in\mathbb{T}^d,\ t>0,\end{aligned} \tag{9}$$

where we assume that $\int_{\mathbb{T}^d}\phi dx$ is fixed, with initial conditions

$$h(\cdot,0) = h_I^\varepsilon, \quad v(\cdot,0) = v_I^\varepsilon \quad \text{in } \mathbb{T}^d. \tag{10}$$

In order to derive the limiting system when $\varepsilon \to 0$, we substitute the formal expansions

$$h = h^0 + \varepsilon h^1 + \varepsilon^2 h^2 + \cdots,\ v = v^0 + \varepsilon v^1 + \varepsilon^2 v^2 + \cdots,\ \phi = \phi^0 + \varepsilon\phi^1 + \varepsilon^2\phi^2 + \cdots$$

in the system (9) and equate equal powers of ε. The lowest-order terms satisfy the equations

$$(\partial_t + v^0\cdot\nabla)h^0 + p'(n(h^0))\nabla\cdot v^0 = 0, \quad \nabla(h^0 - \phi^0) = 0, \quad \Delta\phi^0 = n(h^0) - N. \tag{11}$$

The second equation implies that $h^0 - \phi^0$ is a function of time only. Combining this fact with the third equation, we find that h^0 solves $\Delta h^0 = n(h^0) - N$. It is not difficult to see that $\phi^0 = 0$ and $h^0 = n^{-1}(N)$ are the unique solutions of the corresponding equations. Particularly, the first equation in (11) becomes $\nabla\cdot v^0 = 0$. The first-order terms satisfy

$$\nabla(h^1 - \phi^1) = 0, \quad \Delta\phi^1 = n'(h^0)h^1.$$

The solutions $h^1 = \phi^1 = 0$ are consistent with these equations. At second order, we find

$$(\partial_t + v^0\cdot\nabla)v^0 + v^0 = \nabla(\phi^2 - h^2), \quad \Delta\phi^2 = n'(h^0)h^2. \tag{12}$$

From $\nabla\cdot v^0 = 0$ and the first equation, v^0 and $\phi^2 - h^2$ can be found. Then, h^2 is the solution of the third equation, written in the form of $\Delta h^2 = n'(h^0)h^2 - \Delta(\phi^2 - h^2)$, and finally, ϕ^2 is given by $\phi^2 = h^2 + (\phi^2 - h^2)$. These considerations motivate to choose the initial data as

$$h_I^\varepsilon = h_I^0 + \varepsilon^2 h_I^2, \quad v_I^\varepsilon = v_I^0 + \varepsilon v_I^1. \tag{13}$$

The formal analysis shows that the zero-electron-mass limit has some similarities with the low-Mach-number limit in the compressible Euler system [21]. It is possible to use ideas from Klainerman and Majda [19, 20] to deal with the term $\varepsilon^{-1}\nabla h$ in (7) (after division by ε). However, we have another singularity from $\varepsilon^{-1}\nabla\phi$ which cannot be fixed by their method. Our idea is to control this term by a careful use of the mass conservation and the Poisson equation.

We will show in the following two sections the results and the main ideas of the proof in well and ill prepared initial data.

2 Limit with well prepared initial data

2.1 Main results

We introduce as in [21] the following notations:

$$\|\cdot\|_s = \|\cdot\|_{H^s(\Omega)}, \quad |||\cdot|||_{s,T} = \sup_{0<t<T} \|\cdot\|_s \quad \text{for } s \in \mathbb{R}, \quad \|\cdot\|_\infty = \|\cdot\|_{L^\infty(\Omega)}.$$

Theorem 1. $\Omega = \mathbb{T}^d$. *Let n be a smooth strictly increasing function and let $N > 0$. Furthermore, let $s > d/2 + 1$ and let the initial data $(h_I^\varepsilon, v_I^\varepsilon)$ satisfy $v_I^\varepsilon = v_I^0$ and*

$$\left\| \frac{h_I^\varepsilon - h^0}{\varepsilon} \right\|_s + \|v_I^\varepsilon\|_s \le M_0,$$

where $h^0 = n^{-1}(N)$ and $M_0 > 0$ is a constant independent of ε. Then there exist constants $T_0 > 0$ and $M_0' > 0$, independent of ε, and $\varepsilon_0(M_0) > 0$ such that for all $0 < \varepsilon < \varepsilon_0(M_0)$, the problems (9)–(10) have a classical solution $(h^\varepsilon, v^\varepsilon, \phi^\varepsilon)$ in $[0, T_0]$ satisfying

$$\left|\left|\left| \frac{h^\varepsilon - h^0}{\varepsilon} \right|\right|\right|_{s,T_0} + |||v^\varepsilon|||_{s,T_0} + \left|\left|\left| \frac{\nabla\phi^\varepsilon}{\varepsilon} \right|\right|\right|_{s,T_0} \le M_0'.$$

Theorem 2. *Let the assumptions of Theorem 1 hold with $\nabla \cdot v_I^0 = 0$ and*

$$\left\| \frac{h_I^\varepsilon - h^0}{\varepsilon^2} \right\|_s \le M_1, \tag{14}$$

and let $(h^\varepsilon, v^\varepsilon, \phi^\varepsilon)$ be a classical solution to (9)–(10) in $[0, T_0]$ with $T_0 > 0$ independent of ε. Then, as $\varepsilon \to 0$,

$$h^\varepsilon \to h^0, \ \nabla\phi^\varepsilon \to 0 \text{ strongly in } L^\infty(0, T_0; H^\alpha(\mathbb{T}^d)) \cap C^{0,1}([0, T_0]; L^2(\mathbb{T}^d)),$$
$$v^\varepsilon \to v^0 \text{ strongly in } C^0([0, T_0]; H^\alpha(\mathbb{T}^d)) \quad \text{for all } \alpha < s,$$

where v^0 is the (unique) classical solution of the following incompressible Euler equations with damping,

$$\nabla \cdot v^0 = 0, \quad (\partial_t + v^0 \cdot \nabla)v^0 + v^0 = \nabla\pi, \quad x \in \mathbb{T}^d, \ t > 0, \tag{15}$$
$$v^0(\cdot, 0) = v_I^0 \quad \text{in } \mathbb{T}^d,$$

and π is the limit of

$$\nabla\left(\frac{\phi^\varepsilon - h^\varepsilon}{\varepsilon}\right) \rightharpoonup^* \nabla\pi \qquad \text{weakly* in } L^\infty(0, T_0; L^2(\mathbb{T}^d)).$$

Remark 2. *The above results on well prepared initial data are still true when $\Omega = \mathbb{R}^d$, but in this case, we will use the electric field E instead of $\nabla\phi$ since $\phi = \Gamma * (n_e - N)$ might not be well defined for not good enough $n_e - N$ in the whole space $\mathbb{R}^d$ case, while E is defined for $n_e - N \in H^s(\mathbb{R}^d)$. The the corresponding results are for problems (7)–(8).*

2.2 Ideas

To describe the idea more precisely, we introduce the new variables

$$\tilde{h} = \frac{h - h^0}{\varepsilon}, \quad \tilde{\phi} = \frac{\phi - \phi^0}{\varepsilon}$$

as in [21, Ch. 2.4], where ϕ^0 is any constant fixed by $\int \phi dx$ (if $\int \phi dx = 0$ then $\phi^0 = 0$). The system (7) can be written as

$$A(\varepsilon\tilde{h})(\partial_t + v\cdot\nabla)\tilde{h} + \frac{1}{\varepsilon}\nabla\cdot v = 0, \tag{16}$$

$$(\partial_t + v\cdot\nabla)v + \frac{1}{\varepsilon}\nabla\tilde{h} = \frac{1}{\varepsilon}\nabla\tilde{\phi} - v, \tag{17}$$

$$\Delta\tilde{\phi} = \frac{1}{\varepsilon}(n(\varepsilon\tilde{h} + h^0) - n(h^0)), \qquad x\in\mathbb{T}^d,\ t>0, \tag{18}$$

where $A(\varepsilon\tilde{h}) = 1/p'(\varepsilon\tilde{h} + h^0)$.

For the proof of the limit $\varepsilon \to 0$, it is enough to prove the following lemma by the classical continuity argument on hyperbolic system, which means that we need to derive uniform estimates up to sth-order derivatives, where $s > d/2 + 1$.

Lemma 3. *Suppose that it holds, for some $T^* > 0$ and $M > 0$,*

$$\|\tilde{h}\|_{L^\infty(0,T^*;W^{1,\infty}(\mathbb{T}^d))} + \|v\|_{L^\infty(0,T^*;W^{1,\infty}(\mathbb{T}^d))} \le M. \tag{19}$$

Then there exist $\varepsilon_0 > 0$ and $c(M) > 0$ (depending on M) such that for all $0 < \varepsilon < \varepsilon_0$, it holds

$$|||\tilde{h}|||_{s,T^*} + |||v|||_{s,T^*} + |||\nabla\tilde{\phi}|||_{s,T^*} \le e^{c(M)T^*}(M_0 + c(M)T^*). \tag{20}$$

Here, we will describe only how to derive the lowest-order estimates, which is sufficient to illustrate the idea. As in the assumption of the lemma, there are $L^\infty(0,T;W^{1,\infty}(\mathbb{T}^d))$ estimates for $\tilde{h}$ and v. Friedrich's energy estimate for symmetric hyperbolic systems and integration by parts yield

$$\frac{d}{dt}\int_{\mathbb{T}^d}(A(\varepsilon\tilde{h})|\tilde{h}|^2+|v|^2)dx+\int_{\mathbb{T}^d}|v|^2dx$$
$$\le c\int_{\mathbb{T}^d}(|\tilde{h}|^2+|v|^2)dx-\frac{1}{\varepsilon}\int_{\mathbb{T}^d}\tilde{\phi}\nabla\cdot v dx,$$

where the constant $c>0$ depends on the $L^\infty(0,T;W^{1,\infty}(\mathbb{T}^d))$ bounds for $\tilde{h}$ and v. Replacing the term $\varepsilon^{-1}\nabla\cdot v$ by the mass conservation equation, we are left to control the integrals

$$\int_{\mathbb{T}^d}\tilde{\phi}A(\varepsilon\tilde{h})\tilde{h}_t dx+\int_{\mathbb{T}^d}\tilde{\phi}A(\varepsilon\tilde{h})v\cdot\nabla\tilde{h}dx.$$

The second integral can be easily controlled (after integration by parts) by the $L^\infty(0,T;\, W^{1,\infty}(\mathbb{T}^d))$ bounds for $\tilde{h}$ and v. In order to deal with the first integral we employ the Poisson equation,

$$\Delta\tilde{\phi}_t=n'(\varepsilon\tilde{h}+h^0)\tilde{h}_t.$$

Then we arrive at

$$\int_{\mathbb{T}^d}\tilde{\phi}A(\varepsilon\tilde{h})\tilde{h}_t dx=\int_{\mathbb{T}^d}\frac{A(\varepsilon\tilde{h})}{n'(\varepsilon\tilde{h}+h^0)}\Delta\tilde{\phi}_t\tilde{\phi}dx.$$

Again after integration by parts, we obtain an integral with a "good" sign, i.e.

$$\begin{aligned}I_1&=-\frac{1}{2}\int_{\mathbb{T}^d}\frac{A(\varepsilon\tilde{h})}{n'(\varepsilon\tilde{h}+h^0)}\partial_t|\nabla\tilde{\phi}|^2dx-\int_{\mathbb{T}^d}\partial_t\nabla\tilde{\phi}\cdot\nabla\left(\frac{A(\varepsilon\tilde{h})}{n'(\varepsilon\tilde{h}+h^0)}\right)\tilde{\phi}dx\\&=-\frac{1}{2}\frac{d}{dt}\int_{\mathbb{T}^d}\frac{A(\varepsilon\tilde{h})}{n'(\varepsilon\tilde{h}+h^0)}|\nabla\tilde{\phi}|^2dx+\frac{1}{2}\int_{\mathbb{T}^d}\partial_t\left(\frac{A(\varepsilon\tilde{h})}{n'(\varepsilon\tilde{h}+h^0)}\right)|\nabla\tilde{\phi}|^2dx\\&\quad-\int_{\mathbb{T}^d}\partial_t\nabla\tilde{\phi}\cdot\nabla\left(\frac{A(\varepsilon\tilde{h})}{n'(\varepsilon\tilde{h}+h^0)}\right)\tilde{\phi}dx=H_1+H_2+H_3.\end{aligned}\tag{21}$$

The integral with "good" sign is $-\partial_t\|\nabla\tilde{\phi}\|_{L^2}^2$, and other integrals can be estimated by $\|\varepsilon\nabla\tilde{\phi}_t\|_{L^2}$ and $\|\partial_t(n'(\varepsilon\tilde{h}+h^0))\|_{L^2}$. Using the Poisson equation to bound the first expression, it can be seen that both terms contain the derivative $\varepsilon\tilde{h}_t$ as above including the factor ε. Indeed, by (16), we are now able to control this expression in some norm in terms of the $L^\infty(0,T;W^{1,\infty}(\mathbb{T}^d))$ estimates for $\tilde{h}$ and v.

For higher-order derivatives, we need to take care of the nonlinear terms arising from the partial derivatives, but finally, we end up with estimates for $\tilde{h}$ and v which are appropriate for employing the standard continuation argument (see below for details).

In the case of $\Omega = \mathbb{R}^d$, since we will use electric field E instead of electric potential ϕ, we need to use smoothing technics to get similar results as in (21).

3 Limit with ill prepared initial data

3.1 Main results

Our result on ill prepared initial data is in the case of $\Omega = \mathbb{R}^d$. The problem for $\Omega = \mathbb{T}^d$ is still open.

First we got the uniform local existence

Theorem 4. *Let $s > d/2 + 1$ and $N > 0$. The initial data $(n_I^\varepsilon, v_I^\varepsilon)$ satisfy $n_I^\varepsilon - N \in L^1(\mathbb{R}^d)$ and*

$$\left\| \frac{n_I^\varepsilon - N}{\varepsilon} \right\|_s + \|v_I^\varepsilon\|_s \le M_0,$$

with $M_0 > 0$ being a constant independent of ε. Then there exist constants $T_0 > 0$ and $M_0' > 0$, independent of ε, and $\varepsilon_0(M_0) > 0$ such that for all $0 < \varepsilon < \varepsilon_0(M_0)$, the problems (5)–(6) have a classical solution $(n^\varepsilon, v^\varepsilon, E^\varepsilon)$ in $[0, T_0]$ satisfying

$$\left|\left|\left| \frac{n^\varepsilon - N}{\varepsilon} \right|\right|\right|_{s,T_0} + |||v^\varepsilon|||_{s,T_0} + \left|\left|\left| \frac{E^\varepsilon}{\varepsilon} \right|\right|\right|_{s,T_0} \le M_0'.$$

Zero mass limit for ill prepared initial data is as follows:

Theorem 5. *Let the assumptions of Theorem 4 hold and let $d \ge 3$, $(n^\varepsilon, v^\varepsilon, \phi^\varepsilon)$ be a classical solution to (5)–(6) in $[0, T_0]$ with $T_0 > 0$ independent of ε. Then, as $\varepsilon \to 0$,*

$$\begin{aligned} n^\varepsilon \to n^0, \quad \nabla\phi^\varepsilon &\to 0 \ \text{ strongly in } L^\infty(0, T_0; H^s(\mathbb{R}^d)), \\ v^\varepsilon &\rightharpoonup v^0 \ \text{ weakly* in } L^\infty(0, T_0; H^s(\mathbb{R}^d)), \\ v^\varepsilon &\to v^0 \ \text{ strongly in } C^0_{loc}((0, T_0] \times \mathbb{R}^d), \end{aligned}$$

where $v^0 \in L^\infty([0, T_0]; H^s(\mathbb{R}^d))$ is the unique solution of the following incompressible Euler equations with damping,

$$\nabla \cdot v^0 = 0, \quad (\partial_t + v^0 \cdot \nabla)v^0 + v^0 = \nabla\pi, \quad x \in \mathbb{R}^d, \ t > 0, \qquad (22)$$
$$v^0(\cdot, 0) = Pv_I \quad \text{in } \mathbb{R}^d,$$

for some $\pi \in L^\infty([0, T_0]; H^s(\mathbb{R}^d))$. P is the orthogonal projection of H^s onto the subspace $\{v \in H^s : \nabla \cdot v = 0\}$.

In getting the limit for ill prepared initial data case, we need also the scaling of n, i.e. $\tilde{n} = \frac{n-N}{\varepsilon}$. The system (5) can be written as

$$\partial_t \tilde{n} + \nabla \cdot (\tilde{n}v) + \frac{N}{\varepsilon}\nabla \cdot v = 0,$$

$$\partial_t v + v \cdot \nabla v + v + h'(N)\frac{\nabla \tilde{n}}{\varepsilon} + \frac{\tilde{E}}{\varepsilon} = \frac{h'(N) - h'(\varepsilon\tilde{n} + N)}{\varepsilon}\nabla \tilde{n},$$

$$\tilde{E} = \nabla(\Gamma * \tilde{n}),$$

3.2 Ideas

Our idea to get the uniform local existence is similar to the case of $\Omega = \mathbb{T}^d$ (see Remark 2). We will show in this part the main technics to deal with ill prepared initial data.

To have more estimates in the ill prepared initial data case. we borrowed some of the ideas by Ukai [29] and Grenier [9] to rewrite the system into a linearized form,

$$\begin{aligned} \partial_t \tilde{n} + \frac{1}{\varepsilon}\nabla \cdot v &= G_1^\varepsilon, \\ \partial_t v + v + \frac{1}{\varepsilon}(h'(N) + (-\Delta)^{-1})\nabla \tilde{n} &= G_2^\varepsilon, \end{aligned} \tag{23}$$

where $G_1^\varepsilon = -\nabla \cdot (\tilde{n}v)$, $G_2^\varepsilon = -v \cdot \nabla v + \dfrac{h'(N) - h'(\varepsilon\tilde{n} + N)}{\varepsilon}\nabla \tilde{n}$.

Let $\mathbb{L}^\varepsilon(t)$ be the group generated by the linearized operator

$$L = \begin{pmatrix} 0 & \frac{1}{\varepsilon}\nabla\cdot \\ \frac{1}{\varepsilon}(h'(N) + (-\Delta)^{-1})\nabla & I_d \end{pmatrix},$$

i.e. $U(x,t) = \mathbb{L}^\varepsilon(t)U_0(x)$ solves the following Cauchy problem

$$U_t + \begin{pmatrix} 0 & \frac{1}{\varepsilon}\nabla\cdot \\ \frac{1}{\varepsilon}(a + (-\Delta)^{-1})\nabla & I_d \end{pmatrix} U = 0, \;\; U(x,0) = U_0(x),$$

where $a = h'(N)$. Then the system (23) can be rewritten into

$$\begin{pmatrix} \tilde{n} \\ v \end{pmatrix}_t = \mathbb{L}^\varepsilon(t)\begin{pmatrix} \tilde{n}_I \\ v_I \end{pmatrix} + \int_0^t \mathbb{L}^\varepsilon(t-\tau)\begin{pmatrix} G_1^\varepsilon \\ G_2^\varepsilon \end{pmatrix} d\tau.$$

Then we need to study the linear operator $\mathbb{L}^\varepsilon(t)$.

Obviously, $U(x,t) = \mathbb{L}^\varepsilon(t)U_0(x) = \mathcal{F}^{-1}e^{-tB(\varepsilon,\xi)}\mathcal{F}U_0$ with

$$B(\varepsilon,\xi) = \begin{pmatrix} 0 & \frac{1}{\varepsilon}i\xi \\ \frac{i}{\varepsilon}\Big(a + \frac{1}{|\xi|^2}\Big)\xi^t & I_d \end{pmatrix}.$$

where $\mathcal{F}$ is the Fourier transform. B has eigenvalues and eigenvectors

$$\lambda = 1 \ (d-1 \text{ multiple}) \quad e_i(\xi) = (0,*)^t, i = 1,\cdots,d-1,$$
$$\lambda_\pm = \frac{1}{2} \pm \frac{i}{2\varepsilon}\sqrt{4(a|\xi|^2+1)-\varepsilon^2},$$
$$e_\pm = \Big(\frac{2i|\xi|}{\varepsilon \pm i\sqrt{4(a|\xi|^2+1)-\varepsilon^2}}, \frac{\xi^t}{|\xi|}\Big), \quad |e_\pm|^2 = \frac{|\xi|^2}{a|\xi|^2+1} + 1.$$

$\mathbb{L}^\varepsilon(t)$ has the orthogonal decomposition $\mathbb{L}_1(t)$ and $\mathbb{L}_2^\varepsilon(t)$ according to the eigenvalue 1 and $\lambda_\pm$ in the following sense,

$$\mathbb{L}_1(t)U_0 = e^{-t}\mathcal{F}^{-1}(e_j(\xi)\cdot \hat{U_0}(\xi)e_j(\xi)) = e^{-t}\mathbb{L}_0 U_0$$
$$\mathbb{L}_2^\varepsilon(t)U_0 = e^{-\frac{t}{2}}\mathcal{F}^{-1}e^{\mp\frac{it}{2\varepsilon}\sqrt{4(a|\xi|^2+1)-\varepsilon^2}}(e_\pm(\xi)\cdot\hat{U_0}(\xi)e_\pm(\xi))$$

where we have used the summation convention on j and $+,-$. Since $e_j(\xi) = (0,\tilde{e}_j(\xi))^t$, then $\mathbb{L}_0U_0 = (0, Pv_0)$, $Pv_0 = \mathcal{F}^{-1}(\tilde{e}_j(\xi)\cdot\hat{v_0}(\xi)\tilde{e}_j(\xi))$ for $U_0 = (h_0, v_0)^t$. By Parseval theorem, one gets

$$\|Pv_0\|_s \le \|v_0\|_s, \quad \|\mathbb{L}_2^\varepsilon U_0\|_s \le C\|U_0\|_s.$$

By using the properties of $\mathbb{L}^\varepsilon(t)$, we can decompose the solution $U^\varepsilon = \begin{pmatrix}\tilde{n}^\varepsilon \\ v^\varepsilon\end{pmatrix}$ into two parts by

$$U^\varepsilon = U_1^\varepsilon + U_2^\varepsilon,$$
$$U_1^\varepsilon = (0, v_1^\varepsilon)^t, \quad v_1^\varepsilon = e^{-t}\Big(Pv_I + \int_0^t e^\tau PG_2^\varepsilon(\tau)d\tau\Big),$$
$$U_2^\varepsilon = \mathbb{L}_2^\varepsilon(t)U_0 + \int_0^t \mathbb{L}_2^\varepsilon(t-\tau)(G_1^\varepsilon(\tau), G_2^\varepsilon(\tau))^t d\tau.$$

Now in order to get convergence, we have the following facts with the second being much harder.

- Step 1. By the orthogonal decomposition and $\partial_t v_1^\varepsilon = -v_1^\varepsilon + PG_2^\varepsilon$, uniform energy estimates give $\|v_1^\varepsilon\|_s + \|\partial_t v_1^\varepsilon\|_{s-1} \le C$.
- Step 2. It is needed to deal with U_2^ε in one sense. In fact we have the following convergence, $\forall\tau > 0$,

$$\sup_{\tau\le t}\|U_2^\varepsilon\|_\infty \to 0, \quad as \quad \varepsilon \to 0.$$

We will use the following facts to get the convergence in Step 2. The key estimates in the following is omitted here (see [2] for details).

With $s \ge \frac{d}{2}+1$, $U_0 = (h_0, v_0)$ and for any fixed $\tau > 0$, we have

1. $\exists \varepsilon_0$ s.t. when $\varepsilon < \varepsilon_0$, $t \geq \tau$, the following estimate for $U_0 \in H^s(\mathbb{R}^d) \cap L^(\mathbb{R}^d)$ holds

$$\|\mathbb{L}_2^\varepsilon(t)U_0\|_\infty \leq Ce^{-\frac{t}{2}}\left|\frac{t}{\varepsilon}\right|^{-\sigma}\|U_0\|_s^{1-\sigma}\|U_0\|_{L^1}^\sigma, \quad \sigma = \frac{s-d/2}{s+d/2-1}.$$

2. For $U_0 \in H^s(\mathbb{R}^d)$, it holds that

$$\sup_{\tau\leq t}\|\mathbb{L}_2^\varepsilon(t)U_0\|_\infty \to 0, \quad as \quad \varepsilon \to 0.$$

By uniform energy estimates, for any $\tau \in [0, T_0]$, there exists C independent of ε such that

$$\|G_k^\varepsilon\|_{L^1} + \|G_k^\varepsilon\|_{s-1} \leq C.$$

Then for any fixed $0 < \tau_0$, choosing $0 < \tau' < \tau_0$, $\forall t > \tau_0$, we decompose the integral as

$$U_2^\varepsilon = \mathbb{L}_2^\varepsilon(t)U_0 + \Big(\int_{t-\tau'}^t + \int_0^{t-\tau'}\Big)\mathbb{L}_2^\varepsilon(t-\tau)(G_1^\varepsilon(\tau), G_2^\varepsilon(\tau))^t d\tau.$$

So the L^∞ estimate of U_2^ε is (with $t-\tau \geq \tau'$ for $\tau \in [0, t-\tau']$)

$$\begin{aligned}\|U_2^\varepsilon\|_\infty &\leq \|\mathbb{L}_2^\varepsilon(t)U_0\|_\infty + \tau'\|\mathbb{L}_2^\varepsilon(t-\tau)(G_1^\varepsilon(\tau), G_2^\varepsilon(\tau))^t\|_\infty \\ &\quad + C\int_0^{t-\tau'} e^{\tau-t}\left|\frac{t-\tau}{\varepsilon}\right|^{-\delta}\|G_1^\varepsilon, G_2^\varepsilon\|_s^{1-\sigma}\|G_1^\varepsilon, G_2^\varepsilon\|_{L^1}^\sigma d\tau \\ &\leq \|\mathbb{L}_2^\varepsilon(t)U_0\|_\infty + \tau'\|G_1^\varepsilon, G_2^\varepsilon\|_{s-1} + C\varepsilon^\delta\|G_1^\varepsilon, G_2^\varepsilon\|_s^{1-\sigma}\|G_1^\varepsilon, G_2^\varepsilon\|_{L^1}^\sigma.\end{aligned}$$

By letting $\varepsilon \to 0$, we have

$$\sup_{\tau_0\leq t}\|U_2^\varepsilon\|_\infty \to C\tau'.$$

Since τ' could be arbitrarily small, the conclusion of the last point is drawn.

References

[1] G. Alì, L. Chen, A. Jüngel and Y.-J. Peng. The zero-electron-mass limit in the hydrodynamic model for plasmas. *Preprint*, (2008).

[2] G. Alì, L. Chen, A. Jüngel and Y.-J. Peng. The zero-electron-mass limit in the hydrodynamic model for plasmas. *Preprint*, (2008).

[3] K. Asano. On the incompressible limit of the compressible Euler equations. *Japan J. Appl. Math.* 4 (1987), 455–488.

[4] L. Chen, X. Chen and C. Zhang, Vanishing electron mass limit in the bipolar Euler-Poisson system *Preprint*, (2008).

[5] S. Cordier and E. Grenier, Quasineutral limit of an Euler-Poisson system arising from plasma physics, *Commun. Part. Diff. Eqs.* 25 (2000), 1099–1113.

[6] P. Degond and P. A. Markowich, On a one-dimensional steady-state hydrodynamic model for semiconductors, *Appl. Math. Letters*, 3, 1990, 25–29.

[7] I. M. Gamba, Stationary transonic solutions of a one-dimensional hydrodynamic model for semiconductor, *Comm. P. D. E.*, 17, 1992, 553–577.

[8] I. Gasser and P. Marcati. A quasi-neutral limit in a hydrodynamic model for charged fluids. *Monatsh. Math.* 138 (2003), 189–208.

[9] E. Grenier. Oscillations in quasineutral plasmas. *Commun. Partial Diff. Eqs.* 21 (1996), 363–394.

[10] T. Goudon, A. Jüngel, and Y.-J. Peng. Zero-electron-mass limits in hydrodynamic models for plasmas. *Appl. Math. Lett.* 12 (1999), 75–79.

[11] D. Hewett. Low-frequency electro-magnetic (Darwin) applications in plasma simulation. *Comput. Phys. Commun.* 84 (1994), 243–277.

[12] L. Hsiao and T. Yang, Asymptotics of initial boundary value problems for hydrodynamic and drift diffusion models for semiconductors, *J. Diff. Eqns.* 170, 2001, no.2, 472–493.

[13] L. Hsiao and K. J. Zhang, The relaxation of the hydrodynamic model for semiconductors to the drift-diffusion equations, *J. Diff. Eqns.* 165, 2000, no. 2, 315–354.

[14] A. Jüngel. *Quasi-Hydrodynamic Semiconductor Equations.* Birkhäuser, Basel, 2001.

[15] A. Jüngel and Y.-J. Peng. A hierarchy of hydrodynamic models for plasmas: zero-electron-mass limits in the drift-diffusion equations. *Ann. Inst. H. Poincarè, Anal. non lin.* 17 (2000), 83–118.

[16] H. L. Li, P. Markowich and M. Mei, Asymptotic behavior of solutions of the hydrodynamic model of semiconductors, *Proc. Roy. Soc. Edingburgh Sect. A*, 132, 2002, no.2, 359–378.

[17] T. Luo, R. Natalini and Z. P. Xin, Large-time behavior of the solutions to a hydrodynamic model for semiconductors, *SIAM J. Appl. Math.*, 59, 1998, 810–830.

[18] F. Kazeminezhad, J. Dawson, J. Lebœuf, R. Sydora, and D. Holland. A Vlasov ion zero mass electron model for plasma simulations. *J. Comput. Phys.* 102 (1992), 277–296.

[19] S. Klainerman and A. Majda. Singular perturbations of quasilinear hyperbolic systems with large parameters and the incompressible limit of compressible fluids. *Commun. Pure Appl. Math.* 34 (1981), 481–524.

[20] S. Klainerman and A. Majda. Compressible and incompressible fluids. *Commun. Pure Appl. Math.* 35 (1982), 629–651.

[21] A. Majda. *Compressible Fluid Flow and Systems of Conservation Laws in Several Space Variables.* Springer, New York, 1984.

[22] P. Marcati and R. Natalini. Weak solutions to a hydrodynamic model for semiconductors and relaxation to the drift diffusion equations. *Arch. Rat. Mech. Anal.* 129 (1995), 129–145.

[23] P. Markowich, C. Ringhofer, and C. Schmeiser. *Semiconductor Equations.* Springer, Vienna, 1990.

[24] G. Métivier and S. Schochet. The incompressible limit of the non-isentropic Euler equations. *Arch. Rat. Mech. Anal.* 158 (2001), 61–90.

[25] Y.-J. Peng and Y.-G. Wang. Boundary layers and quasi-neutral limit in steady state Euler-Poisson equations for potential flows. *Nonlinearity* 17 (2004), 835–849.

[26] Y.-J. Peng and Y.-G. Wang. Convergence of compressible Euler-Poisson equations to incompressible type Euler equations. *Asympt. Anal.* 41 (2005), 141–160.

[27] J. Simon, Compact sets in the space $L^p(0,T;B)$, *Ann. Math. Pura Appl.* 146 (1987), 65–96.

[28] M. Slemrod and N. Sternberg. Quasi-neutral limit for Euler-Poisson system. *J. Nonlinear Sci.* 11 (2001), 193–209.

[29] S. Ukai. The incompressible limit and the initial layer of the compressible Euler equation. *J. Math. Kyoto Univ.* 26 (1986), 323–331.

[30] S. Wang. Quasineutral limit of the Euler-Poisson system with and without viscosity. *Commun. Part. Diff. Eqs.* 29 (2004), 419–456.

[31] S. Wang, Z. Xin, and P. Markowich. Quasineutral limit of the drift-diffusion model for semiconductors: the case of general sign-changing doping profile. Submitted for publication, 2004.

[32] B. Zhang, Convergence of the Godunov scheme for the simplifies one-dimensional hydrodynamic model for semiconductor devices, *Comm. Math. Phys.*, 157 (1993), 1–22.

Modeling and Simulation of Fluid-Particles Flows

Thierry Goudon, Pauline Lafitte, Mathias Rousset
Project Team SIMPAF
INRIA Lille Nord Europe Research Centre
Park Plazza, 40 avenue Halley
59650 Villeneuve d'Ascq cedex, France
& Laboratoire Paul Painlevé UMR 8524 CNRS–USTLille
Email: thierry.goudon@inria.fr
pauline.lafitte@math.univ-lille1.fr
mathias.rousset@inria.fr

Abstract

In this paper, we review a few aspects of two phase flows where a disperse phase — the particles — interacts with a dense fluid. We are thus led to consider kinetic equations where the leading term is due to the drag force exerted by the fluid on the particles. We discuss several asymptotic questions and present a numerical scheme which is able to treat the multiscale features of the problem.

1 Introduction

This paper is concerned with models describing disperse particles interacting with a fluid. This work is motivated by the transport of pollutants [14, 39], the dispersion of smokes and dust [18], the modeling of biomedical flows [7, 5] as well as combustion theory, with applications to Diesel engines or propulsors [1, 2, 20, 34, 40].

The basis of the models we are interested in assumes that the leading effect is due to the drag force exerted by the fluid on the particles. As a warm-up, let us explain what is going on with the very simple example of a single particle, spherically shaped with radius a and mass density ρ_{p}, dropped in a fluid. The fluid is characterized by its mass density ρ_{f}, its velocity u, and its dynamic viscosity μ. The drag force is supposed to be proportional to the relative velocity between the fluid and the particle so that the motion of the particle is described by the ODE system

$$\frac{\mathrm{d}}{\mathrm{d}t}X = V, \qquad \frac{4}{3}\pi a^3 \rho_{\mathrm{p}} \frac{\mathrm{d}}{\mathrm{d}t}V = 6\pi\mu a \left(u(t,X) - V\right) + \frac{4}{3}\pi a^3 \rho_{\mathrm{p}} \mathcal{F} \quad (1.1)$$

which defines the evolution of the position X and velocity V of the particle. In (1.1), $\mathcal{F}$ represents the density of external forces applied to the particle; for instance, considering gravity and buoyancy forces, we have

$$\mathcal{F} = \mathrm{g}\Big(\frac{\rho_{\mathrm{f}}}{\rho_{\mathrm{p}}} - 1\Big),$$

with $\mathrm{g} > 0$ the gravitational acceleration. Let us set

$$\mathcal{T}_{St} = \frac{2a^2\rho_P}{9\mu}.$$

When $u = 0$ in the gravity driven case, the velocity tends to the limit value $\mathcal{T}_{St}\mathrm{g}(\rho_{\mathrm{f}}/\rho_{\mathrm{p}} - 1)$, the so-called Stokes settling velocity, and $\mathcal{T}_{St}$, the Stokes settling time, clearly appears as a relaxation time, characterizing how the friction decelerates the particle. In complex mixtures, the disperse phase can be seen as an ensemble of particles, which is described by means of a particle distribution function $f \geq 0$ depending on time and on the phase space variable (x, v) where x stands for a space variable and v a velocity variable:

$$\int_{\Omega}\int_{\mathcal{V}} f(t,x,v)\,\mathrm{d}v\,\mathrm{d}x$$

is the number of particles that can be found at time $t \geq 0$, in the domain $\Omega \times \mathcal{V} \subset \mathbb{R}^N \times \mathbb{R}^N$ of the phase space. Therefore, the evolution of the density f is governed by the following Vlasov-type equation

$$\partial_t f + \nabla_x \cdot (vf) + \frac{1}{\mathcal{T}_{St}} \nabla_v \cdot \big((u-v)f\big) + \nabla_v \cdot \big(\mathcal{F}f\big) = 0. \qquad (1.2)$$

Particularly, if we have P independent particles described by their position-velocity pair (X_j, V_j), $j \in \{1, \ldots, P\}$, obeying (1.1), then $f(t,x,v) = \sum_{j=1}^{P} \delta(x = X_j(t)) \otimes \delta(v = V_j(t))$ satisfies (1.2). It clarifies the connection between the statistical and the particles viewpoints. The questions we address can be summarized as follows:

- Modeling issues are crucial for applications. A fundamental question is concerned with the drag force: the linear expression of the drag force we used above — that is the Stokes law — looks reasonable for low Reynolds numbers; otherwise, a more complex and non linear relation should be used, which will make the mathematical analysis harder. The role of the density of the fluid, which does not appear in the expression of the Stokes drag force above, should be also discussed. Furthermore, additional effects could be important depending on the flow and need to be incorporated in the model: added mass effect, Basset force, lift force $\cdots$ (see [19, 28, 31]). The models can also account for more complex interactions between the particles due to collision effects and shape or size variation through coagulation and fragmentation phenomena (see [2, 4, 34]).

- The surrounding fluid is considered as "turbulent" which, roughly speaking, means that the velocity u has fast and high variations. Hence, one seeks some averaging procedures which is allowed to derive useful and simple models that account for these turbulent effects. In this spirit, we mention [13, 17, 24], and on more physical grounds [41].
- Another viewpoint consists in coupling the evolution of the particles to hydrodynamic equations that describe the behavior of the dense phase. We are thus led to nonlinear systems of PDEs and we address the questions of existence, uniqueness, stability properties of the solutions, as well as we aim at designing efficient numerical schemes able to handle the multiscale features of the problem. We refer in particular to [3, 6, 26, 27, 34, 36] for well-posedness analysis of such coupled fluid-kinetic models. Asymptotic problems and stability properties are investigated in [8, 9, 10, 21, 22, 30, 34, 37]. Concerning numerical methods, of course the key reference is [1]; we also mention [4, 34] and below we shall describe the method introduced in [11].

In this paper, we will focus on some of these questions. In Section 2, we propose a possible modeling of "turbulence" by assuming that u is a time dependent random field. Discussing the scaling of the equations of motion with respect to the relaxation time associated with the Markov properties of the velocity, we identify several relevant asymptotic regimes for which we are able to describe limit effective equations. Section 3 deals with coupled kinetic/hydrodynamic equations and we show how efficient numerical schemes can be designed, based on the dissipative and asymptotic properties of the model.

2 Particles in turbulent flows

In this section, we neglect the external forces ($\mathcal{F} = 0$) and we consider the velocity field as given, but we wish to investigate the behavior of the solutions depending on parameters that characterize the flow. We will assume time randomness intending to mimic some "turbulence effects" and we will derive averaged models.

2.1 Modeling of turbulent flow

The function u is thought of as the velocity field of a "turbulent" flow which is therefore modeled through a time dependent random field. We model the time dependence of the velocity by a Markov process (with, say, dimension 1) which is consistent with the rough idea of an exponentially fast decay of the time-correlations. Let us detail the model we have in mind. The velocity at time t and position x is defined by

$$u(t,x) = \mathcal{U}\ U(x/\ell, Q_{t/\mathcal{T}_M})$$

where:

- $\mathcal{U}$ is the amplitude of the velocity,
- $U : \mathbb{R}^N \times \mathbb{R}$ is a smooth, divergence free and bounded (dimensionless) function in which the modeling of turbulence is embodied, with ℓ being a typical length scale of the variation of the velocity,
- $t \mapsto Q_{t/\mathcal{T}_M}$ is a stationary Markov process modeling the randomness of turbulence with typical decorrelation time $\mathcal{T}_M$.

Typically, one expects some translational invariance property of the field, i.e.:

$$\text{Law}(U(x/\ell, Q_{t/\mathcal{T}_M})) = \text{Law}(U(x/\ell + n, Q_{t/\mathcal{T}_M})),$$

for all $n \in \mathbb{R}^N$ or at least for all $n \in \mathbb{Z}^N$. The asymptotic regimes studied in this paper will typically lead to non-trivial evolution equations as soon as the following time auto-correlations are non-vanishing:

$$\int_0^{+\infty} \int_{(0,1)^N} \mathbb{E}\left(U(y, Q_0) \otimes U(y + vt, Q_{t/\mathcal{T}_M})\right) \,\mathrm{d}y\,\mathrm{d}t \neq 0,$$

for any $(y, v) \in \mathbb{R}^{2N}$.

To simplify the formal computations and the analysis, we restrict ourselves to the following framework:

- The velocity field satisfies

$$y \longmapsto U(y,q) \quad \text{is } (0,1)^N = \mathbb{Y}\text{−periodic}, (Y_t^\star, Q_t^\star) \qquad (2.1)$$

$$\sup_{y\in\mathbb{Y},\ q\in\mathbb{R}} |U(y,q)| \leq C < \infty, \qquad (2.2)$$

$$\nabla_y \cdot U(y,q) = 0 \quad \text{for a. e. } q. \qquad (2.3)$$

- $t \mapsto Q_t \in \mathbb{R}$ is the Markov process at hand. It is described by an operator $\mathcal{Q}$, the Markov generator. We assume long time mixing properties characterized by a stationary probability distribution $\mathcal{M}(q)\,\mathrm{d}q$, with $\mathcal{M}$ being a normalized positive function:

$$\mathcal{M}(q) > 0, \qquad \int_{\mathbb{R}} \mathcal{M}\,\mathrm{d}q = 1, \qquad \mathcal{Q}(\mathbb{1}) = 0, \qquad \mathcal{Q}^\star(\mathbb{1}) = 0,$$

where, here and below, we denote by $\mathcal{Q}^\star$ the adjoint operator defined for the inner product in $L^2(\mathbb{R}, \mathcal{M}(q)\,\mathrm{d}q)$:

$$(F,G) = \int_{\mathbb{R}} F(q)G(q)\ \mathcal{M}(q)\,\mathrm{d}q.$$

The parameter $\mathcal{T}_M$ then appears as a relaxation time.

The mixing requirement on the Markov operators $\mathcal{Q}^\star/\mathcal{Q}$ can be embodied in the spectral gap assumption

$$\begin{cases} \text{There exists } \sigma > 0 \text{ such that} \\ -\displaystyle\int_{\mathbb{R}} \mathcal{Q}(F)\ F\mathcal{M}(q)\,\mathrm{d}q \geq \sigma \int_{\mathbb{R}} \Big|F(q) - \int_{\mathbb{R}} F(q')\mathcal{M}(q')\,\mathrm{d}q'\Big|^2 \mathcal{M}(q)\,\mathrm{d}q \geq 0. \end{cases} \tag{2.4}$$

A typical example uses the Fokker-Planck operator

$$\mathcal{Q}(F) = \frac{1}{\mathcal{M}(q)}\partial_q\big(\mathcal{M}\partial_q F\big) \tag{2.5}$$

with $\mathcal{M}(q) = e^{-q^2/2}/\sqrt{2\pi}$. In this case, the evolution of a particle is then governed by the following set of differential equations

$$\begin{aligned} \mathrm{d}X &= V\,\mathrm{d}t, \\ \mathrm{d}V &= \frac{1}{\mathcal{T}_{St}}\big(\mathcal{U}\ U(X/\ell, Q_{t/\mathcal{T}_M}) - V\big)\,\mathrm{d}t, \\ \mathrm{d}Q &= \partial_q\mathcal{M}(q)\,\mathrm{d}t + \sqrt{2}\,\mathrm{d}W_t \end{aligned}$$

with W_t being a Brownian motion. Another example relies on a jump process, associated with the generator

$$\mathcal{Q}(F) = \int_{\mathbb{R}} F(q_\star)\mathcal{M}(q_\star)\,\mathrm{d}q_\star - F. \tag{2.6}$$

For technical purposes, we need further assumptions involving the velocity field U and the generator $\mathcal{Q}$. We will assume that U has null average

$$\int_{\mathbb{Y}\times\mathbb{R}} U(y,q)\ \mathcal{M}(q)\,\mathrm{d}q\,\mathrm{d}y = 0. \tag{2.7}$$

Occasionally, this assumption will be strengthened with the following pointwise centering condition

$$\text{For a. e. } y \in \mathbb{Y}, \qquad \int_{\mathbb{R}} U(y,q)\ \mathcal{M}(q)\,\mathrm{d}q = 0. \tag{2.8}$$

We shall also need an ergodic property which states

$$\begin{cases} \text{For a. e. } q \in \mathbb{R}, \text{ the solutions of } U(y,q)\cdot\nabla_y f = 0 \\ \text{are constants with respect to } y, \end{cases} \tag{2.9}$$

and the non-degeneracy condition

$$\begin{cases} \text{The matrix } A(y) = \int_{\mathbb{R}} U(y,q) \otimes U(y,q)\ \mathcal{M}(q)\,\mathrm{d}q \\ \text{is positively definite, and its coefficients belong to } W^{1,\infty}(\mathbb{Y}). \end{cases} \tag{2.10}$$

The expression of the time correlations of the space average of the field we obtained depends on the scales separation, that is on the ordering between the different physical parameters involved in the equation. In what follows, it will be given by the following formula:

$$\mathbb{D}_2 = \int_{\mathbb{R}} \int_{\mathbb{Y}} U(y,q)\,\mathrm{d}y \otimes (-\mathcal{Q}^\star)^{-1}\Big(\int_{\mathbb{Y}} U(y,q)\,\mathrm{d}y\Big)\ \mathcal{M}(q)\,\mathrm{d}q\,\mathrm{d}y, \tag{2.11}$$

or by the following two cases when particle transport operates at the same time scale as the random process: the first case will appear when dealing with the "Fine Particle" regime:

$$\mathbb{D}_1 = \int_{\mathbb{Y}\times\mathbb{R}} U(y,q) \otimes (U(y,q)\nabla_y - \mathcal{Q}^\star)^{-1}(U)(y,q)\ \mathcal{M}(q)\,\mathrm{d}q\,\mathrm{d}y \tag{2.12}$$

and for the "High Inertia" regime, we get

$$\mathbb{D}_0(v) = \int_{\mathbb{Y}\times\mathbb{R}} U(y,q) \otimes (v\cdot\nabla_y - \mathcal{Q}^\star)^{-1}(U)(y,q)\ \mathcal{M}(q)\,\mathrm{d}q\,\mathrm{d}y. \tag{2.13}$$

The definition of the inverse operators involved in these formulae is an issue and the arguments, which are of Fredholm alternative type, should be discussed carefully.

Instead of considering the particle distribution function in phase space, it is therefore convenient to deal with the density in $\mathbb{R}^{2N} \times \mathbb{R}$ of the probability distribution of the random variable (X, V, Q). In other words, we introduce $F(t,x,v,q) \geq 0$ such that, at time $t \geq 0$, for any measurable sets $\Omega \subset \mathbb{R}^N$, $\mathcal{V} \subset \mathbb{R}^N$ and $\mathcal{K} \subset \mathbb{R}$, we have,

$$\int_{\Omega\times\mathcal{V}\times\mathcal{K}} F(t,x,v,q)\ \mathcal{M}(q)\,\mathrm{d}q\,\mathrm{d}v\,\mathrm{d}x = \mathrm{Proba}\big(\{(X_t,V_t,Q_t) \in \Omega\times\mathcal{V}\times\mathcal{K}\}\big).$$

Accordingly, we are led to the following evolution PDE on densities

$$\partial_t F + v\cdot\nabla_x F + \frac{1}{\mathcal{T}_{St}}\nabla_v\cdot\big((\mathcal{U}\ U(x/\ell,q) - v)F\big) = \frac{1}{\mathcal{T}_M}\mathcal{Q}^\star(F). \tag{2.14}$$

Our goal is:

– first to identify some relevant asymptotic regimes, depending on the values of the stochastic and physical parameters $\mathcal{T}_M$, $\mathcal{U}$, ℓ and $\mathcal{T}_{St}$, compared to typical time and length scales of observation,

– second, to establish the limit equations which correspond to these regimes.

These questions have been addressed with a different viewpoint in [13], as well as in [41] with a more physical insight. Our approach is also strongly inspired from [24]: there the modeling of turbulence relies on finite time decorrelations of the velocity field. It is allowed to perform the asymptotic analysis by using the method introduced in [38]. We revisit this analysis by replacing this finite time decorrelation by suitable mixing properties of the Markov generator. In turn, the kinetic viewpoint applied to (2.14) is based on the relaxation property of the generator, with methods reminiscient to the analysis of hydrodynamic limits in gas dynamics. Such an approach also appears in [17] where an additional variable is introduced to the usual (position-velocity) phase space in order to describe the carrier flow turbulent velocity encoutered by a particle along its path. We finally refer to the technical developments in [25] which have to be adapted to the two-phase flow context.

2.2 Dimension analysis and asymptotic regimes

We introduce time and length units, denoted by T and L respectively. Then, we define dimensionless variables and unknowns as follows:

$$\begin{gathered} t \to t\mathrm{T}, \qquad x \to x\mathrm{L}, \qquad v \to \tfrac{\mathrm{L}}{\mathrm{T}} v \\ F(t,x,v,q) \to \mathrm{L}^3 \Big(\frac{\mathrm{L}}{\mathrm{T}}\Big)^3 F(t,x,v,q). \end{gathered}$$

We are finally led to

$$\partial_t F + v \cdot \nabla_x F + \frac{1}{\tau} \nabla_v \cdot \big[(\eta U(x/\lambda, q) - v) F \big] = \frac{1}{\varepsilon} \mathcal{Q}^\star(F) \tag{2.15}$$

which is governed by the following four dimensionless parameters

$$\begin{aligned} \varepsilon &= \frac{\mathcal{T}_M}{\mathrm{T}}, \qquad & \tau &= \frac{\mathcal{T}_{St}}{\mathrm{T}} \\ \eta &= \frac{\mathrm{UT}}{\mathrm{L}}, \qquad & \lambda &= \frac{\ell}{\mathrm{L}}. \end{aligned}$$

We shall investigate the behavior of the solutions with respect to the parameters. Let us start with some a priori estimates. Since $\int Q(f) \mathcal{M} \, \mathrm{d}q = 0$, we can reproduce easily the argument in [24] which proves the following estimate on the momentum and kinetic energy.

Proposition 2.1. *Let the initial data $F_0 \geq 0$ satisfy*

$$\begin{aligned} \int_{\mathbb{R}^{2N} \times \mathbb{R}} F_0 \, \mathcal{M} \, \mathrm{d}q \, \mathrm{d}v \, \mathrm{d}x &= M_0 < \infty, \\ \int_{\mathbb{R}^{2N} \times \mathbb{R}} |v|^2 F_0 \, \mathcal{M} \, \mathrm{d}q \, \mathrm{d}v \, \mathrm{d}x &= M_2 < \infty. \end{aligned} \tag{2.16}$$

Then, for any time $t \geq 0$, the solution $F(t, x, v, q)$ of (2.15) verifies $\int_{\mathbb{R}^{2N}\times\mathbb{R}} F(t)\ \mathcal{M}\,\mathrm{d}q\,\mathrm{d}v\,\mathrm{d}x = M_0$ and there exists a constant $C > 0$, depending only on M_0, M_2 and K such that, for any $t \geq 0$,

$$\int_{\mathbb{R}^{2N}\times\mathbb{R}} |v|F(t)\ \mathcal{M}\,\mathrm{d}q\,\mathrm{d}v\,\mathrm{d}x \leq C\eta, \qquad \int_{\mathbb{R}^{2N}\times\mathbb{R}} |v|^2 F(t)\ \mathcal{M}\,\mathrm{d}q\,\mathrm{d}v\,\mathrm{d}x \leq C\eta^2.$$

We can also obtain estimates that use the dissipative properties of the Markov generator. However, these estimates are useful only when τ does not go to 0.

Proposition 2.2. *Let the initial data $F_0 \geq 0$ satisfy (2.16) and*

$$\int_{\mathbb{R}^{2N}\times\mathbb{R}} |F_0|^2\ \mathcal{M}\,\mathrm{d}q\,\mathrm{d}v\,\mathrm{d}x = M_e < \infty. \tag{2.17}$$

Then, for any time $0 \leq t \leq T < \infty$, there exists a constant $C(T/\tau) > 0$, which blows up as $\tau \to 0$, such that

$$\int_{\mathbb{R}^{2N}\times\mathbb{R}} |F(t)|^2\ \mathcal{M}\,\mathrm{d}q\,\mathrm{d}v\,\mathrm{d}x \leq C(T/\tau),$$

$$\frac{1}{\epsilon}\int_0^t \int_{\mathbb{R}^{2N}\times\mathbb{R}} \left|F - \int F(q_\star)\ \mathcal{M}(q_\star)\,\mathrm{d}q_\star\right|^2 \mathcal{M}\,\mathrm{d}q\,\mathrm{d}v\,\mathrm{d}x\,\mathrm{d}s \leq C(T/\tau).$$

Proof. By using integration by parts, we have

$$\begin{aligned}
&\frac{\mathrm{d}}{\mathrm{d}t}\int_{\mathbb{R}^{2N}\times\mathbb{R}} F^2\mathcal{M}\,\mathrm{d}q\,\mathrm{d}v\,\mathrm{d}x - \frac{1}{\varepsilon}\int_{\mathbb{R}^{2N}\times\mathbb{R}} \mathcal{Q}^\star(F)F\mathcal{M}\,\mathrm{d}q\,\mathrm{d}v\,\mathrm{d}x \\
&\quad = -\frac{1}{\tau}\int_{\mathbb{R}^{2N}\times\mathbb{R}} v\cdot\nabla_v\Big(\frac{F^2}{2}\Big)\mathcal{M}\,\mathrm{d}q\,\mathrm{d}v\,\mathrm{d}x = \frac{N}{2\tau}\int_{\mathbb{R}^{2N}\times\mathbb{R}} F^2\mathcal{M}\,\mathrm{d}q\,\mathrm{d}v\,\mathrm{d}x.
\end{aligned}$$

We conclude by using (2.4) and the Gronwall lemma. □

In the spirit of [24], we shall distinguish "High Inertia Particles Regimes" where τ is kept fixed and "Fine Particles Regimes" where the drag force is the leading term within the equation. The former asymptotically lead to diffusion equations where the diffusion operates on velocity while the latter lead to asymptotic diffusion operating on position. As explained below, the Fine Particles Regimes are much harder to analyze due to concentration phenomena (see [29, 30]). Interestingly, looking into an "Over-Damped Regime" ($\tau \to 0$) yet in the "High Inertia-Particles Regimes" leads to an asymptotic diffusion evolution on position similar to the "Fine Particles Regimes", but based on a different physical background mechanism.

2.2.1 High-inertia-particles regime

The high-inertia regimes investigated in [24] correspond to assume

$$\eta = 1/\sqrt{\varepsilon}, \qquad \lambda = \varepsilon, \qquad \varepsilon \to 0, \qquad \tau \text{ fixed.} \tag{2.18}$$

Furthermore, an adaptation of [25] leads to considering the regime

$$\eta = 1/\sqrt{\varepsilon}, \qquad \lambda = \varepsilon^{\alpha}, \qquad \varepsilon \to 0, \qquad \tau \text{ fixed}, \qquad \alpha > 1. \tag{2.19}$$

We will take $\alpha = 3/2$ in the formal analysis, but the method seems to work just the same for other $\alpha > 1$. Both regimes (2.18) and (2.19) yield Fokker-Planck like equations in phase space. It means that the limiting behavior as $\varepsilon \to 0$ can be described by the PDE

$$\partial_t G + v \cdot \nabla_x G = \nabla_v \cdot \left(\frac{v}{\tau} G + \frac{1}{\tau^2} \mathbb{D}_n \nabla_v G \right)$$

where the unknown $G = G(t, x, v)$ is a distribution function in phase space. For the regime (2.18), the effective diffusion matrix is $\mathbb{D}_0$ as defined in (2.13), whereas for the regime (2.19), it is $\mathbb{D}_2$ given by (2.11).

Considering classically the "Over-Damped" regime $\tau \to 0$ for regime (2.19) leads to a diffusion on position:

$$\partial_t G = \nabla_x \cdot \left(\mathbb{D}_2 \nabla_x G \right),$$

where $G = G(t, x)$ now depends only on time and space. The "Over-Damped" regime could be obtained directly from (2.19) by taking $\sqrt{\varepsilon} \ll \tau_\varepsilon \ll 1$. The case of regime (2.18) with such an over damping would also lead to a similar diffusion on position but solution of a more intricate diffusion homogeneization problem that lies beyond the scope of this paper.

2.2.2 Fine-particles regime

In these regimes the scaled Stokes settling time τ goes to 0 very fast. Let us first discuss the asymptotic properties of the model, without making precisely the relation between the parameters; we only assume that $0 < \tau \ll 1$ sufficiently fast. We can expect that

$$f(t, x, v, q) \sim \rho(t, x, q) \; \delta_{v = \eta u(x/\lambda, q)}, \tag{2.20}$$

with the macroscopic density ρ of order 1. It fixes the dependence with respect to the variable v. It is therefore convenient to use the moment equations associated with (2.14). We set

$$\rho(t, x, q) = \int f \, \mathrm{d}v, \quad J(t, x, q) = \int v f \, \mathrm{d}v, \quad \mathbb{P}(t, x, q) = \int v \otimes v f \, \mathrm{d}v. \tag{2.21}$$

Integration of (2.14) with respect to the velocity variable yields

$$\partial_t \rho(t,x) + \mathrm{div}_x J = \frac{1}{\varepsilon} \mathcal{Q}^\star(\rho), \tag{2.22}$$

$$\partial_t J + \mathrm{Div}_x \mathbb{P} = \frac{1}{\tau}(\eta\ \rho U(x/\lambda, q) - J) + \frac{1}{\varepsilon}\mathcal{Q}^\star(J). \tag{2.23}$$

Due to (2.20), we have

$$\eta \rho U(x/\lambda, q) - J) \to 0,$$

as well as

$$\rho = \mathcal{O}(1), \qquad J = \mathcal{O}(\eta), \qquad \mathbb{P} = \mathcal{O}(\eta^2), \tag{2.24}$$

and all of the above terms remain bounded when $\tau \to 0$. Hence, we can use (2.23) to rewrite (2.22) as follows:

$$\begin{aligned}\partial_t \rho(t,x) &+ \mathrm{div}_x\big(\rho\eta U(x/\lambda, q)\big) - \frac{1}{\varepsilon}\mathcal{Q}^\star(\rho) \\ &= \tau \mathrm{div}_x \Big[\partial_t J + \mathrm{Div}_x \mathbb{P}\Big] - \frac{\tau}{\varepsilon} \mathrm{div}_x \mathcal{Q}^\star(J).\end{aligned}$$

The scaling assumptions $\eta\tau/\varepsilon \ll 1$ and $\eta^2\tau \ll 1$ will guarantee that the right hand side goes to 0, owing to (2.24). Hence, we can expect the problem is close to the more familiar one

$$\partial_t \rho(t,x) + \mathrm{div}_x\big(\eta\rho U(x/\lambda, q)\big) - \frac{1}{\varepsilon}\mathcal{Q}^\star(\rho) = 0$$

which is a standard model describing the passive transport of tracer particles in the flow described by u (see [32]). Note that the above formal computation could also be carried out using a Hilbert expansion on the dual problem (with operators acting on test functions instead of densities).

Let us now specify the scaling. Below, the analysis will be carried out by assuming either

$$\begin{cases} \eta = 1/\sqrt{\varepsilon}, \qquad \lambda = \sqrt{\varepsilon}, \qquad \varepsilon \to 0, \\ \tau = \varepsilon^k \text{ with } k > 3/2. \end{cases} \tag{2.25}$$

or

$$\begin{cases} \eta = 1/\sqrt{\varepsilon}, \qquad \lambda = \varepsilon^\alpha, \qquad \varepsilon \to 0, \\ \tau = \varepsilon^k \text{ with } k > 3/2, \qquad \alpha > 1/2. \end{cases} \tag{2.26}$$

As a consequence of the combination of the hydrodynamic limit to the random homogenization effects, we obtain a macroscopic diffusion equation

$$\partial_t \rho - \nabla_x \cdot (\mathbb{D}_n \nabla_x \rho) = 0$$

with $n = 1$ for (2.25) and $n = 2$ for (2.26) with $\alpha = 3/2$. Referring to [24], these scalings correspond to the "Small Stokes Number-Fine Particles Regimes" (the Stokes number is the ratio τ/ε). However (2.25) and (2.26) are not treated in [24] where so fast space oscillations are excluded (with $\eta = 1/\sqrt{\varepsilon}$, the result in [24] assumes λ is fixed, a case which can be treated easily by the techniques exposed here).

2.3 Derivation of the effective equations in the fine particles regimes

2.3.1 Analysis of the regime (2.25)

Let us assume that (2.25) holds, so that we address the question of the behavior for small ε's of

$$\partial_t \rho_\varepsilon(t,x) + \frac{1}{\sqrt{\varepsilon}} \mathrm{div}_x\big(\rho_\varepsilon U(x/\sqrt{\varepsilon}, q)\big) - \frac{1}{\varepsilon} \mathcal{Q}^\star(\rho_\varepsilon) = 0. \tag{2.27}$$

We insert the ansatz

$$\rho_\varepsilon(t,x,q) = \rho_0(t,x,x/\sqrt{\varepsilon},q) + \sqrt{\varepsilon}\rho_1(t,x,x/\sqrt{\varepsilon},q) + \varepsilon\rho_2(t,x,x/\sqrt{\varepsilon},q) + \dots$$

At leading order we obtain

$$U(y,q)\cdot\nabla_y \rho_0 - \mathcal{Q}^\star(\rho_0) = 0.$$

The $\mathcal{O}(1/\sqrt{\varepsilon})$ equation reads

$$U(y,q)\cdot\nabla_y \rho_1 - \mathcal{Q}^\star(\rho_1) = -U(y,q)\cdot\nabla_x\rho_0,$$

and, finally, $\mathcal{O}(1)$ terms yield

$$U(y,q)\cdot\nabla_y \rho_2 - \mathcal{Q}^\star(\rho_2) = -\partial_t\rho_0 - U(y,q)\cdot\nabla_x\rho_1.$$

We are thus led to investigating the cell equation

$$U(y,q)\cdot\nabla_y \rho - \mathcal{Q}^\star(\rho) = h$$

completed with periodic boundary conditions. Clearly, the equation can make sense only when the right hand side fulfills the compatibility condition

$$\int_{\mathbb{Y}\times\mathbb{R}} h\ \mathcal{M}(q)\,\mathrm{d}q\,\mathrm{d}y = 0.$$

This is a necessary condition; it can be shown to be also sufficient as summarized in the following claim (We analyze the relaxation operator (2.6) only, but it is possible to extend the result to the operator (2.5)).

Lemma 2.3. *Let $\mathcal{Q}^\star$ be a bounded operator on $L^2(\mathbb{R}, \mathcal{M}\,\mathrm{d}q)$, verifying (2.4). Let U satisfy (2.1)–(2.3) and (2.9)–(2.10). Then for any $h \in L^2(\mathbb{Y} \times \mathbb{R}, \mathcal{M}\,\mathrm{d}q\,\mathrm{d}y)$ verifying $\int h\ \mathcal{M}(q)\,\mathrm{d}q\,\mathrm{d}y = 0$ there exists a unique solution $\rho \in L^2(\mathbb{Y} \times \mathbb{R}, \mathcal{M}\,\mathrm{d}q\,\mathrm{d}y)$ of $U(y,q) \cdot \nabla_y \rho - \mathcal{Q}^\star(\rho) = h$ such that $\int_{\mathbb{Y}\times\mathbb{R}} \rho\ \mathcal{M}(q)\,\mathrm{d}q\,\mathrm{d}y = 0$.*

Proof. Let us start by studying the problem when $h = 0$. Integrating with respect to both y and q, using the periodic boundary condition and (2.4), we get

$$\int_{\mathbb{Y}\times\mathbb{R}} \left| \rho(y,q) - \int_{\mathbb{R}} \rho(y,q_\star)\mathcal{M}(q_\star)\,\mathrm{d}q_\star \right|^2 \mathcal{M}(q)\,\mathrm{d}q\,\mathrm{d}y = 0$$

and we infer first that the solution $\rho(y,q) = \rho(y) \in \mathrm{Ker}(\mathcal{Q}^\star)$ does not depend on q. Then, the equation simply becomes $U(y,q) \cdot \nabla_y \rho = 0$. The ergodic condition (2.9) implies that ρ does not depend on the fast variable y. Therefore imposing the solution has a null average force $\rho = 0$.

Now we suppose $h \neq 0$ and we justify the existence of solutions by a regularization argument. Let us consider the sequence $\big(\rho_\lambda\big)_{\lambda>0}$ of solutions to

$$\lambda\rho_\lambda + U \cdot \nabla_y \rho_\lambda - \mathcal{Q}^\star(\rho_\lambda) = h. \tag{2.28}$$

If ρ_λ is bounded in $L^2(\mathbb{Y} \times \mathbb{R})$ letting $\lambda \to 0$ for a suitable (weakly) convergent subsequence yields the desired existence statement. Hence, we assume that $\|\rho_\lambda\|_{L^2(\mathbb{Y}\times\mathbb{R})} = 1$ and the ρ_λ's verifies (2.28) with a right hand side h_λ that tends to 0. We shall show that we are led to a contradiction. Indeed, let us denote $\langle\rho_\lambda\rangle = \int_{\mathbb{R}} \rho_\lambda \mathcal{M}\,\mathrm{d}q$. By using (2.4), we get

$$\int_{\mathbb{Y}\times\mathbb{R}} \left| \rho_\lambda(y,q) - \langle\rho_\lambda\rangle(y) \right|^2 \mathcal{M}(q)\,\mathrm{d}q\,\mathrm{d}y \xrightarrow[\lambda\to 0]{} 0.$$

Next, since $\langle\rho_\lambda\rangle \in \mathrm{Ker}(\mathcal{Q}^\star)$, we can write

$$U \cdot \nabla_y \langle\rho_\lambda\rangle = h_\lambda - \lambda\rho_\lambda + \big(\mathcal{Q}^\star - U \cdot \nabla_y\big)(\rho_\lambda - \langle\rho_\lambda\rangle).$$

We denote by $S_\lambda(y,q)$ the right hand side and we set

$$A(y) = \langle U(y,q) \otimes U(y,q) \rangle.$$

By (2.10) this matrix is invertible and $A(y)^{-1}$ belongs to $W^{1,\infty}(\mathbb{Y})$. We deduce that

$$\nabla_y \langle\rho_\lambda\rangle = A(y)^{-1} \langle U S_\lambda(y,q) \rangle$$

belongs to a compact set of $H^{-1}(\mathbb{Y})$. Accordingly, since $\int h_\lambda \mathcal{M}\,\mathrm{d}q\,\mathrm{d}y = 0$ implies $\int \langle\rho_\lambda\rangle\,\mathrm{d}y = 0$, we deduce by a standard Fourier argument that $\langle\rho_\lambda\rangle$ belongs to a compact set in $L^2(\mathbb{Y})$. Finally, we conclude that

$\rho_\lambda = (\rho_\lambda - \langle \rho_\lambda \rangle) + \langle \rho_\lambda \rangle$ is (strongly) compact in $L^2(\mathbb{Y} \times \mathbb{R}, \mathcal{M}\,\mathrm{d}q\,\mathrm{d}y)$. Extracting a subsequence if necessary, let us denote ρ the limit of the ρ_λ's. It verifies

$$U \cdot \nabla_y \rho - \mathcal{Q}^\star(\rho) = 0, \qquad \int \rho \mathcal{M}\,\mathrm{d}q\,\mathrm{d}y = 0,$$

thus $\rho = 0$, a contradiction. □

Remark 2.4. *In fact, the above cell problem has a probabilistic interpretation that will lead to similar existence/uniqueness result but which can be extended to the space of continuous and bounded functions. Introducing*

$$t \mapsto Q_t^\star,$$

the Markov process with generator $\mathcal{Q}^\star$, and the process solution of

$$\dot{Y}_t^\star = -U(Y_t^\star, Q_t^\star),$$

the solution of the cell problem $U(y,q) \cdot \nabla_y \rho - \mathcal{Q}^\star(\rho) = h$, we can write down:

$$\rho(q,y) = \int_0^{+\infty} -\mathbb{E}\left(h(Y_t^\star, Q_t^\star) | (Y_0^\star = y, Q_0^\star = q)\right) \mathrm{d}t,$$

which will be well defined as soon as the process $t \mapsto (Y_t^\star, Q_t^\star)$ is mixing and h is centered with respect to the invariant measure $\mathcal{M}(q)\,\mathrm{d}q\,\mathrm{d}y$. If $\mathcal{Q}^\star$ is a Fokker-Planck operator, a sufficient condition to get mixing is the hypoellipticity of the operator $U(y,q) \cdot \nabla_y \rho - \mathcal{Q}^\star$ (the usual Hörmander sense) (see [35]).

This statement already proves that $\rho_0(t,x,y,q) = \rho_0(t,x)$ does not depend on neither q nor y. Let us introduce $\chi = (\chi_1, \ldots, \chi_N)$ solution of

$$U(y,q) \cdot \nabla_y \chi - \mathcal{Q}^\star(\chi) = U(y,q)$$

which makes sense thanks to (2.7). Then we get

$$\rho_1(t,x,y,q) = -\chi(y,q) \cdot \nabla_x \rho_0(t,x).$$

Therefore, the compatibility condition for the $\mathcal{O}(1)$ equation leads to

$$\partial_t \rho_0 - \nabla_x \cdot \left(\int_{\mathbb{Y} \times \mathbb{R}} U(y,q) \otimes \chi(y,q)\ \mathcal{M}(q)\,\mathrm{d}q\,\mathrm{d}y\ \nabla_x \rho_0 \right) = 0.$$

The diffusion matrix (2.12) writes down:

$$\mathbb{D}_1 = \int_{\mathbb{Y} \times \mathbb{R}} U(y,q) \otimes \chi(y,q)\ \mathcal{M}(q)\,\mathrm{d}q\,\mathrm{d}y.$$

Note it is indeed non negative since we have, for any $\xi \in \mathbb{R}^N \setminus \{0\}$,

$$\begin{aligned}\mathbb{D}_1 v \cdot v &= \int_{\mathbb{Y}\times\mathbb{R}} (U \cdot \nabla_y - \mathcal{Q}^\star)(\chi \cdot \xi)\ \chi \cdot \xi \,\mathrm{d}y\,\mathrm{d}q \\ &= - \int_{\mathbb{Y}\times\mathbb{R}} \mathcal{Q}^\star(\chi \cdot \xi)\ \chi \cdot \xi\ \mathcal{M}(q)\,\mathrm{d}q\,\mathrm{d}y \\ &\geq \sigma \int_{\mathbb{Y}\times\mathbb{R}} |\chi \cdot \xi|^2\ \mathcal{M}(q)\,\mathrm{d}q\,\mathrm{d}y > 0\end{aligned}$$

by (2.3) and (2.4). Moreover, it cannot vanish due to (2.10). This formal development can be achieved rigorously by adapting the arguments in [25] and we conclude with the following statement.

Theorem 2.5. *Let U verify (2.1)–(2.7) and (2.9)–(2.10). We assume that (2.4) is fulfilled. We suppose that the initial condition verifies*

$$\sup_{\varepsilon>0} \int_{\mathbb{R}^N\times\mathbb{R}} |\rho^\varepsilon_{\mathrm{Init}}(x,q)|^2 \mathcal{M}(q)\,\mathrm{d}q\,\mathrm{d}x \leq C < \infty. \tag{2.29}$$

Then, up to a subsequence, ρ^ε solutions of (2.27) associated with $\rho^\varepsilon_{\mathrm{Init}}$ converges weakly in $L^2((0,T)\times\mathbb{R}^N\times\mathbb{R};\mathcal{M}(q)\,\mathrm{d}q\,\mathrm{d}x)$ and in $C^0([0,T], L^2(\mathbb{R}^N\times\mathbb{R},\mathcal{M}\,\mathrm{d}q\,\mathrm{d}x) - \text{weak})$ to $\rho(t,x)$, where ρ is the solution of

$$\begin{cases}\partial_t \rho = \nabla_v \cdot (\mathbb{D}_1 \nabla_v \rho), \\ \rho(t=0,x) = \text{weak-}\lim_{\varepsilon\to 0} \int_{\mathbb{R}} \rho^\varepsilon_{\mathrm{Init}}(x,q)\mathcal{M}(q)\,\mathrm{d}q,\end{cases} \tag{2.30}$$

with $\mathbb{D}_1$ defined by (2.12).

The statement only deals with Eq. (2.27), which is already an approximation of the moment system coming form (2.14). Taking account of the additional (small) terms in (2.27) leads to technical difficulties due to the functional framework: we have only the L^1 bounds at hand (see Propositions 2.1 and 2.2) and the ugly terms make derivatives with respect to space and time appearing.

2.3.2 Analysis of the regime (2.26)

Let us assume that (2.26) holds, so that we address the question of the behavior for small ε's of

$$\partial_t \rho_\varepsilon(t,x) + \frac{1}{\sqrt{\varepsilon}} \mathrm{div}_x\big(\rho_\varepsilon U(x/\varepsilon^{3/2},q)\big) - \frac{1}{\varepsilon}\mathcal{Q}^\star(\rho_\varepsilon). \tag{2.31}$$

We insert the ansatz

$$\begin{aligned}\rho_\varepsilon(t,x,q) &= \rho_0(t,x,x/\varepsilon^{3/2},q) + \varepsilon^{3/2}\rho_1(t,x,x/\varepsilon^{3/2},q) \\ &\quad + \varepsilon^3 \rho_2(t,x,x/\varepsilon^{3/2},q) + \dots\end{aligned}$$

At leading order we obtain

$$U(y,q)\cdot\nabla_y\rho_0=0.$$

The ergodic condition (2.9) implies that $\rho_0(t,x,y,q)=\rho_0(t,x,q)$ does not depend on the fast variable. Next, we get

$$U(y,q)\cdot\nabla_y\rho_1=\mathcal{Q}^\star(\rho_0).$$

Integrating with y and bearing in mind that ρ_0 does not depend on y, we obtain $\rho_0\in\operatorname{Ker}(\mathcal{Q}^\star)$, so that ρ_0 does not depend on q anymore. Therefore, the equation for the corrector becomes $U\cdot\nabla_y\rho_1=0$ implying that $\rho_1(t,x,y,q)=\rho_1(t,x,q)$. Then, we arrive at

$$U(y,q)\cdot\nabla_y\rho_2=\mathcal{Q}^\star(\rho_1)-U(y,q)\cdot\nabla_x\rho_0(t,x).$$

Integration over $\mathbb{Y}$ yields

$$\mathcal{Q}^\star(\rho_1)=\int_{\mathbb{Y}}U(y,q)\,\mathrm{d}y\cdot\nabla_x\rho_0(t,x).$$

When U verifies the pointwise centering condition (2.8) we can appeal to the Fredholm alternative for $\mathcal{Q}^\star$ and find the auxilliary function $\chi(q)$ such that

$$\mathcal{Q}^\star(\chi)=-\int_{\mathbb{Y}}U(y,\cdot)\,\mathrm{d}y.$$

We thus write $\rho_1(t,x,q)=-\chi(q)\cdot\nabla_x\rho_0(t,x)$. We plug this formula into the $\mathcal{O}(1)$ equation, which after integration leads to

$$\partial_t\rho_0-\nabla_x\cdot\big(\mathbb{D}_2\nabla_x\rho_0(t,x)\big)=0, \tag{2.32}$$

with

$$\mathbb{D}_2=\int_{\mathbb{Y}\times\mathbb{R}}U(y,q)\otimes\chi(q)\mathcal{M}(q)\,\mathrm{d}q\,\mathrm{d}y.$$

Referring to [25] again, we obtain the following statement (note that the restriction (2.8) can be relaxed by using a fully probabilistic proof).

Theorem 2.6. *Let U verify (2.1)–(2.3) and (2.8)–(2.9). We assume that (2.4) is fulfilled. We suppose that the initial condition verifies (2.29). Then, up to a subsequence, ρ^ε solutions of (2.31) associated with $\rho^\varepsilon_{\mathrm{Init}}$ converge weakly in $L^2((0,T)\times\mathbb{R}^N\times\mathbb{R};\mathcal{M}(q)\,\mathrm{d}q\,\mathrm{d}x)$ and in $C^0([0,T],L^2(\mathbb{R}^N\times\mathbb{R},\mathcal{M}\,\mathrm{d}q\,\mathrm{d}x)-\mathrm{weak})$ to $\rho(t,x)$, where ρ is the solution of (2.32) with coefficient (2.11).*

2.4 Derivation of the effective equations in the high-inertia particles regimes

By the same token, we can discuss the effective equations arising with the regimes (2.18) and (2.19). In these situations where we do not have concentration effect, a complete proof can be designed by adapting directly the argument in [25].

Assuming (2.18), we are concerned with the behavior as $\varepsilon \to 0$ of

$$\partial_t F_\varepsilon + v \cdot \nabla_x F_\varepsilon + \nabla_v \cdot \Big[\Big(\frac{1}{\sqrt{\varepsilon}} U(x/\varepsilon, q) - v\Big) F_\varepsilon\Big] = \frac{1}{\varepsilon} \mathcal{Q}^\star(F_\varepsilon). \quad (2.33)$$

A double-scale ansatz leads to the auxilliary equation

$$v \cdot \nabla_y \chi^\star - \mathcal{Q}^\star \chi^\star = U(y, q)$$

which makes sense under the centering assumption (2.7) and we define the (non-negative) matrix by (2.11), namely

$$\mathbb{D}_0(v) = \int_{\mathbb{R}\times\mathbb{Y}} U(y,q) \otimes \chi^\star(v,y,q)\,\mathrm{d}q\,\mathrm{d}y.$$

Assuming (2.19), we are concerned with the behavior as $\varepsilon \to 0$ of

$$\partial_t F_\varepsilon + v \cdot \nabla_x F_\varepsilon + \nabla_v \cdot \Big[\Big(\frac{1}{\sqrt{\varepsilon}} U(x/\varepsilon^{3/2}, q) - v\Big) F_\varepsilon\Big] = \frac{1}{\varepsilon} \mathcal{Q}^\star(F_\varepsilon). \quad (2.34)$$

Then, we are led to the auxilliary equation

$$-\mathcal{Q}^\star \chi^\star = \int_{\mathbb{Y}} U(y,\cdot)\,\mathrm{d}y$$

which makes sense under the centering assumption (2.8). We recall (2.11), in this context:

$$\mathbb{D}_2 = \int_{\mathbb{R}\times} \int_{\mathbb{Y}} U(y,q)\,\mathrm{d}y \otimes \chi^\star(q)\,\mathrm{d}q.$$

The results are summarized as follows:

Theorem 2.7. *Let U verify (2.1)–(2.7). For the scaling (2.19), we assume the strengthened condition (2.8). We assume that (2.4) is fulfilled. We suppose that the initial condition verifies*

$$\sup_{\varepsilon>0} \int_{\mathbb{R}^N\times\mathbb{R}^N\times\mathbb{R}} |F^\varepsilon_{\mathrm{Init}}(x,q)|^2 \mathcal{M}(q)\,\mathrm{d}q\,\mathrm{d}v\,\mathrm{d}x \le C < \infty.$$

Then, up to a subsequence, F^ε solutions of (2.33) (resp. (2.34)) associated with $F^\varepsilon_{\mathrm{Init}}$ converge weakly in $L^2((0,T)\times\mathbb{R}^N\times\mathbb{R}^N\times\mathbb{R}; \mathcal{M}(q)\,\mathrm{d}q\,\mathrm{d}x)$ and in $C^0([0,T], L^2(\mathbb{R}^N\times\mathbb{R}^N\times\mathbb{R}, \mathcal{M}\,\mathrm{d}q\,\mathrm{d}x)-\text{weak})$ to $G(t,x,v)$, solution of

$$\partial_t.G + v\cdot\nabla_x G = \nabla_v \cdot (vG + \mathbb{D}_n \nabla_v G)$$

with $\mathbb{D}_n$ given by (2.13) ($n = 0$) (resp. by (2.11) ($n = 2$)).

3 Numerical schemes for coupled fluid/particles models

This section is devoted to models which take account of the back-reaction of the particles on the dense phase. Therefore, the velocity field u involved in the drag force is defined by an evolution equation depending on the particle density. More precisely, the dense phase is described by the mass density $n(t,x)$ and the velocity field $u(t,x)$. These quantities obey the Euler or Navier-Stokes system. We take the following into account:

- the drag force exerted by each phase on the other; as explained in the Introduction it depends on the relative velocity $v - u(t,x)$;
- the Brownian motion of the particles which leads to diffusion with respect to the velocity variable;
- the effect of external forces embodied into a potential field $\Phi(x)$.

Accordingly, the fluid-particles flow is governed by the PDEs

$$\partial_t f + v\cdot\nabla_x f - \alpha\nabla_x\Phi\cdot\nabla_v f = \frac{9\mu}{2a^2\rho_{\mathrm{p}}}\,\nabla_v\cdot\Big((v-u)f + \frac{k\theta_0}{m_{\mathrm{p}}}\,\nabla_v f\Big), \quad (3.1)$$

$$\partial_t n + \nabla_x\cdot(nu) = 0, \quad (3.2)$$

$$\rho_{\mathrm{f}}\Big(\partial_t(nu) + \mathrm{Div}_x(nu\otimes u) + n\nabla_x\Phi\Big) + \nabla_x p(n) = 6\pi\mu a\int_{\mathbb{R}^3}(v-u)f\,\mathrm{d}v. \quad (3.3)$$

In (3.3), ρ_{f} is a typical mass density of the fluid, k stands for the Boltzmann constant, and $\theta_0 > 0$ denotes the temperature of the fluid, assumed to be constant; $p(n)$ is a general pressure law, for instance, $p(n) = C_\gamma\, n^\gamma$, with $\gamma \geq 1$, $C_\gamma > 0$. The parameter $\alpha \in \mathbb{R}$ is a dimensionless parameter which indicates that the external force can have a different strength and direction on the two phases. Throughout this paper we have in mind the case of gravity forces where

$$\Phi(t,x) = \mathrm{g}x_3,$$
$$\alpha = (1 - \rho_{\mathrm{f}}/\rho_{\mathrm{f}})\frac{\mathrm{U}}{\sqrt{3k\theta_0/4\pi a^3\rho_{\mathrm{p}}}},$$

with U being a typical value of the fluid velocity. Hence, α accounts for the buoyancy and gravity forces. The equation is completed by initial and boundary conditions. We suppose the standard homogeneous condition

$$u\cdot\nu(x) = 0$$

for the dense phase and for the particles the specular reflection boundary condition

$$f(t,x,v) = f(t,x,v-(v\cdot\nu(x))\nu(x))$$

for any $(x, v) \in \partial\Omega \times \mathbb{R}^N$ such that $v \cdot \nu(x) < 0$, where $\nu(x)$ stands for the outer normal vector at the point $x \in \partial\Omega$. Obviously the boundary condition guarantees mass conservation.

Establishing the well-posedness of such a nonlinear system is a tough piece of analysis; we refer to [3, 6, 27, 26, 36] for such existence results in different functional frameworks. Next, relevant asymptotic regimes can be identified and investigated, depending on the mass density ratio, the Stokes settling time, the typical velocity of the particles compared to those of the fluid $\cdots$ We refer to [10, 21, 22, 37] for such discussion and analysis.

Here, we are concerned with the asymptotic regime $\varepsilon \to 0$ in the following rescaled version of (3.1)–(3.3)

$$\begin{cases} \partial_t f_\varepsilon + \dfrac{1}{\sqrt{\epsilon}}\Big(v \cdot \nabla_x f_\varepsilon + \nabla_x \Phi \cdot \nabla_v f_\varepsilon\Big) = \dfrac{1}{\epsilon}\nabla_v \cdot \Big(\big(v - \sqrt{\epsilon}u_\varepsilon\big) f_\varepsilon + \nabla_v f_\varepsilon\Big), \\[2ex] \partial_t n_\varepsilon + \nabla_x \cdot (n_\varepsilon u_\varepsilon) = 0, \\[2ex] \partial_t (n_\varepsilon u_\varepsilon) + \mathrm{Div}_x(n_\varepsilon u_\varepsilon \otimes u_\varepsilon) + \nabla_x p(n_\varepsilon) + \eta_\epsilon\, n_\varepsilon \nabla_x \Phi = J_\varepsilon - \rho_\varepsilon u_\varepsilon, \end{cases} \tag{3.4}$$

where we use the notation

$$\rho_\varepsilon(t,x) = \int_{\mathbb{R}^3} f_\varepsilon(t,x,v)\,\mathrm{d}v, \qquad J_\varepsilon(t,x) = \frac{1}{\sqrt{\epsilon}} \int_{\mathbb{R}^3} v\ f_\varepsilon(t,x,v)\,\mathrm{d}v.$$

It is referred to as the "Bubbling regime" in [10] and it relies on the following scaling assumptions

$$\text{Stokes velocity} = \frac{2\rho_{\mathrm{p}} a^2}{9\mu}\mathrm{g}|1 - \rho_{\mathrm{f}}/\rho_{\mathrm{p}}|$$
$$\ll \text{Typical velocity of the fluid} \simeq \text{Thermal velocity} = \sqrt{\frac{k\theta_0}{m_{\mathrm{p}}}},$$

while the ratio $\rho_{\mathrm{p}}/\rho_{\mathrm{f}}$ is of order ϵ. Finally, we suppose that $\eta_\epsilon \to \eta_\star \in (0, \infty)$ (precisely, for gravity forces it reads $\eta_\epsilon = (1-\varepsilon)^{-1}$). The analysis of the asymptotic behavior $\epsilon \to 0$ relies on the dissipative properties of the system (3.4), which are summarized in the following claim.

Proposition 3.1 (Entropy Dissipation Property). *We set*

$$\mathcal{F}_{\mathrm{p}}(t) = \int_{\mathbb{R}^3}\int_{\mathbb{R}^3} \Big(f \ln(f) + \frac{v^2}{2} f - \Phi f\Big)\,\mathrm{d}v\,\mathrm{d}x,$$
$$\mathcal{F}_{\mathrm{f}}(t) = \int_{\mathbb{R}^3} \Big(n\frac{|u|^2}{2} + \Pi(n) + \eta_\epsilon \Phi n\Big)\,\mathrm{d}x,$$

where $\Pi : \mathbb{R}^+ \longrightarrow \mathbb{R}^+$ *is defined by* $s\Pi''(s) = p'(s)$. *Then, we have*

$$\frac{\mathrm{d}}{\mathrm{d}t}\Big(\mathcal{F}_{\mathrm{p}} + \mathcal{F}_{\mathrm{f}}\Big) + \frac{1}{\varepsilon}\int_{\mathbb{R}^3}\int_{\mathbb{R}^3} \big|(v - \sqrt{\varepsilon}u)\sqrt{f} + 2\nabla_v\sqrt{f}\big|^2 \,\mathrm{d}v\,\mathrm{d}x \le 0. \quad (3.5)$$

Accordingly, we guess that, as ϵ goes to 0,

$$f_\varepsilon \simeq \rho(t,x)\ M(v), \qquad M(v) = (2\pi)^{-N/2}e^{-v^2/2},$$

and the asymptotic dynamics is embodied into the behavior of the macroscopic density $\rho(t,x)$. The formal expansion

$$f_\varepsilon = f^{(0)} + \sqrt{\varepsilon} f^{(1)} + \varepsilon f^{(2)} + \dots \quad (3.6)$$

is allowed to go into a step further. Identifying terms with the same power of ϵ, we find the following equation for the corrector

$$Lf^{(1)} = v\cdot\nabla_x f^{(0)} + (u + \nabla_x\Phi)\nabla_v f^{(0)} = v\ M(v)\ \big(\nabla_x\rho - (u+\nabla_x\Phi)\rho\big),$$

with L being the standard Fokker-Planck operator

$$Lf = \nabla_v\cdot(vf + \nabla_v f).$$

We obtain

$$f^{(1)}(t,x,v) = -vM(v)\ \big(\nabla_x\rho - (u+\nabla_x\Phi)\rho\big).$$

We use this information in the mass conservation relation

$$\partial_t \int f_\varepsilon \,\mathrm{d}v + \nabla_x\cdot\int \frac{v}{\sqrt{\varepsilon}} f_\varepsilon \,\mathrm{d}v = 0$$

which becomes

$$\begin{aligned}&\partial_t \int f^{(0)}\,\mathrm{d}v + \nabla_x\cdot\int v f^{(1)}\,\mathrm{d}v = 0\\ &= \partial_t\rho + \nabla_x\cdot\big(\rho(u+\nabla_x\Phi) - \nabla_x\rho\big) = 0.\end{aligned} \quad (3.7)$$

Similarly, in the fluid equation, we get

$$J_\varepsilon - \rho_\varepsilon u_\varepsilon \simeq -(\nabla_x\rho - \rho\nabla_x\Phi).$$

Hence, in the limit system, (3.7) is completed by

$$\begin{cases}\partial_t n + \mathrm{div}_x(nu) = 0,\\ \partial_t(nu) + \mathrm{Div}_x(nu\otimes u) + \nabla_x(p(n)+\rho) + (\eta_\star n - \rho)\nabla_x\Phi = 0.\end{cases} \quad (3.8)$$

Imposing the reflection law leads to the following Robin condition

$$\big(\nabla_x\rho - (u+\nabla_x\Phi)\rho\big)\cdot\nu(x) = 0 \qquad \text{on } \partial\Omega, \quad (3.9)$$

which completes (3.7) and also preserves mass for the limit system. We wish to design a numerical scheme specifically dedicated to treat the asymptotic regime.

3.1 Asymptotic preserving numerical methods

The previous discussion suggests that the solution expands into

$$f_\varepsilon(t,x,v) = \rho_\varepsilon(t,x)M(v) + \sqrt{\varepsilon} r_\varepsilon(t,x,v) \tag{3.10}$$

where Proposition 3.1 might give an estimate on the remainder r_ε. Then we rewrite (3.4) as follows:

$$\partial_t f_\varepsilon + v\cdot\nabla_x r_\varepsilon + (u_\varepsilon + \nabla_x\Phi)\cdot\nabla_v r_\varepsilon = \frac{1}{\varepsilon}Lf_\varepsilon + \frac{1}{\sqrt{\varepsilon}}M(v)S_\varepsilon(t,x,v), \tag{3.11}$$

with

$$S_\varepsilon(t,x,v) = -v\cdot\nabla_x\rho_\varepsilon - v\cdot(u_\varepsilon(t,x)+\nabla_x\Phi)\rho_\varepsilon.$$

We also have

$$\begin{aligned}\partial_t r_\varepsilon =& \frac{1}{\varepsilon}Lr_\varepsilon + \frac{1}{\varepsilon}MS_\varepsilon \\ &- \frac{1}{\sqrt{\varepsilon}}\left[v\cdot\nabla_x r_\varepsilon + (u_\varepsilon+\nabla_x\Phi)\nabla_v r_\varepsilon - M\nabla_x\cdot\left(\int_{\mathbb{R}^3} v_\star r_\varepsilon \,\mathrm{d}v_\star\right)\right].\end{aligned} \tag{3.12}$$

To derive the numerical scheme, we use a splitting algorithm to compute the evolution of both the density f_ε and its fluctuations r_ε by using (3.11) and (3.12). This approach is inspired by [23]. More precisely, the scheme works as follows: Giving n^k, u^k, f^k, r^k, approximation of n, u, f, r at time $k\Delta t$,

- *Step 0.* Solve the Euler equations for the fluid density n and velocity u. The source term is treated explicitly by plugging

  $$\int_{\mathbb{R}^3} vr^k \,\mathrm{d}v - u^k\int_{\mathbb{R}^3} f^k\,\mathrm{d}v.$$

  We use a numerical method which preserves with accuracy the shock structure of the hyperbolic system, applying directly the scheme designed in [15, 16, 33]. It defines the updated density n^{k+1} and velocity u^{k+1}.

- *Step 1.* Solve the stiff equations

  $$\partial_t f = \frac{1}{\varepsilon}Lf, \qquad \partial_t r = \frac{1}{\varepsilon}Lr + \frac{1}{\varepsilon}MS,$$

  where

  $$S = -v\cdot\nabla_x\rho + v\cdot(u^{k+1}+\nabla_x\Phi)\rho.$$

  Note that we get rid of the $\mathcal{O}(1/\sqrt{\varepsilon})$ terms in (3.11) and (3.12). The crucial point is that $\rho = \int f\,\mathrm{d}v$ is not modified during this step: $\rho^{k+1/2} = \int f^{k+1/2}\,dv = \rho^k$ so that the source term in the second equation can be treated as constant in time. Accordingly, the updated quantities read

$$\begin{cases} f^{k+1/2} = e^{\Delta t L/\varepsilon} f^k, \\ r^{k+1/2} = e^{\Delta t L/\varepsilon} r^k + (1 - e^{\Delta t L/\varepsilon}) M\ S^k. \end{cases} \tag{3.13}$$

- *Step 2.* Solve the transport-like part

$$\partial_t f + v \cdot \nabla_x r + (u^{k+1} + \nabla_x \Phi) \cdot \nabla_v r = 0, \qquad \partial_t r = 0,$$

which defines f^{k+1} and $\rho^{k+1} = \int f^{k+1}\,dv$. Particularly, the convection term is of characteristic speed v other than $v/\sqrt{\varepsilon}$).

Let us comment further on the proposed scheme.

1. *CFL and Sub-cycling.* Since the limit equation for the particles density is a diffusion equation, it involves a different typical time scale compared with the Euler equations. Accordingly, the stability constraints in Step 0 and Steps 1–2 are different. Therefore, given the space mesh size Δx, we define a "parabolic" and a "hyperbolic" time steps, $\Delta t_p = \mathcal{O}(\Delta x^2)$ and $\Delta t_h = \mathcal{O}(\Delta x)$) respectively (with $\Delta t_p < \Delta t_h$). Then, we perform several sub-cycles (Step 1–Step 2) above at time intervals $(k\Delta t_p, (k+1)\Delta t_p)$ and only Step 0 at the time interval $(k\Delta t_h, (k+1)\Delta t_h)$.
2. *Approximation of the Fokker-Planck semi-group.* Formulae (3.13) involve the operator e^{sL}, with L being the Fokker-Planck operator, but the expression is not explicit enough to be incorporated in a numerical subroutine, and a further approximation is needed. The method we proposed is based on the expression of the semi-group by means of convolution with the fundamental solution associated with the Fokker-Planck operator (see [12]). Using the fact that we are concerned with the regime $0 < \epsilon \ll 1$, we derive the following expression used in Step 1

$$\begin{cases} f^{k+1/2}(v) = M(v)\left(\rho^k + e^{-\Delta t/\varepsilon} v \displaystyle\int_{\mathbb{R}^3} v_\star f^k\,dv_\star\right), \\ r^{k+1/2}(v) = e^{-\Delta t/\varepsilon} M(v)\left(v \displaystyle\int_{\mathbb{R}^3} v_\star r^k\,dv_\star\right) + (1 - e^{-\Delta t/\varepsilon})\, M(v) S^k. \end{cases} \tag{3.14}$$

3. *Fundamental properties of the scheme.* The numerical scheme can be shown to fulfill many interesting requirements. First of all, it is asymptotic preserving in the sense that letting ϵ run to 0, we obtain a stable and consistent scheme for the limit systems (3.7)–(3.8). Second of all, the scheme is well balanced which means that

it preserves the equilibrium state. Finally, up to some reasonable care in the space/velocity discretization as well as in the definition of the numerical boundary conditions, the scheme conserves mass. We can check on numerics that the entropy dissipation property is also preserved.

We refer to [11] for further details and comments. We only show below a sample of the simulations, restricting ourselves to a one-dimensional situation (which is a toy-model, for instance, for describing the dispersion of pollutants emitted from ground sources). Initially, the dense phase is at rest $u(0, x) = 0$ with constant density $n(0, x) = 1$ while the distribution of particles is the following centered Maxwellian:

$$f(0, x, v) = 0.5 \ \mathbb{1}_{[a,b]}(x) \ \frac{e^{-v^2/2}}{\sqrt{2\pi}},$$

with $0 \leq a \leq b \leq 4$. The influence of ϵ can be discussed by looking at Figs. 3.1–3.4 and Figs. 3.5–3.8, where in both cases the adiabatic constant is $\gamma = 1.4$. The smoothing effect of the limit $\varepsilon \to 0$ appears clearly and both the fluid unknowns (n, u) and the macroscopic density of particles are smoother for small values of ε.

In Figs. 3.9–3.12, we show the solution at time $T = 20$ for different values of the adiabatic constant. The simulations illustrate the stability of sedimentation profiles, as conjectured from [10]. These profiles depend on the pressure law, with a change of convexity for the critical exponent $\gamma = 2$.

The discussion leaves several important questions open. Particularly, assuming a constant temperature in the systems (3.1)–(3.3) might be questionable. An extension of the model that includes an energy equation and energy exchanges has been proposed and studied in [8]. Another important question relies on the approximation of the Fokker-Planck semi-group, for which different approaches deserve to be discussed.

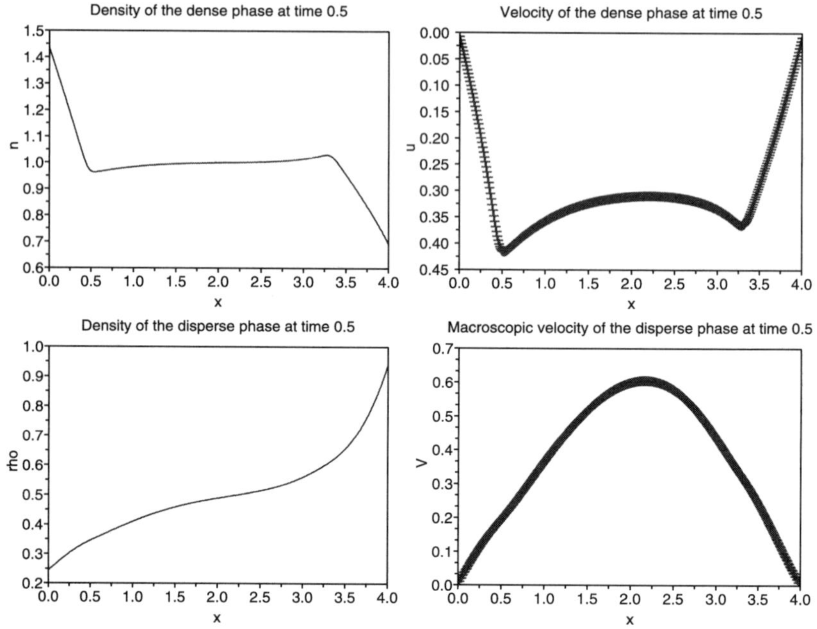

Figure 3.1 Bubbling Regime: Evolution for $\varepsilon = 0.1$, $T = 0.5$.

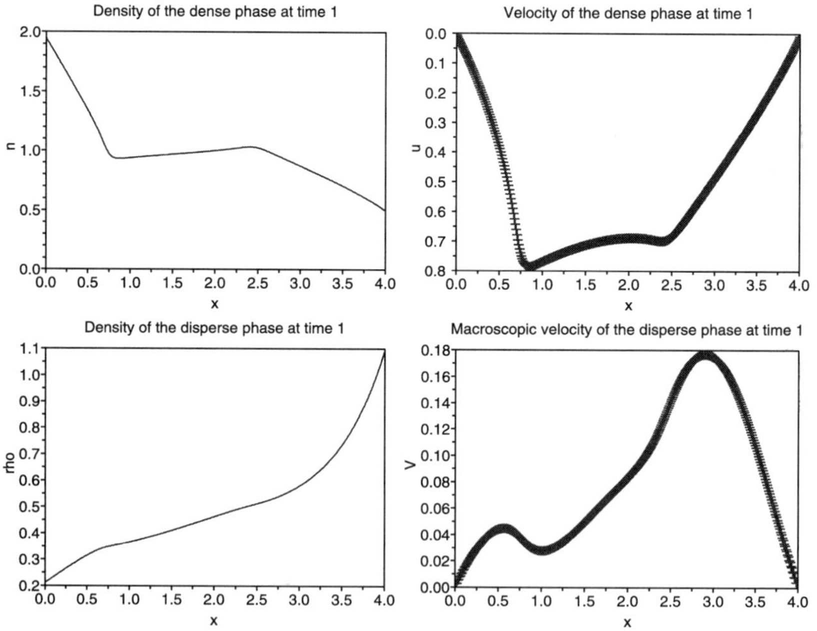

Figure 3.2 Bubbling Regime: Evolution for $\varepsilon = 0.1$, $T = 1$.

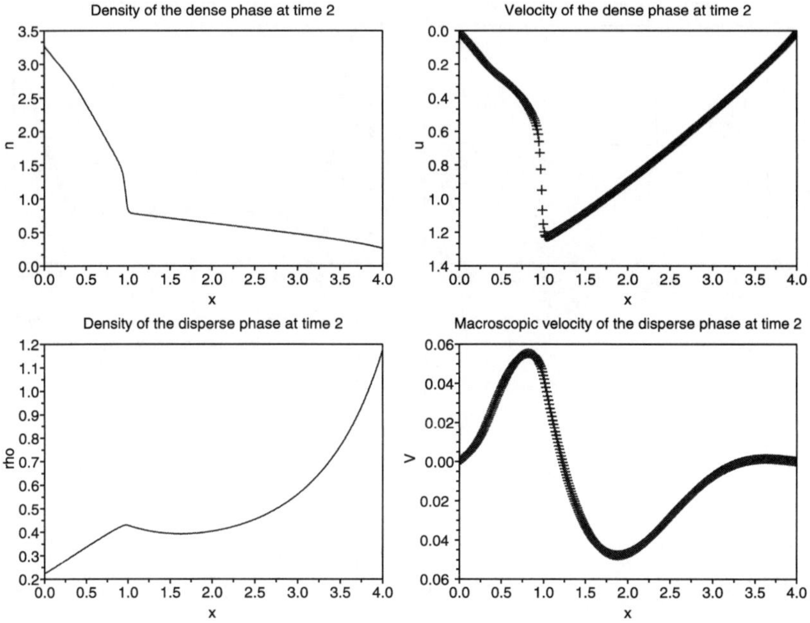

Figure 3.3 Bubbling Regime: Evolution for $\varepsilon = 0.1$, $T = 2$.

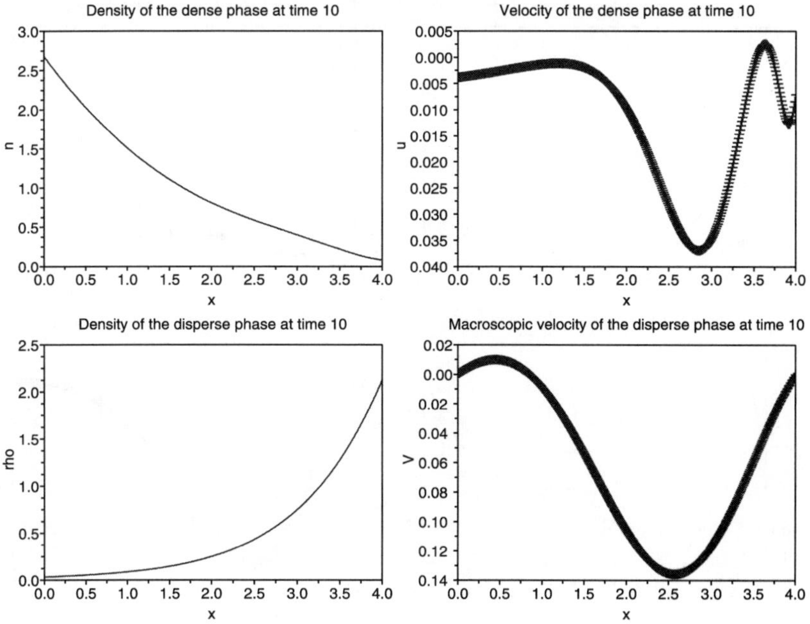

Figure 3.4 Bubbling Regime: Evolution for $\varepsilon = 0.1$, $T = 10$.

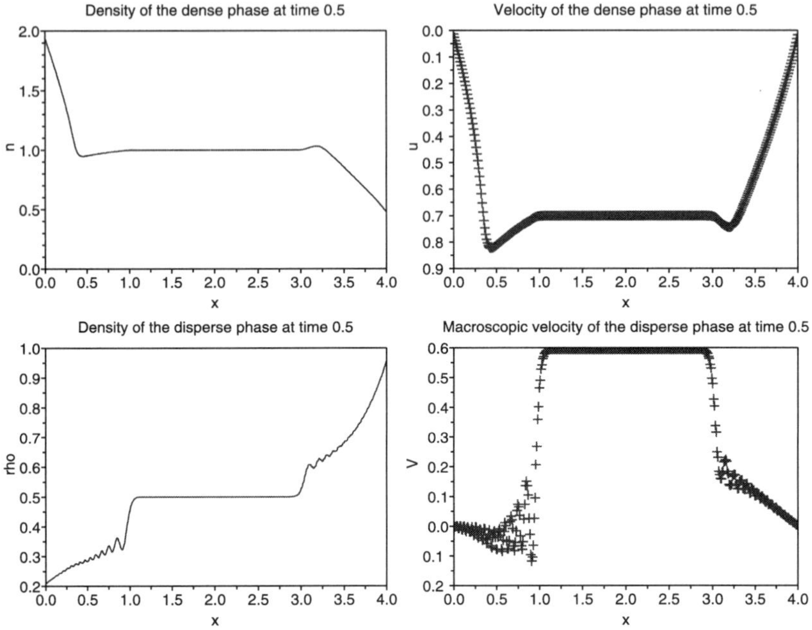

Figure 3.5 Bubbling Regime: Evolution for $\varepsilon = 0.5$, $T = 0.5$.

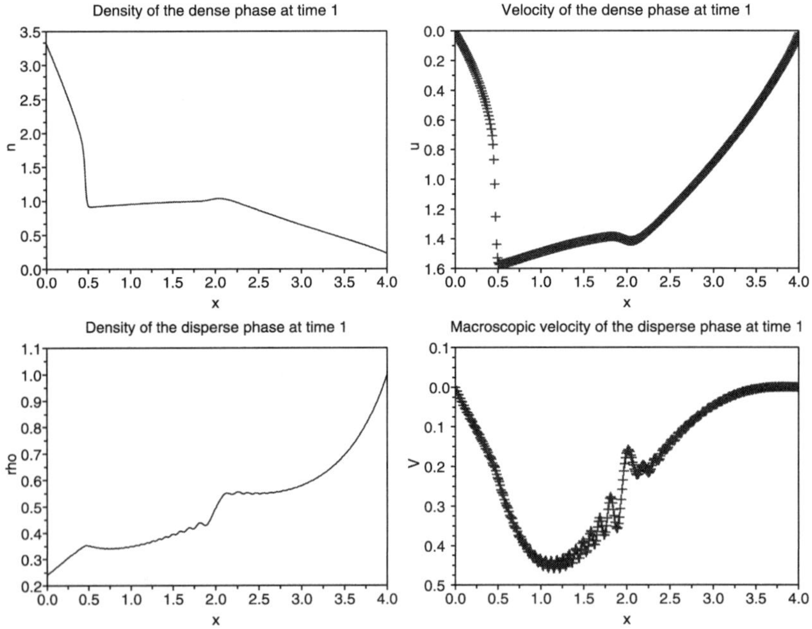

Figure 3.6 Bubbling Regime: Evolution for $\varepsilon = 0.5$, $T = 1$.

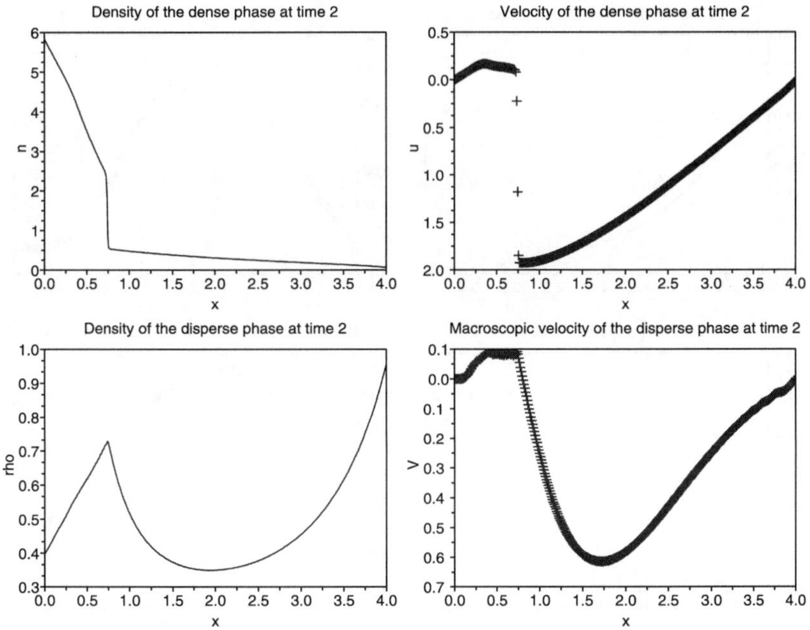

Figure 3.7 Bubbling Regime: Evolution for $\varepsilon = 0.5$, $T = 2$.

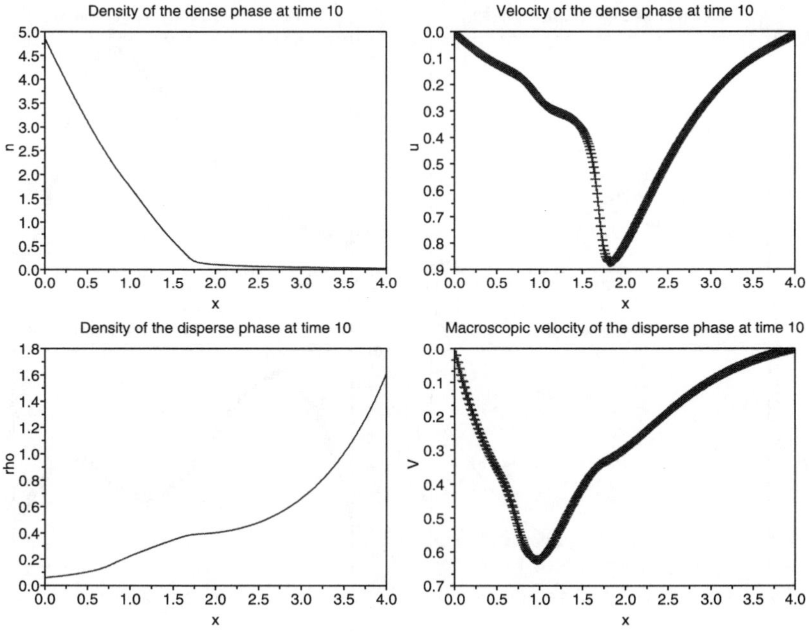

Figure 3.8 Bubbling Regime: Evolution for $\varepsilon = 0.5$, $T = 10$.

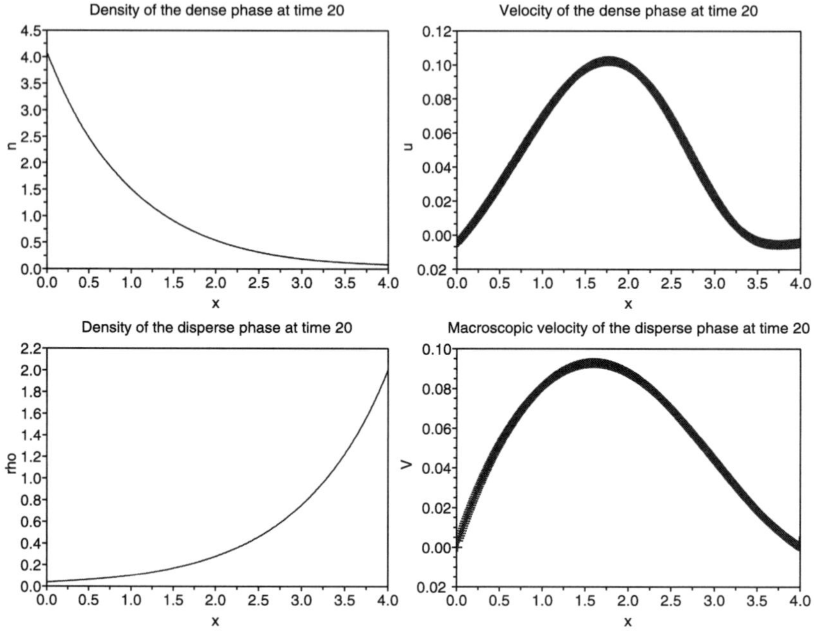

Figure 3.9 $\gamma = 1$.

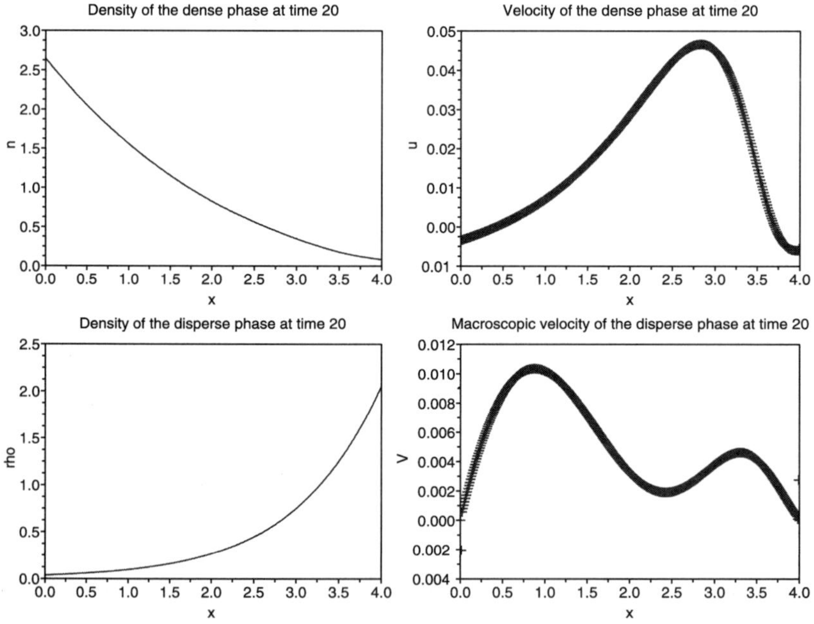

Figure 3.10 $\gamma = 1.4$.

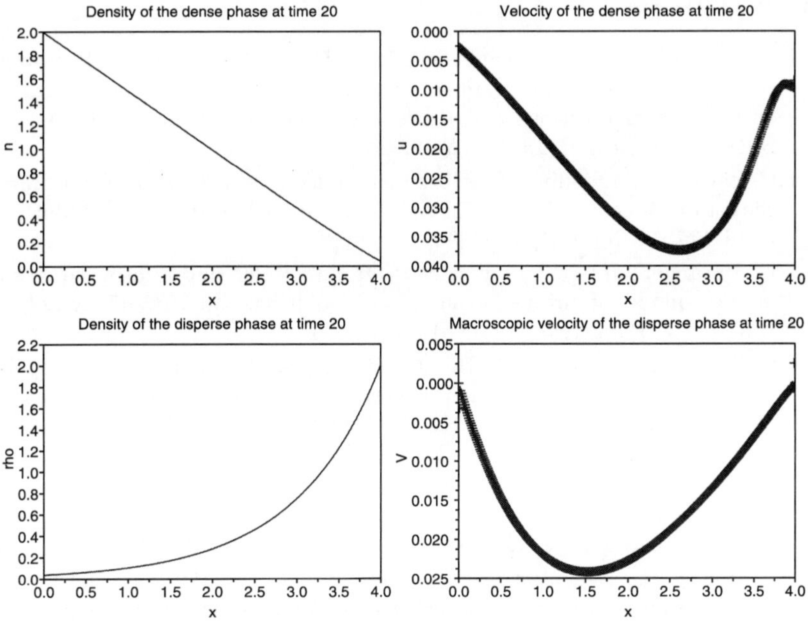

Figure 3.11 $\gamma = 2$.

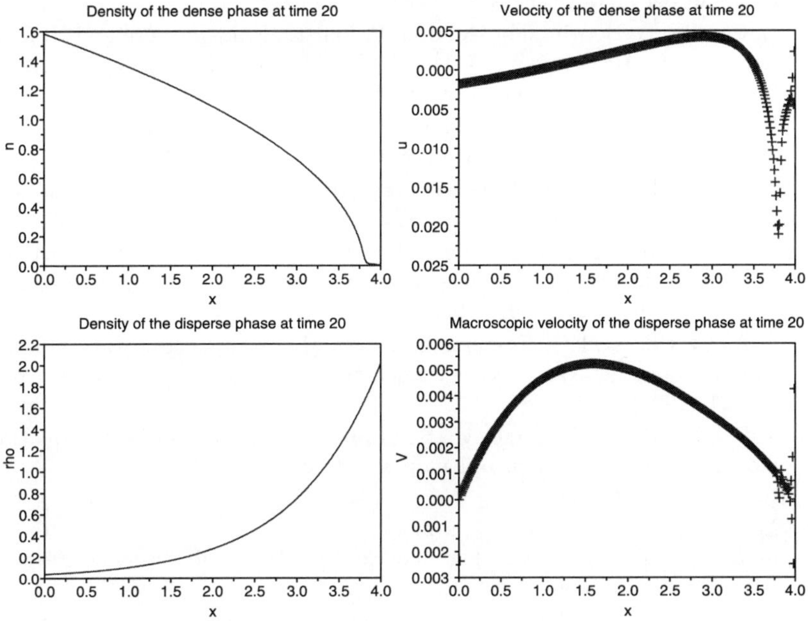

Figure 3.12 $\gamma = 3$.

References

[1] A. A. Amsden, J. D. Ramshaw, P. J. O'Rourke and J. K. Dukowicz. KIVA: A computer program for two- and three-dimensional fluid flows with chemical reactions and fuel sprays. Technical Report, Los Alamos National Laborator (1985) LA-10245-MS.

[2] C. Baranger. Modelling of oscillations, breakup and collisions for droplets: the establishment of kernels for the T.A.B. model. Math. Models Methods Appl. Sci. 14, 5 (2004) 775–794.

[3] C. Baranger and L. Desvillettes. Coupling Euler and Vlasov equations in the context of sprays: local smooth solutions. Journal of Hyperbolic Differential Equations, 3 (2006) no. 1, 1–26.

[4] C. Baranger, G. Baudin, L. Boudin, B. Després, F. Lagoutière, E. Lapébie and T. Takahashi. Liquid jet generation and break-up. In Numerical Methods for Hyperbolic and Kinetic Equations, S. Cordier, Th. Goudon, M. Gutnic, E. Sonnendrucker Eds., IRMA Lectures in Mathematics and Theoretical Physics, vol. 7, EMS Publ. House, 2005.

[5] C. Baranger, L. Boudin, P.-E. Jabin and S. Mancini. A modeling of biospray for the upper airways. ESAIM:Proc, 14 (2005) 41–47.

[6] L. Boudin, L. Desvillettes, C. Grandmont and A. Moussa. Global existence of solutions for the coupled Vlaosov and Noavier-Stokes equations. Preprint 2008.

[7] L. Boudin and L. Weynans. Spray impingement on a wall in the context of the upper airways. ESAIM: Proc, 23 (2008) 1–9.

[8] L. Boudin, B. Boutin, Th. Goudon, P. Lafitte, F. Lagoutière and B. Merlet. Fluid-Particles Flows: A Thin Spray Model with Energy Exchanges. To appear in Special issue of ESAIM:Proc., Cemracs 2008.

[9] R. Caflisch and G. Papanicolaou. Dynamic theory of suspensions with Brownian effects. SIAM J. Appl. Math., 43 (1983) 885–906.

[10] J.-A. Carrillo and Th. Goudon. Stability and asymptotics analysis of a fluid-particles interaction model. Comm. PDE, 31 (2006) 1349–1379.

[11] J.A. Carrillo, Th. Goudon and P. Lafitte, Simulation of fluid & particles flows: Asymptotic preserving schemes for bubbling and flowing regimes. J. Comp. Phys. 227 (2008) 7929–7951.

[12] J.-A. Carrillo and G. Toscani. Exponential convergence toward equilibrium for homogeneous Fokker-Planck-type equations. Math. Meth. Appl. Sci. 21 (1998) 1269–1286.

[13] J.-F. Clouet and K. Domelevo, Solutions of a kinetic stochastic equation modeling a spray in a turbulent gas flow. Math. Models Methods Appl. Sci., 7 (1997) 239–263.

[14] M. De Luca. Contribution à la modélisation de la pulvérisation d'un liquide phytosanitaire en vue de réduire les pollutions. Ph Thesis, University Aix-Marseille 2, 2007.

[15] B. Després and F. Lagoutière. Contact discontinuity capturing schemes for linear advection and compressible gas dynamics. J. Sci. Comput. 16, 4 (2002) 479–524.

[16] B. Després and F. Lagoutière. Generalized Harten formalism and longitudinal variation diminishing schemes for linear advection on arbitrary grids. M2AN Math. Model. Numer. Anal. 35, 6 (2001) 1159–1183.

[17] K. Domelevo and P. Villedieu, A hierarchy of models for turbulent disperse two-phase flows derived from a kinetic equation for the joint particle-gas. Comm. Math. Sci., 5 (2007) 331–353.

[18] T. Elperin, N. Kleeorin, I. Rogachevskii and D. Sokoloff. Turbulent transport of atmospheric aerosols and formation of large-scale structures. Phys Chem Earth A25 (2000) 797–803.

[19] S. Gavrilyuck and V. Teshukhov. Kinetic model for the motion of compressible bubbles in a perfect fluid. Eur. J. Mech. B/Fluids, 21 (2002) 469–491.

[20] M. Golomboka and D. B. Pyea. Droplet dynamics in a diesel injection spray. Fuel, 69, 9 (1990) 1200–1202.

[21] Th. Goudon, P.-E. Jabin and A. Vasseur. Hydrodynamic limit for the Vlasov-Navier-Stokes equations. I. Light particles regime. Indiana Univ. Math. J., 53 (2004) no. 6, 1495–1515.

[22] Th. Goudon, P.-E. Jabin and A. Vasseur. Hydrodynamic limit for the Vlasov-Navier-Stokes equations. II. Fine particles regime. Indiana Univ. Math. J., 53 (2004) no. 6, 1517–1536.

[23] Th. Goudon and P. Lafitte. Splitting schemes for the simulation of non equilibrium radiative flows. Preprint, 2007.

[24] Th. Goudon and F. Poupaud. On the modeling of the transport of particles in turbulent flows. M2AN Math. Model. Numer. Anal., 38 (2004) 673–690.

[25] Th. Goudon and M. Rousset. Stochastic acceleration in an inhomogenous time random force field. Preprint 2008.

[26] Th. Goudon, L. He, A. Moussa and P. Zhang. The Navier-Stokes-Vlasov-Fokker-Planck system near equilibrium. Preprint 2009.

[27] K. Hamdache. Global existence and large time behaviour of solutions for the Vlasov-Stokes equations. Japan J. Indust. Appl. Math. 15 (1998) 51–74.

[28] H. Herrero, B. Lucquin-Desreux and B. Perthame. On the motion of dispersed balls in a potential flow: a kinetic description of the added mass effect. SIAM J. Appl. Math., 60 (1999) 61–83.

[29] P.-E. Jabin. Large time concentrations for solutions to kinetic equations with energy dissipation. Comm. PDE., 25 (2000) 541–557.

[30] P.-E. Jabin. Macroscopic limit of Vlasov type equations with friction. Ann. IHP Anal. Non Linéaire, 17 (2000) 651–672.

[31] P.-E. Jabin and B. Perthame. Notes on mathematical problems on the dynamics of dispersed particles interacting through a fluid. In Modeling in applied sciences, a kinetic theory approach, N. Bellomo, M. Pulvirenti Eds., Birkhäuser (2000), 111–147.

[32] H. Kesten and G. C. Papanicolaou. A limit theorem for turbulent diffusion. Commun. Math. Phys. 65 (1979) 97–128.

[33] F. Lagoutière. A non-dissipative entropic scheme for convex scalar equations via discontinuous cell-reconstruction. C. R. Math. Acad. Sci. Paris 338, 7 (2004) 549–554.

[34] J. Mathiaud. Etude de systèmes de type gaz-particules. PhD Thesis ENS Cachan, Sept. 2006.

[35] J. C. Mattingly, A. M. Stuart and D. J. Higham. Ergodicity for SDEs and approximations: locally Lipschitz vector fields and degenerate noise. Stochastic Process. Appl. 101 (2002) 185–232.

[36] A. Mellet and A. Vasseur. Global weak solutions for a Vlasov-Fokker-Planck/Navier-Stokes system of equations. Math. Mod. Meth. Appl. Sci 17 (2007) no. 7, 1039–1063.

[37] A. Mellet and A. Vasseur. Asymptotic analysis for a Vlasov-Fokker-Planck/Compressible Navier-Stokes system of equations. To appear in Comm. Math. Phys.

[38] F. Poupaud and A. Vasseur, Classical and quantum transport in random media, J. Math. Pures et Appl., 82 (2003) 711–748.

[39] I. Vinkovic. Dispersion et mélange turbulents de particules solides et de gouttelettes par une simulation des grandes échelles et une modélisation stochastique lagrangienne. Application à la pollution de l'atmosphère. PhD thesis, Ecole Centrale de Lyon, 2005.

[40] F. A. Williams. Combustion theory (2nd ed.). Benjamin Cummings Publ., 1985.

[41] L. I. Zaichik. A statistical model of particle transport and heat transfer in turbulent shear flows. Phys. Fluids, 11 (1999) 1521–1534.

Well-Posedness and Stability of Quantum Hydrodynamics for Semiconductors in $\mathbb{R}^3$*

Feimin Huang
Institute of Applied Mathematics, AMSS, Academia Sinica
Beijing 100190, China
Email: fhuang@amt.ac.cn

Hailiang Li
Department of Mathematics, Capital Normal University
Beijing 100037, China
Email: hailiang.li.math@gmail.com

Akitaka Matsumura
Graduate School of Information Science and Technology
Osaka University, Toyonaka 560-0043, Japan
Email: akitaka@math.sci.osaka-u.ac.jp

Shinji Odanaka
Cybermedia Center, Computer Assisted Science Division
Osaka University, Toyonaka 560-0043, Japan
Email: odanaka@math.sci.osaka-u.ac.jp

Abstract

The well-posedness and stability of multi-dimensional quantum hydrodynamic equations for charge transport in ultra-small electronic devices like semiconductors where quantum effects (like particle tunnelling through potential barriers and built-up in quantum wells which can not be simulated by classical hydrodynamic models) take place are considered in $\mathbb{R}^3$. The local existence and uniqueness of classical solutions subject to general regular initial data are proven in terms of an extended system with which the original problem under investigation is consistent as a special case. Particularly, the nonlinear dispersive term appears mainly in the

*F.H. is supported in part by NSFC Grant No. 10471138, NSFC-NSAF Grant No. 10676037 and 973 Project of China, Grant No. 2006CB805902. H.L. is supported partially by the NNSFC No.10431060, the Beijing Nova Program 2005B48, the NCET support of the Ministry of Education of China, and the Huo Ying Dong Foundation No.111033. A.M. is supported by the Grants-in-Aid of JSPS for Scientific Research No.15340043.

form of a fourth-order wave type operator. Then, we establish the existence, uniqueness and exponential stability of steady-state under a stability condition viewed as a quantized version of classical subsonic condition.

1 Introduction and main results

In the modelling of semiconductor devices in nano-size, for instance, HEMT's, MOSFET's and RTD's where quantum effects (like particle tunnelling through potential barriers and built-up in quantum wells [10, 14, 32] which can not be simulated by classical hydrodynamic models) take place, the quantum hydrodynamical equations are important and dominative in the description of the motion of electron or hole transport in the self-consistent electric field. The basic observation regarding quantum hydrodynamics lies on that the energy density consists of additional new quantum correction term of the order $O(\hbar)$, the planck constant, introduced first by Wigner [49] in 1932, and that the stress tensor contains also additional quantum correction part [2, 3] related to the quantum Bohm potential [5]

$$Q(\rho) = -\frac{\hbar^2}{2m} \frac{\Delta\sqrt{\rho}}{\sqrt{\rho}}, \tag{1.1}$$

with observable $\rho > 0$ the density, m the charge mass, and $\hbar$ the Planck constant. It is no wonder, however, since the original idea initialized by Madelung [39] in 1927 to derive quantum fluid-type equations has already described such possible relation. To have an intuition let us just consider the motion of an electron in a potential field described by the linear Schrödinger equation

$$i\hbar\partial_t\phi = -\frac{\hbar^2}{2m}\Delta\phi - qU(x)\phi$$

with Δ the Laplacian operator on $\mathbb{R}^3$, q the electron charge, and $U(x)$, $x \in \mathbb{R}^3$ the given potential field. To understand the dynamics of physical observables like density ρ, momentum J and so on, we apply the Madelung's transformation $\phi = \sqrt{\rho}e^{iS/\hbar}$ of the wave function to the above linear Schrödinger equation to obtain the quantum fluid equations for potential flow away from vacuum

$$\partial_t\rho + \frac{1}{m}\nabla\cdot(\rho\nabla S) = 0, \tag{1.2}$$

$$\partial_t S + \frac{1}{2m}|\nabla S|^2 - qU(x) - \frac{\hbar^2}{2m}\frac{\Delta\sqrt{\rho}}{\sqrt{\rho}} = 0. \tag{1.3}$$

Based on such an idea, one is able to derive quantum fluid type equations from the pure-state (nonlinear) Schrödinger equation [19, 26, 18].

A practicable and rigorous approach to derive quantum hydrodynamic equations for semiconductor device at nano-size is based on the moment method applied to Wigner-Boltzmann (or quantum Liouville) equation. The kinetic structure behind the Schrödinger Hamiltonian was justified through Wigner transformation [49]. In fact, the action of Wigner transformation on the wave function of the Schrödinger equation gives rise to a quantum Liouville equation, the Wigner-Boltzmann equation [42, 44]. Start with the Wigner-Boltzmann equation

$$W_t + \xi\cdot\nabla_x W + \frac{q}{m}\mathbb{P}[V]W = [W_t]_c \tag{1.4}$$

where $W = W(x,\xi,t), (x,\xi,t) \in \mathbb{R}^3 \times \mathbb{R}^3 \times \mathbb{R}_+$, and $\mathbb{P}$ denotes the pseudo-differential operator defined by

$$\mathbb{P}[V]W = \frac{im}{(2\pi)^N}\iint \frac{V(x+\frac{\hbar}{2m}\eta) - V(x-\frac{\hbar}{2m}\eta)}{\hbar} e^{i\eta\cdot(\xi-\xi')}W(x,\xi',t)d\eta d\xi'.$$

The electrostatic potential $V = V(x,t)$ is self-consistent through Poisson equation

$$\lambda_0\,\Delta V = q\left(\int W d\xi - \mathcal{C}\right)$$

with $\lambda_0 > 0$ the characteristic of permitivity and $\mathcal{C} = \mathcal{C}(x) > 0$ the given doping profile, and $[W_t]_c$ refers to the quantum collision operator [44]. Applying moment method to the above Wigner-Boltzmann equation (1.4) near the "momentum-shifted quantum Maxwellian" [49] together with appropriate closure assumption [14, 20] and modified Baccarani-Wordeman type relaxation time approximation of electron scattering effects [1, 14, 43], we can obtain the *quantum hydrodynamic equations* [14, 26]. For more derivation from and references to the modelling of quantum hydrodynamical or quantum moment models, one refers to [44, 19, 15, 8, 9] and references therein.

The advantage of the macroscopic quantum hydrodynamical models is such that they are able to describe directly the dynamic evolution of physical observables and is convenient to simulate quantum phenomena. Moreover, in the semiclassical (or the zero dispersion) limit the macroscopic quantum quantities, like density, momentum, and temperature, are shown to converge in one sense to the of Newtonian fluid-dynamical quantities [18]. Similar macroscopic quantum models are also used in other physical areas such as superfluid [38] and superconductivity [11].

In the present paper, we consider the initial value problem (IVP) of the *quantum hydrodynamic* model in $\mathbb{R}^3$ (QHD) under re-scaling in $\mathbb{R}^3$

as follows:

$$\partial_t \rho + \nabla \cdot (\rho \mathbf{u}) = 0, \tag{1.5}$$

$$\partial_t(\rho \mathbf{u}) + \nabla \cdot (\rho \mathbf{u} \otimes \mathbf{u}) + \nabla P = \rho E + \frac{1}{2}\varepsilon^2 \rho \nabla \left(\frac{\Delta \sqrt{\rho}}{\sqrt{\rho}} \right) - \frac{\rho \mathbf{u}}{\tau}, \tag{1.6}$$

$$\lambda^2 \nabla \cdot E = \rho - \mathcal{C}(x), \ \nabla \times E = 0, \quad E(x) \to 0, \quad |x| \to +\infty, \tag{1.7}$$

$$(\rho, \mathbf{u})(x, 0) = (\rho_1, \mathbf{u}_1)(x), \quad (\rho_1, \mathbf{u}_1)(x) \to (\bar{\rho}, 0), \quad |x| \to +\infty, \tag{1.8}$$

where $\rho > 0$ is a constant, $\varepsilon > 0$ the scaled Planck constant, $\tau > 0$ the scaled momentum relaxation time, $\lambda > 0$ the scaled Debye length, and $P = P(\rho)$ the pressure-density function. The electric field $E = -\nabla V$ is a gradient vector field of the electrostatic potential V. Note here that the nonlinear dispersive term $Q_2 = \frac{1}{2}\varepsilon^2 \rho \nabla \left(\frac{\Delta \sqrt{\rho}}{\sqrt{\rho}} \right)$ requires the strict positivity of density for classical solution.

Recently, important progress has been made on the analysis about the QHD (1.5)–(1.7). The existence and uniqueness of (classical) steady-state solutions to the QHD (1.5)–(1.7) for current density $J = 0$ (thermal equilibrium) were studied in one and high dimension bounded domain with density and electrostatic potential imposed at boundary [4, 17, 48], the existence of unique stationary solutions of QHD (1.5)–(1.7) for $J_0 > 0$ (non-thermal equilibrium) was proven in [13, 21, 25, 50] for monotonous increasing pressure functions, and in [27] for general pressure functions $P(\rho)$, and the asymptotic properties of stationary or dynamical solutions with respect to small physical parameter like Planck constant, relaxation time, and Debye length were investigated [12, 21, 34, 35]. For dynamical system, an analysis of special travelling wave solution was also made [16], the local and global in-time existence of classical solution was obtained in one-dimensional bounded domain [28, 22, 23] subject to different boundary conditions and in real line [24], in multi-dimensional bounded domain in terms of Schroödinger-Poisson type description via Schrödinger semigroup for potential flow [31] and multi-dimensional torus $\mathbb{T}^n$ for ir-rotational (potential) flow [36]. However, the well-posedness theory of multi-dimensional QHD for general rotational initial data is still not known yet. The readers can refer to one review paper [29] and above-mentioned papers for more references.

The aim of the present paper is to investigate the well-posedness theory of classical solutions of the QHD (1.5)–(1.7) in $\mathbb{R}^3$ for general regular *large* (*rotational*) initial data, and the existence, uniqueness and stability of steady-state. The mathematical analysis on these topics is absolutely nontrivial due to the strong coupling of nonlinear dispersion and Euler-Poisson. The quantum systems (1.5)–(1.7) can be (for smooth solutions) re-written mathematically as the same form as the classical hydrodynamic system for hot electron transport

$$\partial_t \rho + \nabla\cdot(\rho \mathbf{u}) = 0, \tag{1.9}$$

$$\partial_t(\rho\mathbf{u}) + \nabla\cdot(\rho\mathbf{u}\otimes\mathbf{u} + \bar{\mathbf{T}}) = \rho E - \frac{\rho\mathbf{u}}{\tau}, \tag{1.10}$$

$$\lambda^2 \nabla\cdot E = \rho - \mathcal{C}, \tag{1.11}$$

with stress tensor $\bar{\mathbf{T}} = P(\rho)I_{n\times n}$ since for quantum hydrodynamic systems (1.5)–(1.7) the quantized stress tensor $\mathbf{P}$ of classical one $\bar{\mathbf{T}}$ satisfies

$$\nabla\cdot\mathbf{P} = \nabla P - \frac{1}{2}\varepsilon^2\rho\nabla\Big(\frac{\Delta\sqrt{\rho}}{\sqrt{\rho}}\Big), \quad \mathbf{P} = P(\rho)I_{n\times n} - \frac{1}{4}\varepsilon^2\rho\nabla\otimes\nabla\log\rho.$$

However, the standard Kato-Lax-Friedrich's theory of quasilinear symmetric hyperbolic system [33, 40], which works for hydrodynamic systems (1.9)–(1.11) about well-posedness theory in Sobolev space for arbitrarily large initial data, does not apply here for quantum hydrodynamic systems (1.5)–(1.7) due to the influence of nonlinear dispersion operator which as we know is not symmetrizable for general charge fluid and requires positivity and enough regularity of charge density for classical solutions.

It is convenient to make use of the variable transformation $\rho = \psi^2$ in (1.5)–(1.8). Then, we derive the corresponding IVP for $(\psi, \mathbf{u}, E)$:

$$2\psi\cdot\partial_t\psi + \nabla\cdot(\psi^2\mathbf{u}) = 0, \tag{1.12}$$

$$\partial_t\mathbf{u} + (\mathbf{u}\cdot\nabla)\mathbf{u} + \nabla h(\psi^2) + \frac{\mathbf{u}}{\tau} = E + \frac{\varepsilon^2}{2}\nabla\Big(\frac{\Delta\psi}{\psi}\Big), \tag{1.13}$$

$$\nabla\cdot E = \psi^2 - \mathcal{C}, \quad \nabla\times E = 0, \ E(x)\to 0, \ |x|\to\infty, \tag{1.14}$$

$$\psi(x,0) = \psi_1(x) := \sqrt{\rho_1(x)}, \ \mathbf{u}(x,0) = \mathbf{u}_1(x), \tag{1.15}$$

with $\rho h'(\rho) = P'(\rho)$, and $(\psi_1, \mathbf{u}_1)(x) \to (\sqrt{\bar{\rho}}, 0)$ as $|x|\to\infty$. Note here that the two problems (1.5)–(1.8) and (1.12)–(1.15) are equivalent for classical solutions. For convenience in the rest part of the paper we just consider the initial value problem (1.12)–(1.15) on $\mathbb{R}^3$ for general (quantum) flow, and show how to prove both the local and global existence and investigate the long-time behavior of classical solutions.

First of all, based on sharp observations about the quantum hydrodynamic system (1.5)–(1.7) to be explained later, we are able to prove the local existence and uniqueness theorem:

Theorem 1.1. *Suppose that $P(\rho)\in C^5(0,+\infty)$. Assume $(\psi_1 - \sqrt{\bar{\rho}}, \mathbf{u}_1) \in H^6(\mathbb{R}^3)\times\mathcal{H}^5(\mathbb{R}^3)$ satisfying $\psi_* =: \inf_{x\in\mathbb{R}^3}\psi_1(x) > 0$. Then, there is a short time $T_{**} > 0$ such that the unique solution $(\psi, \mathbf{u}, E)$ with $\psi > 0$ of the IVP (1.12)–(1.15) exists for $t\in[0, T_{**}]$ and satisfies*

$$\psi - \sqrt{\bar{\rho}} \in C^i([0,T_{**}]; H^{6-2i}(\mathbb{R}^3))\bigcap C^3([0,T_{**}]; L^2(\mathbb{R}^3)), \quad i = 0,1,2;$$

$$\mathbf{u} \in C^i([0, T_{**}]; \mathcal{H}^{5-2i}(\mathbb{R}^3)), \ i = 0, 1, 2; \quad E \in C^1([0, T_{**}]; \mathcal{H}^3(\mathbb{R}^3)).$$

Hereafter $\mathcal{H}^k(\mathbb{R}^3) = \{f \in L^6(\mathbb{R}^3), \ Df \in H^{k-1}(\mathbb{R}^3)\}, \ k \geq 1.$

Remark 1.2. Though the above local existence result is proven when the "boundary" value of density (or doping profile) at spatial infinity is a positive constant $\bar{\rho}$, we claim that it is applicable to more general case with some modification on the proof.

The proof of Theorem 1.1 for the quantum hydrodynamics (1.12)–(1.14) is absolutely nontrivial due to a strong coupling between hyperbolicity, ellipticity, and nonlinear dispersion in particular. The nonlinear dispersion term is a nonlinear third-order differential operator with respect to space variables. It requires strict positivity and higher order regularity of density for time-dependent classical solutions. However, there is no maximum principle applicable to obtaining the a-priori bounds on density, and it is not obvious how to keep higher order regularity for density directly from Eqs. (1.12)–(1.14). We have to establish the short time existence of classical solutions through another way. Fortunately, the information behind the conservation law of mass and the balance law of momentum helps us finally to overcome the above difficulties. In fact, the conservation law of mass describes in one sense the transport property of charge particles and implies at least for short time the positivity of density. Combining the balance law of momentum with conservation law of mass, one can observe dispersive effect induced from the nonlinear dispersion operator — a fourth-order wave operator acting on density as main part, this can help to keep the same regularity as the initial data of density as expected [28, 36]. These sharp observations are, however, far from the real program to prove Theorem 1.1. In fact, inspired by the above ideas, we still need to introduce new unknown variables and construct an extended system, derived based on Eqs. (1.12)–(1.14), for $(\mathbf{v}, \mathbf{z}, \varphi, \psi, \mathbf{u}, E)$ where the dispersive term appears in the form of a fourth-order wave type operator, and prove the short time existence of classical solutions $(\mathbf{v}, \mathbf{z}, \varphi, \psi, \mathbf{u}, E)$ of the initial value problems for this extended system. The key point is that the local in-time classical solutions $(\mathbf{v}, \mathbf{z}, \varphi, \psi, \mathbf{u}, E)$ of the extended system shall be equivalent to these of IVP (1.12)–(1.15) for classical solution so long as $\mathbf{u} = \mathbf{v} + \mathbf{z}$ and $\psi = \varphi$ initially, which is different from [28, 36] where the solution of original system is located on the invariant sub-manifolds of extended system in phase space and makes the construction much more difficult. The expected positivity and higher order regularity of density ψ follow (see Section 3 for details).

With the help of local existence theory, we can investigate the existence, uniqueness and stability of steady-state $(\bar{\psi}, \bar{\mathbf{u}}, \bar{E})$ for the quantum hydrodynamic equations (1.5)–(1.8). To this end, we first need to establish the well-posedness of stationary state $(\bar{\psi}, \bar{\mathbf{u}}, \bar{E})$ of Eqs. (1.12)–(1.14) in $\mathbb{R}^3$. As a starting point we consider a special stationary state $(\bar{\psi}, \bar{\mathbf{u}}, \bar{E}) = (\widetilde{\psi}, 0, \widetilde{E})$ with $\widetilde{\psi}$ a small perturbation of the state $\sqrt{\bar{\rho}}$.

Theorem 1.3. *Suppose that $P(\rho) \in C^4(0, +\infty)$. Let $\bar{\rho}$ be a positive constant and*

$$\mathcal{C}(x) = \bar{\rho} + f(x), \quad f \in H^2(\mathbb{R}^3), \quad \delta_0 = \|f\|_{H^2}. \tag{1.16}$$

Assume

$$\varepsilon\sqrt{\bar{\rho}} + P'(\bar{\rho}) > 0. \tag{1.17}$$

Then there exists a constant $m_0 > 0$ such that for any $\delta_0 \leq m_0$, there exists a unique steady solution $(\widetilde{\psi}, 0, \widetilde{E})$, with $\inf_{x\in\mathbb{R}^3} \psi(x) > 0$, of (1.12)–(1.14) satisfying

$$\|\widetilde{\psi} - \sqrt{\bar{\rho}}\|_{H^4} + \|\widetilde{E}\|_{H^3} \leq C_0\delta_0, \tag{1.18}$$

where $C_0 = C_0(\bar{\rho}, \varepsilon) > 0$ is a constant independent of δ_0.

The proof can be completed in terms of the a-priori estimates and the fixed point theorem, as used in [24], we omit the details.

Remark 1.4. (1) The condition (1.17) can be viewed as a quantum correction of the subsonic condition for classical fluids [6, 7] in the sense that it is (formally) equivalent to the subsonic condition as re-scaled Planck constant ε goes to zero. When $\varepsilon > 0$ and $P'(\rho) > 0$, the "sound" speed $\tilde{c}(\bar{\rho}) = \sqrt{\varepsilon\sqrt{\bar{\rho}} + P'(\bar{\rho})}$ is bigger than the sound speed $c(\bar{\rho}) = \sqrt{P'(\bar{\rho})}$ for classical fluids. Moreover, for a general steady-state $(\widetilde{\psi}, \widetilde{\mathbf{u}}, \widetilde{E})$ of BVP (1.12)–(1.14) to exist and be stable in multi-dimension, the *quantum subsonic* condition is

$$\inf_{x\in\mathbb{R}^3}(\varepsilon\widetilde{\psi} + P'(\widetilde{\psi}^2) - |\widetilde{\mathbf{u}}|^2) > 0. \tag{1.19}$$

Here we should mention that the condition (1.19) generalizes the one proposed before by authors in [36] which is related to the spatial structure of the semiconductor device under consideration, while the *quantum subsonic* condition (1.19) is generic and independent of the geometric structure of the domain.

(2) It is known [37] that classical solutions of hydrodynamical model for semiconductors (i.e., (1.5)–(1.7) with $\varepsilon = 0$) may only exist in subsonic regime [6, 7]. When dispersive regularity is involved in (1.17), however, the classical (strong) solutions of (1.12)–(1.15) exist even in the transonic or supersonic regime so long as the stability (1.17) holds. Note that the condition (1.19) also admits the existence of strong solution for the case $\inf_{x\in\mathbb{R}^3}(\varepsilon\widetilde{\psi} + P'(\widetilde{\psi}^2) - |\widetilde{\mathbf{u}}|^2) > 0 > \sup_{x\in\mathbb{R}^3}(P'(\widetilde{\psi}^2) - |\widetilde{\mathbf{u}}|^2)$, i.e.,

the existence of strong solution in transonic regime with the thickness of order $O(\varepsilon)$. For more consideration on the influence of quantum correction to the existence of solutions, one can refer to [21, 24, 30] for further analysis.

Finally, we obtain the global-in-time existence of classical solutions and exponential stability of steady-state.

Theorem 1.5. *Assume that* (1.16) *holds. Let* $(\widetilde{\psi}, 0, \widetilde{E})$ *the steady state solution obtained in Theorem 1.3. Suppose that* $P(\rho) \in C^5(0,+\infty)$ *with* (1.17) *satisfied. Assume* $(\psi_1 - \widetilde{\psi}, \mathbf{u}_1) \in H^6(\mathbb{R}^3) \times \mathcal{H}^5(\mathbb{R}^3)$. *Then, there is* $m_1 > 0$ *such that if* $\|\psi_1 - \widetilde{\psi}\|_{H^6(\mathbb{R}^3)} + \|\mathbf{u}_1\|_{\mathcal{H}^5(\mathbb{R}^3)} + \delta_0 \le m_1$, *solution* $(\psi, \mathbf{u}, E)$ *of the IVP* (1.12)–(1.15) *exists globally in time and satisfies*

$$\|(\psi - \widetilde{\psi})(t)\|_{H^6(\mathbb{R}^3)} + \|\mathbf{u}(t)\|_{\mathcal{H}^5(\mathbb{R}^3)} + \|(E - \widetilde{E})(t)\|_{\mathcal{H}^3(\mathbb{R}^3)} \le C_1 \delta_1 e^{-\Lambda_0 t}$$

for $t \ge 0$. *Here* $C_1 > 0$, $\Lambda_0 > 0$ *are constants and*

$$\delta_1 = \|\psi_1 - \widetilde{\psi}\|_{H^6(\mathbb{R}^3)} + \|D\mathbf{u}_1\|_{H^4(\mathbb{R}^3)}. \tag{1.20}$$

Remark 1.6. (1) Theorems 1.1–1.5 can be extended to multi-dimension $\mathbb{R}^N$, $N \ge 3$, for IVP (1.12)–(1.15) for smooth (large) initial data. The proof can be done within the same framework.

(2) Once we prove the local existence (resp. global existence) of solutions $(\psi, \mathbf{u}, E)$ of IVP (1.12)–(1.15), we can obtain the local existence (resp. global existence) of solutions $(\rho, \mathbf{u}, E)$ of IVP (1.5)–(1.8) by setting $\rho = \psi^2$.

This paper is arranged as follows. In Section 2, we present necessary results on divergence equation, vorticity equation, and a semilinear fourth-order wave equation on $\mathbb{R}^3$; then list some known calculus inequalities. In Section 3, we prove Theorem 1.1. After the construction of an extended system to be dealt with in Section 3.1, we make the approximate solution series, derive uniform estimates, and prove the Theorem 1.1 in Section 3.2. Section 4 is devoted to the proof of Theorem 1.5. After the reformulation of original problem in Section 4.1, we establish the a-priori estimates on the local solutions in Section 4.2, and prove their global existence and large time behavior finally.

Notation. C always denotes generic positive constant. $L^p(\mathbb{R}^3)$, $1 \le p \le \infty$, is the space of p-powers integral functions on $\mathbb{R}^3$ with the norm $\|\cdot\|_{L^p}$. Particularly, the norm of the space of square integral functions on $\mathbb{R}^3$ is denoted by $\|\cdot\|$. $H^k(\mathbb{R}^3)$ with integer $k \ge 1$ denotes the usual Sobolev space of function f, satisfying $\partial_x^i f \in L^2$ $(0 \le i \le k)$, with norm

$$\|f\|_k = \sqrt{\sum_{0 \le |l| \le m} \|D^l f\|^2},$$

here and after $D^\alpha = \partial_1^{\alpha_1}\partial_2^{\alpha_2}\cdots\partial_n^{\alpha_n}$ for $|\alpha| = \alpha_1 + \alpha_2 + \cdots + \alpha_n$ and $\partial_j = \partial_{x_j}$, $j = 1, 2, ..., n$, for abbreviation. Especially, $\|\cdot\|_0 = \|\cdot\|$. Moreover, $W^{k,p}$, with $k \geq 1, p \geq 1$, denotes the space of functions with $D^l f \in L^p$, $0 \leq |l| \leq k$, and $\mathcal{H}^k(\mathbb{R}^3)$ denotes the subspace of $L^6(\mathbb{R}^3)$ with $Df \in H^{k-1}(\mathbb{R}^3)$. Let $T > 0$ and $\mathcal{B}$ be a Banach space. $C^k(0, T; \mathcal{B})$ ($C^k([0, T]; \mathcal{B})$ resp.) denotes the space of $\mathcal{B}$-valued k-times continuously differentiable functions on $(0, T)$ (or $[0, T]$ resp.), $L^2([0, T]; \mathcal{B})$ the space of $\mathcal{B}$-valued L^2-functions on $[0, T]$, and $H^k([0, T]; \mathcal{B})$ the spaces of $f(x, t)$ with $\partial_t^i f \in L^2([0, T]; \mathcal{B})$, $1 \leq i \leq k$, $1 \leq p \leq \infty$.

2 Preliminaries

In this section, we list the existence and uniqueness of solutions of divergence equation and vorticity equation in $\mathbb{R}^3$ without proof, mention the orthogonal decomposition of velocity vector field, and then turn to prove the well-posedness for an abstract second-order semi-linear wave equation. Finally, some useful calculus inequalities are listed.

First, we have the theorem on the divergence operator and vorticity operator on $\mathbb{R}^3$:

Theorem 2.1. *Let $f \in H^s(\mathbb{R}^3)$, $s \geq 3/2$. There is a unique solution u of the divergence equation*

$$\nabla\cdot\mathbf{u} = f, \quad \nabla\times\mathbf{u} = 0, \quad \mathbf{u}(x) \to 0, \quad |x| \to \infty, \tag{2.1}$$

satisfying

$$\|\mathbf{u}\|_{L^6(\mathbb{R}^3)} \leq C\|f\|_{L^2(\mathbb{R}^3)}, \quad \|D\mathbf{u}\|_{H^s(\mathbb{R}^3)} \leq C\|f\|_{H^s(\mathbb{R}^3)}. \tag{2.2}$$

Theorem 2.2. *Let $f \in H^s(\mathbb{R}^3)$, $s \geq 3/2$. There is a unique solution u of the vorticity equation*

$$\nabla\times\mathbf{u} = f, \quad \nabla\cdot\mathbf{u} = 0, \quad \mathbf{u}(x) \to 0, \quad |x| \to \infty, \tag{2.3}$$

satisfying

$$\|\mathbf{u}\|_{L^6(\mathbb{R}^3)} \leq C\|f\|_{L^2(\mathbb{R}^3)}, \quad \|D\mathbf{u}\|_{H^s(\mathbb{R}^3)} \leq C\|f\|_{H^s(\mathbb{R}^3)}. \tag{2.4}$$

Theorem 2.3. *For $\mathbf{u} \in \mathcal{H}^s(\mathbb{R}^3)$, $s \geq 3/2$, it has a unique decomposition consisting of the gradient vector field $\mathbf{v} \in \mathcal{H}^s(\mathbb{R}^3)$ and the divergence free vector field $\mathbf{z} \in \mathcal{H}^s(\mathbb{R}^3)$, i.e.,*

$$\mathbf{u} = \mathbf{v} + \mathbf{z} = \mathcal{Q}\mathbf{u} + \mathcal{P}\mathbf{u}, \quad \mathcal{Q} = I - \mathcal{P}, \ \nabla\cdot\mathcal{P} = 0.$$

Proof. The proof of the above theorems can be made by using the standard arguments and the Riesz's potential theory. The reader can refer to [41, 47], we omit the details here. □

Based on Theorem 2.1, we obtain the initial electric field $E(x,0) = E_1(x)$ through (1.14) in view of initial density:

$$\nabla\cdot E_1 = \psi_1^2 - \mathcal{C}, \quad \nabla\times E_1 = 0, \quad E_1(x) \to 0, \quad |x| \to \infty. \tag{2.5}$$

By $\psi_1 - \sqrt{\bar\rho} \in H^6(\mathbb{R}^3)$ and $\mathcal{C} - \bar\rho \in H^3(\mathbb{R}^3)$, we obtain $E_1 \in \mathcal{H}^3(\mathbb{R}^3)$, satisfying

$$\|E_1\|_{\mathcal{H}^4(\mathbb{R}^3)} \le C\|\psi_1 - \sqrt{\bar\rho}\|_{H^3(\mathbb{R}^3)} + C\|\mathcal{C} - \bar\rho\|_{H^3(\mathbb{R}^3)}. \tag{2.6}$$

Finally, let us turn to consider an abstract initial value problem in Hilbert space $L^2(\mathbb{R}^3)$:

$$u'' + \frac{1}{\tau}u' + Au + \mathcal{L}u' = F(t), \tag{2.7}$$

$$u(0) = u_0, \quad u'(0) = u_1. \tag{2.8}$$

where u' denotes $\frac{du}{dt}$, and the operator A is given by

$$Au = \nu_0\Delta^2 u + \nu_1 u, \tag{2.9}$$

with Δ the Laplace operator on $\mathbb{R}^3$, and $\nu_0, \nu_1 > 0$ constants. The domain of linear operator A is $D(A) = H^4(\mathbb{R}^3)$. Related to the operator A, we define a continuous and symmetric bilinear form $a(u,v)$ on $H^2(\mathbb{R}^3)$

$$a(u,v) = \int_{\mathbb{R}^3} (\nu_0 \Delta u \Delta v + \nu_1 uv)dx, \quad \forall\, u,v \in H^2(\mathbb{R}^3), \tag{2.10}$$

which is coercive, i.e.,

$$\exists\, \nu > 0, \quad a(u,u) \ge \nu\|u\|_2, \quad \forall\, u \in H^2(\mathbb{R}^3). \tag{2.11}$$

Related to $\mathcal{L}u$ and $F(t)$, we have

$$< \mathcal{L}u, v >= \int_{\mathbb{R}^3} (b(x,t)\cdot\nabla u)v dx, \quad u,\ v \in H^2(\mathbb{R}^3), \tag{2.12}$$

$$< F(t), v >= \int_{\mathbb{R}^3} f(x,t) v dx, \quad v \in\ H^2(\mathbb{R}^3), \tag{2.13}$$

where $b : \mathbb{R}\times[0,T] \to \mathbb{R}^3$ and $f : \mathbb{R}\times[0,T] \to \mathbb{R}$ are measurable functions.

Note that the space $H^4(\mathbb{R}^3)$ is separable and has a complete basis $\{r_j\}_{j\ge1}$. Applying the Faedo-Galerkin method [36, 51], we can obtain the existence of solutions of (2.7)–(2.8).

Theorem 2.4. *Let $T > 0$ and assume that*

$$F \in C^1([0,T]; L^2(\mathbb{R}^3)), \quad b \in C^1([0,T]; \mathcal{H}^3(\mathbb{R}^3)). \tag{2.14}$$

Then, if $u_0 \in H^4(\mathbb{R}^3)$ and $u_1 \in H^2(\mathbb{R}^3)$, the solution of (2.7)–(2.8) *exists and satisfies*

$$u \in C^i([0,T]; H^{4-2j}(\mathbb{R}^3)) \bigcap C^2([0,T]; L^2(\mathbb{R}^3)), \ j = 0, 1. \tag{2.15}$$

Moreover, assume

$$F \in C^1([0,T]; H^2(\mathbb{R}^3)). \tag{2.16}$$

Then, if $u_0 \in H^6(\mathbb{R}^3)$ and $u_1 \in H^4(\mathbb{R}^3)$, it holds

$$u \in C^i([0,T]; H^{6-2j}(\mathbb{R}^3)) \bigcap C^3([0,T]; L^2(\mathbb{R}^3)), \ j = 0, 1, 2. \tag{2.17}$$

Proof. The (2.17) follows from (2.15) with some modification when we consider the similar problem for new variable $v = D^\alpha u$, $|\alpha| = 2$. The (2.15) can be proven by applying the Faedo-Galerkin method. We omit the details here. □

Finally, we list below the Moser-type calculus inequalities [33, 40]:

Lemma 2.5. 1). Let $f, g \in L^\infty \cap H^s(\mathbb{R}^3)$, $s \geq 3/2$. Then, it holds

$$\|D^\alpha(fg)\| \leq C\|g\|_{L^\infty}\|D^\alpha f\| + C\|f\|_{L^\infty}\|D^\alpha g\|, \tag{2.18}$$

$$\|D^\alpha(fg) - fD^\alpha g\| \leq C\|g\|_{L^\infty}\|D^\alpha f\| + C\|f\|_{L^\infty}\|D^{\alpha-1} g\|, \tag{2.19}$$

for $1 \leq |\alpha| \leq s$.

2). Let $\mathbf{u} \in \mathcal{H}^1(\mathbb{R}^3)$, then it holds

$$\|\mathbf{u}\|_{L^6} \leq C\|D\mathbf{u}\|. \tag{2.20}$$

3 Local-in-time existence

This section is concerned with the proof of Theorem 1.1. Instead, we shall prove the well-posedness for a new extended problem, derived based on (1.12)–(1.14), for $U = (\mathbf{v}, \mathbf{z}, \varphi, \psi, \mathbf{u}, E)$

$$\nabla\cdot \mathbf{v} = r(t), \quad \nabla\times\mathbf{v} = 0, \quad \mathbf{v}(x,t) \to 0, \ |x| \to \infty, \tag{3.1}$$

$$\begin{cases} \mathbf{z} = B_0 \int_{\mathbb{R}^3} |x-y|^{-3}(x-y) \times \omega(y,t)\, dy, \\ \omega' + \omega + (\mathbf{v}+\mathbf{z})\cdot\nabla\omega + \omega\nabla\cdot\mathbf{v} - (\omega\cdot\nabla)[\mathbf{v}+\mathbf{z}] = 0, \\ \omega(x,0) = \omega_1(x) =: \nabla\times\mathbf{u}_1(x), \end{cases} \tag{3.2}$$

$$\begin{cases} \varphi' + \frac{1}{2}(\nabla\cdot \mathbf{v})\varphi + \mathbf{u}\cdot\nabla\psi = 0, \\ \varphi(x,0) = \psi_1(x), \end{cases} \tag{3.3}$$

$$\begin{cases} \psi'' + \psi' + \nu\Delta^2\psi + \nu\psi + k(t)\cdot\nabla\psi' = h(t), \\ \psi(x,0) = \psi_1(x),\ \psi'(x,0) = \psi_0 =: -\frac{1}{2}\psi_1\nabla\cdot \mathbf{u}_1 - \mathbf{u}_1\cdot\nabla\psi_1, \end{cases} \tag{3.4}$$

$$\begin{cases} \mathbf{u}' + \mathbf{u} = g(t), \\ \mathbf{u}(x,0) = \mathbf{u}_1(x), \end{cases} \tag{3.5}$$

$$\nabla\cdot E = q =: \psi^2 - \mathcal{C}, \quad \nabla\times E = 0, \quad E(x,t) \to 0,\ |x| \to \infty \tag{3.6}$$

where $\nu = \frac{1}{4}\varepsilon^2$, and

$$r(t) = r(x,t) = -\frac{2(\psi' + \mathbf{u}\cdot\nabla\psi)}{\varphi}, \tag{3.7}$$

$$k(t) = k(x,t) = \mathbf{u}_p(x,t) + \mathbf{v}(x,t) + \mathbf{z}(x,t), \tag{3.8}$$

$$\begin{aligned} h(t) = h(x,t) =& \frac{1}{\varphi}\psi'(\psi' + \mathbf{u}\cdot\nabla\psi) + \frac{\varepsilon^2}{4}\frac{|\Delta\psi|^2}{\varphi} - \frac{1}{2}\psi\nabla\cdot E - \nabla\psi\cdot E \\ &+ \frac{1}{2}\frac{\Delta P(\psi^2)}{\varphi} + \nu\psi + \frac{1}{2}\nabla\psi\cdot\nabla(|\mathbf{v}+\mathbf{z}|^2) - \frac{1}{2}\psi|\,\omega|^2 \\ &- [\mathbf{v}+\mathbf{z}]\cdot\nabla(\mathbf{u}\cdot\nabla\psi) + \frac{1}{\varphi}(\psi' + \mathbf{u}\cdot\nabla\psi)(\mathbf{v}\cdot\nabla\psi) \\ &- \nabla\psi\cdot([\mathbf{v}+\mathbf{z}]\times\omega) + \frac{1}{2}\psi\nabla(\mathbf{v}+\mathbf{z}):\nabla(\mathbf{v}+\mathbf{z}), \end{aligned} \tag{3.9}$$

$$\begin{aligned} g(t) = g(x,t) =& E - \frac{1}{2}\nabla\Big(|\mathbf{v}+\mathbf{z}|^2\Big) + [\mathbf{v}+\mathbf{z}]\times\omega - \nabla h(\psi^2) \\ &+ \frac{1}{2}\varepsilon^2\Big(\frac{\nabla\Delta\psi}{\varphi} - \frac{\Delta\psi}{\varphi^2}\nabla\psi\Big), \end{aligned} \tag{3.10}$$

where $\mathbf{u} = (u^1, u^2, u^3)$ and $\mathbf{v} = (v^1, v^2, v^3)$. The most important fact which will be belown in Section 3.2 is to note that the above extended system for $U = (\mathbf{v}, \mathbf{z}, \varphi, \psi, \mathbf{u}, E)$ is equivalent to the equations (1.12)–(1.14) of $(\psi, \mathbf{u}, E)$ for classical solutions when $\mathbf{u} = \mathbf{v} + \mathbf{z}$ and $\psi = \varphi > 0$.

The main result in this section is:

Theorem 3.1. *Assume that* $P \in C^5(0,\infty)$ *and* $(\psi_1 - \sqrt{\bar{\rho}}, \mathbf{u}_1) \in H^6(\mathbb{R}^3)\times \mathcal{H}^5(\mathbb{R}^3)$ *satisfying*

$$\psi^* = \sup_{x\in\mathbb{R}^3}\psi_1(x), \quad \psi_* =: \inf_{x\in\mathbb{R}^3}\psi_1(x) > 0. \tag{3.11}$$

Then, there is a uniform time T_{**} *such that there exists a solution series* $U = (\mathbf{v}, \mathbf{z}, \varphi, \psi, \mathbf{u}, E)$ *which solves uniformly the systems* (3.1)–(3.6) *for* $t \in [0, T_{**}]$ *and satisfies*

$$\begin{cases} \mathbf{v} \in C^j([0,T_{**}];\mathcal{H}^{4-j}(\mathbb{R}^3)) \cap C^2([0,T_*];\mathcal{H}^1(\mathbb{R}^3)), \\ \mathbf{z} \in C^l([0,T_{**}];\mathcal{H}^{4-l}(\mathbb{R}^3)), \ \ \omega \in C^l([0,T_1];H^{3-l}(\mathbb{R}^3)), \\ \mathbf{u} \in C^1([0,T_{**}];\mathcal{H}^3(\mathbb{R}^3)) \cap C^2([0,T_{**}];\mathcal{H}^1(\mathbb{R}^3)), \\ \varphi - \sqrt{\bar{\rho}} \in C^1([0,T_*];H^3(\mathbb{R}^3)) \cap C^2([0,T_{**}];H^2(\mathbb{R}^3)) \\ \qquad\qquad \cap\, C^3([0,T_{**}];L^2(\mathbb{R}^3)), \\ \psi - \sqrt{\bar{\rho}} \in C^k([0,T_*];H^{6-2k}(\mathbb{R}^3)) \cap C^3([0,T_{**}];L^2(\mathbb{R}^3)), \\ E \in C^1([0,T_{**}];\mathcal{H}^3(\mathbb{R}^3)), \end{cases} \tag{3.12}$$

where $j = 0, 1$, $l = 0, 1, 2$, $k = 0, 1, 2$, *and we recall that* $\mathcal{H}^m = \{f \in L^6(\mathbb{R}^3);\ Df \in H^{m-1}(\mathbb{R}^3)\}$, $m \geq 1$.

We will show the construction of the extended systems (3.1)–(3.10) based on (1.12)–(1.14) in Section 3.1. Then we define an iterative scheme of approximate solution sequence of the extended system and obtain the uniform estimates, and then prove Theorem 1.1 in Section 3.2. For simplicity, we set $\tau = 1$ and $\lambda = 1$.

3.1 Construction of new problems

We construct the extended systems (3.1)–(3.10) based on Eqs. (1.12)–(1.14) by modifying the main idea in [28, 36]. For general smooth fluid-dynamics, the velocity vector filed can be (uniquely) decomposed into the gradient field and the divergence free vector field:

$$\mathbf{u} = \mathbf{v} + \mathbf{z} = \mathcal{Q}\mathbf{u} + \mathcal{P}\mathbf{u} = \nabla S + \mathbf{z}, \quad \nabla \cdot \mathbf{z} = 0. \tag{3.13}$$

Equation (1.13) for the velocity vector field $\mathbf{u}$ can be rewritten as

$$\partial_t \mathbf{u} + \mathbf{u} + \frac{1}{2}\nabla(|\mathbf{u}|^2) - \mathbf{u} \times (\nabla \times \mathbf{u}) + \nabla h(\psi^2) = E + \frac{\varepsilon^2}{2}\nabla\Big(\frac{\Delta\psi}{\psi}\Big), \tag{3.14}$$

where we use the relation of the convection term

$$(\mathbf{u}\cdot\nabla)\mathbf{u} = \frac{1}{2}\nabla(|\mathbf{u}|^2) - \mathbf{u} \times (\nabla \times \mathbf{u}). \tag{3.15}$$

Taking *curl* on (3.14) and letting $\omega = \nabla \times \mathbf{u}$, we have

$$\partial_t \omega + \omega + \mathbf{u}\cdot\nabla\omega + \omega\nabla\cdot\mathbf{u} - (\omega\cdot\nabla)\mathbf{u} = 0. \tag{3.16}$$

For smooth $\psi > 0$ Eq. (1.12) is equivalent to

$$2\partial_t \psi + 2\mathbf{u}\cdot\nabla\psi + \psi\nabla\cdot\mathbf{u} = 0. \tag{3.17}$$

Based on Eqs. (3.13)–(3.17), we show the ideas on how to construct the extended systems (3.1)–(3.10) for $U = (\mathbf{v}, \mathbf{z}, \varphi, \psi, \mathbf{u}, E)$ to be dealt with on the basis of Section 3.2. Given $\mathbf{u}$ and ψ, we can introduce new equations for "density" $\varphi > 0$ and gradient velocity vector field $\mathbf{v}$ in terms of divergence free vector field $\mathbf{z}$ as

$$\partial_t\varphi + \frac{1}{2}\varphi\nabla\cdot\mathbf{v} + \mathbf{u}\cdot\nabla\psi = 0, \quad \varphi(x,0) = \psi_1(x) > 0, \tag{3.18}$$

$$\nabla\cdot\mathbf{v} = -\frac{2(\partial_t\psi + \mathbf{u}\cdot\nabla\psi)}{\varphi}, \quad \nabla\times\mathbf{v} = 0, \quad \mathbf{v}(x,t) \to 0,\ |x| \to \infty. \tag{3.19}$$

The new divergence free vector field $\mathbf{z}$ is represented by its vorticity (which will still be denoted by ω) $\omega = \nabla\times\mathbf{z}$ as

$$\mathbf{z}(x,t) = B_0 \int_{\mathbb{R}^3} |x-y|^{-3}(x-y) \times \omega(y,t)\, dy \tag{3.20}$$

where B_0 is a constant matrix, and the vorticity vector field $\omega = \nabla\times\mathbf{z}$ solves the following equation

$$\partial_t\omega + \omega + (\mathbf{v}+\mathbf{z})\cdot\nabla\omega + \omega\nabla\cdot\mathbf{v} - (\omega\cdot\nabla)[\mathbf{v}+\mathbf{z}] = 0, \tag{3.21}$$

$$\omega(x,0) = \nabla\times\mathbf{u}_1(x), \tag{3.22}$$

which is obtained by taking *curl* on (3.16) after replacing $\mathbf{z}+\mathbf{v}$ for $\mathbf{u}$.

We need to pay attention to, however, that we should be able to determine $\mathbf{u}$ and ψ again so long as we can solve the above equations for $\mathbf{v}$, $\mathbf{z}$ and φ. Namely, we will propose the corresponding two equations for $\mathbf{u}$ and ψ based on (3.18), (3.19), and (3.20) as follows. In fact, we can construct the expected equation for the velocity $\mathbf{u}$ as

$$\partial_t\mathbf{u} + \mathbf{u} + \frac{1}{2}\nabla(|\mathbf{v}+\mathbf{z}|^2) - (\mathbf{v}+\mathbf{z})\times\omega + \nabla h(\psi^2)$$
$$= E + \frac{\varepsilon^2}{2}\left(\frac{\nabla\Delta\psi}{\varphi} - \frac{\Delta\psi\nabla\psi}{\varphi^2}\right), \tag{3.23}$$

$$\mathbf{u}(x,0) = \mathbf{u}_1(x), \tag{3.24}$$

which is derived from (3.14) by substituting $\mathbf{v}+\mathbf{z}$ for $\mathbf{u}$ into the convection term through the relation (3.15), and by using the equality

$$\nabla\left(\frac{\Delta\psi}{\psi}\right) = \left(\frac{\nabla\Delta\psi}{\psi} - \frac{\Delta\psi\nabla\psi}{\psi^2}\right) \tag{3.25}$$

and replacing $\frac{1}{\psi}$ by $\frac{1}{\varphi}$ on the right hand side terms of (3.25). And we construct the equation for the density ψ as

$$\psi_{tt} + \psi_t + \frac{1}{4}\varepsilon^2\Delta^2\psi - \frac{1}{4}\varepsilon^2\frac{|\Delta\psi|^2}{\varphi} - \frac{1}{2\varphi}\Delta P(\psi^2) + \frac{1}{2}\psi\Delta V + \nabla\psi\cdot E$$

$$+ (\mathbf{u}+\mathbf{v}+\mathbf{z})\cdot\nabla\psi_t - \frac{\psi_t}{\varphi}(\psi_t + \mathbf{u}\cdot\nabla\psi) - \frac{1}{2}\nabla\psi\cdot\nabla(|\mathbf{v}+\mathbf{z}|^2)$$
$$+ \nabla\psi\cdot([\mathbf{v}+\mathbf{z}]\times\omega) - \frac{1}{2}\psi\nabla(\mathbf{v}+\mathbf{z}):\nabla(\mathbf{v}+\mathbf{z}) + \frac{1}{2}\psi|\omega|^2$$
$$+ (\mathbf{v}+\mathbf{z})\cdot\nabla(\mathbf{u}\cdot\nabla\psi) - \frac{1}{\varphi}(\psi_t + \mathbf{u}\cdot\nabla\psi)([\mathbf{v}+\mathbf{z}]\cdot\nabla\psi) = 0 \tag{3.26}$$

with initial data

$$\psi(x,0) = \psi_1,\ \psi_t(x,0) = \psi_0 =: -\frac{1}{2}\psi_1\nabla\cdot\mathbf{u}_1 - \mathbf{u}\cdot\nabla\psi_1, \tag{3.27}$$

where $\mathbf{v} = (v^1, v^2, ..., v^n)$ and

$$\nabla\mathbf{v}:\nabla\mathbf{v} = \sum_{i,j}|\partial_j v^i|^2.$$

In fact, by differentiating (3.17) with respect to time, replacing $\mathbf{u}_t$ in terms of (3.14) where the unknown $\mathbf{u}$ of the convection term is substituted by $\mathbf{v}+\mathbf{z}$, using(3.15) and replacing the term $\frac{1}{2}\psi[\frac{1}{2}\Delta(|\mathbf{v}+\mathbf{z}|^2) - \nabla\cdot([\mathbf{v}+\mathbf{z}]\times\omega)]$ by

$$\frac{1}{2}\psi\nabla(\mathbf{v}+\mathbf{z}):\nabla(\mathbf{v}+\mathbf{z}) - \frac{1}{2}\psi|\omega|^2 - (\mathbf{v}+\mathbf{z})\cdot\nabla\psi_t$$
$$- (\mathbf{v}+\mathbf{z})\cdot\nabla(\mathbf{u}\cdot\nabla\psi) + \frac{1}{\psi}(\psi_t + \mathbf{u}\cdot\nabla\psi)([\mathbf{v}+\mathbf{z}]\cdot\nabla\psi),$$

and by using the relation

$$\nabla\cdot\left[\psi^2\nabla\Big(\frac{\Delta\psi}{\psi}\Big)\right] = \psi\left[\Delta^2\psi - \frac{|\Delta\psi|^2}{\psi}\right],$$

and by replacing all $\frac{1}{\psi}$ in the resulting equation by $\frac{1}{\varphi}$, we can obtain Eq. (3.26). Finally, from (1.14) we still solve E from the same divergence equation on $\mathbb{R}^3$:

$$\nabla\cdot E = \psi^2 - \mathcal{C}, \quad \nabla\times E = 0, \quad E(x,t) \to 0,\ |x| \to \infty. \tag{3.28}$$

So far, we have constructed the extended and closed systems (3.1)–(3.10) for $U = (\mathbf{v}, \mathbf{z}, \varphi, \psi, \mathbf{u}, E)$ which consists of two O.D.E.s (3.18) for φ and (3.23) for $\mathbf{u}$, a wave type equation (3.26) for ψ, two divergence equations (3.19) and (3.28) for $\mathbf{v}$ and E, and a formula (3.20) for $\mathbf{z}$ in terms of ω which solves a hyperbolic equation (3.21).

3.2 Iteration scheme and local existence

We define an iterative scheme of approximate solution sequence of the extended system and obtain the uniform estimates, and then prove Theorem 3.1 and Theorem 1.1. We consider the corresponding problem for an approximate solution sequence $\{U^i\}_{i=1}^{\infty}$ with

$$U^p = (\mathbf{v}_p, \mathbf{z}_p, \varphi_p, \psi_p, \mathbf{u}_p, E_p)$$

based on the extended systems (3.1)–(3.10) constructed in Section 3.1. We construct an iterative scheme for the solution

$$U^{p+1} = (\mathbf{v}_{p+1}, \mathbf{z}_{p+1}, \varphi_{p+1}, \psi_{p+1}, \mathbf{u}_{p+1}, E_{p+1}), \quad p \geq 1,$$

on $\mathbb{R}^3$ by solving the following problems

$$\nabla\cdot \mathbf{v}_{p+1} = r_p(t), \quad \nabla\times\mathbf{v}_{p+1} = 0, \ \mathbf{v}_{p+1}(x,t) \to 0, \ |x| \to \infty, \tag{3.29}$$

$$\begin{cases} \mathbf{z}_{p+1} = B_0 \int_{\mathbb{R}^3} |x-y|^{-3}(x-y) \times \omega_{p+1}(y,t)\, dy, \\ \omega'_{p+1} + \omega_{p+1} + (\mathbf{v}_p + \mathbf{z}_p)\cdot\nabla\omega_{p+1} \\ \qquad\qquad + \omega_{p+1}\nabla\cdot \mathbf{v}_p - (\omega_{p+1}\cdot\nabla)[\mathbf{v}_p + \mathbf{z}_p] = 0, \\ \omega_{p+1}(x,0) = \omega_1(x) =: \nabla\times\mathbf{u}_1(x), \end{cases} \tag{3.30}$$

$$\begin{cases} \varphi'_{p+1} + \frac{1}{2}(\nabla\cdot \mathbf{v}_p)\varphi_{p+1} + \mathbf{u}_p\cdot\nabla\psi_p = 0, \\ \varphi_{p+1}(x,0) = \psi_1(x), \end{cases} \tag{3.31}$$

$$\begin{cases} \psi''_{p+1} + \psi'_{p+1} + \nu\Delta^2\psi_{p+1} + \nu\psi_{p+1} + k_p(t)\cdot\nabla\psi'_{p+1} = h_p(t), \\ \psi_{p+1}(x,0) = \psi_1(x), \\ \psi'_{p+1}(x,0) = \psi_0 =: -\frac{1}{2}\psi_1\nabla\cdot \mathbf{u}_1 - \mathbf{u}_1\cdot\nabla\psi_1, \end{cases} \tag{3.32}$$

$$\begin{cases} \mathbf{u}'_{p+1} + \mathbf{u}_{p+1} = g_p(t), \\ \mathbf{u}_{p+1}(x,0) = \mathbf{u}_1(x), \end{cases} \tag{3.33}$$

$$\begin{cases} \nabla\cdot E_{p+1} = q_p =: \psi_p^2 - \mathcal{C}, \quad \nabla\times E_{p+1} = 0, \\ E_{p+1}(x,t) \to 0, \ |x| \to \infty, \end{cases} \tag{3.34}$$

where $\nu = \frac{1}{4}\varepsilon^2$, and

$$r_p(t) = r_p(x,t) = -\frac{2(\psi'_p + \mathbf{u}_p\cdot\nabla\psi_p)}{\varphi_p}, \tag{3.35}$$

$$k_p(t) = k_p(x,t) = \mathbf{u}_p(x,t) + \mathbf{v}_p(x,t) + \mathbf{z}_p(x,t), \tag{3.36}$$

$$\begin{aligned} h_p(t) = h_p(x,t) = {} & \frac{1}{\varphi_p}\psi'_p(\psi'_p + \mathbf{u}_p\cdot\nabla\psi_p) + \frac{\varepsilon^2}{4}\frac{|\Delta\psi_p|^2}{\varphi_p} - \frac{1}{2}\psi_p\nabla\cdot E_p \\ & - \nabla\psi_p\cdot E_p + \frac{1}{2}\frac{\Delta P(\psi_p^2)}{\varphi_p} + \nu\psi_p + \frac{1}{2}\nabla\psi_p\cdot\nabla(|\mathbf{v}_p + \mathbf{z}_p|^2) \end{aligned}$$

$$
\begin{aligned}
&-\frac{1}{2}\psi_p|\omega_p|^2-[\mathbf{v}_p+\mathbf{z}_p]\cdot\nabla(\mathbf{u}_p\cdot\nabla\psi_p)\\
&+\frac{1}{\varphi_p}\big(\psi_p'+\mathbf{u}_p\cdot\nabla\psi_p\big)(\mathbf{v}_p\cdot\nabla\psi_p)-\nabla\psi_p\cdot([\mathbf{v}_p+\mathbf{z}_p]\times\omega_p)\\
&+\frac{1}{2}\psi_p\nabla(\mathbf{v}_p+\mathbf{z}_p):\nabla(\mathbf{v}_p+\mathbf{z}_p),
\end{aligned}
\tag{3.37}
$$

$$
\begin{aligned}
g_p(t)=&g_p(x,t)=E_p-\frac{1}{2}\nabla\Big(|\mathbf{v}_p+\mathbf{z}_p|^2\Big)+[\mathbf{v}_p+\mathbf{z}_p]\times\omega_p-\nabla h(\psi_p^2)\\
&+\frac{1}{2}\varepsilon^2\left(\frac{\nabla\Delta\psi_p}{\varphi_p}-\frac{\Delta\psi_p}{\varphi_p^2}\nabla\psi_p\right),
\end{aligned}
\tag{3.38}
$$

where $\mathbf{u}_p=(u_p^1,u_p^2,u_p^3)$ and $\mathbf{v}_p=(v_p^1,v_p^2,v_p^3)$. Here, we also note that functions $r_p(0)$,$k_p(0)$, $h_p(0)$, $g_p(0)$, $q_p(0)$ only depend on initial data $(\psi_1,\mathbf{u}_1)$.

The main result in this subsection is

Lemma 3.2. *Let the assumption of Theorem 3.1 hold. Then, there is a uniform time T_* such that there exists a solution series $\{U^p\}_{p=1}^\infty$ which solves uniformly the systems* (3.29)–(3.34) *for $t\in[0,T_*]$ and satisfies*

$$
\begin{cases}
\mathbf{v}_p\in C^j([0,T_*];\mathcal{H}^{4-j}(\mathbb{R}^3))\bigcap C^2([0,T_*];\mathcal{H}^1(\mathbb{R}^3)),\\
\mathbf{z}_p\in C^k([0,T_*];\mathcal{H}^{4-k}(\mathbb{R}^3)),\ \ \omega_{p+1}\in C^k([0,T_1];H^{3-k}(\mathbb{R}^3)),\\
\mathbf{u}_p\in C^1([0,T_*];\mathcal{H}^3(\mathbb{R}^3))\bigcap C^2([0,T_*];\mathcal{H}^1(\mathbb{R}^3)),\\
\varphi_p-\sqrt{\bar\rho}\in C^1([0,T_*];H^3(\mathbb{R}^3))\bigcap C^2([0,T_*];H^2(\mathbb{R}^3))\\
\qquad\qquad\bigcap C^3([0,T_*];L^2(\mathbb{R}^3)),\\
\psi_p-\sqrt{\bar\rho}\in C^l([0,T_*];H^{6-2l}(\mathbb{R}^3))\bigcap C^3([0,T_*];L^2(\mathbb{R}^3)),\\
E_p\in C^1([0,T_*];\mathcal{H}^3(\mathbb{R}^3)),
\end{cases}
\tag{3.39}
$$

where $j=0,1$, $k=0,1,2$, $l=0,1,2$, and we recall that $\mathcal{H}^k=\{f\in L^6(\mathbb{R}^3);\ Df\in H^{k-1}(\mathbb{R}^3)\}$, $k\geq 1$. Furthermore, the solution series $\{U^p\}_{p=1}^\infty$ is bounded uniformly for each $p\geq1$ by

$$
\begin{cases}
\|D(\mathbf{v}_p',\mathbf{u}_p,\mathbf{u}_p',E_p,E_p')(t)\|_{H^2}^2+\|D\mathbf{v}_p(t)\|_{H^3}^2+\|D(\mathbf{v}_p'',\mathbf{u}_p'')(t)\|^2\leq M_*,\\
\|(\omega_p,\omega_p',\omega_p'')(t)\|_{H^3\times H^2\times H^1}^2+\|D(\mathbf{z}_p,\mathbf{z}_p',\mathbf{z}_p'')(t)\|_{H^3\times H^2\times H^1}^2\leq M_*,\\
\|(\psi_p-\sqrt{\bar\rho},\psi_p',\psi_p'',\psi_p''')(t)\|_{H^6\times H^4\times H^2\times L^2}^2\leq M_*,\\
\|(\varphi_p-\sqrt{\bar\rho},\varphi_p',\varphi_p'',\varphi_p''')(t)\|_{H^3\times H^3\times H^2\times L^2}^2\leq M_*,
\end{cases}
\tag{3.40}
$$

with M_* a positive constant independent of U^p, $p\geq1$.

Remark 3.3. Here, note that by (2.20) we can obtain automatically the L^6-norm of the unknown $\mathbf{v}_{p+1},\mathbf{z}_{p+1},E_{p+1},\mathbf{u}_{p+1}$ and their time derivatives so long as (3.40) holds, i.e.,

$$
\|(\mathbf{v}_{p+1},\mathbf{v}_{p+1}',\mathbf{v}_{p+1}'')\|_{L^6}\leq C\|D(\mathbf{v}_{p+1},\mathbf{v}_{p+1}',\mathbf{v}_{p+1}'')\|_{L^2},
\tag{3.41}
$$

$$\|(\mathbf{z}_{p+1},\mathbf{z}'_{p+1},\mathbf{z}''_{p+1})\|_{L^6}\le C\|D(\mathbf{z}_{p+1},\mathbf{z}'_{p+1},\mathbf{z}''_{p+1})\|_{L^2}, \tag{3.42}$$

$$\begin{aligned}&\|(\mathbf{u}_{p+1},\mathbf{u}'_{p+1},\mathbf{u}''_{p+1},E_{p+1},E'_{p+1})\|_{L^6}\\ \le& C\|D(\mathbf{u}_{p+1},\mathbf{u}'_{p+1},\mathbf{u}''_{p+1},E_{p+1},E'_{p+1})\|_{L^2}.\end{aligned} \tag{3.43}$$

Below, we will often use this fact without mentioning it.

Proof. We prove Lemma 3.2 and verify the a-priori estimates (3.40) in terms of energy method and induction argument as follows:

First of all, we consider the case $p=1$. We choose

$$U^1=(\mathbf{v}_1,\mathbf{z}_1,\varphi_1,\psi_1,\mathbf{u}_1,E_1)=(\mathcal{Q}\mathbf{u}_1,\mathcal{P}\mathbf{u}_1,\psi_1,\psi_1,\mathbf{u}_1,E_1),$$

which obviously satisfies (3.39)–(3.40) for $t\in[0,T_1]$ with M_* replaced by a constant $B_1>0$, where E_1 is determined by (2.5). Starting with $U^1=(\mathcal{Q}\mathbf{u}_1,\mathcal{P}\mathbf{u}_1,\psi_1,\psi_1,\mathbf{u}_1,E_1)$ and solving the problems (3.29)–(3.34) for $p=1$, we shall prove the (local in time) existence of solution $U^2=(\mathbf{v}_2,\mathbf{z}_2,\psi_2,\varphi_2,\mathbf{u}_2,E_2)$ which also satisfies (3.39)–(3.40) for $t\in[0,T_1]$ with $T_1>0$ and with M_* replaced by another constant $\widetilde{M_2}>0$. In fact, for $U^1=(\mathcal{Q}\mathbf{u}_1,\mathcal{P}\mathbf{u}_1,\psi_1,\psi_1,\mathbf{u}_1,E_1)$ the functions r_1, k_1, h_1, g_1, q_1 depend only on the initial data $(\psi_1,\mathbf{u}_1)$, i.e.,

$$\begin{gathered}r_1(x,t)=\tilde r_1(x),\ k_1(x,t)=\tilde k_1(x),\ h_1(x,t)=\tilde h_1(x),\\ g_1(x,t)=\tilde g_1(x),\ q_1(x,t)=\tilde q_1(x),\end{gathered}$$

satisfying

$$\|\tilde r_1\|_2^2+\|\tilde k_1\|_{L^6}^6+\|D\tilde k_1\|_3^2+\|\tilde h_1\|_3^2+\|D\tilde g_1\|_3^2+\|\tilde g_1\|_{L^6}^6+\|\tilde q_1\|_2^2\le NI_0^4, \tag{3.44}$$

here and after $N>0$ denotes a generic constant independent of $U^p,p\ge 1$,

$$I_0=\|(\psi_1-\sqrt{\bar\rho})\|^2+\|\nabla\psi_1\|_5^2+\|\mathbf{u}_1\|_5^2+\|\mathcal{C}-\bar\rho\|_3^2. \tag{3.45}$$

The systems (3.29)–(3.34) with $p=1$ are linear with $U^2=(\mathbf{v}_2,\mathbf{z}_2,\psi_2,\varphi_2,\mathbf{u}_2,E_2)$. It can be solved with the help of the estimates (3.44) about the right hand side terms as follows. Applying Theorem 2.1 to the divergence equations (3.29) with $r_1(x,t)$ replaced by $\tilde r_1(x)$ and (3.34) with $q_1(x,t)$ replaced by $\tilde q_1(x)$, we obtain the existence of solution

$$\mathbf{v}_2\in C^j([0,T_1];\mathcal{H}^{4-j}(\mathbb{R}^3))\cap C^2([0,T_*];\mathcal{H}^1(\mathbb{R}^3)),\ j=0,1,$$

and $E_2\in C^1([0,T_1];\mathcal{H}^3(\mathbb{R}^3))$. Making use of the theory of linear hyperbolic system, we show the existence of $\omega_2\in C^j([0,T_1];H^{3-j}(\mathbb{R}^3))$, $j=0,1,2$, of Eq. $(3.30)_{2,3}$, which together with $(3.30)_1$ and (2.4) implies the existence of $\mathbf{z}_2\in C^j([0,T_1];\mathcal{H}^{4-j}(\mathbb{R}^3))$, $j=0,1,2$. By the theory of linear O.D.E. system, we prove the existence of solution $\mathbf{u}_2\in$

$C^1([0,T_1];\mathcal{H}^3(\mathbb{R}^3)) \cap C^2([0,T_*];\mathcal{H}^1(\mathbb{R}^3))$ of (3.33) for $g_1(x,t) = \tilde{g}_1(x)$, and solution of (3.31) satisfying

$$\varphi_2 - \sqrt{\bar{\rho}} \in C^1([0,T_1];H^3(\mathbb{R}^3)) \cap C^2([0,T_1];H^2(\mathbb{R}^3)) \cap C^3([0,T_1];L^2(\mathbb{R}^3)).$$

Finally, applying Theorem 2.4 with $b(x,t) = 2\mathbf{u}_1(x)$ in (2.12) and $f(x,t) = \tilde{h}_1(x)$ in (2.13) to the semilinear wave equation (3.32), we conclude the existence of solution $\psi_2 - \sqrt{\bar{\rho}} \in C^j([0,T_1];H^{6-2j})$, $j = 0,1,2$. Moreover, based on the estimates (3.44), we conclude that there is a constant $\widetilde{M}_2$ such that U^2 satisfies (3.40) where $p = 2$, $M_* = \widetilde{M}_2$ and $T_* = T_1$.

Next, let us prove the estimates for $p \geq 2$. Assume that $\{U^i\}_{i=1}^p$ ($p \geq 2$) exists uniformly for $t \in [0,T_1]$, solves the systems (3.29)–(3.34), and satisfies (3.39)–(3.40) with (the constant M_* replaced by) the upper bound $\widetilde{M}_p$ ($\geq \max_{1\leq j\leq p-1}\{\widetilde{M}_j\}$). We shall prove that it still holds for U^{p+1} for $t \in [0,T_1]$. In fact, the systems (3.33)–(3.34) are linear with $U^{p+1} = (\mathbf{v}_{p+1}, \mathbf{z}_{p+1}, \varphi_{p+1}, \psi_{p+1}, \mathbf{u}_{p+1}, E_{p+1})$ for given U^p. In analogy, the application of Theorem 2.1 to divergence equations (3.29) for $\mathbf{v}_{p+1}$ and (3.34) for E_{p+1}, theory of linear O.D.E. system to Eq. (3.31) for φ_{p+1} and Eq. (3.33) for $\mathbf{u}_{p+1}$, and Theorem 2.4 to wave equation (3.32) for ψ_{p+1} with $f(x,t) = h_p(t)$ and $b(x,t) = k_p(t)$, shows that $U^{p+1} = (\mathbf{v}_{p+1}, \mathbf{z}_{p+1}, \psi_{p+1}, \varphi_{p+1}, \mathbf{u}_{p+1}, E_{p+1})$ exists for $t \in [0,T_1]$ and satisfies

$$\begin{cases} \mathbf{v}_{p+1} \in C^j([0,T_1];\mathcal{H}^{4-j}(\mathbb{R}^3)) \cap C^2([0,T_*];\mathcal{H}^1(\mathbb{R}^3)), \quad j = 0,1, \\ \mathbf{z}_{p+1} \in C^j([0,T_1];\mathcal{H}^{4-j}(\mathbb{R}^3)),\ \omega_{p+1} \in C^j([0,T_1];H^{3-j}(\mathbb{R}^3)),\ j = 0,1,2, \\ \mathbf{u}_{p+1} \in C^1([0,T_1];\mathcal{H}^3(\mathbb{R}^3)) \cap C^2([0,T_*];\mathcal{H}^1(\mathbb{R}^3)), \\ \varphi_{p+1} - \sqrt{\bar{\rho}} \in C^1([0,T_1];H^3(\mathbb{R}^3)) \cap C^2([0,T_1];H^2(\mathbb{R}^3)) \\ \qquad\qquad \cap C^3([0,T_*];L^2(\mathbb{R}^3)), \\ \psi_{p+1} - \sqrt{\bar{\rho}} \in C^l([0,T_1];H^{6-2l}(\mathbb{R}^3)) \cap C^3([0,T_*];L^2(\mathbb{R}^3)), \quad l = 0,1,2, \\ E_{p+1} \in C^1([0,T_1];\mathcal{H}^3(\mathbb{R}^3)). \end{cases}$$

What we do next is to obtain a uniform upper bound of U^{j+1}, $1 \leq j \leq p$, for a fixed time period.

Let us verify the L^2 norm of the initial value of $\psi_{p+1}, \psi'_{p+1}, \psi''_{p+1}$ first, where the initial value $\hat{\psi}$ of ψ''_{p+1} is obtained through $(3.32)_1$ at $t = 0$ with ψ_{p+1} and ψ'_{p+1} replaced by initial data ψ_1, ψ_0:

$$\hat{\psi} = -\psi_0 - \nu\Delta^2\psi_0 - \nu\psi_1 - 2\mathbf{u}_1\cdot\nabla\psi_0 + \tilde{h}(0) \tag{3.46}$$

with $\tilde{h}(0) = h_p(0)$ only depending on $(\psi_1, \mathbf{u}_1)$. Hence, the initial values of $\psi_{p+1}, \psi'_{p+1}, \psi''_{p+1}$ only depend on $(\psi_1, \mathbf{u}_1)$. Obviously, there is a constant $M_2 > 0$ such that the initial values of $\psi_{p+1}, \psi'_{p+1}, \psi''_{p+1}$ for $p \geq 1$ are bounded by

$$\max\left\{ \|\psi_1 - \sqrt{\bar{\rho}}\|_2^2,\ \|\psi_0\|_2^2,\ \|\hat{\psi}\|_2^2,\ \|\mathcal{C} - \bar{\rho}\|_3^2,\ \|D\mathbf{u}_1\|_4^2 \right\} \leq M_2 I_0. \tag{3.47}$$

Here, we recall that I_0 is defined by (3.45).

Set

$$\left.\begin{aligned} M_0 &= 16NI_0 \cdot \max\{1, \nu^{-1}\}, \\ M_1 &= 2N(I_0+1+M_0)^7 \cdot \max\{1, \nu^{-2}\}, \end{aligned}\right\} \tag{3.48}$$

$$\psi^* =: \sup_{x\in\mathbb{R}^3} \psi_1(x) > 0, \ \psi_* =: \inf_{x\in\mathbb{R}^3} \psi_1(x) > 0 \tag{3.49}$$

and choose

$$T_* = \min\left\{1, \ \frac{\psi_*}{4M_0}, \ \frac{2I_0}{M_3}, \ \frac{\ln 2}{NM_4}\right\}, \tag{3.50}$$

where

$$M_3 = (I_0+1+M_0+M_1)^{14}, \ M_4 = 2(I_0+1+M_0+M_1)^8, \tag{3.51}$$

and we choose the generic constant $N > M_2$ independent of $U^p, p \geq 1$.

The main hard work in this part is to apply energy method to the coupled system (3.29)–(3.34) for $U^{p+1} = (\mathbf{v}_{p+1}, \mathbf{z}_{p+1}, \varphi_{p+1}, \psi_{p+1}, \mathbf{u}_{p+1}, E_{p+1})$ in a similar argument to the above step for U^2, and obtain after a tedious but straightforward computation (We omit the details here, one can also refer to [36] for the main steps for irrotaional case) the following statement:

If it holds for $\{U^j\}_{j=1}^p$ $(p \geq 2)$, *solving problems* (3.29)–(3.34), *that*

$$\begin{cases} \|(D\mathbf{z}_j, \omega_j)(t)\|_3^2 + \|D\mathbf{u}_j(t)\|_2^2 \leq M_0, \\ \|(\psi_j - \sqrt{\bar\rho}, \psi_j')(t)\|_4^2 + \|\psi_j''(t)\|_2^2 \leq M_0, \\ \|D\mathbf{v}_j(t)\|_3^2 + \|D\Delta\psi_j(t)\|_1^2 \leq M_1, \end{cases} \tag{3.52}$$

for $1 \leq j \leq p$ *and* $t \in [0, T_*]$, *then it also holds for* U^{p+1} *that*

$$\begin{cases} \|(D\mathbf{z}_{p+1}, \omega_{p+1})(t)\|_3^2 + \|D\mathbf{u}_{p+1}(t)\|_2^2 \leq M_0, \\ \|(\psi_{p+1} - \sqrt{\bar\rho}, \psi_{p+1}')(t)\|_4^2 + \|\psi_{p+1}''(t)\|_2^2 \leq M_0, \\ \|D\mathbf{v}_{p+1}(t)\|_3^2 + \|D\Delta\psi_{p+1}(t)\|_1^2 \leq M_1, \end{cases} \tag{3.53}$$

for $t \in [0, T_*]$. *Here* M_0 *and* M_1 *are given by* (3.48) *and* (3.49).

Furthermore, with the help of the above statement, we can conclude, by a direct complicated computation, that the approximate solution sequence $U^{p+1} = (\mathbf{v}_{p+1}, \mathbf{z}_{p+1}, \varphi_{p+1}, \psi_{p+1}, \mathbf{u}_{p+1}, E_{p+1}), p \geq 1$, which solves (3.29)–(3.34) uniformly for $t \in [0, T_*]$ with T_* defined by (3.50), is uniformly bounded for $t \in [0, T_*]$, and satisfies (3.39), (3.40) with the constant $M_* > 0$ chosen by

$$M_* = N(I_0+1+M_0+M_1)^{15}. \tag{3.54}$$

In addition, for given $(x,t) \in \mathbb{R}^3 \times [0, T_*]$ using the standard argument of O.D.E., we obtain φ_{j+1} from (3.31), that is,

$$\begin{cases} \varphi_{j+1}(x,t) = & (\psi_1(x) - \int_0^t e^{\frac{1}{2}\int_0^s \nabla\cdot\mathbf{v}_j(x,\xi)d\xi}\mathbf{u}_j\cdot\nabla\psi_j(x,s)ds) \\ & e^{-\frac{1}{2}\int_0^t \nabla\cdot\mathbf{v}_j(x,s)ds}, \\ \varphi_{j+1} - \sqrt{\bar{\rho}} \in & C^1([0,T_1]; H^3(\mathbb{R}^3)) \cap C^2([0,T_1]; H^2(\mathbb{R}^3)) \\ & \cap C^3([0,T_1]; L^2(\mathbb{R}^3)), \end{cases} \tag{3.55}$$

which together with (3.52) give rise to the uniform positivity for $(x,t) \in \mathbb{R}^3 \times [0, T_*]$

$$\frac{1}{4}\psi_* \le \varphi_{j+1}(x,t) \le 2(\psi^* + \psi_*). \tag{3.56}$$

Recall here that M_0, M_1, ψ^*, and ψ_* are defined by (3.48) and (3.49) respectively and $N > 0$ is a generic constant, which are independent of U^{p+1}, $p \ge 1$. Thus, the proof of Lemma 3.2 is completed. □

Proof of Theorem 3.1. By Lemma 3.2, we obtain an approximate solution sequence $\{U^i\}_{i=1}^{\infty}$ satisfying (3.39)–(3.40). To prove the uniform convergence of the whole sequence, we need to estimate the difference of the approximate solution sequence

$$Y^{p+1} = (\bar{\mathbf{v}}_{p+1}, \bar{\varphi}_{p+1}, \bar{\psi}_{p+1}, \bar{\mathbf{u}}_{p+1}, \bar{E}_{p+1}) =: U^{p+1} - U^p, p \ge 1,$$

based on Lemma 3.2. In fact, let

$$\begin{aligned} \bar{\mathbf{v}}_{p+1} &= \mathbf{v}_{p+1} - \mathbf{v}_p, & \bar{\mathbf{z}}_{p+1} &= \mathbf{z}_{p+1} - \mathbf{z}_p, & \bar{\varphi}_{p+1} &= \varphi_{p+1} - \varphi_p, \\ \bar{\psi}_{p+1} &= \psi_{p+1} - \psi_p, & \bar{\mathbf{u}}_{p+1} &= \mathbf{u}_{p+1} - \mathbf{u}_p, & \bar{E}_{p+1} &= E_{p+1} - E_p, \end{aligned}$$

then by using Lemma 3.2 and repeating the similar arguments as above, we can show after a tedious computation that there is a time $0 < T_{**} \le T_*$ such that the difference Y^{p+1}, $p \ge 1$, of the approximate solution sequence satisfies the following estimates

$$\begin{aligned} \sum_{p=1}^{\infty} (\|(\bar{\mathbf{u}}_{p+1}, \bar{E}_{p+1})(t)\|^2_{\mathcal{H}^3(\mathbb{R}^3)} &+ \|\bar{\mathbf{u}}'_{p+1}(t)\|^2_{\mathcal{H}^2(\mathbb{R}^3)} \\ &+ \|\bar{E}'_{p+1}(t)\|^2_{\mathcal{H}^3(\mathbb{R}^3)}) \le C_*, \end{aligned} \tag{3.57}$$

$$\sum_{p=1}^{\infty} \Big(\|(\bar{\mathbf{v}}_{p+1}, \bar{\mathbf{z}}_{p+1})(t)\|^2_{\mathcal{H}^4} + \|(\bar{\mathbf{v}}'_{p+1}, \bar{\mathbf{z}}'_{p+1})(t)\|^2_{\mathcal{H}^3(\mathbb{R}^3)} \Big) \le C_*, \tag{3.58}$$

$$\begin{aligned} \sum_{p=1}^{\infty} (\|\bar{\psi}_{p+1}\|^2_{C^i([0,T_{**}];H^{6-2i})} &+ \|\bar{\varphi}_{p+1}\|^2_{C^1([0,T_{**}];H^4)} \\ &+ \|\bar{\varphi}''_{p+1}\|^2_{C([0,T_{**}];H^2)}) \le C_*, \end{aligned} \tag{3.59}$$

where $i = 0, 1, 2$, and $C_* = C_*(N, M_*)$ denotes a positive constant depending on N and M_*. Here we recall that $\mathcal{H}^k = \{f \in L^6(\mathbb{R}^3);\ Df \in H^{k-1}(\mathbb{R}^3)\},\ k \geq 1$.

By applying Ascoli-Arzela Theorem (to time variable) and Rellich-Kondrachov theorem (to spatial variable) [45], we can prove by the standard argument [40] that there exists a (unique) $U = (\mathbf{v}, \mathbf{z}, \varphi, \psi, \mathbf{u}, E)$ such that as $p \to \infty$

$$\begin{cases} \mathbf{v}_p \to \mathbf{v} \ \text{ in } \ C^i([0, T_{**}]; \mathcal{H}^{4-i-\sigma}(\mathbb{R}^3)), \\ \mathbf{z}_p \to \mathbf{z} \ \text{ in } \ C^i([0, T_{**}]; \mathcal{H}^{4-i-\sigma}(\mathbb{R}^3)), \\ \mathbf{u}_p \to \mathbf{u} \ \text{ in } \ C^i([0, T_{**}]; \mathcal{H}^{3-\sigma}(\mathbb{R}^3)), \\ E_p \to E \ \text{ in } \ C^i([0, T_{**}]; \mathcal{H}^{3-\sigma}(\mathbb{R}^3)), \\ \varphi_p \to \varphi \ \text{ in } \ C^1([0, T_{**}]; H^{3-\sigma}(\mathbb{R}^3)) \cap C^2([0, T_{**}]; H^{2-\sigma}(\mathbb{R}^3)), \\ \psi_p \to \psi \ \text{ in } \ C^j([0, T_{**}]; H^{6-2j-\sigma}(\mathbb{R}^3)) \end{cases}$$

with $i = 0, 1$, $j = 0, 1, 2$, and $0 < \sigma < 1/2$. Moreover, by (3.56) it holds

$$\varphi(x, t) \geq \frac{1}{4}\psi_* > 0, \quad (x, t) \in \mathbb{R}^3 \times [0, T_{**}]. \tag{3.60}$$

Passing into limit $p \to \infty$ in (3.29)–(3.34), we obtain the (short time) existence and uniqueness of classical solution of the extended systems (3.1)–(3.6) constructed in Section 3.1. The proof of Theorem 3.1 is completed.

Proof of Theorem 1.1. By Theorem 3.1, we have the existence and uniqueness of short time strong solution $(\mathbf{v}, \mathbf{z}, \psi, \varphi, \mathbf{u}, E)$ of the extended systems (3.1)–(3.6) with initial data

$$(\mathbf{v}, \mathbf{z}, \varphi, \psi, \mathbf{v}, E)(x, 0) = (\mathcal{Q}\mathbf{u}_1, \mathcal{P}\mathbf{u}_1, \psi_1, \psi_1, \mathbf{u}_1, E_1)(x).$$

The most important fact we show below is that the extended systems (3.1)–(3.6) for $U = (\mathbf{v}, \mathbf{z}, \varphi, \psi, \mathbf{u}, E)$ are equivalent to Eqs. (1.12)–(1.14) for $(\psi, \mathbf{u}, E)$ for classical solutions. That is, it holds

$$\psi = \varphi, \ \mathbf{u} = \mathbf{v} + \mathbf{z}, \ t \geq 0, \quad \text{so long as} \quad [\varphi - \psi](0) = 0, \ [\mathbf{u} - \mathbf{v} - \mathbf{z}](0) = 0. \tag{3.61}$$

In fact, erasing the common term $\mathbf{u}\cdot\nabla\psi$ in the ODE equation (3.3) for φ and the divergence equation for $\mathbf{v}$, i.e.,

$$\varphi_t + \mathbf{u}\cdot\nabla\psi + \frac{1}{2}\varphi\nabla\cdot\mathbf{v} = 0, \quad \nabla\cdot\mathbf{v} = -\frac{2(\psi_t + \mathbf{u}\cdot\nabla\psi)}{\varphi}, \tag{3.62}$$

we obtain

$$(\varphi - \psi)_t(x, t) = 0, \quad t \in [0, T_{**}]. \tag{3.63}$$

By (3.63), (3.60) and the fact

$$\varphi(x, 0) = \psi(x, 0) = \psi_1(x) \quad \Rightarrow \quad (\varphi - \psi)(x, 0) = 0,$$

we obtain

$$\psi(x,t) = \varphi(x,t) \geq \frac{1}{4}\psi_* > 0, \qquad (x,t) \in \mathbb{R}^3 \times [0, T_{**}], \tag{3.64}$$

$$\psi_t + \mathbf{u}\cdot\nabla\psi + \frac{1}{2}\psi\nabla\cdot\mathbf{v} = 0, \qquad (x,t) \in \mathbb{R}^3 \times [0, T_{**}]. \tag{3.65}$$

With the help of $(3.2)_1$ which gives $\omega = \nabla\times\mathbf{z}$, (3.64) and (3.25), we obtain from Eq. (3.2) for $\mathbf{u}$ the following equation

$$\partial_t\mathbf{u} + \frac{1}{2}\nabla(|\mathbf{v}+\mathbf{z}|^2) - (\mathbf{v}+\mathbf{z})\times(\nabla\times\mathbf{z}) + \nabla h(\psi^2) + \mathbf{u} = E + \frac{\varepsilon^2}{2}\nabla\Big(\frac{\Delta\psi}{\psi}\Big). \tag{3.66}$$

Taking *curl* to Eq. (3.66) and making the difference between the resulted equation and Eq. $(3.2)_2$ for $\omega = \nabla\times\mathbf{z}$, we have

$$\nabla\times(\mathbf{u}-\mathbf{z})_t + \nabla\times(\mathbf{u}-\mathbf{z}) = 0, \quad t \geq 0. \tag{3.67}$$

Similarly, recombining the various terms in Eq.(3.26) for ψ we can obtain, with the help of (3.64) and (3.25), that

$$\begin{aligned} \psi_{tt} + \psi_t - \frac{1}{4\psi}\nabla\cdot\Big(\psi^2\frac{1}{2}\nabla(|\mathbf{v}+\mathbf{z}|^2) - \psi^2(\mathbf{v}+\mathbf{z})\times(\nabla\times\mathbf{z})\Big) \\ -\frac{1}{2\psi}\Delta P(\psi^2) + \mathbf{u}\cdot\nabla\psi_t + \frac{1}{2}\psi_t(\nabla\cdot\mathbf{v}) \\ +\frac{1}{2\psi}\nabla\cdot(\psi^2 E) + \frac{1}{4\psi}\varepsilon^2\nabla\cdot\Big(\psi^2\nabla\Big(\frac{\Delta\psi}{\psi}\Big)\Big) = 0. \end{aligned} \tag{3.68}$$

From (3.65) we have $\psi_t = -\mathbf{u}\cdot\nabla\psi - \frac{1}{2}\psi\nabla\cdot\mathbf{v}$. Substituting it into (3.68) and representing $\mathbf{u}_t$ by (3.66), we obtain after a computation that

$$\nabla\cdot(\mathbf{u}-\mathbf{v})_t + \nabla\cdot(\mathbf{u}-\mathbf{v}) = 0. \tag{3.69}$$

Since

$$\nabla\cdot(\mathbf{u}-\mathbf{v})(x,0) = 0, \quad \nabla\times(\mathbf{u}-\mathbf{z})(x,0) = 0,$$

it follows from (3.67), (3.69) that

$$\nabla\cdot(\mathbf{u}-\mathbf{v})(x,t) = 0, \quad \nabla\times(\mathbf{u}-\mathbf{z})(x,t) = 0, \quad (x,t) \in \mathbb{R}^3\times[0, T_{**}]. \tag{3.70}$$

After decomposing the velocity $\mathbf{u}$ into $\mathbf{u}_g + \mathbf{u}_r =: \mathcal{Q}\mathbf{u} + \mathcal{P}\mathbf{u}$ and using Theorems 2.1–2.3, we conclude from (3.70) that

$$\mathbf{u} = \mathbf{v} + \mathbf{z}, \quad \mathcal{Q}\mathbf{u} = \mathbf{v}, \ \mathcal{P}\mathbf{u} = \mathbf{z}, \quad (x,t) \in \mathbb{R}^3 \times [0, T_{**}]. \tag{3.71}$$

Thus, by (3.71), (3.66), and (3.15), we recover the equation for $\mathbf{u}$ which is exactly Eq. (3.14). Multiplying (3.65) with ψ and using (3.71) we recover the equation for ψ (which is exactly Eq. (1.12))

$$\partial_t(\psi^2) + \nabla\cdot(\psi^2\mathbf{u}) = 0. \tag{3.72}$$

Finally, by (3.6) and Theorem 2.1 we show that $E \in C^1([0,T_{**}];\mathcal{H}^3(\mathbb{R}^3))$ is the unique solution of the divergence equation:

$$\nabla\cdot E = \psi^2 - \mathcal{C}, \quad \nabla\times E = 0, \quad E(x,t) \to 0,\ |x| \to \infty.$$

Therefore $(\psi, \mathbf{u}, E)$ with $\psi \geq \frac{1}{2}\psi_* > 0$ is the unique local (in time) solution of IVP (1.12)–(1.15). Again by a straightforward computation once more, we can find

$$\psi \in C^i([0,T_{**}];H^{6-2i}(\mathbb{R}^3)) \cap C^3([0,T_{**}];L^2(\mathbb{R}^3)), \quad i = 0,1,2;$$
$$\mathbf{u} \in C^i([0,T_{**}];\mathcal{H}^{5-2i}(\mathbb{R}^3)),\ i = 0,1,2; \quad E \in C^1([0,T_{**}];\mathcal{H}^3(\mathbb{R}^3)).$$

The proof of Theorem 1.1 is completed. □

4 Global existence and large time behavior

We establish the uniform a-priori estimates for local classical solutions $(\psi, \mathbf{u}, E)$ of the IVP (1.12)–(1.15) for any fixed $T > 0$ when $(\psi, \mathbf{u}, E)$ is around the asymptotic state $(\widetilde{\psi}, 0, \widetilde{E})$.

4.1 Reformulation of original problems

We reformulate the original problem (1.12)–(1.15) for classical solutions. For simplicity, we still set $\tau = 1$ and $\lambda = 1$.

Set

$$w = \psi(x,t) - \widetilde{\psi}(x), \quad \theta = E - \widetilde{E}.$$

By (1.12), (1.14), (3.26) we have the following systems for $(w, \mathbf{u}, \theta)$

$$\mathbf{u}_t + (\mathbf{u}\cdot\nabla)\mathbf{u} + \mathbf{u} = f_1(x,t), \tag{4.1}$$

$$w_{tt} + w_t + \frac{1}{4}\varepsilon^2\Delta^2 w + \widetilde{\psi}^2 w + 2\mathbf{u}\cdot\nabla w_t - \nabla\cdot(P'(\widetilde{\psi}^2)\nabla w) = f_2, \tag{4.2}$$

$$\nabla\cdot\theta = (2\widetilde{\psi} + w)w, \tag{4.3}$$

and the corresponding initial values are

$$w(x,0) = w_1(x),\ w_t(x,0) = w_2(x),\ \mathbf{u}(x,0) = \mathbf{u}_1(x), \tag{4.4}$$

with

$$w_1(x) =: \psi_1 - \widetilde{\psi},\ w_2(x) =: -\mathbf{u}_1\cdot\nabla(\widetilde{\psi}+w_1) - \frac{1}{2}(\widetilde{\psi}+w_1)\nabla\cdot\mathbf{u}_1. \quad (4.5)$$

Here

$$f_1(x,t) = \theta - \nabla(h((\widetilde{\psi}+w)^2) - h(\widetilde{\psi}^2)) \quad (4.6)$$

$$+ \frac{1}{2}\varepsilon^2\nabla\left(\frac{\Delta(w+\widetilde{\psi})}{w+\widetilde{\psi}} - \frac{\Delta\widetilde{\psi}}{\widetilde{\psi}}\right), \quad (4.7)$$

$$\begin{aligned} f_2(x,t) = &- \frac{w_t^2}{w+\widetilde{\psi}} - \frac{1}{2(\widetilde{\psi}+w)}\nabla\cdot\Big((\widetilde{\psi}+w)^2(\widetilde{E}+\theta)\Big) + \frac{1}{2\widetilde{\psi}}\nabla\cdot\Big(\widetilde{\psi}^2\widetilde{E}\Big) \\ &+ \widetilde{\psi}^2 w + \frac{\varepsilon^2}{4}\frac{|\Delta(\widetilde{\psi}+w)|^2}{(\widetilde{\psi}+w)} - \frac{\varepsilon^2}{4}\frac{|\Delta\widetilde{\psi}|^2}{\widetilde{\psi}} \\ &+ \frac{1}{2(\widetilde{\psi}+w)}\Delta P((\widetilde{\psi}+w)^2) - \frac{1}{2\widetilde{\psi}}\Delta P(\widetilde{\psi}^2) - \nabla\cdot\Big(P'(\widetilde{\psi}^2)\nabla w\Big) \\ &+ \frac{1}{2(\widetilde{\psi}+w)}\nabla^2\cdot\Big([\widetilde{\psi}+w]^2\mathbf{u}\otimes\mathbf{u}\Big) + 2\mathbf{u}\cdot\nabla w_t. \end{aligned} \quad (4.8)$$

The derivatives of w and $\mathbf{u}$ satisfy

$$2w_t + 2\mathbf{u}\cdot\nabla(\widetilde{\psi}+w) + (\widetilde{\psi}+w)\nabla\cdot\mathbf{u} = 0. \quad (4.9)$$

4.2 The a-priori estimates

For $T > 0$ define the workspace for $(w, \mathbf{u}, \theta)$ of the IVP (4.2)–(4.4) as

$$X(T) = \{(w,\mathbf{u},\theta)(t) \in H^6(\mathbb{R}^3)\times\mathcal{H}^5(\mathbb{R}^3)\times\mathcal{H}^3(\mathbb{R}^3), \quad 0 \le t \le T\},$$

and assume that it holds

$$\delta_T = \max_{0\le t\le T}(\|w(t)\|_{H^6(\mathbb{R}^3)} + \|\mathbf{u}(t)\|_{\mathcal{H}^5(\mathbb{R}^3)}) \ll 1. \quad (4.10)$$

It is easy to verify that under the assumption (4.10) it holds that

$$\frac{1}{2}\sqrt{\bar{\rho}} \le w + \widetilde{\psi} \le \frac{3}{2}\sqrt{\bar{\rho}}. \quad (4.11)$$

First, by Theorem 2.1 we derive the following estimates.

Lemma 4.1. *It holds for* $(w,\mathbf{u},\theta) \in X(T)$

$$\begin{cases} |\theta| + \|\theta\|_{L^6} + \|D\theta\|_3 \le C\|w\|_3, \\ |\theta_t| + \|\theta_t\|_{L^6} + \|D\theta_t\|_2 \le C\|w_t\|_2, \end{cases} \quad (4.12)$$

provided that $\delta_T + \delta_0 \ll 1$.

The estimates (4.12) together with (4.10) give

$$|\theta| + |\theta_t| + \|(\theta, \theta_t)\|_{L^6}^6 + \|D(\theta, \theta_t)\|_2^2 \le C\delta_T. \tag{4.13}$$

Next, we have the following basic estimates:

Lemma 4.2. *It holds, for $(w, \mathbf{u}, E) \in X(T)$, that*

$$\|(w, \nabla w, \Delta w, w_t, \nabla\times\mathbf{u})(t)\|^2 \le C(\|w_1\|_2^2 + \|D\mathbf{u}_1\|^2)e^{-\beta_1 t}, \tag{4.14}$$

$$\|(w, \Delta w, w_t)(t)\|_2^2 + \|(w_t, \Delta w_t, w_{tt})(t)\|_2^2 + \|\nabla\times\mathbf{u}\|_4^2 \le C(\|w_1\|_6^2 + \|D\mathbf{u}_1\|_4^2)e^{-\beta_2 t}, \tag{4.15}$$

with $\beta_1, \beta_2 > 0$ two constants, provided that δ_T and δ_0 are small enough.

Proof. Let us prove (4.14) first. Take inner product between (4.2) and $w + 2w_t$ and integrate it by parts over $\mathbb{R}^3$, we obtain by Cauchy' inequality, (4.10), (4.11), Lemma 4.1, that

$$\begin{aligned}
&\frac{d}{dt}\int_{\mathbb{R}^3}\left(\frac{1}{2}w^2 + ww_t + w_t^2 + \widetilde{\psi}^2 w^2 + \frac{1}{4}\varepsilon^2|\Delta w|^2 + P'(\widetilde{\psi}^2)|\nabla w|^2\right)dx \\
&+ \int_{\mathbb{R}^3}\left(w_t^2 + \widetilde{\psi}^2 w^2 + \frac{1}{4}\varepsilon^2|\Delta w|^2 + P'(\widetilde{\psi}^2)|\nabla w|^2\right)dx \\
&\le C(\delta_T + \delta_0)\|(w_t, w, \nabla w, \Delta w)\|^2 + C(\delta_T + \delta_0)\|\nabla\times\mathbf{u}\|^2.
\end{aligned} \tag{4.16}$$

It is easy to verify under the condition (1.17) that there is an $a_0 > 0$ so that

$$\int_{\mathbb{R}^3}\left(\widetilde{\psi}^2 w^2 + \frac{1}{4}\varepsilon^2|\Delta w|^2 + P'(\widetilde{\psi}^2)|\nabla w|^2\right)dx \ge a_0(\|w\|^2 + \|\Delta w\|^2) \tag{4.17}$$

due to the facts that the above inequality holds when $P'(\bar{\rho}) > 0$ since $P'(\widetilde{\psi}^2) = P'(\bar{\rho}) + O(1)\delta_0 > 0$ when δ_0 is small, and if $P'(\bar{\rho}) \le 0$ and (1.17) holds, we know that there is a positive constant k_0 with $0 < 1 - k_0 < 1$ such that $\sqrt{\bar{\rho}}k_0\varepsilon + P'(\bar{\rho}) > 0$.

To estimate the L^2-norm of $\nabla\times\mathbf{u}$, we take $\nabla\times$ on Eq. (4.1) to obtain

$$\partial_t\nabla\times\mathbf{u} + \nabla\times\mathbf{u} + \mathbf{u}\cdot\nabla(\nabla\times\mathbf{u}) + (\nabla\times\mathbf{u})\nabla\cdot\mathbf{u} - ([\nabla\times\mathbf{u}]\cdot\nabla)\mathbf{u} = 0. \tag{4.18}$$

Taking inner product between (4.18) and $\nabla\times\mathbf{u}$ and integrating it by parts, we have

$$\frac{1}{2}\frac{d}{dt}\|\nabla\times\mathbf{u}\|^2 + (1 - C\delta_T)\|\nabla\times\mathbf{u}\|^2 \le C\delta_T\|(w_t, w, \nabla w, \Delta w)\|^2. \tag{4.19}$$

Making summation of (4.16) and (4.19) and applying Gronwall's lemma and Nirenberg inequality, we get finally the expected estimate (4.14) for a constant $\beta_1 > 0$, provided that δ_T and δ_0 are small enough.

The higher order estimate (4.15) for w and $\mathbf{u}$ can be obtained in a similar fashion, we omit the details. Finally, since we can estimate $D\mathbf{u}$ by

$$\|D\mathbf{u}\|_4^2 \leq C\|\nabla\times\mathbf{u}\|_4^2 + \|\nabla\cdot\mathbf{u}\|_4^2, \tag{4.20}$$

and express $D^5 w$ and $D^6 w$ through Eq.(4.2), we have, by Lemmas 4.1–4.2, and (2.20), that

Theorem 4.3. *It holds, for $(w, \mathbf{u}, \theta) \in X(T)$, that*

$$\begin{aligned}\|w(t)\|_{H^6(\mathbb{R}^3)} + \|(\mathbf{u},\theta)(t)\|_{L^6(\mathbb{R}^3)} &+ \|D\mathbf{u}(t)\|_{H^4(\mathbb{R}^3)} \\ &+ \|D\theta(t)\|_{H^2(\mathbb{R}^3)} \leq C\delta_1 e^{-\beta_3 t},\end{aligned} \tag{4.21}$$

provided that $\delta_T + \delta_0 \ll 1$. Here $\beta_3 = \min\{\beta_2,\ \beta_1\}$ and δ_1 is given by (1.20).

Proof of Theorem 1.5. Based on Theorem 4.3, we can prove that (4.10) is true for the classical solution existing locally in time so long as $\delta_1 = ||\psi_1 - \widetilde{\psi}||_6 + ||D\mathbf{u}_1||_4$ is small enough such that $C\delta_1 \ll 1$. Then, the continuity argument together with the uniform a-priori bounds (4.21) gives the global existence, and the time-asymptotic behavior of the global solutions follows. □

References

[1] G. Baccarani and W.R. Wordeman, An investigation of steady-state velocity overshoot effects in Si and GaAs devices, *Solid State Electronics* **28** (1985), 407–416.

[2] M.G. Ancona and H.F. Tiersten, Microscopic physics of the Silicon inversion layer, *Physical Review B*, **35** (1987), 7959–7965.

[3] M.G. Ancona and G.I. Iafrate, Quantum correction to the equation of state of an electron gas in a semiconductor, *Physical Review B*, **39** (1989), 9536–9540.

[4] F. Brezzi, I Gasser, P. Markowich, and C. Schmeiser, Thermal equilibrium state of the quantum hydrodynamic model for semiconductor in one dimension, *Appl. Math. Lett.*, **8** (1995), 47–52.

[5] D. Bohm, A suggested interpretation of the quantum theory in terms of "hidden" valuables: I; II, *Phys. Rev.* **85** (1952), 166–179;180–193.

[6] R. Courant and K. O. Friedrichs, *Supersonic flow and shock waves*, AMS, Vol. **21** Springer-Verlag, New York-Heidelberg, 1976.

[7] C. Dafermos, *Hyperbolic conservation law in continous mechanics*, Grundlehren der mathematischen Wissenschaften Vol. 325, Springer, Berlin, 2000.

[8] P. Degond, and C. Ringhofer, Quantum moment hydrodynamics and the entropy principle, *J. Statist. Phys.* **112** (2003), no. 3–4, 587–628.

[9] P. Degond, F. Méhats and C. Ringhofer, Quantum hydrodynamic models derived from the entropy principle, *Nonlinear partial differential equations and related analysis* 107–131, Contemp. Math., 371, Amer. Math. Soc., Providence, RI, 2005.

[10] D.K. Ferry and J.-R. Zhou, Form of the quantum potential for use in hydrdynamic equations for semiconductor device modeling, *Phys. Rev. B*, **48** (1993), 7944–7950.

[11] R. Feynman, *Statistical mechanics, a set of lectures*, New York: W.A. Benjamin, 1972.

[12] I. Gamba and A. Jüngel, Asymptotic limits in quantum trajectory models, *Commun. PDEs.*, **27** (2002), no. 3–4, 669–691.

[13] I. Gamba and A. Jüngel, Positive solutions to singular second and third order differential equations for quantum fluids, *Arch. Rat. Mech. Anal.*, **156** (2001), 183–203.

[14] C. Gardner, The quantum hydrodynamic model for semiconductors devices, *SIAM J. Appl. Math.*, **54** (1994), 409–427.

[15] C. Gardner and C. Ringhofer, Dispersive/hyperbolic models for transport in semiconducotr devices, accepted for publication in IMA Volumes in Mathematics and its Applications.

[16] I. Gasser, Traveling wave solutions for a quantum hydrodynamic model, *Appl. Math. Lett.* **14** (2001), no. 3, 279–283.

[17] I. Gasser and A. Jüngel, The quantum hydrodynamic model for semiconductors in thermal equilibrium, *Z. Angew. Math. Phys.*, **48** (1997), 45–59.

[18] I. Gasser, C.-K. Lin and P. Markowich, A review of dispersive limits of the (non)linear Schrödinger-type equation, *Taiwanese J. of Math.*, **4** (2000), 501–529.

[19] I. Gasser and P. Markowich, Quantum hydrodynamics, Wigner transforms and the classical limit, *Asymptotic Anal.*, **14** (1997), 97–116.

[20] I. Gasser, P. A. Markowich, and C. Ringhofer, Closure conditions for classical and quantum moment hierarchies in the small temperature limit, *Transp. Theory Stat. Phys.*, **25** (1996), 409–423.

[21] M. T. Gyi and A. Jüngel, A quantum regularization of the one-dimensional hydrodynamic model for semiconductors, *Adv. Diff. Eqs.*, **5** (2000), 773–800.

[22] C.-C Hao, Y.-L. Jia, and H.-L. Li, Quantum Euler-Poisson system: local existence, *J. Partial Differential Equations*, **16** (2003), no. 4, 306–320.

[23] Y.-L. Jia and H.-L. Li, Large-time behavior of solutions of quantum hydrodynamic model for semiconductors, *Acta Math. Sci. Ser. B* **26** (2006), no. 1, 163–178.

[24] F. Huang, H.-L. Li, and A. Matsumura, Existence and stability of steady-state of one-dimensional quantum hydrodynamic system for semiconductors, *J. Differential Equations*, **225** (2006), no. 1, 440–464

[25] A. Jüngel, A steady-state potential flow Euler-Poisson system for charged quantum fluids, *Comm. Math. Phys.*, **194** (1998), 463–479.

[26] A. Jüngel, *Quasi-hydrodynamic semiconductor equations*, Progress in Nonlinear Differential Equations, Birkhäuser, Basel (2001).

[27] A. Jüngel and H.-L. Li, Quantum Euler-Poisson systems: existence of stationary states, *Archivum Mathematicum*, **40** (2004), no. 4, 435–456.

[28] A. Jüngel and H.-L. Li, Quantum Euler-Poisson systems: global existence and exponential decay, *Quarterly Appl. Math.*, **62** (2004), no. 3, 569–600.

[29] A. Jüngel, H.-L. Li, P. Markowich, and S. Wang, Recent progress on quantum hydrodynamic models for semiconductors, *Hyperbolic problems: theory, numerics, applications*, 217–226, Springer, Berlin, 2003.

[30] A. Jüngel, H.-L. Li, A. Matsumura, Stability and relaxation asymptotic of quantum hydrodynamics model in R^3, *J. Differential Equations* **225** (2006), 1–25

[31] A. Jüngel, M. C. Mariani and D. Rial, Local existence of solutions to the transient quantum hydrodynamic equations, *Math. Models Methods Appl. Sci.* **12** (2002), no. 4, 485–495.

[32] N. Klusdahl, A. Kriman, D. Ferry, and C. Ringhofer, Self-consistent study of the resonant-tunneling diode, *Phys. Rev. B*, **39** (1989), 7720–7735.

[33] T. Kato, Quasi-linear equations of evolution, with applications to partial differential equations, *Lecture Notes in Math.*, **448** (1975), 25–70, Springer, Berlin.

[34] H.-L. Li and C.-K. Lin, Semiclassical limit and well-posedness of nonlinear Schrödinger-Poisson, *Electronic Journal of Differential Equations*, Vol. 2003(2003), No. 93, 1–17.

[35] H.-L. Li and C.-K. Lin, Zero Debye length asymptotic of the quantum hydrodynamic model for semiconductors, *Comm. Math. Phys.* **256** (2005), no. 1, 195–212.

[36] H.-L. Li and P. Marcati, Existence and asymptotic behavior of multi-dimensional quantum hydrodynamic model for semiconductors, *Comm. Math. Phys.*, **245** (2004), no. 2, 215–247.

[37] H.-L. Li and P. Markowich, A review of hydrodynamical models for semiconductors: asymptotic behavior, *Bol. Soc. Brasil. Mat.(N.S.)*, **32** (2001), no. 3, 321–342.

[38] M. Loffredo and L. Morato, On the creation of quantum vortex lines in rotating HeII. *Il nouvo cimento*, **108B** (1993), 205–215.

[39] E. Madelung, Quantentheorie in hydrodynamischer Form, *Z. Physik*, **40** (1927), 322.

[40] A. Majda, *Compressible Fluid Flow and Systems of Conservation Laws in Several Space Variables*, Springer-Verlag 1984.

[41] A. Majda and A. Bertozzi, *Vorticity and incompressible flow*, Cambridge University Press, Cambridge, 2002.

[42] P. Markowich, On the equivalentce of the Schrödinger and the quantum Liouville equations, *Math. Meth. Appl. Sci.*, **11** (1989), 459–469.

[43] P. Markowich and C.Ringhofer, Quantum hydrodynamics for semiconductors in the high-field case, *Appl. Math. Lett.* **7** (1994), no. 5, 37–41.

[44] P. A. Markowich, C. Ringhofer and C. Schmeiser, *Semiconductor Equations*, Springer, Wien 1990.

[45] J. Simon, Compact sets in the space $L^p(0,T;B)$, *Ann. Math. Pura. Appl.*, **146** (1987), 65–96.

[46] J. Smoller, *Shock waves and reaction-diffusion equations*, Berlin, Heidelberg, New York, Springer 1982.

[47] E. M. Stein, *Singular integrals and differentiability properties of functions*, Princeton University Press, New Jersey 1970.

[48] A. Unterreiter, The thermal equilibrium solution of a generic bipolar quantum hydrodynamic model, *Comm. Math. Phys.*, **188** (1997), 69–88.

[49] E. Wigner, On the quantum correction for thermodynamic equilibrium, *Phys. Rev.*, **40** (1932), 749–759.

[50] B. Zhang and W. Jerome, On a steady-state quantum hydrodynamic model for semiconductors, *Nonlinear Anal., TMA*, **26** (1996), 845–856.

[51] E. Zeidler, *Nonlinear functional analysis and its applications*, **II:** Nonlinear monotone operators, Springer-Verlag 1990.

Bloch Decomposition-Based Method for High Frequency Waves in Periodic Media*

Zhongyi Huang
Dept. of Mathematical Sciences, Tsinghua University
Beijing 100084, China
Email: zhuang@math.tsinghua.edu.cn

Shi Jin
Dept. of Mathematics, University of Wisconsin-Madison
Madison, WI 53706, USA
Email: jin@math.wisc.edu

Peter A. Markowich
DAMTP, University of Cambridge
Wilberforce Road, Cambridge CB3 0WA, UK
Email: P.A.Markowich@damtp.cam.ac.uk

Christof Sparber
DAMTP, University of Cambridge
Wilberforce Road, Cambridge CB3 0WA, UK
Email: c.sparber@damtp.cam.ac.uk

Abstract

In this paper, we introduce the Bloch decomposition-based time splitting spectral method to conduct numerical simulations of the (non)linear dispersive wave equations with periodic coefficients. We first consider the numerical simulations of the dynamics of nonlinear Schrödinger equations subject to periodic and confining potentials. We consider this system as a two-scale asymptotic problem with different scalings of the nonlinearity. Particularly we discuss (nonlinear) mass transfer between different Bloch bands and also present three-dimensional simulations for lattice Bose-Einstein condensates in the superfluid regime. We further

*This work was partly supported by the NSFC (10676017) and the National Basic Research Program of China (2005CB321701), the Austrian-Chinese Technical-Scientific Cooperation Agreement, the NSF grant No. DMS-0608720.

estimate the stability of our scheme in the presence of random perturbations and give numerical evidence for the well-known phenomenon of Anderson's localization.

0 Introduction

In this paper, we consider the propagation of (non)linear *high frequency* waves in the heterogeneous media with *periodic microstructures.* Such problems arise, e.g., in the studies of

- Bose-Einstein condensates (BECs) in optical lattices,
- composite materials,
- photonic crystals,
- laser optics,
- plasma physics,
- ······

We are interested in the case where the *typical wavelength* is comparable to the *period of the medium*, both of which are assumed to be *small* on the *length-scale of the considered physical domain.* This consequently leads us to a problem involving *two-scales* where from now on we shall denote by $0 < \varepsilon \ll 1$ the small dimensionless parameter describing the *microscopic/macroscopic scale ratio.*

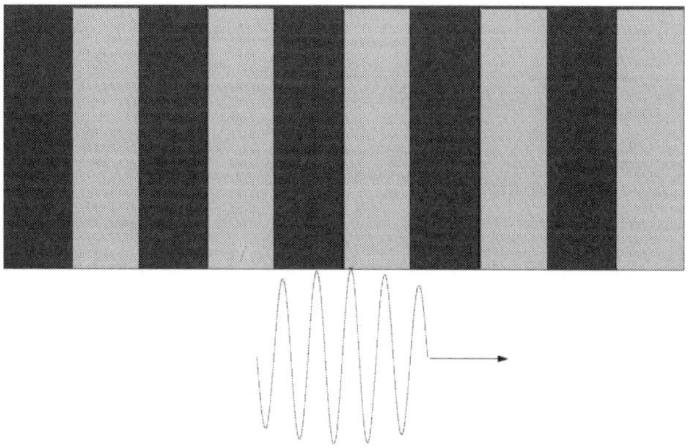

Figure 0.1

Here we shall mainly consider two kinds of problems: the *dynamics of lattice BECs* and the *acoustic waves in composite materials.* Therefore, we will study the *Schrödinger equation* and the *Klein-Gordon equation.*

Recently there is a growing interest in the theoretical description and the experimental realization of *Bose-Einstein condensates* (BECs) under the influence of the so-called *optical lattices*, *cf.* [8, 25, 28, 30]. In such a system there are two extreme situations one needs to distinguish: the *superfluid*, or *Gross-Pitaevskii* (GP) regime and the so-called *Mott insulator.* The two regimes are essentially induced by the strength of the optical lattice, experimentally generated via intense laser fields. In the following we shall focus solely on the superfluid regime, corresponding to situations where the optical lattice potential is of order $\mathcal{O}(1)$ in amplitude. The BEC is usually modelled by the *Gross-Pitaevskii equation*, a cubically nonlinear Schrödinger equation (NLS), given by [30]

$$i\hbar\partial_t\psi = -\frac{\hbar^2}{2m}\Delta\psi + V(x)\,\psi + U_0(x)\psi + N\alpha|\psi|^2\psi, \quad x\in\mathbb{R}^3, t\in\mathbb{R}, \tag{0.1}$$

where m is the *atomic mass*, $\hbar$ is the *Planck constant*, N is the number of atoms in the condensate and $\alpha = 4\pi\hbar^2 a/m$, with $a\in\mathbb{R}$ denoting the characteristic scattering length of the particles.

The external potential $U_0(x)$ is being confined in order to describe the electromagnetic trap needed for the experimental realization of a BEC. Typically it is assumed to be of harmonic form

$$U_0(x) = m\omega_0^2\frac{|x|^2}{2}, \quad \omega_0\in\mathbb{R}. \tag{0.2}$$

$V(x)$ is a *periodic* external potential induced by the applied laser field. A particular example for the periodic potentials used in physical experiments is then given by [11, 30]

$$V(x) = s\sum_{\ell=1}^{3}\frac{\hbar^2\xi_\ell^2}{m}\sin^2(\xi_\ell x_\ell), \quad \xi_\ell\in\mathbb{R}, \tag{0.3}$$

where $\xi = (\xi_1, \xi_2, \xi_3)$ denotes the wave vector of the applied laser field and $s > 0$ is a dimensionless parameter describing the depth of the optical lattice (expressed in terms of the recoil energy).

After appropriate scaling, *cf.* [7, 33], we therefore arrive at the following nonlinear Schrödinger equation in a *semiclassical* asymptotic scaling, *i.e.*

$$\begin{cases} i\varepsilon\partial_t\psi = -\dfrac{\varepsilon^2}{2}\Delta\psi + V_\Gamma\left(\dfrac{x}{\varepsilon}\right)\psi + U(x)\psi + \beta|\psi|^2\psi, & x\in\mathbb{R}^d, \\ \psi\big|_{t=0} = \psi_{\text{in}}(x), \end{cases} \tag{0.4}$$

where $0 < \varepsilon \ll 1$ denotes the small *semiclassical parameter* describing the microscopic/macroscopic scale ratio. The dimensionless equation (0.4) describes the motion of the electrons on the macroscopic scale induced by the external potential $U(x)$. The highly oscillating *lattice-potential* $V_\Gamma(y)$ is assumed to be *periodic* w.r.t some *regular lattice* Γ. For definiteness we shall assume that

$$V_\Gamma(y+2\pi) = V_\Gamma(y) \quad \forall y \in \mathbb{R}, \tag{0.5}$$

i.e. $\Gamma = 2\pi\mathbb{Z}$. It is well known that (0.4) preserves *mass*

$$M\big(\psi(t)\big) = \int_{\mathbb{R}^d} |\psi|^2 dx = M\big(\psi(0)\big),$$

and *energy*

$$E\big(\psi(t)\big) = \int_{\mathbb{R}^d} \left[\frac{\varepsilon^2}{2}|\nabla\psi|^2 + (U + V_\Gamma)|\psi|^2 + \frac{\beta}{2}|\psi|^4\right] dx = E\big(\psi(0)\big).$$

The mathematically precise asymptotic description of $\psi(t)$, solution to (0.4), as $\varepsilon \to 0$, has been intensively studied in [5, 14, 19, 29], relying on different analytical tools. On the other hand the numerical literature on these issues is not so abundant [16, 17, 18].

Generally speaking, for the traditional numerical methods, one needs the restriction $\triangle x = \mathcal{O}(\varepsilon)$ and $\triangle t = \mathcal{O}(\varepsilon)$ to achieve a satisfactory numerical result.

We want to develop an efficient numerical approach with *high accuracy* and *reasonable computational cost*, especially for non-smooth lattice potentials and/or non-smooth external potentials. Furthermore, we need that

- it has *high accuracy* even in the case of *heterogeneous media* and/or *with discontinuities*,
- and it has *uniform convergence* w.r.t the parameter ε.

This paper is organized as follows:
1. Description of the Bloch decomposition-based algorithm,
2. Numerical implementation for lattice BECs,
3. Random coefficients: Stability tests and Anderson localization,
4. Conclusions.

1 Description of the Bloch decomposition-based algorithm

First, let us introduce some notation used throughout this paper, respectively recall some basic definitions used when dealing with periodic Schrödinger operators [2, 5, 34, 36].

With V_Γ obeying (0.5) we have:

- The fundamental domain of our lattice $\Gamma = 2\pi\mathbb{Z}$, is $\mathcal{C} = (0, 2\pi)$.
- The *dual lattice* Γ^* can then be defined as the set of all wave numbers $k \in \mathbb{R}$, for which plane waves of the form $\exp(ikx)$ have the same periodicity as the potential V_Γ. This yields $\Gamma^* = \mathbb{Z}$ in our case.
- The fundamental domain of the dual lattice, *i.e.* the (first) *Brillouin zone*, $\mathcal{B} = \mathcal{C}^*$ is the set of all $k \in \mathbb{R}$ closer to zero than to any other dual lattice point. In our case, that is $\mathcal{B} = \left(-\frac{1}{2}, \frac{1}{2}\right)$.

1.1 Recapitulation of Bloch's decomposition method

One of our main points in all what follows is that the dynamical behavior of (0.4) is mainly governed by the periodic part of the Hamiltonian, for $\varepsilon \ll 1$ in particular. Thus it will be important to study its spectral properties. To this end consider the periodic *Hamiltonian*, where for the moment we set $y = x/\varepsilon$ for simplicity,

$$H = -\frac{1}{2}\,\partial_{yy} + V_\Gamma\,(y)\,, \tag{1.1}$$

which will be focused here only on $L^2(\mathcal{C})$. This is possible due to the periodicity of V_Γ which is allowed since then to cover all of $\mathbb{R}$ by simple translations. More precisely, for $k \in \overline{\mathcal{B}} = \left[-\frac{1}{2}, \frac{1}{2}\right]$ we equip the operator H with the following *quasi-periodic* boundary conditions

$$\begin{cases} \psi(t, y + 2\pi) = \mathrm{e}^{2ik\pi}\psi(t, y) \quad \forall\, y \in \mathbb{R},\ k \in \overline{\mathcal{B}}, \\ \partial_y\psi(t, y + 2\pi) = \mathrm{e}^{2ik\pi}\partial_y\psi(t, y) \quad \forall\, y \in \mathbb{R},\ k \in \overline{\mathcal{B}}. \end{cases} \tag{1.2}$$

It is well known [36] that under very mild conditions on V_Γ, the operator H admits a complete set of eigenfunctions $\varphi_m(y, k), m \in \mathbb{N}$, providing, for each fixed $k \in \overline{\mathcal{B}}$, an orthonormal basis in $L^2(\mathcal{C})$. Correspondingly there exists a countable family of real-valued eigenvalues which can be ordered according to $E_1(k) \leq E_2(k) \leq \cdots \leq E_m(k) \leq \cdots$, $m \in \mathbb{N}$, including the respective multiplicity. The set $\{E_m(k) \,|\, k \in \mathcal{B}\} \subset \mathbb{R}$ is called the mth *energy band* of the operator H and the eigenfunctions $\varphi_m(\cdot, k)$ are usually called *Bloch function.* In the following the index $m \in \mathbb{N}$ will *always* denote the *band index.* Concerning the dependence on $k \in \mathcal{B}$, it has been shown [36] that for any $m \in \mathbb{N}$ there exists a closed subset $\mathcal{A} \subset \mathcal{B}$ such that: $E_m(k)$ is analytic and $\varphi_m(\cdot, k)$ can be chosen to be real analytic function for all $k \in \overline{\mathcal{B}}\backslash\mathcal{A}$. Moreover

$$E_{m-1} < E_m(k) < E_{m+1}(k) \quad \forall\, k \in \overline{\mathcal{B}}\backslash\mathcal{A}. \tag{1.3}$$

If this condition indeed holds for all $k \in \mathcal{B}$ then $E_m(k)$ is called an *isolated Bloch band* [34]. Moreover, it is known that

$$\operatorname{meas}\mathcal{A} = \operatorname{meas}\{k \in \overline{\mathcal{B}} \mid E_n(k) = E_m(k),\ n \neq m\} = 0. \tag{1.4}$$

In this set of measure zero one encounters so called *band crossings*. Note that due to (1.2) we can rewrite $\varphi_m(y,k)$ as

$$\varphi_m(y,k) = \mathrm{e}^{iky}\chi_m(y,k) \quad \forall m \in \mathbb{N}, \tag{1.5}$$

for some 2π-periodic function $\chi_m(\cdot,k)$. In terms of $\chi_m(y,k)$ the *Bloch eigenvalue problem* reads

$$\begin{cases} H(k)\chi_m(y,k) = E_m(k)\chi_m(y,k), \\ \chi_m(y+2\pi,k) = \chi_m(y,k) \quad \forall k \in \mathcal{B}, \end{cases} \tag{1.6}$$

where $H(k)$ denotes the shifted Hamiltonian

$$H(k) := \frac{1}{2}(-\mathrm{i}\partial_y + k)^2 + V_\Gamma(y). \tag{1.7}$$

By solving the eigenvalue problem (1.6), the Bloch decomposition allows us to decompose the Hilbert space $\mathcal{H} = L^2(\mathbb{R})$ into a direct sum of orthogonal *band spaces* [26, 31, 36], i.e.

$$L^2(\mathbb{R}) = \bigoplus_{m=1}^{\infty} \mathcal{H}_m, \quad \mathcal{H}_m := \left\{ f_m(y) = \int_{\mathcal{B}} g(k)\,\varphi_m(y,k)\,\mathrm{d}k, \ g \in L^2(\mathcal{B}) \right\}.$$

This consequently allows us to write

$$\forall f \in L^2(\mathbb{R}): \quad f(y) = \sum_{m\in\mathbb{N}} f_m(y), \quad f_m \in \mathcal{H}_m. \tag{1.8}$$

The corresponding projection of f onto the m-th band space is given by [26]

$$\begin{aligned} f_m(y) &\equiv (\mathbb{P}_m f)(y) \\ &= \int_{\mathcal{B}} \left(\int_{\mathbb{R}} f(\zeta)\overline{\varphi}_m(\zeta,k)\,\mathrm{d}\zeta \right) \varphi_m(y,k)\,\mathrm{d}k. \end{aligned} \tag{1.9}$$

In what follows, we will denote by

$$C_m(k) := \int_{\mathbb{R}} f(\zeta)\overline{\varphi}_m(\zeta,k)\,\mathrm{d}\zeta \tag{1.10}$$

the coefficient of the Bloch decomposition. The main use of the Bloch decomposition is that it reduces an equation of the form

$$\mathrm{i}\partial_t\psi = -\frac{1}{2}\partial_{yy}\psi + V_\Gamma(y)\psi, \quad \psi\big|_{t=0} = \psi_{\mathrm{in}}(y), \tag{1.11}$$

into countably many, exactly solvable problems on $\mathcal{H}_m$. Indeed in each band space one simply obtains

$$\mathrm{i}\partial_t\psi_m = E_m(-\mathrm{i}\partial_y)\psi_m, \quad \psi_m\big|_{t=0} = (\mathbb{P}_m\psi_{\mathrm{in}})(y), \tag{1.12}$$

where $E_m(-i\varepsilon\partial_y)$ denotes the Fourier-multiplier corresponding to the symbol $E_m(k)$. Using the Fourier transformation $\mathcal{F}$, equation (1.12) is easily solved by

$$\psi_m(t,y) = \mathcal{F}^{-1}\left(e^{-iE_m(k)t}(\mathcal{F}(\mathbb{P}^\varepsilon_m\psi_{in}))(k)\right). \tag{1.13}$$

Here the energy band $E_m(k)$ is understood to be periodically extended to all of $\mathbb{R}$.

1.2 The Bloch decomposition(BD)-based split-step algorithm

In [20] we introduced a new numerical method, based on the Bloch decomposition described above. In order to make the paper self-contained, we shall recall here the most important steps of our algorithm and then show how to generalize it to more than one spatial dimension.

As a necessary preprocessing, we first need to calculate the energy bands $E_m(k)$ as well as the eigenfunction $\varphi_m(y,k)$ from (1.2) (or, likewise from (1.6)). In $d = 1$ dimension this is rather easy to do with an acceptable numerical cost as described in [20] (see also [17] for an analogous treatment). We shall therefore not go into details here and only remark that the numerical cost for this preprocessing does not depend on the spatial grid to be chosen later on and is therefore almost *negligible* when compared to the costs spent in the evolutionary algorithms below.

For the convenience of the computations, we consider equation (0.4), for $d = 1$, on a bounded domain $\mathcal{D} = [0, 2\pi]$ with *periodic boundary conditions.* This represents an approximation of the (one-dimensional) whole-space problem, as long as the observed wave function does not touch the boundaries $x = 0, 2\pi$. Then, for some $N \in \mathbb{N}$, $t > 0$, let the time step be

$$\triangle t = \frac{t}{N}, \quad \text{and } t_n = n\triangle t, \ n = 1, \cdots, N.$$

Suppose that there are $L \in \mathbb{N}$ lattice cells of Γ within the computational domain $\mathcal{D}$, and that there are $R \in \mathbb{N}$ grid points in each lattice cell, which yields the following discretization

$$\begin{cases} k_\ell = -\dfrac{1}{2} + \dfrac{\ell - 1}{L}, & \text{where } \ell = \{1, \cdots, L\} \subset \mathbb{N}, \\ y_r = \dfrac{2\pi(r-1)}{R}, & \text{where } r = \{1, \cdots, R\} \subset \mathbb{N}. \end{cases} \tag{1.14}$$

Thus, for any time-step t_n, we evaluate $\psi(t_n, \cdot)$, the solution of (0.4), at the grid points

$$x_{\ell,r} = \varepsilon(2\pi(\ell - 1) + y_r). \tag{1.15}$$

Now we introduce the following unitary transformation of $f \in L^2(\mathbb{R})$

$$(Tf)(y,k) \equiv \widetilde{f}(y,k) := \sum_{\gamma \in \mathbb{Z}} f(\varepsilon(y + 2\pi\gamma))\, \mathrm{e}^{-\mathrm{i}2\pi k\gamma}, \quad y \in \mathcal{C},\ k \in \mathcal{B}, \tag{1.16}$$

such that $\widetilde{f}(y + 2\pi, k) = \mathrm{e}^{2\mathrm{i}\pi k}\widetilde{f}(y,k)$ and $\widetilde{f}(y, k+1) = \widetilde{f}(y,k)$. In other words $\widetilde{f}(y,k)$ admits the same periodicity properties w.r.t. k and y as the Bloch eigenfunction $\varphi_m(y,k)$. Thus we can decompose $\widetilde{f}(y,k)$ as a linear combination of such eigenfunctions $\varphi_m(y,k)$. We introduce the transform T instead of the traditional Bloch transform, in order to solely use FFT in (1.25) and (1.29) below. Note that the following inversion formula holds

$$f(\varepsilon(y + 2\pi\gamma)) = \int_{\mathcal{B}} \widetilde{f}(y,k)\, \mathrm{e}^{\mathrm{i}2\pi k\gamma} \mathrm{d}k. \tag{1.17}$$

Moreover one easily sees that the Bloch coefficient, defined in (1.10), can be equivalently written as

$$C_m(k) = \int_{\mathcal{C}} \widetilde{f}(y,k)\overline{\varphi}_m(y,k)\, \mathrm{d}y, \tag{1.18}$$

which, in view of (1.5), resembles a Fourier integral.

We are now in position to set up the time-splitting algorithm. To this end, we first set $d = 1$, for simplicity. We then solve (0.4) in two steps.

Step 1. First, we solve the equation

$$\mathrm{i}\varepsilon\partial_t\psi = -\frac{\varepsilon^2}{2}\partial_{xx}\psi + V_\Gamma\left(\frac{x}{\varepsilon}\right)\psi, \quad x \in \mathbb{R}, \tag{1.19}$$

on a fixed time-interval $\triangle t$. To do so we consider for each fixed $t \in \mathbb{R}$, the corresponding transformed solution $(T\psi(t,\cdot)) \equiv \widetilde{\psi}(t,y,k)$, where T is defined in (1.16) and $y = x/\varepsilon$. Note that if we would not use T here, the solution $\psi(t,\cdot)$ in general would not satisfy the same periodic boundary conditions (w.r.t. y) as the eigenfunctions $\varphi_m(y,k)$. After applying T we can decompose $\widetilde{\psi}(t,y,k)$ according to

$$\widetilde{\psi}(t,y,k) = \sum_{m \in \mathbb{N}} (\mathbb{P}_m\widetilde{\psi}) = \sum_{m \in \mathbb{N}} C_m(t,k)\varphi_m(y,k). \tag{1.20}$$

Of course, we have to truncate this summation at a certain fixed $M \in \mathbb{N}$. Numerical experiments on the band mixing (see also the next section) give us enough experience to choose M large enough, typically $M = 32$, in order to maintain mass conservation up to a sufficiently high accuracy.

By (1.12), this consequently yields the following evolution equation for the coefficient $C_m(t,k)$

$$i\varepsilon\partial_t C_m(t,k) = E_m(k)\, C_m(t,k), \tag{1.21}$$

which yields

$$C_m(t,k) = C_m(0,k)e^{-iE_m(k)t/\varepsilon}. \tag{1.22}$$

Step 2. We solve the ordinary differential equation

$$i\varepsilon\partial_t\psi = \left(U(x) + \beta|\psi|^2\right)\psi, \tag{1.23}$$

on the same time-interval as before, where the solution obtained in Step 1 serves as initial condition for Step 2. Again, we easily obtain the exact solution for this equation by

$$\psi(t,x) = \psi(0,x)\, e^{-i\left(U(x)+\beta|\psi|^2\right)t/\varepsilon}. \tag{1.24}$$

Note that this splitting conserves the total particle number $\|\psi(t,x)\|_{L^2}$ also on the fully discrete level and is thus *unconditionally stable* in the sense used by Iserles in [23] (w.r.t. to the discrete L^2-norm). Clearly, the algorithm given above is the first order in time. But we can easily obtain a second order scheme by the Strang's splitting method, i.e. perform Step 1 with time-step $\triangle t/2$, then Step 2 with $\triangle t$ and finally once again Step 1 with $\triangle t/2$. Indeed, this is what we should do when we implement the algorithm. Step 1 consequently consists of several intermediate steps:

Step 1.1. We compute $\widetilde{\psi}$, *cf.* (1.16), at time t^n by

$$\widetilde{\psi}(t_n, x_{\ell,r}, k_\ell) = \sum_{j=1}^{L} \psi(t_n, x_{j,r})\, e^{-i2\pi k_\ell\cdot(j-1)}, \tag{1.25}$$

where $x_{\ell,r}$ is as in (1.15).

Step 1.2. Next, we compute the coefficient $C_m(t_n, k_\ell)$ via (1.18),

$$C_m(t_n,k_\ell) \approx \frac{2\pi}{R}\sum_{r=1}^{R} \widetilde{\psi}(t_n, x_{\ell,r}, k_\ell)\overline{\chi_m}(y_r, k_\ell)\, e^{-ik_\ell y_r}. \tag{1.26}$$

Step 1.3. The obtained Bloch coefficients are then evolved up to t^{n+1} as given by (1.22),

$$C_m(t_{n+1},k_\ell) = C_m(t_n,k_\ell)\, e^{-iE_m(k_\ell)\triangle t/\varepsilon}. \tag{1.27}$$

Step 1.4. Then we obtain $\widetilde{\psi}$ at the time t_{n+1} by summing up all band contributions

$$\widetilde{\psi}(t_{n+1}, x_{\ell,r}, k_\ell) = \sum_{m=1}^{M} C_m(t_{n+1},k_\ell)\chi_m(y_r,k_\ell)\, e^{ik_\ell y_r}. \tag{1.28}$$

Step 1.5. Finally, we perform the inverse transformation (1.17),

$$\psi(t_{n+1}, x_{\ell,r}, k_\ell) \approx \frac{1}{L} \sum_{j=1}^{L} \widetilde{\psi}(t_{n+1}, x_{j,r}, k_j)\, \mathrm{e}^{\mathrm{i}2\pi k_j(\ell-1)}. \tag{1.29}$$

This concludes the numerical procedure performed in Step 1.

In our algorithm, we compute the dominant effects from the dispersion and the periodic lattice potential in one step, maintaining their strong interaction, and treat the non-periodic potential as a perturbation. Because the split-step error between the periodic and non-periodic parts is relatively small, the time-steps can be chosen to be *considerably larger* than these for a conventional time-splitting algorithm [3, 4], see [20] for more details.

Moreover, an extension of the above given algorithm to more than one spatial dimension is straightforward, if the periodic potential V_Γ is of the following form

$$V_\Gamma(y) = \sum_{j=1}^{d} V_j(x_j), \text{ such that: } V_j(x_j + \gamma_j) = V_j(x_j). \tag{1.30}$$

In other words, V_Γ is given by the sum of one-dimensional lattice periodic potentials V_j. In this case, Step 1 consequently generalizes to the task of solving an equation of the form (1.19) for each spatial direction $x_j \in \mathbb{R}$ separately. Since our new algorithm allows for much larger time-steps and much coarser spatial grid, compared with a conventional time-splitting code, we can apply it to such multi-dimensional problems with reasonable computational complexity.

Remark 1.1. Note that the separability property (1.30) is necessary in order to easily compute the Bloch bands as a preparatory step. If V_Γ does not obey (1.30), the computational treatment of (1.2) is in itself a formidable task. For the main application we have in mind, namely lattice BECs, the separability condition (1.30) holds, since there V_Γ is typically given by (0.3).

1.3 The classical time splitting pseudo spectral (TS) method

Often finite difference methods are used to simulate the (non)linear Schrödinger equations. However, the results of [27] show that these methods disqualify in the semiclassical regime from a practical point of view, since they require exceedingly small temporal and spatial mesh sizes. By contrast, time-splitting spectral schemes have performed very well in such cases, *cf.* [3, 4]. In the present setting, however, the fast varying periodic potential V_Γ introduces additional difficulties. In [20] we compared our Bloch-decomposition-based algorithm with a time-splitting method which splits the dispersion from all potential terms (the approach used in [17]). Even in the linear, one-dimensional case, this method is *not* comparable in efficiency with our Bloch decomposition approach. To complete the picture we shall now present a comparison with a method, invoking the same time-splitting as above but with a trigonometric pseudo-spectral discretization of the periodic Hamiltonian.

More precisely, the classical pseudo-spectral method consists of the following steps:

Step 1. We solve the equation

$$\mathrm{i}\varepsilon\partial_t\psi = -\frac{\varepsilon^2}{2}\,\partial_{xx}\psi + V_\Gamma\left(\frac{x}{\varepsilon}\right)\psi, \quad x \in \mathbb{R}, \tag{1.31}$$

on a fixed time-interval $\triangle t$. Denoting by "$\widehat{\cdot}$" the Fast Fourier Transform (FFT) we then solve the ordinary differential equation

$$\mathrm{i}\varepsilon\partial_t\widehat{\psi}(t,\xi) = \frac{\varepsilon^2|\xi|^2}{2}\,\widehat{\psi} + \widehat{(V_\Gamma\psi)}. \tag{1.32}$$

From here we consequently obtain $\psi(t,x)$ by invoking an inverse FFT.

Step 2. We solve, as before, the ordinary differential equation (1.23) at the same time-interval, where the solution obtained in Step 1 serves as initial condition for Step 2. The splitting of the equation (0.4) is therefore as the above, only the numerical approach for solving (1.31) differs. Again we shall implement this pseudo-spectral method by using Strang's splitting to gain a second order scheme in time.

1.4 Application to lattice BEC in 3D

Now we turn to the application to lattice BEC in 3D, *i.e.*

$$\begin{cases} \mathrm{i}\varepsilon\partial_t\psi = -\dfrac{\varepsilon^2}{2}\,\Delta\psi + V_\Gamma\left(\dfrac{x}{\varepsilon}\right)\psi + U(x)\psi + \beta|\psi|^2\psi, \quad x \in \mathbb{R}^3, \\ \psi\big|_{t=0} = \psi_{\mathrm{in}}(x). \end{cases} \tag{1.33}$$

In practice, we usually consider the lattice potential V_Γ as follows:

$$V_\Gamma(y) = V_1(y_1) + V_2(y_2) + V_3(y_3), \qquad \text{for } y = (y_1, y_2, y_3) \in \mathbb{R}^3. \tag{1.34}$$

For example, $V_\Gamma(y) = \sum_{j=1}^{3} \sin^2(\xi_j y_j)$ with some constants $\xi_j \in \mathbb{R}$.

In this case (1.34), we can split **Step 1** in our BD algorithm into three steps:

Step 1.1. First, we solve the equation

$$\mathrm{i}\varepsilon\partial_t\psi = -\frac{\varepsilon^2}{2}\,\partial_{x_1x_1}\psi + V_1\left(\frac{x_1}{\varepsilon}\right)\psi, \tag{1.35}$$

at the time interval Δt.

Step 1.2. Then, we solve the equation

$$\mathrm{i}\varepsilon\partial_t\psi = -\frac{\varepsilon^2}{2}\,\partial_{x_2x_2}\psi + V_2\left(\frac{x_2}{\varepsilon}\right)\psi. \tag{1.36}$$

Step 1.3. Third, we solve the equation

$$\mathrm{i}\varepsilon\partial_t\psi = -\frac{\varepsilon^2}{2}\,\partial_{x_3x_3}\psi + V_3\left(\frac{x_3}{\varepsilon}\right)\psi. \tag{1.37}$$

In each substep given above, we can use the Bloch decomposition based algorithm given in Section 1.2.

2 Numerical implementation

We shall use several examples to show the efficiency of our algorithm. In order to compare the different numerical algorithms we denote by

- $\psi^{\mathrm{ts}}(t,x)$ — the solution gained from the *time-splitting spectral method*,
- $\psi^{\mathrm{bd}}(t,x)$ — the solution obtained via the new method base on *Bloch's decomposition*,
- $\psi^{\mathrm{ex}}(t,x)$ — the *"exact"* solution obtained using a *very fine spatial grid and time step.*

We also consider the following errors

$$\begin{aligned}\Delta_\infty^{\mathrm{bd/ts}}(t) &:= \left\| \psi^{\mathrm{ex}}(t) - \psi^{\mathrm{bd/ts}}(t)\right\|_{L^\infty(\mathbb{R})},\\ \Delta_2^{\mathrm{bd/ts}}(t) &:= \left\| \psi^{\mathrm{ex}}(t) - \psi^{\mathrm{bd/ts}}(t)\right\|_{L^2(\mathbb{R})},\end{aligned} \tag{2.1}$$

between the "exact solution" and the corresponding solutions obtained via our methods.

2.1 Numerical tests for 1D problems ($\beta = 0$)

We choose the initial data $\psi_{\mathrm{in}} \in \mathcal{S}(\mathbb{R})$ of the following form

$$\psi_{\mathrm{in}}(x) = \left(\frac{2\omega}{\pi}\right)^{1/4} \mathrm{e}^{-\omega(x-\pi)^2}. \tag{2.2}$$

Concerning slowly varying, external potentials U, we shall choose,

- a *harmonic oscillator* type potential:

$$U(x) = \frac{|x-\pi|^2}{2}, \tag{2.3}$$

- or an external (non-smooth) *step potential*,

$$U(x) = \begin{cases} 1, \ x \in \left[\frac{\pi}{2}, \frac{3\pi}{2}\right], \\ 0, \ \text{else}. \end{cases} \tag{2.4}$$

Within the setting described above, we shall focus on two particular choices for the lattice potential, namely:

Example 2.1. (**Mathieu's model**)

The so-called *Mathieu's model, i.e.*

$$V_\Gamma(x) = \cos(x), \tag{2.5}$$

as already considered in [17]. (For applications in solid state physics this is rather unrealistic, however it fits quite well with experiments on Bose-Einstein condensates in optical lattices.)

Example 2.2. (**Kronig-Penney's model**)

The so-called *Kronig-Penney's model, i.e.*

$$V_\Gamma(x) = 1 - \sum_{\gamma \in \mathbb{Z}} \mathbf{1}_{x \in \left[\frac{\pi}{2}+2\pi\gamma, \frac{3\pi}{2}+2\pi\gamma\right]}, \tag{2.6}$$

where $\mathbf{1}_\Omega$ denotes the characteristic function of a set $\Omega \subset \mathbb{R}$. In contrast to Mathieu's model this case comprises a non-smooth lattice potential. The corresponding Bloch eigenvalue problem is known to be explicitly solvable (see, *e.g.*, [17]).

Figure 2.1 shows a plot of the first few energy bands, drawn over $\mathcal{B}$.

Here we have some results of Example 2.1 for $\varepsilon = \frac{1}{1024}$: (for TS, we let $\triangle t = \frac{1}{10000}$, $\triangle x = \frac{1}{16384}$; for BD, we let $\triangle t = \frac{1}{10}$, $\triangle x = \frac{1}{8192}$.)

- $U(x) = \frac{|x-\pi|^2}{2}$.

$$\Delta^{\text{ts}}_\infty(t) = 3.04\text{E}-2, \quad \Delta^{\text{ts}}_2(\text{t}) = 1.34\text{E}-2,$$
$$\Delta^{\text{bd}}_\infty(t) = 4.63\text{E}-3, \quad \Delta^{\text{bd}}_2(\text{t}) = 2.94\text{E}-3.$$

- $U(x) = \begin{cases} 1, \ x \in \left[\frac{\pi}{2}, \frac{3\pi}{2}\right], \\ 0, \ \text{else}. \end{cases}$

$$\Delta^{\text{ts}}_\infty(t) = 3.04\text{E}-2, \quad \Delta^{\text{ts}}_2(\text{t}) = 1.34\text{E}-2,$$
$$\Delta^{\text{bd}}_\infty(t) = 8.46\text{E}-4, \quad \Delta^{\text{bd}}_2(\text{t}) = 6.67\text{E}-4.$$

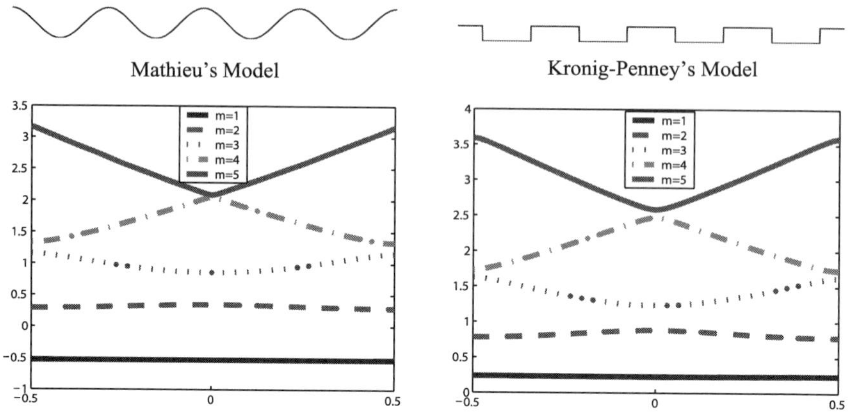

Figure 2.1 $E_m(k)$, $m = 1, \cdots, 5$.

Figure 2.2 shows the comparison of the spatial and temporal discretization error tests of our BD method and TS method.

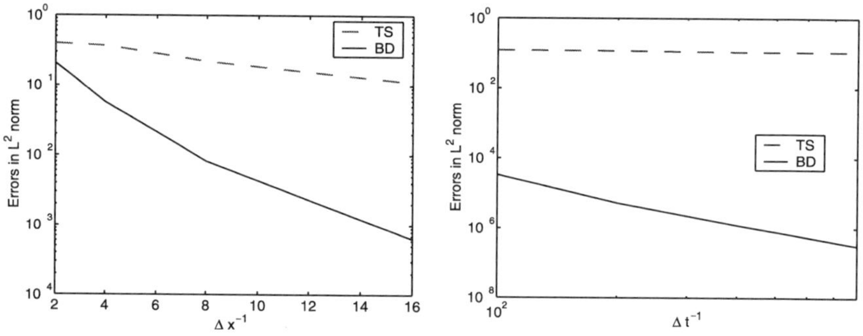

TS: $\triangle t = \frac{1}{100000}$, BD: $\triangle t = \frac{1}{1000}$ TS and BD: $\frac{\triangle x}{\varepsilon} = \frac{1}{128}$

Figure 2.2 The spatial and temporal discretization error test of Example 2.2 with $U(x) = \frac{|x-\pi|^2}{2}$, $\varepsilon = \frac{1}{1024}$, $t = 0.01$.

Here we have some remarks on linear problems.
If $U(x) \equiv 0$:

- We can use only *one step* in time to obtain the numerical solution, because the *Bloch decomposition method* indeed is *"exact"* in this case (independent of ε).
- On the other hand, by using the *usual time-splitting method*, one has to refine the *time steps (depending on ε)* as well as the *mesh*

size in order to achieve the same accuracy.

If $U(x) \neq 0$ and $\varepsilon \ll 1$:

- We can achieve quite good accuracy by using the *Bloch decomposition method* with $\triangle t = \mathcal{O}(1)$ and $\triangle x = \mathcal{O}(\varepsilon)$.
- On the other hand, by using the *time splitting spectral algorithm*, we have to use $\triangle t = \mathcal{O}(\varepsilon^\alpha)$, $\triangle x = \mathcal{O}(\varepsilon^\alpha)$, for some $\alpha \geq 1$. Particularly $\alpha > 1$ is required for the case of a non-smooth lattice potential V_Γ.

2.2 Numerical tests for 1D NLS

Before applying our algorithm to the simulation of three-dimensional lattice BECs we shall first study in more detail the influence of the nonlinearity on the Bloch decomposition. The corresponding numerical experiments are of some interest on their own, since so far the mixing of Bloch bands (i.e. the mass transfer between different bands) due to nonlinear interactions has not been fully clarified. We remark that these tests have to be seen as mathematical experiments which do not necessarily correspond to realistic physical experiments.

Example 2.3. (**Tests for nonlinear band mixing**) The periodic potential is chosen to be (2.5). Figure 2.1 shows a plot of the first few energy bands, drawn over $\mathcal{B}$. For the slowly varying, external potentials $U(x)$, we shall choose a harmonic oscillator type potential centered in the middle of the computational domain (2.3). Obviously, if $U(x) \neq 0$ an exact treatment along the lines of (1.11) – (1.13) is no longer possible for the evolution equation (0.4), even if $\beta = 0$. This is due to the fact that one has to take account of the action of the non-periodic potential $U(x)$, which in general *mixes* all Bloch bands $E_m(k)$. It is well known however, at least in the linear case, that one has a so-called *adiabatic decoupling* of the individual bands, as long as $U(x)$ varies slowly on the scale of the periodic potential which is the case in our scaling. More precisely (see [34] and the references given therein)

$$\sup_{t\in[0,T]} \left\|(\mathbf{1} - \mathbb{P}_m)U^\varepsilon(t)\,\mathbb{P}_m\right\|_{\mathcal{B}(L^2(\mathbb{R}))} \leq \mathcal{O}(\varepsilon), \tag{2.7}$$

where $\mathbb{P}$ is the ε-rescaled projection onto the m-th Bloch band defined in (1.9) and $U^\varepsilon(t) = \mathrm{e}^{-\mathrm{i}H^\varepsilon t/\varepsilon}$ is the unitary group corresponding to the linear Hamiltonian operator

$$H^\varepsilon = -\frac{\varepsilon^2}{2}\,\partial_{xx} + V_\Gamma\left(\frac{x}{\varepsilon}\right) + U(x).$$

In other words, under the influence of $U(x)$ the m-th band is stable, up to errors of order $\mathcal{O}(\varepsilon)$. The estimate (2.7) however only holds for energy bands $E_m(k)$ which are *isolated* from the rest of the spectra, i.e. do not exhibit band-crossings. In the latter case mass transfer of order $\mathcal{O}(1)$ is possible, the so-called *Landau-Zener phenomena* (see [12, 24, 34] and the references given therein). In the nonlinear case, the situation is even more complicated, as the strength (in terms of ε) of the nonlinear coupling λ^ε is expected to play a crucial role. So far, only the case of a *weak* nonlinearity, i.e. $\beta \sim \mathcal{O}(\varepsilon)$, has been treated rigorously in [7, 15]. It has been shown there, that, apart from certain resonance phenomena, an adiabatic decoupling also holds in the weakly nonlinear case.

In the following we shall numerically study such band mixing phenomena. The reason for this is twofold: Firstly, it gives us more experience on how many Bloch bands one has to take into account to guarantee that our numerical algorithm preserves mass with sufficient accuracy. Secondly, we aim to present some qualitative and quantitative studies on the phenomena for band mixing in the nonlinear case, which are of some interest on their own.

Now we start with the initial condition like

$$\psi_{\rm I}(x) = \mathbb{P}_{m_0}\psi_{\rm in}(x)e^{{\rm i}kx}, \tag{2.8}$$

where $\psi_{\rm in}(x)$ is given in (2.2). We'll test the mass transition from one band to others.

Here we have the following results (*cf.* Figures 2.3–2.6).

- The *isolated band* with $m_0 = 1$ is more stable than other bands.
- If m_0 is large, there will be more mass transfers to other bands.
- If the eigenvalue E_{m_0} is not isolated, there will be $\mathcal{O}(1)$ mass transfers to other bands.
- If $\beta = \mathcal{O}(1)$, there will be $\mathcal{O}(1)$ mass transfers to other bands.

2.3 Numerical examples for lattice BEC in 3D

Having gained sufficient insight on the phenomena of band mixing, we shall finally turn to the simulation of three-dimensional lattice BECs described by (0.4). To do so we have to choose physically relevant initial data, having in mind the following experimental situation: We assume that in the first step, the BEC is formed in a trap *without* the lattice potential, i.e. only under the influence of $U(x)$, where $x = (x_1, x_2, x_3)$. Then, in a second step, we assume that the lattice potential $V_\Gamma(x/\varepsilon)$ is switched on and the (nonlinear) dynamics of the BEC under the *combined* influence of $U(x)$ and $V_\Gamma(x/\varepsilon)$ is studied. For definiteness we shall

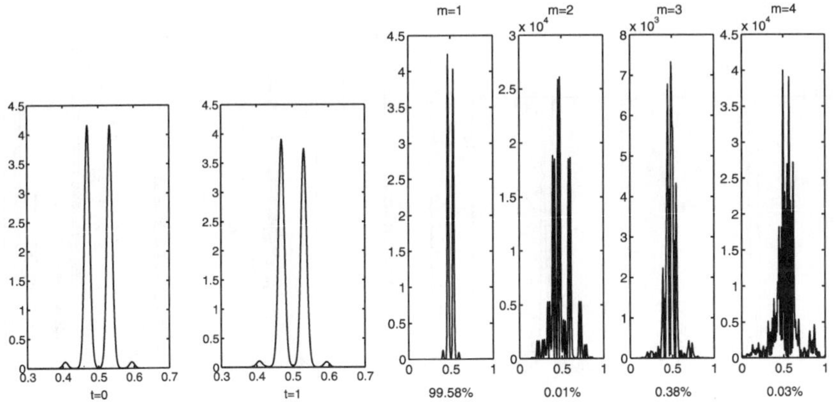

Total mass density $\rho(t, 2\pi x) = |\psi(t, x)|^2$

Mass distribution for different bands at $t = 1$

Figure 2.3 The test of band mixing for Example 2.3 with $U(x) = \frac{|x-\pi|^2}{2}$, $\varepsilon = \frac{1}{16}$, $\beta = \frac{1}{16}$, $m_0 = 1$, $k = 5$.

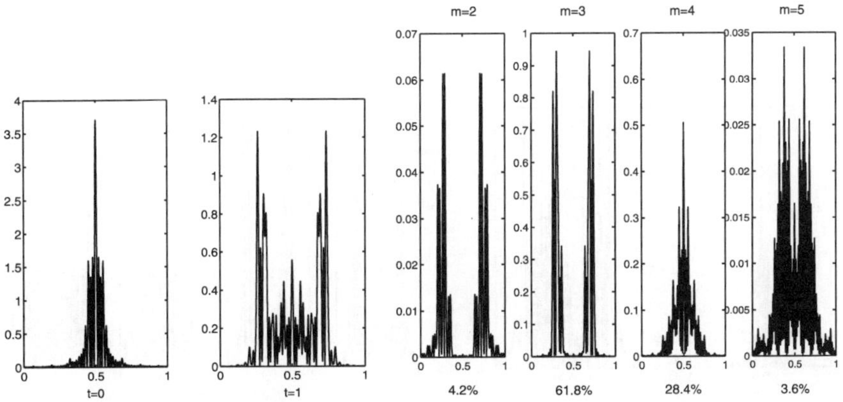

Total mass density $\rho(t, 2\pi x) = |\psi(t, x)|^2$

Mass distribution for different bands at $t = 1$

Figure 2.4 The test of band mixing for Example 2.3 with $U(x) = \frac{|x-\pi|^2}{2}$, $\varepsilon = \frac{1}{16}$, $\beta = \frac{1}{16}$, $m_0 = 4$, $k = 0$.

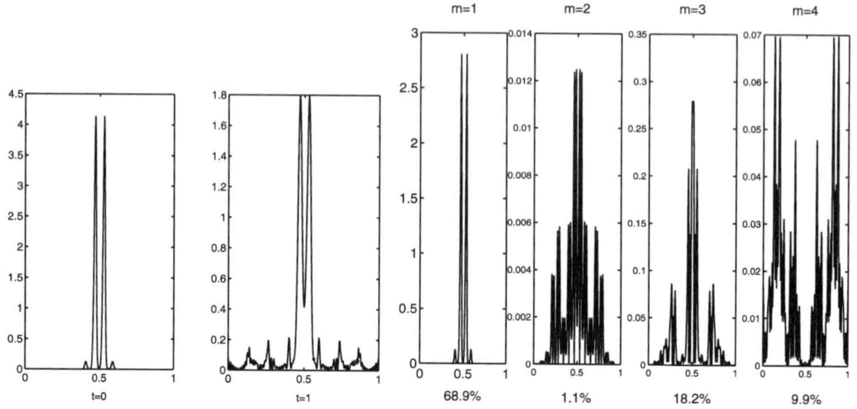

Total mass density
$\rho(t, 2\pi x) = |\psi(t, x)|^2$

Mass distribution for different
bands at $t = 1$

Figure 2.5 The test of band mixing for Example 2.3 with $U(x) = \frac{|x-\pi|^2}{2}$, $\varepsilon = \frac{1}{16}$, $\beta = 1$, $m_0 = 1$, $k = 0$.

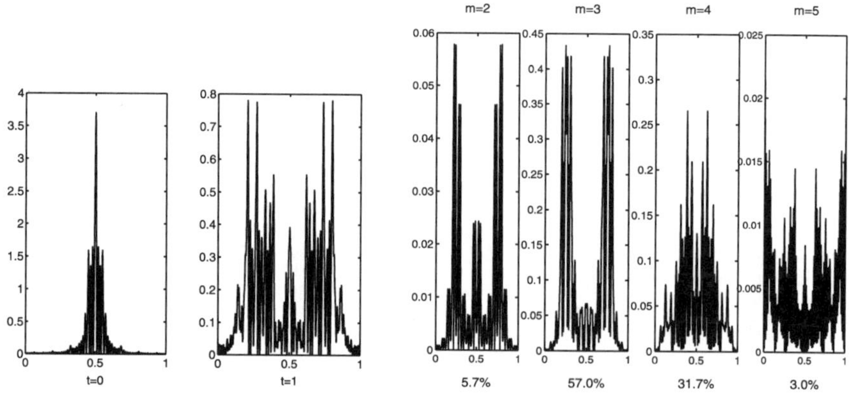

Total mass density
$\rho(t, 2\pi x) = |\psi(t, x)|^2$

Mass distribution for different
bands at $t = 1$

Figure 2.6 The test of band mixing for Example 2.3 with $U(x) = \frac{|x-\pi|^2}{2}$, $\varepsilon = \frac{1}{16}$, $\beta = 1$, $m_0 = 4$, $k = 0$.

from now on consider the following potentials acting on the BEC

$$V_\Gamma(x) = \sum_{\ell=1}^{3} \sin^2(x_\ell), \qquad U(x) = \frac{1}{2}\sum_{\ell=1}^{3} |x_\ell - \pi|^2. \tag{2.9}$$

These choices, obtained from the potentials (0.2)–(0.3) by scaling, are consistent with various physical experiments [1, 6, 8, 10, 25, 30].

Example 2.4. **(Dynamics of lattice BECs)**

We consequently have to set for the initial data of (0.4):

$$\psi\big|_{t=0} = \psi_{\text{in}}(x),$$

where $\psi_{\text{in}}(x)$ is the *ground state* of the nonlinear eigenvalue problem

$$\begin{cases} \mu\phi(x) = -\frac{1}{2}\Delta\phi + U\phi + \beta|\phi|^2\phi \\ \|\phi\|_{L^2} = \int_{\mathbb{R}^d} |\phi|^2(x)dx = 1. \end{cases}$$

Then we will turn on the lattice potential after $t > 0$.

Numerical examples for lattice BEC in 3D, for example, in 3D case, with $U(x)$ given by (2.3) (harmonic oscillator) are listed below.

- *weak interaction*: $|\beta| \ll 1$,

$$\mu_g = \frac{3\varepsilon}{2}, \qquad \phi_g = \frac{1}{(\pi\varepsilon)^{3/4}} e^{-U(x)/\varepsilon};$$

- *strong interaction*: $\beta = \mathcal{O}(1)$,

$$\mu_g^s = \frac{1}{2}\left(\frac{15\beta}{4\pi}\right)^{2/5}, \qquad \phi_g = \begin{cases} \sqrt{(\mu_g^s - U(x))/\beta}, & U(x) < \mu_g^s, \\ 0, & \textit{otherwise}. \end{cases}$$

The comparison of defocusing and focusing cases is given in Figure 2.7. In the defocusing case we see that the density starts to redistribute itself under the influence of the periodic potential and the nonlinearity. In the case where $\beta < 0$ this behavior is countervailed by the typical concentration effects of the focusing nonlinearity, leading to a blow-up for solutions to (0.4) and thus a collapse of the condensate.

3 Random coefficients: stability tests and Anderson localization

In this section, we present numerical studies on the Klein-Gordon equation including *random coefficients*. This describes waves propagating in

$|\beta| = \frac{1}{8}$ and $\varepsilon = \frac{1}{8}$

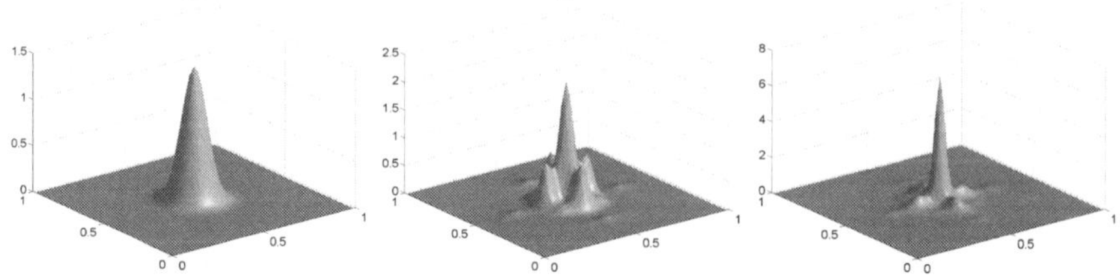

$|\beta| = 1$ and $\varepsilon = \frac{1}{8}$

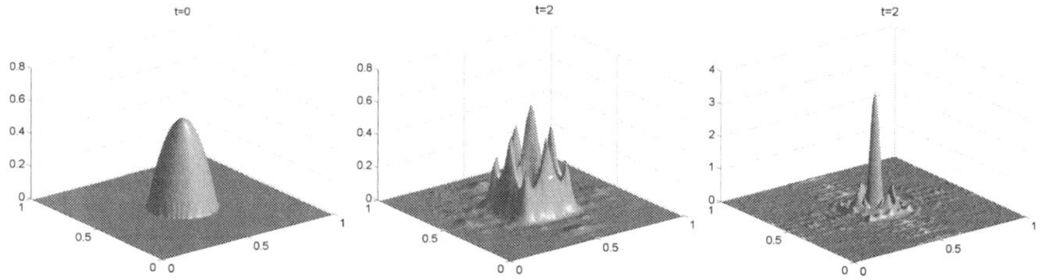

Figure 2.7 Comparison of the initial and final mass densities, evaluated at $x_3 = 0$.

disordered media, a topic of intense physical and mathematical research (*cf.* P. A. Robinson, *Phil. Magazine B* **80**, 2000).

We shall study the following class of (one-dimensional) *Klein-Gordon* type equations

$$\begin{cases} \dfrac{\partial^2 u}{\partial t^2} = \dfrac{\partial}{\partial x}\left(a_\Gamma\left(\dfrac{x}{\varepsilon}\right)\dfrac{\partial u}{\partial x}\right) - \dfrac{1}{\varepsilon^2} W_\Gamma\left(\dfrac{x}{\varepsilon}\right) u + f(x), \quad t > 0, \\ u\big|_{t=0} = u_0(x), \quad \dfrac{\partial u}{\partial t}\Big|_{t=0} = v_0(x), \end{cases} \tag{3.1}$$

with given initial data $u_0(x), v_0(x) \in \mathbb{R}$ and $f(x) \in \mathbb{R}$ describing some slowly varying source terms.

The highly oscillatory coefficients $a_\Gamma(y), W_\Gamma(y) \in \mathbb{R}$ are assumed to be *periodic* with respect to some *regular lattice* $\Gamma \simeq \mathbb{Z}$. Equation (3.1) henceforth describes the propagation of waves on macroscopic length- and time-scales. The *purely periodic coefficients* $a_\Gamma(y)$ and $W_\Gamma(y)$ describe an idealized situation where *no defects* are present within the material. More realistic descriptions for *disordered media* usually rely on the introduction of *random perturbations* within these coefficients and we wish to include such random perturbations also in our numerics.

Since our numerical method relies on $\{\varphi_m(y,k)\}_{m=1}^M$ as basis functions, the *stability* of our method w.r.t. to *perturbation* of these *Bloch functions* is an important question.

3.1 Stability of our BD algorithm

To this end we consider, instead of (1.2), the *randomly perturbed* eigenvalue problem

$$\left(-\frac{\partial}{\partial_y}\left(a_\Gamma(\omega,y)\frac{\partial}{\partial_y}\right)+W_\Gamma(y)\right)\varphi_m(\omega,y,k) = \lambda_m(\omega,k)\varphi_m(\omega,y,k), \tag{3.2}$$

subject to the quasi-periodic boundary condition. Here, the coefficient $a_\Gamma = a_\Gamma(\omega, y)$ is assumed to be a function of a *uniformly distributed random variable* ω with mean zero and variance $\sigma^2 \geq 0$. In the following we shall vary σ in such a way that we do not loose the uniform ellipticity, i.e. we have, as before, that $\lambda_m(\omega, k) \geq 0$ (for every realization of ω) and we consequently set $E_m(\omega, k) = \sqrt{\lambda_m(\omega, k)}$. Note that we do *not* assume any randomness in W_Γ, since this would only result in a shift of the eigenvalues.

In our algorithm, we solve the *random eigenvalue problem* (3.2), for different choices of σ, to obtain the corresponding eigenvalues $\lambda_m(\omega, k)$ and eigenfunctions $\varphi_m(\omega, y, k)$. We shall then take the *average* of them and use these averaged quantities in our Bloch decomposition based algorithm (as described in Section 1.2).

Example 3.1. **(Stability test)** Consider (3.1) with $f(x) \equiv 0$ and initial data

$$u_0(x) = \left(\frac{2}{\pi\varepsilon}\right)^{1/4} \mathrm{e}^{-\frac{(x-\pi)^2}{\varepsilon}}, \quad v_0(x) = 0. \tag{3.3}$$

The random coefficient a_Γ is chosen as

$$a_\Gamma(\omega, y) = a_\Gamma(y) + \omega, \quad a_\Gamma\,(y) = 2.5 + \cos(y), \tag{3.4}$$

i.e. including an *additive noise*, and

$$W_\Gamma(y) = 1 - \sum_{\gamma\in\mathbb{Z}} \mathbf{1}_{x\in\left[\frac{\pi}{2}+2\pi\gamma,\frac{3\pi}{2}+2\pi\gamma\right]}. \tag{3.5}$$

For a given choice of σ we numerically generate $N \in \mathbb{N}$ realization of ω and consequently take the ensemble average. In our examples we usually choose $N = 100$. Figure 3.1 shows the average of the first few Bloch bands, i.e.

$$E_m(k) := \mathbb{E}\{E_m(\omega, k)\} \approx \frac{1}{N}\sum_{\ell=1}^{N} E_m(\omega_\ell, k), \tag{3.6}$$

for different values of σ.

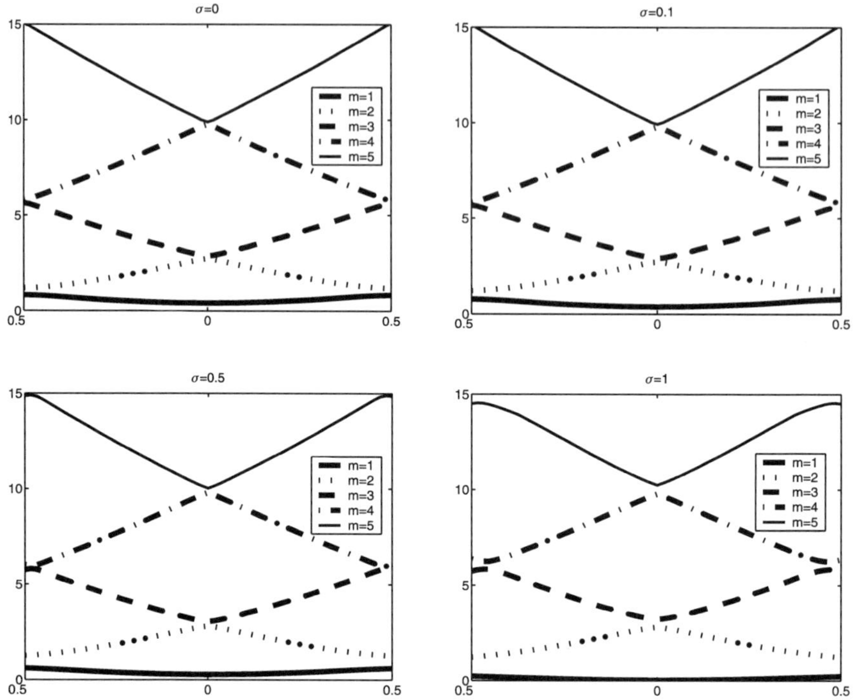

Figure 3.1 The first five averaged Bloch bands $E_m(k) = \mathbb{E}\{E_m^\omega(k)\}$ after random perturbation.

Figure 3.2 shows a comparison between the solution $u^\sigma(t,x)$ with noise and the solution $u(t,x)$ without noise. To this end we consider two different kinds of errors

$$\Delta_\infty^\sigma(t) := \|u(t,\cdot) - u^\sigma(t,\cdot)\|_{L^\infty(\mathbb{R})}, \quad \Delta_2^\sigma(t) := \|u(t,\cdot) - u^\sigma(t,\cdot)\|_{L^2(\mathbb{R})}.$$

Numerically, we find that $\Delta_\infty^\sigma \sim \sigma$, and $\Delta_2^\sigma \sim \sigma\|u(t,\cdot)\|_{L^2(\mathbb{R})}$. That means our BD algorithm is stable with the numerical simulation of the Bloch eigenvalue problem.

3.2 Numerical evidence for the Anderson's localization

The phenomenon of Anderson's localization, also known as the *strong localization*, describes the absence of dispersion for waves in random media with sufficiently *strong random perturbation*. It has been predicted by P. W. Anderson (Philos. Mag. B, **52**, 1985) in the context of (quantum mechanical) electron dynamics but is now regarded as a general

Graphs of the differences: $u(1, 2\pi x) - u^\sigma(1, 2\pi x)$

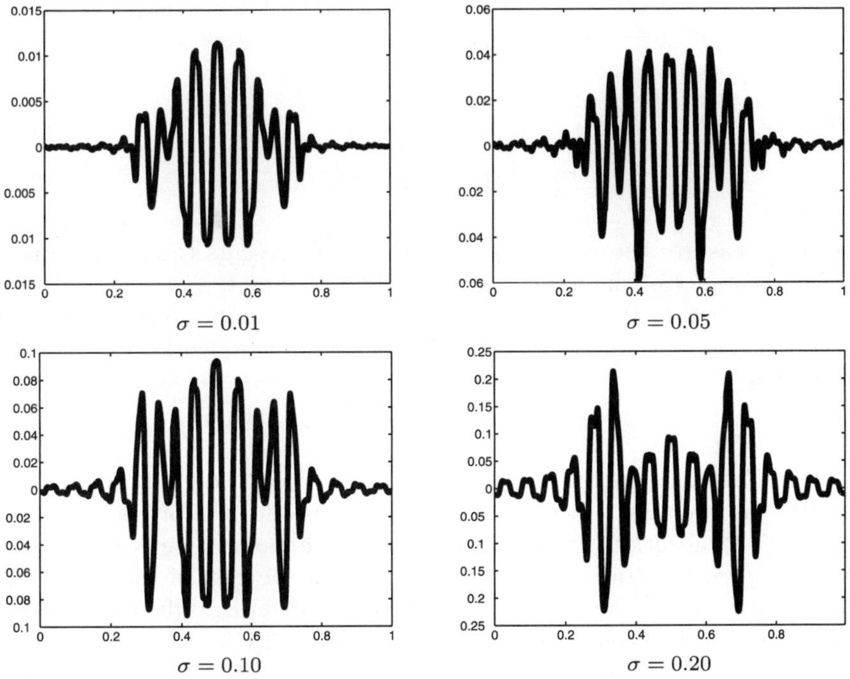

Figure 3.2 Comparison between the solution $u^\sigma(t, x)$ with noise and the solution $u(t, x)$ without noise, where $\varepsilon = \frac{1}{32}$, $\triangle t = \frac{1}{10}$, $\triangle x = \frac{\pi}{512}$.

wave phenomenon applied to the transport of electromagnetic or acoustic waves as well, *cf.* [9, 32, 35].

In the following, we shall again assume that $a_\Gamma = a_\Gamma(\omega, y)$ depends on a uniformly distributed random variable ω with mean zero and variance σ^2. We then study the random Klein-Gordon equation

$$\begin{cases} \dfrac{\partial^2 u^\omega}{\partial t^2} = \dfrac{\partial}{\partial x}\left(a_\Gamma\left(\omega, \dfrac{x}{\varepsilon}\right)\dfrac{\partial u^\omega}{\partial x}\right) - \dfrac{1}{\varepsilon^2} W_\Gamma\left(\dfrac{x}{\varepsilon}\right) u^\omega + f(x), \\ u^\omega\big|_{t=0} = u_0(x), \qquad \dfrac{\partial u^\omega}{\partial t}\Big|_{t=0} = v_0(x), \end{cases} \tag{3.7}$$

which describes the propagation of waves in disordered media.

In order to realize the emergence of these localization phenomena we consider the *local energy density* $e^\omega(t, x)$ of the solution $u^\omega(t, x)$:

$$e^\omega(t, x) := \frac{1}{2}\left(\left|\frac{\partial u^\omega}{\partial t}\right|^2 + a_\Gamma\left(\omega, \frac{x}{\varepsilon}\right)\left|\frac{\partial u^\omega}{\partial x}\right|^2 + \frac{1}{\varepsilon^2} W_\Gamma\left(\frac{x}{\varepsilon}\right)|u^\omega|^2\right).$$

The *total energy* $E_0^\omega(t)$ of $u^\omega(t, x)$ is then given by the zeroth spatial moment of $e^\omega(t,x)$, i.e.

$$E_0^\omega(\omega,t) = \int_{\mathbb{R}} e^\omega(t,x)\mathrm{d}x, \tag{3.8}$$

and we likewise define

$$E_2^\omega(\omega,t) = \int_{\mathbb{R}} x^2 e^\omega(t,x)\mathrm{d}x, \tag{3.9}$$

which measures the *spread of the wave.* It represents the mean square of the distance of the wave from the origin at time t.

Note that in the case, where $f(x) \equiv 0$ (no source term), we have energy conservation, i.e. $E_0^\omega(t) = E_0^\omega(0)$. We consequently consider the function

$$A^\sigma(t) := \frac{\mathbb{E}\{E_2^\omega(t)\}}{\mathbb{E}\{E_0^\omega(t)\}}, \tag{3.10}$$

where $\mathbb{E}$ again denotes the mathematical expectation. The quantity $A^\sigma(t)$ has been introduced in [13] as a measure for the presence of Anderson's localization. In the absence of any random perturbation $A^\sigma(t)$ should grow quadratically in time whereas in the case of the Anderson localization $A^\sigma(t)$ should grow only linearly, indicating diffusive behavior and eventually become a constant in time [13, 32, 35].

Example 3.2. **(Anderson's localization)** Here we also consider (3.7) with $f(x) \equiv 0$ and $a_\Gamma(\omega, y)$ is given by (3.4), the potential W_Γ is given by (3.5) and the initial data are chosen as (3.3). Now we make a different test with random perturbation. We then solve the Klein-Gordon equation (3.7) with 100 kinds of different realization of the random variable ω. Finally we take an ensemble average to obtain $\mathbb{E}\{u^\omega(t,x)\}$, *cf.* Figure 3.3.

We plot the graph of the quantity $A^\sigma(t)$ in Figure 3.4. As we see it first grows almost linearly in t, a typical diffusive behavior, and then, around $t = 2$ it flattens. The latter is a strong indication of *Anderson's localization.*

4 Conclusions

In this paper, we present a new numerical method for accurate computations of solutions to (non)linear dispersive wave equations with periodic coefficients.

- Our approach is based on the classical *Bloch decomposition method* (BD) and it proves to be superior to the mainly used time-splitting pseudo spectral schemes (TS).

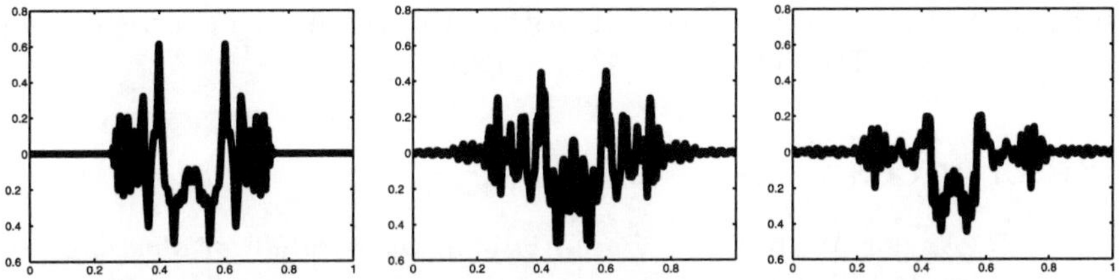

$\mathbb{E}\{u^\omega(1, 1\pi x)\}$, where, from left to right: $\sigma = 0$, 0.5, and 1.0.

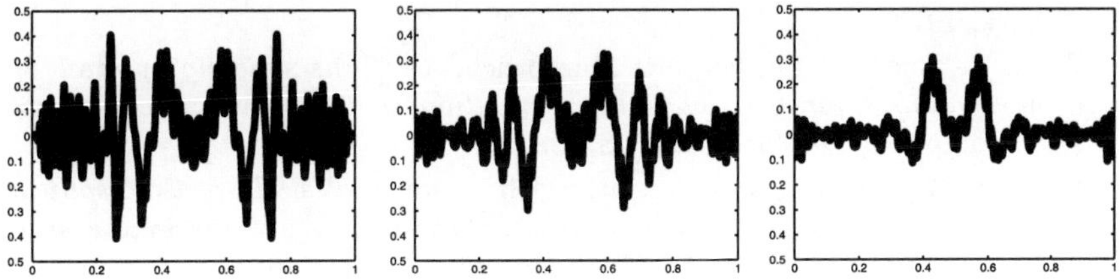

$\mathbb{E}\{u^\omega(2, 1\pi x)\}$, where, from left to right: $\sigma = 0$, 0.5, and 1.0.

Figure 3.3 Averaged solutions at different time ($\varepsilon = \frac{1}{64}$).

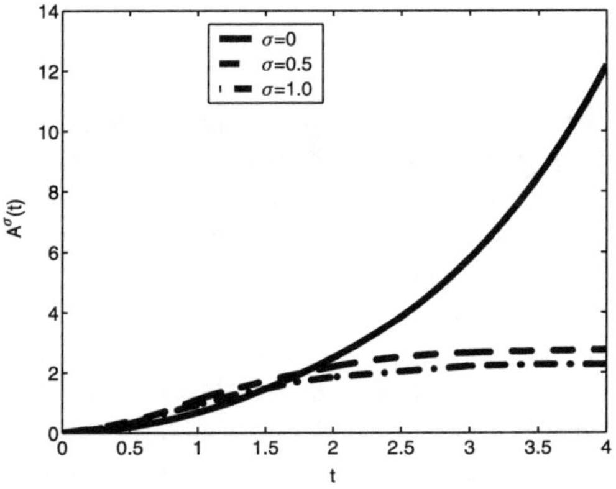

Figure 3.4 The graph of $A^\sigma(t)$ for different σ ($\varepsilon = \frac{1}{64}$)

- It is shown by the given numerical examples that our method is *unconditionally stable*, and has *conserved mass* and *uniform convergence rate* in temporal discretization.
- Our new method allows for *much larger time-steps* and sometimes even a *coarser spatial grid*, to achieve the same accuracy as for the usual *time-splitting pseudo-spectral method*. This is particularly visible in cases, where the lattice potential is *no longer smooth* and $\varepsilon \ll 1$.

Indeed in these cases the BD algorithm turns out to be *considerably faster* than the TS method.

References

[1] J. R. Anglin, W. Ketterle, *Bose-Einstein condensation of atomic gases*, Nature, **416** (2002), 211–218.

[2] N. W. Ashcroft and N. D. Mermin, *Solid state physics*, Saunders New York, 1976.

[3] W. Z. Bao, S. Jin, and P. Markowich, *On time-splitting spectral approximations for the Schrödinger equation in the semiclassical regime*, J. Comp. Phys. **175** (2002), 487–524.

[4] W. Z. Bao, S. Jin, and P. Markowich, *Numerical study of time-splitting spectral discretizations of nonlinear Schrödinger equations in the semi-Classical regime*, SIAM J. Sci. Comp. **25** (2003), 27–64.

[5] A. Bensoussan, J. L. Lions, and G. Papanicolaou, *Asymptotic Analysis for Periodic Structures*, North-Holland Pub. Co. (1978).

[6] S. Burger, F. S. Cataliotti, C. Fort, F. Minardi, and M. Inguscio, *Superfluid and Dissipative Dynamics of a Bose-Einstein Condensate in a Periodic Optical Potential*, Phys. Rev. Lett., **86** (2001), 4447–4450.

[7] R. Carles, P. A. Markowich and C. Sparber, *Semiclassical asymptotics for weakly nonlinear Bloch waves*, J. Stat. Phys. **117** (2004), 369–401.

[8] D. I. Choi and Q. Niu, *Bose-Einstein condensates in an optical lattice*, Phys. Rev. Lett, **82** (1999), 2022–2025.

[9] C. A. Condat and T. R. Kirkpatrick, *Resonant scattering and Anderson localization of acoustic waves*, Phys. Rev. B, **36** (1987), 6782–6793.

[10] S. L. Cornish and D. Cassettari, *Recent progress in Bose-Einstein condensation experiments*, Phil. Trans. R. Soc. Lond. A **361** (2003), 2699–2713.

[11] B. Deconinck, B. A. Frigyik, and J. N. Kutz, *Dynamics and stability of Bose-Einstein condensates: the nonlinear Schrödinger equation with periodic potential*, J. Nonlinear Sci. **12** (2002), 169–205.

[12] C. Fermanian Kammerer and P. Gérard, *A Landau-Zener formula for non-degenerated involutive codimension 3 crossings*, Ann. Henri Poincaré **4** (2003), 513–552.

[13] J. Fröhlich and T. Spencer, *A Rigorous Approach to Anderson Localization*, Physics Reports, **103** (1984), 9–25.

[14] P. Gérard, P. Markowich, N. Mauser, and F. Poupaud, *Homogenization Limits and Wigner transforms*, Comm. Pure and Appl. Math **50** (1997), 323–378.

[15] J. Gianoullis, A. Mielke, and C. Sparber, *Interaction of modulated pulses in the nonlinear Schrödinger equation with periodic potential*, preprint (2007).

[16] L. Gosse, *Multiphase semiclassical approximation of an electron in a one-dimensional crystalline lattice. II. Impurities, confinement and Bloch oscillations*, J. Comput. Phys. **201** (2004), no. 1, 344–375.

[17] L. Gosse and P. A. Markowich, *Multiphase semiclassical approximation of an electron in a one-dimensional crystalline lattice - I. Homogeneous problems*, J. Comput Phys. **197** (2004), 387–417.

[18] L. Gosse and N. Mauser, *Multiphase semiclassical approximation of an electron in a one-dimensional crystalline lattice. III. From ab initio models to WKB for Schrödinger-Poisson*, to appear in J. Comput. Phys. **211** (2006), no. 1, 326–346.

[19] J. C. Guillot, J. Ralston, and E. Trubowitz, *Semiclassical asymptotics in solid-state physics*, Comm. Math. Phys. **116** (1998), 401–415.

[20] Z. Huang, S. Jin, P. Markowich, and C. Sparber, *A Bloch decomposition based split-step pseudo spectral method for quantum dynamics with periodic potentials*, SIAM J. Sci. Comput. **29** (2007), 515–538.

[21] Z. Huang, S. Jin, P. Markowich, and C. Sparber, *Numerical simulation of the nonlinear Schrödinger equation with multi-dimensional periodic potentials*, Multiscale Model. Simul., **7** (2008), 539–564.

[22] Z. Huang, S. Jin, P. Markowich, and C. Sparber, *On the Bloch decomposition based spectral method for wave propagation in periodic media*, Wavemotion **46** (2009), 15–28.

[23] A. Iserles, *A First Course in the Numerical Analysis of Differential Equations*, Cambridge University Press, Cambridge, UK, 1996.

[24] A. Joye, *Proof of the Landau-Zener formula*, Asymptotic Anal. **9** (1994), 209–258.

[25] M. Kraemer, C. Menotti, L. Pitaevskii, and S. Stringari, *Bose-Einstein condensates in 1D optical lattices: compressibility, Bloch bands and elementary excitations*, Eur. Phys. J. D **27** (2003), 247–263.

[26] P. A. Markowich, N. Mauser, and F. Poupaud, *A Wigner-function approach to (semi)classical limits: electrons in a periodic potential*, J. Math. Phys. **35** (1994), 1066–1094.

[27] P. A. Markowich, P. Pietra, and C. Pohl, *Numerical Approximation of Quadratic Observables of Schrödinger-type Equations in the Semi-classical Limit*, Num. Math. **81** (1999), 595–630.

[28] O. Morsch and E. Arimondo, *Ultracold atoms and Bose-Einstein condensates in optical lattices*, in: T. Dauxois, S. Ruffo, E. Arimondo, and M. Wilkens (eds.), *Dynamics and Thermodynamics of Systems with Long Range Interactions*, Lecture Notes in Physics 602, Springer (2002).

[29] G. Panati, H. Spohn, and S. Teufel, *Effective dynamics for Bloch electrons: Peierls substitution and beyond*, Comm. Math. Phys. **242** (2003), 547–578.

[30] L. Pitaevskii and S. Stringari, *Bose-Einstein condensation*, Internat. Series of Monographs on Physics 116, Clarendon Press, Oxford (2003).

[31] M. Reed and B. Simon, *Methods of modern mathematical physics IV. Analysis of operators*, Academic Press (1978).

[32] C. M. Soukoulis, E. N. Economou, G. S. Grest, and M. H. Cohen, *Existence of Anderson Localization of Classical Waves in a Random Two-Component Medium*, Phys. Rev. Lett. **62** (1989), 575–578.

[33] C. Sparber, *Effective mass theorems for nonlinear Schröinger equations*, SIAM J. Appl. Math. **66** (2006), 820–842.

[34] S. Teufel, *Adiabatic perturbation theory in quantum dynamics*, Lecture Notes in Mathematics 1821, Springer (2003).

[35] R. L. Weaver, *Anderson localization in the time domain: Numerical sdudies of waves in two-dimensional disordered media*, Phys. Rev. B, **49** (1994), 5881–5895.

[36] C. H. Wilcox, *Theory of bloch waves*, J. Anal. Math. **33** (1978), 146–167.

Some Results of the Euler-Poisson System for Plasmas and Semiconductors

Ingrid Lacroix-Violet
Laboratoire Paul Painlevé UMR 8524 CNRS-USTLille
Université des Sciences et Technologies de Lille
Cité Scientifique 59655 Villeneuve d'Ascq Cedex, France
Email: ingrid.violet@math.univ-lille1.fr

Abstract

This paper is a review of different results already published concerning the steady state Euler-Poisson system for a potential flow. In a first part we present results of the zero electron mass limit and the quasineutral limit of the system using an asymptotic expansion method. For the quasineutral limit, we consider the case where boundary layers can appear. In a second part, we present some numerical schemes of finite volume type to compute approximate solutions of the system for semiconductors in the unipolar case. Particularly, some numerical simulations are given to illustrate some smallness conditions on given data and parameters in the proof of existence of solutions to the system.

1 Introduction

In this paper we consider the Euler-Poisson system which is a hydrodynamic model widely used in the mathematical modeling and numerical simulation for plasmas [9] and semiconductors [32]. It consists in two nonlinear equations given by the conservation laws of momentum and density, called the Euler equations, plus a Poisson equation for the electrostatic potential. Due to the hyperbolicity of the transient nonlinear Euler equations, the weak solution is only studied in one space dimension. In such a situation, the existence of global weak solution is shown in the set of bounded functions [31].

Here we only consider the unipolar steady-state case for a potential flow. In the scaled variables, the Euler-Poisson system reads then as follows (see [13, 34, 35]):

$$-\text{div}(n\nabla\psi) = 0, \tag{1.1}$$

$$\frac{\varepsilon}{2}|\nabla\psi|^2 + h(n) = V + \frac{\varepsilon\psi}{\tau}, \tag{1.2}$$

$$-\lambda^2\Delta V = n - C. \tag{1.3}$$

This system will be studied in an open and bounded domain Ω in $\mathbb{R}^d$ ($d = 2$ or $d = 3$ in practice) with sufficiently smooth boundary Γ. The unknowns of the system are $n = n(x)$, $\psi = \psi(x)$ and $V = V(x)$ which represent respectively the electron density, the velocity potential and the electrostatic potential. The function $h = h(n)$ corresponds to the enthalpy of the system and is defined by:

$$h'(n) = \frac{p'(n)}{n}, \quad n > 0, \text{ and } h(1) = 0,$$

where $p = p(n)$ is the pressure function, supposed to be sufficiently smooth and strictly increasing for $n > 0$. In practice, the pressure function is typically governed by the γ-law, $p(n) = cn^\gamma$ where $c > 0$ and $\gamma \geq 1$ are constants. The case $\gamma = 1$ corresponds to the isothermal flow, since in this case the temperature is constant. The function $C = C(x)$ stands for the doping profile for a semiconductor and for the ion density for a plasma. The physical scaled parameters λ, ε, τ represent respectively the Debye length, the electron mass and the relaxation time of the system. They are dimensionless and small compared to the characteristic length of physical interest.

In all the following, systems (1.1)–(1.3) will be completed with Dirichlet type boundary conditions. We will see later which ones are exact.

First of all let us say that this system has already been studied a lot. Particularly let us mention [13] where the authors have shown existence and uniqueness of solutions (with all the physical parameters equal to one) under a smallness condition on the data, which implies that the problem is in the subsonic region. In [34], it is shown that the smallness condition on the data can be replaced by a smallness condition on the parameter ε. Then the existence and uniqueness hold for large data provided that ε is small enough. In the same article, the author was also interested in the asymptotic limit of the system when the physical parameters tend, independently, to zero. There are then three limits called respectively the zero electron mass limit (case ε tends to zero), the zero relaxation time limit (case τ tends to zero) and the quasineutral limit (case λ tends to zero). Particularly, in [34], the author obtained the convergence, for the electron mass limit, in $O(\varepsilon)$ for an asymptotic expansion of order zero, and, a convergence in $O(\lambda^2)$ for the quasineutral limit in case of an asymptotic expansion of order zero under a compatibility condition. In [35], the asymptotic expansion is justified up to any order for the zero electron mass limit and the zero relaxation time limit. In [40], the same result is obtained for the quasineutral limit without compatibility condition.

Let us note that the asymptotic limits for the Euler-Poisson system have been studied by a lot of authors. In one-dimensional steady state Euler-Poisson system, the quasineutral limit was performed in [39] for well-prepared boundary data and in [33] for general boundary data. In [12], by using pseudo-differential techniques, the quasineutral limit was studied for local smooth solutions of a one-dimensional and isothermal model for plasmas in which the electron density is described by the Maxwell-Boltzmann relation. This relation can be obtained in the zero electron mass limit of the Euler-Poisson equations which we will discuss below. See also [4] for the study of the quasineutral limit in a semi-linear Poisson equation in which the Maxwell-Boltzmann relation is also used.

The zero relaxation time limit in one-dimensional transient Euler-Poisson system has been investigated in [31] and [26, 27] by the compensated compactness arguments for global weak solutions. The limit system is governed by the classical drift-diffusion model. In multi-dimensional case and for local smooth solutions this limit has been studied in [1].

From a numerical point of view, the hydrodynamic model has essentially been studied in its complete form with the energy balance equation. In [3] the authors provide numerical simulations and show that the model exhibits velocity overshoot. In [19] the authors propose numerical methods of the hydrodynamic model and give numerical results of the ballistic diode. In [18], the author extends the simulations to the case of transonic flow. There exists also a wide literature on the analysis and simulation of the drift-diffusion equations (see [2, 5, 6, 7, 10, 11, 25, 29, 38] and references therein). The steady-state drift-diffusion system, as the steady state Euler-Poisson model, is a fully nonlinear system which is frequently solved with a Gummel map method [24]. In [8] the authors propose iterative schemes to solve a system of linear partial differential equations for the electrostatic and velocity potentials and nonlinear algebraic equation for the density instead of solving a fully nonlinear system of partial differential equations. They consider in their article also the case of the bi-polar system (which means that they consider the two species: electrons and ions). Particularly they can see numerically the smallness condition on the parameter ε for the existence of solutions. They can also obtain some current-voltage characteristics and the case of a ballistic diode.

In this paper we make a review of results of the steady state Euler-Poisson system for a potential flow obtained by the author and co-authors. Particularly we will present the construction and justification of an asymptotic expansion up to any order, and in the multi-dimensional case, for the zero electron mass limit and the quasineutral limit. Let us note that for the quasineutral limit, we will consider a case without compatibility condition, which means that boundary layers can appear. These two results were the objects of two previous papers [35, 40] and will be here presented in Section 2. Moreover we will be interested in numerical simulation for systems (1.1)–(1.3). As mentioned above, in [8] the authors propose two numerical schemes of finite volume type with reconstruction of the gradient appearing in (1.2). We will present them in Section 3.

2 Asymptotic limits

In this section we are interested in two asymptotic limits: the zero electron mass limit and the quasineutral limit. We will just give the main ideas of the results and we refer to [35] and [40] for more details. In all this section we take $\tau = 1$.

First of all, as mentioned in the introduction, we complete the systems (1.1)–(1.3) with Dirichlet type boundary conditions on the density and the velocity potential:

$$n = n_D, \quad \psi = \psi_D, \quad \text{on } \Gamma. \tag{2.1}$$

By eliminating V of (1.2) and (1.3) and using (1.1) we have:

$$\begin{aligned} -\Delta h(n) &+ \frac{\varepsilon}{n}\sum_{i,j=1}^{d}\frac{\partial\psi}{\partial x_i}\frac{\partial\psi}{\partial x_j}\frac{\partial^2 n}{\partial x_i\partial x_j} - \frac{\varepsilon}{\tau n}\nabla\psi.\nabla n - \frac{\varepsilon}{n^2}(\nabla\psi.\nabla n)^2 \\ &+ \frac{\varepsilon}{n}\sum_{i,j=1}^{d}\frac{\partial\psi}{\partial x_i}\frac{\partial^2\psi}{\partial x_i\partial x_j}\frac{\partial n}{\partial x_j} + n - C(x) = Q(\psi), \end{aligned} \tag{2.2}$$

where Q is given by

$$Q(\psi) = \sum_{i,j=1}^{d}\left(\frac{\partial^2\psi}{\partial x_i\partial x_j}\right)^2. \tag{2.3}$$

For $n > 0$ it is easy to see that (n, ψ, V) is a smooth solution to the systems (1.1)–(1.3) if and only if (n, ψ) is a smooth solution to (1.1) and (2.2). Moreover, for ψ given, equation (2.2) is elliptic if and only if the flow is subsonic, i.e., the condition $|\nabla\psi| < \sqrt{p'(n)/\varepsilon}$ holds.

The first goal of this part is to construct asymptotic expansions in the case of the zero electron mass limit and the quasineutral limit. The second one is to justify them, which means that we can obtain the existence and uniqueness of each profile, and estimates of the difference between the exact solution and the asymptotic expansion.

2.1 Zero electron mass limit

Here we are interested in the construction and justification of an asymptotic expansion for the zero electron mass limit i.e. for ε tends to zero. Then we assume in all this part $\lambda = 1$ and we note $(n_\varepsilon, \psi_\varepsilon, V_\varepsilon)$ the solution of (1.1)–(1.3) supplemented with the boundary conditions:

$$n_\varepsilon = \sum_{k=0}^{m} \varepsilon^k \overline{n}_k + n_{D,\varepsilon}^{m+1}, \quad \psi_\varepsilon = \sum_{k=0}^{m} \varepsilon^k \overline{\psi}_k + \psi_{D,\varepsilon}^{m+1} \quad \text{sur } \Gamma, \tag{2.4}$$

where $n_{D,\varepsilon}^{m+1}$ and $\psi_{D,\varepsilon}^{m+1}$ are smooth enough and defined in $\overline{\Omega}$ such that $n_{D,\varepsilon}^{m+1} = O(\varepsilon^{m+1})$ and $\psi_{D,\varepsilon}^{m+1} = O(\varepsilon^{m+1})$ uniformly in ε.

Let us recall that for fixed ε, the existence and uniqueness of solutions to the system have been already shown in the space

$$B \stackrel{\text{def}}{=} \mathcal{W}^{2,q}(\Omega) \times \mathcal{C}^{2,\delta}(\overline{\Omega}) \times \mathcal{C}^{1,\delta}(\overline{\Omega})$$

for small boundary data [13] or on a smallness condition on ε [34] which guarantee that the problem is located in the subsonic region.

2.1.1 Construction of the asymptotic expansion

Let us first explain how to derive the profile equations. We assume that:

- (A1) Ω is a bounded and convex domain of $\mathbb{R}^d$ with $\Gamma \in \mathcal{C}^{2,\delta}$, $\delta \in]0,1[$,
- (A2) $p \in \mathcal{C}^{m+4}(\mathbb{R}^+)$, $m \in \mathbb{N}$, $p'(n) > 0 \ \forall \ n > 0$,
- (A3) $C \in L^\infty(\Omega), 0 < \underline{C} \leq C(x)$,
- (A4) $\overline{n}_k \in \mathcal{W}^{2,q}(\Omega)$ for $q > \frac{d}{1-\delta}$ and $\forall \ 0 \leq k \leq m$, $0 < \underline{n} \leq \overline{n}_0(x) \ \ \forall x \in \Gamma$,
- (A5) $\overline{\psi}_k \in \mathcal{C}^{2,\delta}(\overline{\Omega}), \quad \forall \ 0 \leq k \leq m$,
- (A6) the sequence $(\varepsilon^{-(m+1)} n_{D,\varepsilon}^{m+1})_{\varepsilon > 0}$ is bounded in $\mathcal{W}^{2,q}(\Omega)$,
- (A7) the sequence $(\varepsilon^{-(m+1)} \psi_{D,\varepsilon}^{m+1})_{\varepsilon > 0}$ is bounded in $\mathcal{C}^{2,\delta}(\overline{\Omega})$.

Let $(n_{a,\varepsilon}, \psi_{a,\varepsilon}, V_{a,\varepsilon})$ be defined by the following ansatz :

$$n_{a,\varepsilon} = \sum_{k \geq 0} \varepsilon^k n_k, \ \psi_{a,\varepsilon} = \sum_{k \geq 0} \varepsilon^k \psi_k, \ V_{a,\varepsilon} = \sum_{k \geq 0} \varepsilon^k \phi_k \ \text{ in } \Omega, \tag{2.5}$$

with the boundary conditions :

$$n_{a,\varepsilon} = \sum_{k \geq 0} \varepsilon^k \overline{n}_k, \ \psi_{a,\varepsilon} = \sum_{k \geq 0} \varepsilon^k \overline{\psi}_k \quad \text{on } \Gamma. \tag{2.6}$$

Plugging expression (2.5) into the systems (1.1)–(1.3), using the Taylor's formula to develop

$$h\left(\sum_{k\geq 0}\varepsilon^k n_k\right),$$

and by identification of the power of ε, we obtain the system for each (n_k, ψ_k, V_k), $k \geq 0$. More precisely, the first order (n_0, ψ_0, V_0) satisfies the nonlinear problem in Ω:

$$-\text{div}(n_0 \nabla \psi_0) = 0, \tag{2.7}$$

$$h(n_0) = V_0, \tag{2.8}$$

$$-\Delta V_0 = C(x) - n_0, \tag{2.9}$$

with the following boundary conditions :

$$n_0 = \overline{n}_0, \quad \psi_0 = \overline{\psi}_0 \quad \text{on } \Gamma. \tag{2.10}$$

For all $k \geq 1$, (n_k, ψ_k, V_k) is obtained by induction on k in the following linear problem in Ω:

$$-\text{div}(n_0 \nabla \psi_k) = \sum_{i=1}^{k} \text{div}(n_i \nabla \psi_{k-i}), \tag{2.11}$$

$$h'(n_0) n_k - V_k = f_k, \tag{2.12}$$

$$-\Delta V_k = -n_k, \tag{2.13}$$

with the boundary conditions :

$$n_k = \overline{n}_k, \quad \psi_k = \overline{\psi}_k \quad \text{on } \Gamma, \tag{2.14}$$

where

$$f_k = \psi_{k-1} - \frac{1}{2}\sum_{i=0}^{k-1} \nabla\psi_{k-1-i}.\nabla\psi_i - \overline{h}_k((n_i)_{0\leq i\leq k-1}). \tag{2.15}$$

Remark 2.1. Equation (2.8) expresses a Maxwell-Boltzmann type relation. Indeed for the isothermal plasma, the pressure is a linear function. Then $p(n) = a^2 n$ with $a > 0$. This implies from the definition of h that $h(n) = a^2 \log n$ and hence, from (2.8) $n_0 = \exp(V_0/a^2)$. This is the classical Maxwell-Boltzmann relation which has been used in [4, 12, 37] for the study of the quasineutral limit.

2.1.2 Justification of the asymptotic expansion

To justify the asymptotic expansion there are two necessary steps. First, we have to show that each profile exists and is unique. Then we have to obtain estimates for the difference between the exact solution and the asymptotic expansion in the good spaces.

Using classical results we can prove

Theorem 2.2. *Let assumptions (A1)–(A5) hold. The problem* (2.7)–(2.10) *has a unique solution* (n_0, ψ_0, V_0) *in* $\mathcal{W}^{2,q}(\Omega) \times \mathcal{C}^{2,\delta}(\overline{\Omega}) \times \mathcal{C}^{1,\delta}(\overline{\Omega})$ *which satisfies*

$$n_0(x) \geq \min(\underline{C}, \underline{n}) > 0, \quad \forall x \in \overline{\Omega}.$$

We refer to [35] for the details of proof. Considering now the problem (2.11)–(2.14), we can prove by induction on k that it has also a unique solution

Theorem 2.3. *Let assumptions (A1)–(A5) hold and* $1 \leq k \leq m$. *The problem* (2.11)–(2.14) *has a unique solution* (n_k, ψ_k, V_k) *in* $\mathcal{W}^{2,q}(\Omega) \times \mathcal{C}^{2,\delta}(\overline{\Omega}) \times \mathcal{C}^{1,\delta}(\overline{\Omega})$.

We refer again to [35] for more details on proof. The two theorems 2.2 and 2.3 give immediately

Theorem 2.4. *Let* $m \in \mathbb{N}$ *and assumptions (A1)-(A5) hold. Then there exists a unique asymptotic expansion* (2.5) *up to order* m, *i.e., for all* $0 \leq k \leq m$, *there exists a unique profile* $(n_k, \psi_k, V_k) \in B$, *solution to the problem* (2.7)–(2.10) *of* $k = 0$ *or* (2.11)–(2.14) *if* $1 \leq k \leq m$.

It remains now obtaining estimates for the difference between a sequence of exact solution and the asymptotic expansion. Let $(n_\varepsilon, \psi_\varepsilon, V_\varepsilon)$ be a smooth solution of (1.1)–(1.3) and (2.4) and $(n^m_{a,\varepsilon}, \psi^m_{a,\varepsilon}, V^m_{a,\varepsilon})$ be approximate solution of order m defined by

$$n^m_{a,\varepsilon} = \sum_{k=0}^{m} \varepsilon^k n_k, \quad \psi^m_{a,\varepsilon} = \sum_{k=0}^{m} \varepsilon^k \psi_k, \quad V^m_{a,\varepsilon} = \sum_{k=0}^{m} \varepsilon^k V_k, \tag{2.16}$$

where $(n_k, \psi_k, V_k)_{0 \leq k \leq m}$ is the unique solution of (2.7)–(2.10) for $k = 0$ and (2.11)–(2.14) for $1 \leq k \leq m$. In [35] it is shown that

Theorem 2.5. *Let* $(n_\varepsilon, \psi_\varepsilon, V_\varepsilon)$ *be the solution of the systems* (1.1)–(1.3) *and* (2.4) *and* $(n^m_{a,\varepsilon}, \psi^m_{a,\varepsilon}, V^m_{a,\varepsilon})$ *be the approximate solution given by the asymptotic expansion* (2.16). *Let assumptions (A1)–(A7) hold. Then there exists* $\varepsilon_0 > 0$ *such that for all* $\varepsilon \in (0, \varepsilon_0]$, *we have the following estimates*

$$\|n_\varepsilon - n^m_{a,\varepsilon}\|_{\mathcal{W}^{2,q}(\Omega)} \leq A_1 \varepsilon^{m+1}, \|\psi_\varepsilon - \psi^m_{a,\varepsilon}\|_{\mathcal{C}^{2,\delta}(\overline{\Omega})} \leq A_1 \varepsilon^{m+1}, \tag{2.17}$$
$$\|V_\varepsilon - V^m_{a,\varepsilon}\|_{\mathcal{C}^{1,\delta}(\overline{\Omega})} \leq A_1 \varepsilon^{m+1},$$

where $A_1 > 0$ is a constant independent of ε.

Here we give only the main steps of the proof of Theorem 2.4 and we refer one more time to [35] for more details.

The first step consists in obtaining the system satisfied by the approximate solution $(n^m_{a,\varepsilon}, \psi^m_{a,\varepsilon}, V^m_{a,\varepsilon})$. To this end we use the problems satisfied by each profile of the asymptotic expansion. Then we subtract the systems verified by the exact solution and the approximate solution to obtain

$$-\operatorname{div}(n_\varepsilon \nabla \psi_\varepsilon) + \operatorname{div}(n^m_{a,\varepsilon} \nabla \psi^m_{a,\varepsilon}) = \varepsilon^{m+1} D_1^\varepsilon, \tag{2.18}$$

$$\frac{\varepsilon}{2}(|\nabla\psi_\varepsilon|^2 - |\nabla\psi^m_{a,\varepsilon}|^2) + h(n_\varepsilon) - h(n^m_{a,\varepsilon}) = V_\varepsilon - V^m_{a,\varepsilon} + \varepsilon(\psi_\varepsilon - \psi^m_{a,\varepsilon}) + \varepsilon^{m+1} D_2^\varepsilon, \tag{2.19}$$

$$-\Delta(V_\varepsilon - V^m_{a,\varepsilon}) = -(n_\varepsilon - n^m_{a,\varepsilon}), \tag{2.20}$$

and

$$n_\varepsilon - n^m_{a,\varepsilon} = n^{m+1}_{D,\varepsilon}, \quad \psi_\varepsilon - \psi^m_{a,\varepsilon} = \psi^{m+1}_{D,\varepsilon} \quad \text{on } \Gamma. \tag{2.21}$$

where

$$D_1^\varepsilon = \sum_{k=m+1}^{2m} \left(\varepsilon^{k-m-1} \sum_{i=k-m}^{m} \operatorname{div}(n_i \nabla \psi_{k-i}) \right),$$

$$D_2^\varepsilon = -\frac{1}{2} \sum_{k=m}^{2m} \left(\varepsilon^{k-m} \sum_{i=k-m}^{m} \nabla\psi_i . \nabla\psi_{k-i} \right) - r_\varepsilon(n) + \psi_m,$$

and

$$r_\varepsilon(n) = \frac{1}{(m+1)!} \frac{d^{m+1} h(n^m_{a,\xi})}{d\varepsilon^{m+1}} \quad \text{with } \xi \in [0, \varepsilon].$$

We eliminate first $V_\varepsilon - V^m_{a,\varepsilon}$ to obtain an elliptic nonlinear system satisfied by $(n_\varepsilon - n^m_{a,\varepsilon}, \psi_\varepsilon - \psi^m_{a,\varepsilon})$. Using different lemmas and uniform boundedness of the sequence of solution $(n_\varepsilon, \psi_\varepsilon, V_\varepsilon)$, we can show the estimates (2.17) for $(n_\varepsilon - n^m_{a,\varepsilon}, \psi_\varepsilon - \psi^m_{a,\varepsilon})$. Then, using (2.19), we obtain easily the last estimate of (2.17) for $V_\varepsilon - V^m_{a,\varepsilon}$.

Remark 2.6. Here, this kind of proof is possible only due to the fact that we have already existence, uniqueness and uniform boundedness of a sequence of solution to the problem (1.1)–(1.3) and (2.1) thanks to [34]. In the following section for the quasineutral limit, the situation will be very different, since without compatibility condition, we don't have anymore existence of solutions for the problem (1.1)–(1.3) and (2.1).

Remark 2.7. In a same way, it is shown in [35] that there exists an analogous result for the zero relaxation time limit. Moreover, as an application of Theorem 2.4 here and Theorem 4.2 in [35], when the boundary data are compatible with the function C, it is possible to obtain the convergence to the incompressible Euler equations via the zero electron mass limit and the zero relaxation time limit.

2.2 Quasineutral limit

Here we are interested in the construction and justification of an asymptotic expansion for the quasineutral limit, i.e. λ tends to zero, without compatibility condition which means that boundary layers can appear. Indeed, if we formally take $\lambda = 0$ in (1.3) and (2.1) we obtain

$$C(x) = n(x), \text{ in } \Omega \text{ and } n = n_D \text{ on } \Gamma.$$

Then, if $n_D \neq C$ on Γ some boundary layers appear.

In all the section we keep $\varepsilon > 0$ as a small parameter independent of λ in the equations and we note $(n_\lambda, \psi_\lambda, V_\lambda)$ a solution of (1.1)–(1.3) supplemented with the boundary conditions

$$n_\lambda = \sum_{j=0}^{m} \lambda^j n_D^j + n_{D,\lambda}^m, \ \psi_\lambda = \sum_{j=0}^{m} \lambda^j \psi_D^j + \psi_{D,\lambda}^m, \text{ on } \Gamma, \tag{2.22}$$

where $n_{D,\lambda}^m$ and $\psi_{D,\lambda}^m$ are smooth enough and defined on $\overline{\Omega}$. Let us note that here since we consider the case without compatibility condition, we don't have existence and uniqueness of solution to (1.1)–(1.3) and (2.22) contrary to the previous. We will see later that it is important to keep ε in the equations since the ellipticity of the system would be equivalent to a smallness condition on ε as before.

2.2.1 Construction of the asymptotic expansion

Let us first explain how to construct the asymptotic expansion. The method used here is the one presented in [36]. We assume that:

- (H1) $C \in C^\infty(\overline{\Omega})$, $0 < \underline{n} \leq C(x) \leq \overline{n}$, $x \in \overline{\Omega}$, $\underline{n}$, $\overline{n} \in \mathbb{R}$,
- (H2) $n_D^j \in C^\infty(\overline{\Omega})$ for $0 \leq j \leq m$,
- (H3) $\psi_D^j \in C^{2,\delta}(\overline{\Omega})$ for $0 \leq j \leq m$,
- (H4) $n_D^0(x) = C(x), n_D^1(x) = 0$, $x \in \overline{\Omega}$,
- (H5) $(\lambda^{-m-1} n_{D,\lambda}^m)_{\lambda>0}$ is bounded in $W^{2,q}(\Omega)$, $q > \frac{d}{1-\delta}$, $\delta \in (0,1)$,
- (H6) $(\lambda^{-m+1} \psi_{D,\lambda}^m)_{\lambda>0}$ is bounded in $C^{2,\delta}(\overline{\Omega})$.

Remark 2.8. The assumption (H4) is a compatibility condition for the first and second order terms. It assures that any boundary layers will not appear in these two terms. The case without compatibility conditions presents some difficulties in which we didn't succeed in the study [40]. We will give more details on it below.

Here due to the boundary layers, we have to construct an asymptotic expansion including an internal expansion and an external expansion.

Internal expansion Let

$$n(x) = \sum_{k\geq 0} \lambda^k n_k(x); \ \psi(x) = \sum_{k\geq 0} \lambda^k \psi_k(x); \ V(x) = \sum_{k\geq 0} \lambda^k V_k(x).$$

Plugging this into (1.1)–(1.3), using the same method as in the previous section, and by identification of the power of λ, we obtain the problems satisfied by (n_k, ψ_k, V_k) for all k. More precisely

$$V_0 = -\frac{\varepsilon}{2}|\nabla\psi_0|^2 - h(n_0) + \varepsilon\psi_0, \tag{2.23}$$
$$\mathrm{div}(n_0\nabla\psi_0) = 0, \tag{2.24}$$
$$n_0 = C(x), \tag{2.25}$$

$$V_1 = -\varepsilon\nabla\psi_0.\nabla\psi_1 - h'(n_0)n_1 + \varepsilon\psi_1, \tag{2.26}$$
$$-\mathrm{div}(n_0\nabla\psi_1) = \mathrm{div}(n_1\nabla\psi_0), \tag{2.27}$$
$$n_1 = 0, \tag{2.28}$$

and for all $k \geq 2$

$$V_k = -\frac{\varepsilon}{2}\sum_{i=0}^{k} \nabla\psi_i.\nabla\psi_{k-i} - h'(n_0)n_k - \overline{h}_k((n_i)_{0\leq i\leq k-1}) + \varepsilon\psi_k \tag{2.29}$$

$$-\mathrm{div}(n_0\nabla\psi_k) = \sum_{i=1}^{k} \mathrm{div}(n_i\nabla\psi_{k-i}), \tag{2.30}$$
$$n_k = \Delta V_{k-2}, \tag{2.31}$$

where $\overline{h}_k$ is smooth and $\overline{h}_1 \equiv 0$ (see [35]).

All the profiles (n_k, ψ_k, V_k) can be determined to become uniquely and sufficiently smooth by induction on k with boundary conditions given later. Then the internal expansion is constructed. For $m \geq 2$ let us denote

$$n^\lambda_{I,m} = \sum_{k=0}^{m} \lambda^k n_k; \ \psi^\lambda_{I,m} = \sum_{k=0}^{m} \lambda^k \psi_k; \ V^\lambda_{I,m} = \sum_{k=0}^{m} \lambda^k V_k.$$

By construction, it is easy to see that if (n_k, ψ_k, V_k) are smooth enough, then the error equations are of order $O(\lambda^{m+1})$. Since $n_k = \Delta V_{k-2}$, for $k \geq 2$, and is not necessarily equal to n^k_D on Γ, a boundary layer can appear.

External expansion We follow the notations in [40]. For $x \in \Omega$, we note $t(x)$ the distance from Γ to x and $s(x)$ the point of Γ nearest to x. For $\theta > 0$, let Ω_θ be the boundary layer of size θ :

$$\Omega_\theta = \{x \in \Omega; |x - y| < \theta, \ y \in \Gamma\}.$$

If θ is small enough, $s(x)$ is defined uniquely for all $x \in \Omega_\theta$. In Ω_θ, we define the fast variable by $\xi(x,\lambda) = t(x)/\lambda$. For $x \in \Omega_\theta$, let $\nu(x) = (\nu_1, ..., \nu_d)$ the unit interior-directional normal vector of Γ passing from x. Then from :

$$t(x) = \|x - s(x)\|, \; x - s(x) = t(x)\nu(x),$$

and due to the fact that for all $i = 1, ..., d$, $\partial s(x)/\partial x_i$ is orthogonal to $\nu(x)$, it is easy to see that $\nabla_x t = \nu(x)$. Hence the partial derivative of a function $w(s(x), \xi(x,\lambda))$ may be decomposed as :

$$\frac{\partial w(s(x), \xi(x,\lambda))}{\partial x_i} = \lambda^{-1}\nu_i \frac{\partial w}{\partial \xi} + D_i w, \tag{2.32}$$

where D_i is a first order differential operator in s defined by : $D_i w = \nabla_s w . \frac{\partial s}{\partial x_i}$. Similarly :

$$\frac{\partial^2 w(s(x), \xi(x,\lambda))}{\partial x_i \partial x_j} = \lambda^{-2}\nu_i \frac{\partial^2 w}{\partial \xi^2} + \lambda^{-1} D_{ji} \frac{\partial w}{\partial \xi} + D_j D_i w + \nabla_s w . \frac{\partial^2 s}{\partial x_i \partial x_j}, \tag{2.33}$$

where $D_{ji} = \nu_i D_j + \nu_j D_i + \partial \nu_i / \partial x_j$. Note that for all i, j we have : $D_{ji} = D_{ij}$.

For each function $w(x)$ defined in Ω_θ the equivalent function of (s,t) is denoted by $\tilde{w}$ i.e. $w(x) = \tilde{w}(s(x), t(x)) = \tilde{w}(s(x), \lambda\xi(x,\lambda))$. We develop $\tilde{w}(s(x), \lambda\xi(x,\lambda))$ formally to obtain

$$\tilde{w}(s(x), \lambda\xi(x,\lambda)) = \tilde{w}(s(x), 0) + O(\lambda).$$

Let $\overline{w}(s) = \tilde{w}(s, 0)$. Then the ansatz of an approximate solution up to order m of (1.1)–(1.3) in Ω_θ is given by

$$\begin{aligned} \tilde{n}^\lambda_{a,m}(x) &= n^\lambda_{I,m}(x) + \tilde{n}^\lambda_{B,m}(s(x), \xi(x,\lambda)), \\ \tilde{\psi}^\lambda_{a,m}(x) &= \psi^\lambda_{I,m}(x) + \tilde{\psi}^\lambda_{B,m}(s(x), \xi(x,\lambda)), \\ \tilde{V}^\lambda_{a,m}(x) &= V^\lambda_{I,m}(x) + \tilde{V}^\lambda_{B,m}(s(x), \xi(x,\lambda)); \end{aligned}$$

where the boundary layers $(\tilde{n}^\lambda_{B,m}, \tilde{\psi}^\lambda_{B,m}, \tilde{\phi}^\lambda_{B,m})$ have the expansion:

$$\tilde{n}^\lambda_{B,m} = \sum_{k=0}^{m} \lambda^k n^b_k, \; \tilde{\psi}^\lambda_{B,m} = \sum_{k=0}^{m+1} \lambda^k \psi^b_k, \; \tilde{V}^\lambda_{B,m} = \sum_{k=0}^{m} \lambda^k V^b_k,$$

in which each term $(n^b_k(s,\xi), \psi^b_k(s,\xi), V^b_k(s,\xi))$ will be chosen to decay exponentially when ξ tends to $+\infty$. They are determined by setting $(\tilde{n}^\lambda_{a,m}, \tilde{\psi}^\lambda_{a,m}, \tilde{V}^\lambda_{a,m})$ in (1.1)–(1.3) and identification of the power of λ. Let $\partial/\partial\nu = \sum_{i=1}^d \nu_i \partial/\partial x_i$. After computation we obtain

$$\psi^b_0 \equiv 0,$$

$$(S_0)\qquad \begin{cases} (\overline{n}_0 + n_0^b)\dfrac{\partial^2 \psi_1^b}{\partial \xi^2} + \left(\dfrac{\partial \overline{\psi}_0}{\partial \nu} + \dfrac{\partial \psi_1^b}{\partial \xi}\right)\dfrac{\partial n_0^b}{\partial \xi} = 0, \\ \dfrac{\varepsilon}{2}\left(\dfrac{\partial \psi_1^b}{\partial \xi}\right)^2 + \varepsilon \dfrac{\partial \psi_1^b}{\partial \xi}\dfrac{\partial \overline{\psi}_0}{\partial \nu} + h(\overline{n}_0 + n_0^b) - V_0^b = \overline{V}_0 + \varepsilon \overline{\psi}_0, \\ \dfrac{\partial^2 V_0^b}{\partial \xi^2} = n_0^b, \end{cases}$$

and for $k \geq 1$:

$$(S_k)\qquad \begin{cases} \varepsilon \dfrac{\partial \psi_{k+1}^b}{\partial \xi}\left(\dfrac{\partial \psi_1^b}{\partial \xi} + \dfrac{\partial \overline{\psi}_0}{\partial \nu}\right) + h'(\overline{n}_0 + n_0^b)n_k^b - V_k^b \\ \qquad\qquad = F_{1,k}(n_l^b, \psi_{l+1}^b, 0 \leq l \leq k-1), \\ (\overline{n}_0 + n_0^b)\dfrac{\partial^2 \psi_{k+1}^b}{\partial \xi^2} + \left(\dfrac{\partial \overline{\psi}_0}{\partial \nu} + \dfrac{\partial \psi_1^b}{\partial \xi}\right)\dfrac{\partial n_k^b}{\partial \xi} + n_k^b \dfrac{\partial^2 \psi_1^b}{\partial \xi^2} \\ \qquad + \dfrac{\partial n_0^b}{\partial \xi}\dfrac{\partial \psi_{k+1}^b}{\partial \xi} = F_{2,k}(n_l^b, \psi_{l+1}^b, 0 \leq l \leq k-1), \\ -\dfrac{\partial^2 V_k^b}{\partial \xi^2} + n_k^b = F_{3,k}(V_l^b, k-1 \leq l \leq k-2), \end{cases}$$

where $F_{i,k},\ i = 1, 2, 3$, are given functions of $(n_l^b, \psi_{l+1}^b)_{0 \leq l \leq k-1}$ for $F_{1,k}$, $F_{2,k}$, and of $(V_l^b)_{k-1 \leq l \leq k-2}$ for $F_{3,k}$.

Hence the approximate solution is constructed in Ω_θ. To complete the definition of the approximate solution in $\overline{\Omega}$, let $\sigma \in C^\infty(0, \infty)$ be a smooth function such that $\sigma(t) = 1$ for $0 \leq t \leq \theta/2$ and $\sigma \equiv 0$ for $t \geq \theta$ and $(n_{B,m}^\lambda(x), \psi_{B,m}^\lambda(x), V_{B,m}^\lambda(x))$ defined by

$$\begin{cases} (\tilde{n}_{B,m}^\lambda(s(x), t(x)/\lambda), \tilde{\psi}_{B,m}^\lambda(s(x)t(x)/\lambda), \tilde{V}_{B,m}^\lambda(s(x)t(x)/\lambda))\sigma(t(x)), \\ \qquad\qquad\qquad\qquad\qquad\qquad\qquad\qquad \text{for } x \in \Omega_\theta, \\ 0, \quad \text{for } x \in \Omega - \Omega_\theta. \end{cases}$$

Then, $(n_{B,m}^\lambda, \psi_{B,m}^\lambda, V_{B,m}^\lambda)$ has the same regularity as $(\tilde{n}_{B,m}^\lambda, \tilde{\psi}_{B,m}^\lambda, \tilde{V}_{B,m}^\lambda)$. For $(n_k^b(s,\xi), \psi_k^b(s,\xi), V_k^b(s,\xi))$ decreasing exponentially when ξ tends to $+\infty$, it is easy to see that the difference between $(n_{B,m}^\lambda, \psi_{B,m}^\lambda, V_{B,m}^\lambda)$ and $(\tilde{n}_{B,m}^\lambda, \tilde{\psi}_{B,m}^\lambda, \tilde{V}_{B,m}^\lambda)$ is uniform of order of $e^{-\mu/\lambda}$ for a constant $\mu > 0$.

Finally, the boundary conditions (2.22) give for $s \in \Gamma$:

$$n_0 = n_D^0, n_1 = n_D^1, n_0^b(s,0) = n_1^b(s,0) = 0, \overline{n}_k(s) + n_k^b(s,0) = n_D^k, k \geq 2, \tag{2.34}$$

$$\begin{aligned} &\psi_0 = \psi_D^0, \psi_1 = \psi_D^1, \psi_2 = \psi_D^2, \\ &\psi_1^b(s,0) = \psi_2^b(s,0) = 0, \overline{\psi}_k(s) + \psi_k^b(s,0) = \psi_D^k,\ k \geq 3. \end{aligned} \tag{2.35}$$

We refer to [36] for the scheme of determination of $(n_k, \psi_k, V_k, n_k^b, \psi_{k+1}^b, V_k^b)$.

Remark 2.9. Due to the assumption (H4),

$$n_0^b = n_1^b = \psi_1^b = \psi_2^b = 0.$$

This means that there are no boundary layers terms of order zero and one for the density and zero, one and two for the velocity potential.

The approximate solution up to order m is now constructed in the form:

$$(n_\lambda^a, \psi_\lambda^a, V_\lambda^a) = (n_{I,m}^\lambda + n_{B,m}^\lambda, \psi_{I,m}^\lambda + \psi_{B,m}^\lambda, V_{I,m}^\lambda + V_{B,m}^\lambda), \text{ in } \overline{\Omega}. \quad (2.36)$$

Moreover, $n_\lambda^a = \sum_{k=0}^m \lambda^k n_D^k$, $\psi_\lambda^a = \sum_{k=0}^m \lambda^k \psi_D^k$ on Γ and,

$$n_\lambda^a = n_0 + \sum_{j=2}^m \lambda^j (n_j + n_j^b), \quad (2.37)$$

$$\psi_\lambda^a = \psi_0 + \lambda\psi_1 + \lambda^2\psi_2 + \sum_{j=3}^m \lambda^j (\psi_j + \psi_j^b) + \lambda^{m+1}\psi_{m+1}^b. \quad (2.38)$$

2.2.2 Justification of the asymptotic expansion

To justify one more time the asymptotic expansion, we have to prove the existence and uniqueness of each profile and to obtain some estimates for the difference between the exact solution and the approximate solution. Moreover, here, since we consider the case where boundary layers can appear, we have also to prove existence of solutions to the problems (1.1)–(1.3) and (2.22). We will give here only the results and we refer to [40] for more details.

Theorem 2.10. *Under the assumptions (H1)–(H6), there exists a unique asymptotic expansion* (2.36) *up to order* m*, sufficiently smooth satisfying* (2.37)–(2.38).

We have already seen that (1.1)–(1.3) and (2.22) are equivalent to (1.1), (2.2) and (2.22). Then

Theorem 2.11. *Let the assumptions (H1)–(H6) hold. For* λ *is small enough there is an* $\varepsilon_0 > 0$ *independent of* λ *such that for all* $\varepsilon \in [0, \varepsilon_0]$*, the problems* (1.1), (2.2), (2.22) *have a unique solution* $(n_\lambda, \psi_\lambda)$ *in* $\mathcal{W}^{2,q}(\Omega) \times \mathcal{C}^{2,\delta}(\overline{\Omega})$ *which satisfy*

$$\|n_\lambda - n_\lambda^a\|_{W^{2,q}(\Omega)} \le A\lambda^{m-1}, \ \|\psi_\lambda - \psi_\lambda^a\|_{C^{2,\delta}(\overline{\Omega})} \le A\lambda^{m-1}, \quad (2.39)$$

where A *is a constant independent of* λ.

Remark 2.12. Using equation (1.2), the continuity of h and estimates (2.39), we can easily obtain, for λ is small enough

$$\|V_\lambda - V_\lambda^a\|_{C^{1,\delta}(\overline{\Omega})} \le A\lambda^{m-1},$$

where A is a constant independent of λ.

The proof of Theorem 2.11 is long and complicated and we refer to [40] for details. Let us just mention that the main idea is to search a solution under the form

$$n_\lambda = n_\lambda^a + \lambda^{m-1} r_\lambda,\ \psi_\lambda = \psi_\lambda^a + \lambda^{m-1} p_\lambda.$$

Then we consider the problem verified by r_λ and p_λ. It is clear that if we obtain the existence and the boundedness of r_λ and p_λ we immediately have the result of Theorem 2.11. The problem for r_λ and p_λ is a nonlinear elliptic problem. To solve it, we use the Schauder fixed point theorem, in which, to obtain the existence, uniqueness and boundedness of solution for the linearized problem, we use another fixed point theorem: the Leray-Schauder fixed point theorem.

3 Numerical simulations

In this section we construct numerical schemes to the systems (1.1)–(1.3). As seen before, from a theoritical point of view, to study these systems one uses (1.1) and (1.3) to eliminate V in (1.2) to obtain a system of two equations of unknowns (n, ψ), supplemented with Dirichlet boundary conditions. Recall that the resulting equation for n, equation (2.2), is elliptic if and only if the flow is subsonic which corresponds to a smallness condition on the data or on the parameter ε. When (n, ψ) are solved one obtains easily V from (1.2). However, equation (2.2) is fully nonlinear and coupled to ψ till its second derivatives, so that the numerical discretization is not an easy task. Let us now recall the systems (1.1)–(1.3):

$$-\text{div}(n\nabla\psi) = 0, \tag{3.1}$$

$$\frac{\varepsilon}{2}|\nabla\psi|^2 + h(n) = V + \frac{\varepsilon\psi}{\tau}, \tag{3.2}$$

$$-\lambda^2\Delta V = n - C, \quad \in \Omega. \tag{3.3}$$

The first and last equations are linear with (ψ, V) and the second one is nonlinear only algebraically with n. This motivates us to make the following iterative scheme: for a given n^m $(m \ge 0)$, we first solve (ψ^m, V^m) by:

$$-\text{div}(n^m\nabla\psi^m) = 0, \tag{3.4}$$

$$-\Delta V^m = C - n^m, \qquad \text{in } \Omega, \tag{3.5}$$

subject to mixed Dirichlet-Neumann boundary conditions:

$$V^m = \overline{V}, \ \psi^m = \overline{\psi}, \quad \text{on } \Gamma_D, \tag{3.6}$$

$$\nabla V^m.\nu = \nabla \psi^m.\nu = 0, \quad \text{on } \Gamma_N, \tag{3.7}$$

where ν is the unit outward normal to $\Gamma = \Gamma_D \cup \Gamma_N$.

Remark 3.1. Let us note that these boundary conditions are physically motivated in the case of a semiconductor. However,usually, for a semiconductor the boundary conditions are given for the electrostatic potential and the electron density. But, since here we need some boundary conditions on ψ and not on n, we choose

$$\overline{\psi} \stackrel{def}{=} (h(n_D) - \overline{V})/\varepsilon. \tag{3.8}$$

With such boundary conditions we are able to obtain Dirichlet type boundary conditions for n on Γ_D (see Remark 2.1 in [8] for more details).

Then, n^{m+1} is computed with the algebraic equation

$$h(n^{m+1}) + \frac{\varepsilon}{2}|\nabla \psi^m|^2 = V^m + \varepsilon \psi^m. \tag{3.9}$$

Equations (3.4) and (3.5) are of elliptic type (provided that n remains positive). There are several numerical methods to solve them. In [8], we choose to use two finite volume schemes. The first scheme is "classical" with a two-point discretization of the fluxes through the edges, see [15]. It leads to piecewise constant approximate solutions and needs to be completed by a reconstruction of the gradients $\nabla \psi^m$, necessary for the computation of n^{m+1} in (3.9). The second scheme is of mixed finite volume type as introduced in [14], in which the construction of the gradients is intrinsic. Here we will only present the first scheme and we refer to [8] for details on the second scheme. Let us just mention that the results obtained with each scheme are really similar. Our first goal in this study is to see numerically the necessary smallness condition on ε for the existence of solutions. But our schemes can also be used to obtain the current-voltage characteristics or to simulate a ballistic diode.

3.1 Mesh and notations

Let us first of all introduce some notations useful in the presentation of the schemes. It concerns the mesh, the initial and boundary data.

A mesh of Ω is given by a family $\mathcal{T}$ of control volumes (open polygonal convex disjoint subsets of Ω), a family $\mathcal{E}$ of edges in 2-d (faces in 3-d) and a set $\mathcal{P}$ of points of Ω indexed by $\mathcal{T}$: $\mathcal{P} = (\mathbf{x}_K)_{K\in\mathcal{T}}$. For a control volume $K \in \mathcal{T}$ we denote by $m(K)$ the measure of K and $\mathcal{E}_K$ the set of edges of K. The (d-1)-dimensional measure of an edge σ is denoted

$m(\sigma)$. In the case where $\sigma \in \mathcal{E}$ such that $\overline{\sigma} = \overline{K} \cap \overline{L}$ with K and L being two neighboring cells, we note $\sigma = K|L$.

The set of interior (resp. boundary) edges is denoted by $\mathcal{E}^{int}$ (resp. $\mathcal{E}^{ext}$), that is $\mathcal{E}^{int} = \{\sigma \in \mathcal{E}; \sigma \not\subset \partial\Omega\}$ (resp. $\mathcal{E}^{ext} = \{\sigma \in \mathcal{E}; \sigma \subset \partial\Omega\}$). We note $\mathcal{E}_D^{ext}$ (resp. $\mathcal{E}_N^{ext}$) the set of $\sigma \subset \Gamma_D$ (resp. $\sigma \subset \Gamma_N$). For all $K \in \mathcal{T}$, we note $\mathcal{E}_K^{ext} = \mathcal{E}_K \cap \mathcal{E}^{ext}$, $\mathcal{E}_{D,K}^{ext}$ (resp. $\mathcal{E}_{N,K}^{ext}$) the edges of K included in Γ_D (resp. Γ_N), and $\mathcal{E}_K^{int} = \mathcal{E}_K \cap \mathcal{E}^{int}$. Finally, for $\sigma \in \mathcal{E}_K$, we denote by $\mathbf{x}_\sigma$ its barycenter and by $\nu_{K,\sigma}$ the exterior unit normal vector to σ.

Given an initial datum n^0 and boundary data $\overline{V}$, $\overline{\psi}$, their approximations on each control volume or on each boundary edge are denoted by

$$\begin{aligned} n_K^0 &= \frac{1}{m(K)} \int_K n^0, \\ \overline{V}_\sigma &= \frac{1}{m(\sigma)} \int_\sigma \overline{V}, \\ \overline{\psi}_\sigma &= \frac{1}{m(\sigma)} \int_\sigma \overline{\psi}. \end{aligned}$$

We also set

$$f_K^m = C_K - n_K^m, \quad \text{with} \quad C_K = \frac{1}{m(K)} \int_K C.$$

3.2 Classical finite volume scheme (VF4-scheme)

Now we are able to present the classical finite volume scheme used in [8] to solve the problem (3.1)–(3.3).

Let us consider an admissible mesh of Ω given by $\mathcal{T}$, $\mathcal{E}$ and $\mathcal{P}$ which satisfy Definition 3.8 in [15]. We recall that the admissibility of $\mathcal{T}$ implies that the straight line between two neighboring centers of cells $(\mathbf{x}_K, \mathbf{x}_L)$ is orthogonal to the edge $\sigma = K|L$. Finally, let us define the transmissibility coefficients:

$$\tau_\sigma = \frac{m(\sigma)}{d(\mathbf{x}_K, \mathbf{x}_L)} \quad \text{if } \sigma = K|L \in \mathcal{E}_K^{int} \quad \text{and} \quad \tau_\sigma = \frac{m(\sigma)}{d(\mathbf{x}_K, \Gamma)} \quad \text{if } \sigma \in \mathcal{E}_K^{ext}, \tag{3.10}$$

and the size of the mesh:

$$H = \max_{K \in \mathcal{T}} \operatorname{diam}(K). \tag{3.11}$$

In all the sequel, we assume that the points $\mathbf{x}_K$ are located inside each control volume. Let $(V_K^m)_{K \in \mathcal{T}}$ and $(\psi_K^m)_{K \in \mathcal{T}}$ be the discrete unknowns.

A finite volume scheme to the mixed Dirichlet-Neumann problem (3.4)–(3.7) is defined by the following set of equations (see [15]) :

$$-\sum_{\sigma\in\mathcal{E}_K} dV^m_{K,\sigma} = m(K) f^m_K, \tag{3.12}$$

$$-\sum_{\sigma\in\mathcal{E}_K} n^m_\sigma d\psi^m_{K,\sigma} = 0, \tag{3.13}$$

where

$$dV^m_{K,\sigma} = \begin{cases} \tau_\sigma(V^m_L - V^m_K), \ \sigma = K|L, \\ \tau_\sigma(\overline{V}_\sigma - V^m_K), \ \sigma \in \mathcal{E}^{ext}_{D,K}, \\ 0, \ \sigma \in \mathcal{E}^{ext}_{N,K}, \end{cases}$$

$$d\psi^m_{K,\sigma} = \begin{cases} \tau_\sigma(\psi^m_L - \psi^m_K), \ \sigma = K|L, \\ \tau_\sigma(\overline{\psi}_\sigma - \psi^m_K), \ \sigma \in \mathcal{E}^{ext}_{D,K}, \\ 0, \ \sigma \in \mathcal{E}^{ext}_{N,K}, \end{cases}$$

$$n^m_\sigma = \begin{cases} \dfrac{n^m_K + n^m_L}{2}, \ \sigma = K|L, \\ n^m_K, \ \sigma \in \mathcal{E}^{ext}_K. \end{cases}$$

The quantities $dV^m_{K,\sigma}$ and $d\psi^m_{K,\sigma}$ are the approximations of the fluxes through each edge for each function, i.e.

$$dV^m_{K,\sigma} \approx \int_\sigma \nabla V^m \cdot \nu_{K,\sigma} \quad \text{and} \quad d\psi^m_{K,\sigma} \approx \int_\sigma \nabla \psi^m \cdot \nu_{K,\sigma}.$$

For given n^m, since equations (3.4)–(3.5) are linear, we obtain the piecewise constant functions ψ^m and V^m, unique solution of (3.12)–(3.13). Then we need to define the gradient of ψ^m. Therefore, we use the reconstruction proposed in [16]; the approximate gradient is a piecewise constant function, defined on each control volume by

$$\mathbf{w}^m_K = \frac{1}{m(K)} \sum_{\sigma\in\mathcal{E}_K} d\psi^m_{K,\sigma}(\mathbf{x}_\sigma - \mathbf{x}_K), \quad \forall K \in \mathcal{T}.$$

Finally, from (3.2) we obtain the piecewise constant function n^{m+1} by:

$$n^{m+1}_K = h^{-1}\Big(V^m_K + \varepsilon\psi^m_K - \frac{\varepsilon}{2}|\mathbf{w}^m_K|^2\Big), \tag{3.14}$$

with h^{-1} being the inverse function of h (see [8] for a discussion on the invertibility of h).

3.3 Numerical results

In [8] the numerical simulations are performed in two space dimensions by taking the domain $\Omega = [0,1] \times [0,1]$. A point $\mathbf{x}$ of Ω is denoted by its coordinates $\mathbf{x} = (x_1, x_2)$ and then the boundary $\Gamma = \Gamma_N \cup \Gamma_D$ is defined by $\Gamma_N = \{(x_1,x_2), x_1 \in [0,1], x_2 \in \{0,1\}\}$ and $\Gamma_D = \Gamma_{D,l} \cup \Gamma_{D,r}$ with

$$\Gamma_{D,l} = \{(x_1,x_2), x_1 = 0, x_2 \in [0,1]\},$$
$$\Gamma_{D,r} = \{(x_1,x_2), x_1 = 1, x_2 \in [0,1]\}.$$

The considered pressure functions are $p(s) = s^\gamma$ with $\gamma = 1$ or $5/3$, which implies for the enthalpy :

$$h(s) = \begin{cases} \ln(s), & \text{if } \gamma = 1, \\ \dfrac{5}{2}(s^{2/3} - 1), & \text{if } \gamma = 5/3. \end{cases}$$

For the case $\gamma = 5/3$, the inverse function of h is defined on all $\mathbb{R}$ by setting

$$h^{-1}(t) = \begin{cases} \left(\dfrac{2}{5}t + 1\right)^{3/2}, & \text{if } t > -5/2, \\ 0, & \text{else.} \end{cases}$$

We refer again to [8] for more details. In all the simulations, the used mesh is a triangular mesh of size 5×10^{-2} and the accuracy of the numerical results is defined as the difference between n^m and n^{m+1} in $L^2(\Omega)$ or $L^\infty(\Omega)$ norm. For results of the validity of the schemes, on the bipolar case, we refer one more time to [8]. Here we present the obtained results of a ballistic diode using either the VF4-scheme or the mixed finite volume scheme (DE-scheme) since they are always very similar.

A ballistic diode is a semiconductor which consists of a weakly doped n-region S between two highly doped n^+-regions Ω/S. It corresponds to the unipolar case since in such devices the charge transport is only due to electrons. In [8] the numerical solution of the systems (3.1)–(3.3) is computed with the doping profile

$$C(\mathbf{x}) = \begin{cases} 10^{-3}, & \text{if } (x_1,x_2) \in S = [1/6, 5/6] \times [0,1], \\ 1, & \text{else.} \end{cases}$$

The considered boundary conditions for the electrostatic potential are the following

$$V^m = 0, \quad \text{on } \Gamma_{D,l} \quad \text{and} \quad V^m = U, \quad \text{on } \Gamma_{D,r}, \quad m \geq 0, \tag{3.15}$$

where U is a given applied voltage. Two kinds of boundary conditions are considered for the velocity potential. First the authors consider the following one:

$$\psi^m = 0, \quad \text{on } \Gamma_{D,l} \text{ and } \psi^m = -U, \quad \text{on } \Gamma_{D,r}, \quad m \geq 0. \tag{3.16}$$

For different values of γ, U and ε, the numerical solutions of the electron density, velocity potential and electrostatic potential are calculated. Note that the smallness condition on ε, for boundary conditions independent of ε (see [34]), which ensures the strict ellipticity of the system, appears clearly in the numerical simulations. Indeed, when ε is not small enough, the gradient of the velocity potential becomes more and more larger in the iteration. Moreover, due to the negative sign before $|\nabla\psi_K^m|^2$ in the formula (3.14), the condition $n > 0$ is not numerically satisfied and the matrix involved in the computation of ψ^m becomes singular. A numerical example in this case is given in Figure 3.1 (the computation is stopped after 4 iterations).

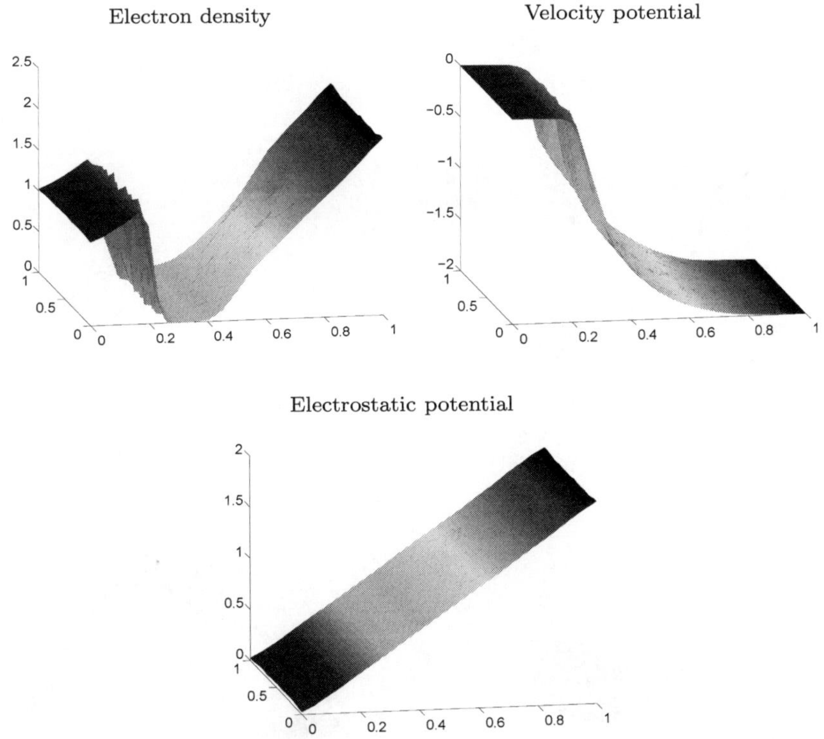

Figure 3.1 Case $\gamma = 1$, $U = 2$, $\varepsilon = 0.6$.

In the case $\gamma = 5/3$, due to the definition of the inverse function of h,

U should satisfy $U > -5/2$. Indeed, for ε is small enough, n_K^{m+1} is nearly given by $h^{-1}(V_K^m)$ according to (3.14). If $\phi_K^m \leq -5/2$, then $h^{-1}(V_K^m) = 0$ and $n_K^{m+1} \approx 0$, so that the matrix involved in the computation of ψ^{m+1} becomes singular.

Note that the usual boundary conditions used for a ballistic diode are on V and n instead of ψ and in general we choose $n = n_D = C(x) = 1$ on Γ_D (see Remark 2.6). Since $h(1) = 0$ by definition, from (3.8) we deduce the following boundary conditions on ψ:

$$\psi^m = 0, \quad \text{on } \Gamma_{D,l} \text{ and } \psi^m = -U/\varepsilon, \quad \text{on } \Gamma_{D,r}, \quad m \geq 0. \quad (3.17)$$

For such boundary conditions, the ellipticity condition depends on the ratio of U and $\sqrt{\varepsilon}$ and not only on ε. This ratio has to be small enough to ensure the ellipticity condition (which remains ensuring that we are in the subsonic region). In Figure 3.2 we show the obtained solution for $\varepsilon = 0.6$ and $U = 0.1$ with the DE-scheme. The required accuracy is of order 10^{-7} in $L^\infty(\Omega)$ norm for stopping the iterations (the iteration number is 8).

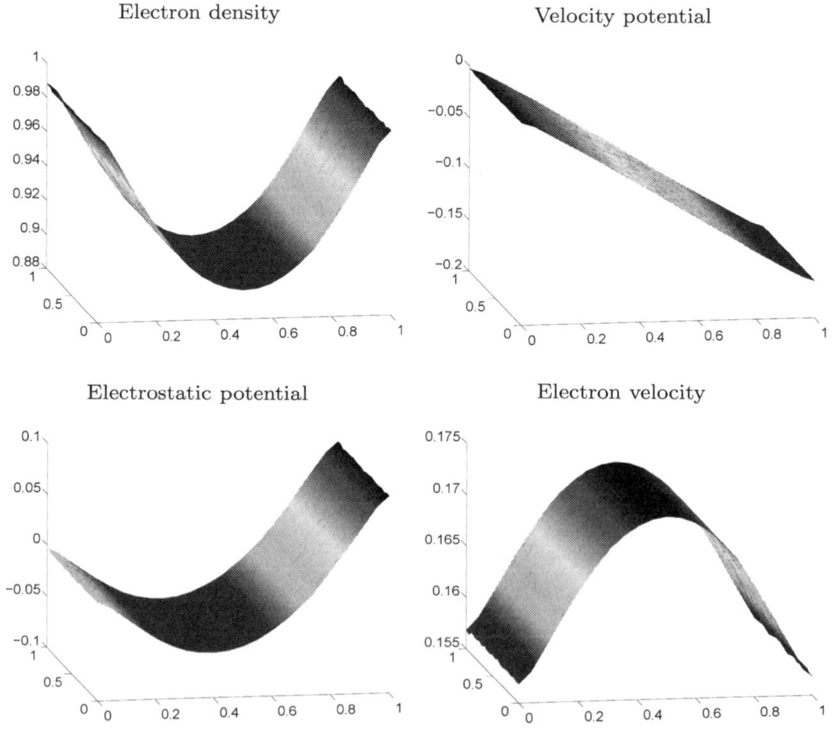

Figure 3.2 Case $\gamma = 1$, $U = 0.1$, $\varepsilon = 0.6$ for a ballistic diode.

Let us now present the current-voltage characteristics for the boundary conditions (3.15) and (3.17). By definition of Γ_N, the second coordinate of the electron current density is vanishing. Then, the problem is reduced to a one-dimensional case and the first coordinate of the electron current density is constant in the device. To obtain the current-voltage characteristics, we compute the electron current density on each control volume and we take the average of these values. Here by definition of the boundary conditions on ψ, the ellipticity condition is satisfied when $|U|/\sqrt{\varepsilon}$ is strictly less than 1. Then for $\varepsilon = 1$ we choose $-0.83 \leq U \leq 0.83$ and for $\varepsilon = 0.6, \quad -0.66 \leq U \leq 0.66$. We show the results in Figure 3.3. The required accuracy in $L^\infty(\Omega)$ norm for stopping the iterations is 10^{-7} and we still present the results of the DE-scheme.

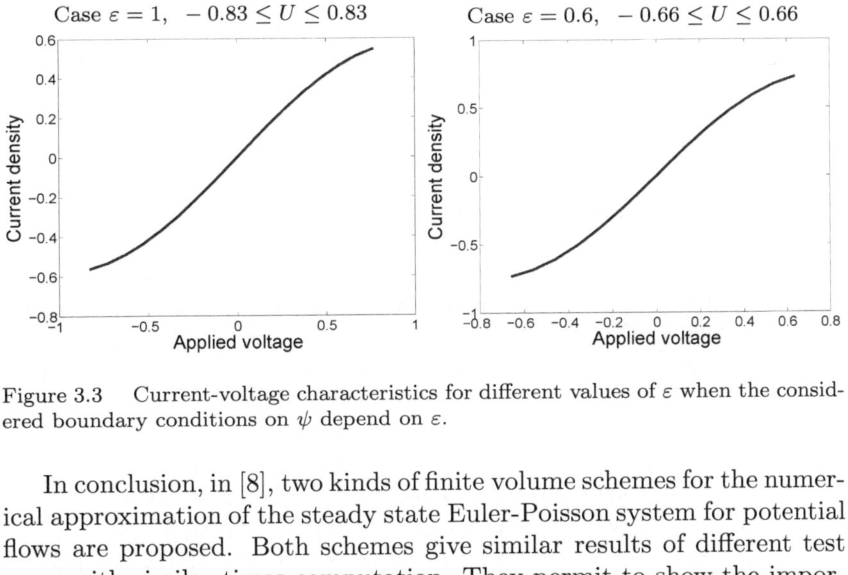

Figure 3.3 Current-voltage characteristics for different values of ε when the considered boundary conditions on ψ depend on ε.

In conclusion, in [8], two kinds of finite volume schemes for the numerical approximation of the steady state Euler-Poisson system for potential flows are proposed. Both schemes give similar results of different test cases with similar times computation. They permit to show the importance of the smallness of ε and the boundary data for the ellipticity of the system.

The VF4-scheme is a bit simpler to implement but it only works on admissible meshes, whereas the DE-scheme enables to treat very general meshes.

References

[1] G. Ali and A. Jüngel, Global smooth solutions to the multi-dimensional hydrodynamic model for two-carrier plasmas, *J. Diff. Eqn.*, 190 (2003).

[2] F. Arimburgo, C. Baiocchi and L. D. Marini, Numerical approximation of the 1-D nonlinear drift-diffusion model in semiconductors, *Nonlinear Ki-*

netic Theory and Mathematical Aspects of Hyperbolic Systems (Rappallo, 1992) (World Scientific, 1992).

[3] G. Baccarani and M. R. Wordeman, An investigation of steady-state velocity overshoot effects in Si and GaAs devices, *Solid State Electronics*, 28 (1985).

[4] H. Brézis, F. Golse and R. Sentis, Analyse asymptotique de l'équation de Poisson couplée à la relation de Boltzmann. Quasi-neutralité des plasmas, *C. R. Acad. Sci. Paris*, 321 (1995).

[5] F. Brezzi, L. D. Marini and P. Pietra, Numerical simulation of semiconductor devices, *Comput. Meth. Appl. Mech. Engrg.* 75 (1989).

[6] C. Chainais-Hillairet, J. G. Liu and Y. J. Peng, Finite volume scheme for multi-dimensional drift-diffusion equations and convergence analysis, *Math. Mod. Anal. Numer.*, 37 (2003).

[7] C. Chainais-Hillairet and Y. J. Peng, Finite volume approximation for degenerate drift-diffusion system in several space dimensions, *Math. Mod. Meth. Appl. Sci.*, 14 (3) (2004).

[8] C. Chainais, Y. J. Peng, I. Violet, Numerical solutions of the Euler-Poisson system for a potential flow, *Applied Numerical Mathematics*, 59 (2009).

[9] F. Chen, *Introduction to plasma physics and controlled fusion*, vol. 1, Plenum Press, New York, (1984).

[10] Z. X. Chen and B. Cockburn, Analysis of a finite element method for the drift-diffusion semiconductor device equations: The multidimensional case, *Numer. Math.*, 71 (1995).

[11] B. Cockburn and I. Triandaf, Error estimates for a finite element method for the drift-diffusion semiconductor device equations: The zero diffusion case, *Math. Comput.*, 63 (1994).

[12] S. Cordier and E. Grenier, Quasi-neutral limit of an Euler-Poisson system arising from plasma physics, *Comm. Part. Diff. Eqs*, 25 (2000).

[13] P. Degond and P. A. Markowich, A steady state potential flow model for semiconductors, *Annali di matematica pura ed applicata*, (IV), Vol CLXV (1993).

[14] J. Droniou and R. Eymard, A mixed finite volume scheme for anisotropic diffusion problems on any grid, *Num. Math.*, 105 (2006).

[15] R. Eymard, T. Gallouët and R. Herbin, Finite volume methods, *Handbook of numerical analysis, North-Holland, Amsterdam*, Vol. VII (2000).

[16] R. Eymard, T. Gallouët and R. Herbin, A cell-centred finite-volume approximation for anisotropic diffusion operators on unstructured meshes in any space dimension, *IMA Journal of Numerical Analysis*, 26(2) (2006).

[17] P. C. Fife, Semilinear elliptic boundary value problems with small parameters, *Arch. Ration. Mech. Anal*, 52 (1973).

[18] C.L. Gardner, Numerical simulation of a steady-state electron shock wave in a submicrometer semiconductor device, *IEEE Trans. Electron Devices*, 28 (1991).

[19] C. L. Gardner, J. W. Jerome and D. J. Rose, Numerical methods for the hydrodynamic device model: subsonic flow, *IEEE Trans. Computer-Aided Design.*, 8 (1989).

[20] I. Gasser, The initial time layer problem and the quasineutral limit in a nonlinear drift-diffusion model for semiconductors, *Nonlinear Differential Equations and Applications*, 8 (2001).

[21] I. Gasser, D. Levermore, P. A. Markowich and C. Schmeiser, The initial time layer problem and the quasineutral limit in the semiconductor drift-diffusion model, *European J. Appl. Math.*, 12 (2001).

[22] I. Gasser and P. Marcati, The combined relaxation and vanishing Debye length limit in the hydrodynamic model for semiconductors, *Math. Mech. Appl. Sci.*, 24 (2001).

[23] D. Gilbarg and N. S. Trudinger, *Elliptic partial differential equations of second order*, Springer, Berlin, New York (1984).

[24] H. K. Gummel, A self-consistent iterative scheme for one-dimensional steady-state transistor calculations, *IEEE Trans. Electron Devices*, 11 (1964).

[25] A. Jüngel, Numerical approximation of a drift-diffusion model for semiconductors with nonlinear diffusion, *Z. Angew. Math. Mech.*, 75 (1995).

[26] A. Jüngel and Y. J. Peng, A hierarchy of hydrodynamic models for plasmas. Zero-relaxation-time limit, *Comm. Part. Diff. Eqs*, 24 (1999).

[27] A. Jüngel and Y. J. Peng, Zero-relaxation-time limits in hydrodynamic models for plasmas revisited, *Z.Angew. Math. Phys.*, 51 (2000).

[28] A. Jüngel and Y. J. Peng, A hierarchy of hydrodynamic models for plasmas. Quasineutral limits in the drift-diffusion equations, *Asymptotic Analysis*, 28 (2001).

[29] A. Jüngel and P. Pietra, A discretization scheme for a quasi-hydrodynamic semiconductor model, *Math. Mod. Meth. Appl. Sci*, 7 (1997).

[30] C. Lattanzio and P. Marcati, The relaxation to the drift-diffusion system for the 3-D isentropic Euler-Poisson model for semiconductors, *Discr. contin. dyn. syst.*, 5 (1999).

[31] P. Marcati and R. Natalini, Weak solutions to a hydrdynamic model for semiconductors and relaxation to the drift-diffusion equation, *Arch. Rat. Mech. Anal.*, 129 (1995).

[32] P. A. Markowich, *The stationnary semiconductor device equations*, Springer, New York (1986).

[33] Y. J. Peng, Asymptotic limits of one-dimensional hydrodynamic models for plasmas and semiconductors, *Chinese Ann. Math.*, 23B (2002).

[34] Y. J. Peng, Some asymptotic analysis in steady state Euler-Poisson equations for potential flow, *Asymptotic Analysis*, 36 (2003).

[35] Y. J. Peng and I. Violet, Asymptotic expansions in a steady state Euler-Poisson system and convergence to incompressible Euler equations, *M3AS*, 15 (2005).

[36] Y. J. Peng and Y. G. Wang, Boundary layers and quasi-neutral limit in steady state Euler-Poisson equations for potential flows, *Nonlinearity*, 17 (2004).

[37] P. Raviart, On singular perturbation problems for the nonlinear Poisson equation: A mathematical approach to electrostatic sheaths and plasma erosion, *Lecture Notes of the Summer school in Ile d'Oléron, France* (1997).

[38] R. Sacco and F. Saleri, Mixed finite volume methods for semiconductor device simulation, *Numer. Meth. Part. Diff. Equations*, 13 (1997).

[39] M. Slemrod and N. Sternberg, Quasi-neutral limit for the Euler-Poisson system, *J.Nonlinear Sciences*, 11 (2001).

[40] I. Violet, High order expansions in quasi-neutral limit of the Euler-Poisson system for a potential flow, *Proceedings of the Royal Society of Edinburgh*, 137A (2007).

Behavior of Discontinuities in Thermoelasticity with Second Sound*

Zheng Li, Yaguang Wang
Department of Mathematics, Shanghai Jiao Tong University
Shanghai 200240, China
Email: {lizheng, ygwang}@sjtu.edu.cn

Abstract

This note is devoted to the study of the asymptotic behavior of discontinuous solutions to the Cauchy problem for linear and semilinear thermoelastic equations with second sound and variable coefficients in one space variable. When the relaxation parameter tends to zero, we obtain that the jump of temperature vanishes while jumps of elastic waves and heat flux are propagated in the speed of elastic waves. Furthermore, it is observed that these jumps decay exponentially when the time goes to infinity, and the decay rates depend on not only the growth rate of the nonlinear source terms and heat conduction coefficient, but also the change rates of variable speed of elastic waves.

1 Introduction

Thermoelastic equations describe the elastic and thermal behavior of elastic heat conductive media. The classical equations in thermoelasticity, based on the Fourier law for heat conduction, are of a hyperbolic-parabolic coupled type ([1, 5]). After introducing a microlocally decoupling idea in [11], there already have been some interesting results of the propagation of singularities in hyperbolic-parabolic coupled systems of thermoelasticity (c.f. [11, 3, 9, 12] and references therein). When the Fourier law is replaced by the Cattaneo law for the heat conduction, the thermoelastic system becomes purely hyperbolic, which shows that the thermal disturbance is transmitted as a "wave-like" pulse with a finite speed ([2, 4]). This kind of equation is the so-called thermoelastic system with second sound. Tarabek in [10] first studied the existence of

*The authors are partly supported by the NSF of China (No. 10531020; No. 10676020).

smooth solutions to linear thermoelastic equations with second sound in one space variable, and recently Racke and Wang in [8] have obtained the well-posedness of its nonlinear problems.

The semilinear version of thermoelastic equations with second sound in one space variable is as follows:

$$\begin{cases} u_{tt} - \alpha^2 u_{xx} + \beta\theta_x = f(u, u_x, u_t, \theta) \\ \theta_t + \gamma q_x + \delta u_{tx} = g(u, u_x, u_t, \theta) \\ \tau q_t + q + \kappa\theta_x = 0 \end{cases} \tag{1.1}$$

where u, θ and q denote the displacement, temperature and heat flux respectively, all coefficients in (1.1) are supposed to be smooth functions of $(t, x) \in [0, \infty) \times I\!R$ with $\alpha \geq \alpha_0 > 0$, $\beta\delta > 0$, $\tau > 0$ and $\kappa\gamma \geq \alpha_0 > 0$ for a constant $\alpha_0 > 0$.

Obviously, equations (1.1) are strictly hyperbolic, and when the relaxation parameter τ goes to zero, they formally converge to the classical thermoelastic equations of hyperbolic-parabolic coupled type. So, one approach to study the behavior of discontinuous solutions to problems for the classical thermoelastic equations of hyperbolic-parabolic type is to investigate the asymptotic behavior of discontinuities as τ vanishes for problems of the thermoelastic equations (1.1). This was first studied by Racke and Wang in [6, 7] for the linear and semilinear thermoelastic equations with second sound and constant coefficients in one and three space variables. In this note, we study this problem for equations with variable coefficients in one space variable. The goal is to investigate the influence of coefficients on the behavior of discontinuities. This observation shall be also helpful to studying fully nonlinear problems, which will be considered in a forth coming work. An interesting phenomenon is observed that the change rate of the variable speed of elastic waves has an explicit influence on the decay of the discontinuities of elastic waves and heat flux.

In §2, we shall present some preliminary facts of equations (1.1). The linear and semilinear problems for (1.1) with discontinuous initial data will be studied in §3 and §4 respectively.

2 Preliminaries

In this section, let us first state some elementary facts related to equations (1.1). Some of them had been given similarly in [6], so we shall only give their main steps of calculations for completeness.

Denote by

$$U := \begin{pmatrix} u_1 \\ u_2 \\ u_3 \\ u_4 \\ u_5 \end{pmatrix} = \begin{pmatrix} u \\ u_t + \alpha u_x \\ u_t - \alpha u_x \\ \theta \\ q \end{pmatrix}. \tag{2.1}$$

From (1.1), we know that $U(t,x)$ satisfies the following equations

$$\partial_t U + B_1 \partial_x U + B_0 U = F(U) \tag{2.2}$$

where $F(U) = (0, f, f, g, 0)^T$,

$$B_1 := \begin{pmatrix} 0 & 0 & 0 & 0 & 0 \\ 0 & -\alpha & 0 & \beta & 0 \\ 0 & 0 & \alpha & \beta & 0 \\ 0 & \frac{\delta}{2} & \frac{\delta}{2} & 0 & \gamma \\ 0 & 0 & 0 & \frac{\kappa}{\tau} & 0 \end{pmatrix},$$

and

$$B_0 := \begin{pmatrix} 0 & -\frac{1}{2} & -\frac{1}{2} & 0 & 0 \\ 0 & \frac{\alpha\alpha_x - \alpha_t}{2\alpha} & \frac{\alpha_t - \alpha\alpha_x}{2\alpha} & 0 & 0 \\ 0 & \frac{\alpha_t + \alpha\alpha_x}{2\alpha} & -\frac{\alpha_t + \alpha\alpha_x}{2\alpha} & 0 & 0 \\ 0 & 0 & 0 & 0 & 0 \\ 0 & 0 & 0 & 0 & \frac{1}{\tau} \end{pmatrix}.$$

It is easy to know ([6]) that the eigenvalues of B_1 are

$$\lambda_1 = 0, \quad \lambda_{2,3} = \mp\alpha\left(1 - \frac{\delta\beta}{2\kappa\gamma}\tau\right) + O(\tau^2), \quad \lambda_{4,5} = \mp\sqrt{\frac{\kappa\gamma}{\tau}} + O(\sqrt{\tau}). \tag{2.3}$$

Denote by r_k and l_k the right and left eigenvectors of B_1 with respect to λ_k for $1 \le k \le 5$, i.e. $(\lambda_k \cdot Id - B_1)r_k = l_k(\lambda_k \cdot Id - B_1) = 0$. As in [6], we have $l_1 = r_1^T = (1,0,0,0,0)$,

$$r_2 = \begin{pmatrix} 0 \\ 1 \\ \frac{\lambda_2+\alpha}{\lambda_2-\alpha} \\ \frac{\lambda_2+\alpha}{\beta} \\ \frac{\kappa(\lambda_2+\alpha)}{\tau\beta\lambda_2} \end{pmatrix}, \quad r_3 = \begin{pmatrix} 0 \\ \frac{\lambda_3-\alpha}{\lambda_3+\alpha} \\ 1 \\ \frac{\lambda_3-\alpha}{\beta} \\ \frac{\kappa(\lambda_3-\alpha)}{\tau\beta\lambda_3} \end{pmatrix}, \quad r_k = \begin{pmatrix} 0 \\ \frac{\beta}{\lambda_k+\alpha} \\ \frac{\beta}{\lambda_k-\alpha} \\ 1 \\ \frac{\kappa}{\tau\lambda_k} \end{pmatrix} \quad \text{for } k = 4,5, \tag{2.4}$$

and

$$\begin{aligned} l_2 &= c_2\left(0,\ 1,\ \frac{\lambda_2+\alpha}{\lambda_2-\alpha},\ \frac{2(\lambda_2+\alpha)}{\delta},\ \frac{2\gamma(\lambda_2+\alpha)}{\lambda_2\delta}\right) \\ l_3 &= c_3\left(0,\ \frac{\lambda_3-\alpha}{\lambda_3+\alpha},\ 1,\ \frac{2(\lambda_3-\alpha)}{\delta},\ \frac{2\gamma(\lambda_3-\alpha)}{\lambda_3\delta}\right) \\ l_k &= c_k\left(0,\ \frac{\delta}{2(\lambda_k+\alpha)},\ \frac{\delta}{2(\lambda_k-\alpha)},\ 1,\ \frac{\gamma}{\lambda_k}\right) \quad \text{for } k = 4,5 \end{aligned} \tag{2.5}$$

with constants $\{c_k\}_{k=2}^5$ satisfying the normalization

$$l_j r_k = \delta_{jk} = \begin{cases} 1, \; j = k \\ 0, \; j \neq k. \end{cases}$$

By a simple computation, one can choose

$$c_2 = 1+O(\tau), \quad c_3 = 1+O(\tau), \quad c_4 = \frac{1}{2}+O(\tau), \quad c_5 = \frac{1}{2}+O(\tau) \tag{2.6}$$

in (2.5).

Denote by

$$L := \begin{pmatrix} l_1 \\ l_2 \\ l_3 \\ l_4 \\ l_5 \end{pmatrix}, \qquad R := (r_1, r_2, r_3, r_4, r_5).$$

From (2.2), we know that $V = L \cdot U$ satisfies

$$\partial_t V + \Lambda \partial_x V + \tilde{B}_0 V = \widetilde{F}(V) \tag{2.7}$$

where

$$\Lambda = \text{diag}\,(\lambda_1, \lambda_2, \lambda_3, \lambda_4, \lambda_5)$$

and

$$\tilde{B}_0 = L B_0 R + L\,(\partial_t R + B_1 \partial_x R)\,. \tag{2.8}$$

Denote by $\{\Sigma_j\}_{j=1}^5$ characteristic curves of the operator $\partial_t + \Lambda\partial_x$ passing the origin $\{t = x = 0\}$, i.e. $\Sigma_j = \{x = \gamma_j(t)\}$ with $\gamma_j(t)$ satisfying

$$\frac{d\gamma_j(t)}{dt} = \lambda_j(t, \gamma_j(t)), \quad \gamma_j(0) = 0.$$

Denote by

$$[u]_{\Sigma_j}(t^*, x^*) = \lim_{\substack{(t,x)\to(t^*,x^*) \\ x>\gamma_j(t)}} u(t,x) - \lim_{\substack{(t,x)\to(t^*,x^*) \\ x<\gamma_j(t)}} u(t,x)$$

the jump of u at (t^*, x^*) across Σ_j.

First, for the diagonal equations (2.7), from [6] we know

Lemma 2.1 *Let V be a bounded solution to equations (2.7). For each fixed $1 \leq j \leq 5$, $V_j(t,x)$ may have jump only on Σ_j.*

For a fixed $1 \leq j \leq 5$, by noting that the vector field $X_j = \partial_t + \lambda_j(t,x)\partial_x$ is tangential to Σ_j, and using Lemma 2.1, we know from (2.7) that the jump of V_j on Σ_j satisfies the following transport equation:

$$(\partial_t + \lambda_j \partial_x)[V_j]_{\Sigma_j} + b_{jj}[V_j]_{\Sigma_j} = [\tilde{F}_j(V)]_{\Sigma_j}. \tag{2.9}$$

Thus, to have the asymptotic behavior of the jump of V_j on Σ_j it is important to compute b_{jj}.

Denote the right and left eigenvectors given in (2.4)(2.5) by

$$r_j = (r_{1j}, \dots, r_{5j})^T, \quad l_j = (l_{j1}, \dots, l_{j5})$$

for $1 \le j \le 5$.

By a direct computation, from (2.8) we have

$$\begin{aligned} b_{jj} &= \tfrac{1}{\tau} l_{j5} r_{5j} + \left(-\tfrac{l_{j1}}{2} - \tfrac{\alpha_t - \alpha\alpha_x}{2\alpha} l_{j2} + \tfrac{\alpha_t + \alpha\alpha_x}{2\alpha} l_{j3}\right) r_{2j} \\ &\quad + \left(-\tfrac{l_{j1}}{2} + \tfrac{\alpha_t - \alpha\alpha_x}{2\alpha} l_{j2} - \tfrac{\alpha_t + \alpha\alpha_x}{2\alpha} l_{j3}\right) r_{3j} + \textstyle\sum_{k=1}^{5} l_{jk} \partial_t r_{kj} \\ &\quad + \left(\tfrac{\delta}{2} l_{j4} - \alpha l_{j2}\right) \partial_x r_{2j} + \left(\tfrac{\delta}{2} l_{j4} + \alpha l_{j3}\right) \partial_x r_{3j} \\ &\quad + \left(\beta\left(l_{j2} + l_{j3}\right) + \tfrac{\kappa}{\tau} l_{j5}\right) \partial_x r_{4j} + \gamma l_{j4} \partial_x r_{5j} \end{aligned} \tag{2.10}$$

for all $1 \le j \le 5$.

Noting that $r_{11} = 1$ and $r_{k1} = 0$ for all $k \ge 2$, we have

$$b_{11} = 0 \tag{2.11}$$

immediately. By using (2.4) and (2.5) in (2.10), it is not difficult to obtain:

Lemma 2.2 *When $\tau \to 0$, we have*

$$\begin{cases} b_{22} = \frac{\beta\delta}{2\kappa\gamma} + \frac{\alpha\alpha_x - \alpha_t}{2\alpha} + O(\tau) \\ b_{33} = \frac{\beta\delta}{2\kappa\gamma} - \frac{\alpha\alpha_x + \alpha_t}{2\alpha} + O(\tau) \\ b_{jj} = \frac{1}{2\tau}(1 + O(\sqrt{\tau})), \quad j = 4, 5 \end{cases} \tag{2.12}$$

3 Linear problems

From now on, let us study the Cauchy problem for equations (1.1) in $\{(t,x) | t > 0, \ x \in \mathbb{R}\}$ with the initial data

$$u(0,x) = u_0(x), \quad u_t(0,x) = u_1(x), \quad \theta(0,x) = \theta_0(x), \quad q(0,x) = q_0(x) \tag{3.1}$$

where u_0 is continuous, and $(u_0', u_1, \theta_0, q_0)$ are continuous away from the origin, and may have jumps at $\{x = 0\}$.

In this section, we study the behavior of discontinuous solutions to a linear problem for (1.1) and (3.1) with $f = g = 0$. In this case, the unknowns

$$V = L \cdot (u, u_t + \alpha u_x, u_t - \alpha u_x, \theta, q)^T \tag{3.2}$$

satisfy the diagonal equations (2.7) with $\tilde{F} = 0$. From Lemma 2.1 and (2.9), we know that for each $1 \le j \le 5$, V_j may have jump only on Σ_j, and $[V_j]_{\Sigma_j}$ satisfies

$$(\partial_t + \lambda_j \partial_x)[V_j]_{\Sigma_j} + b_{jj}[V_j]_{\Sigma_j} = 0. \tag{3.3}$$

Integrating (3.3) along Σ_j, it follows

$$[V_j]_{\Sigma_j(t)} = [V_{0,j}]_{\{0\}} e^{-\int_0^t b_{jj}(s,\gamma_j(s))ds} \tag{3.4}$$

where $[V_j]_{\Sigma_j(t)}$ denotes the jump of V_j at $(t, \gamma_j(t))$ across Σ_j, and $[V_{0,j}]_{\{0\}}$ is the jump of the initial data $V_{0,j}$ at $\{x = 0\}$.

Obviously, we have $\Sigma_1 = \{t > 0, x = 0\}$. Since $u(0,x) = u_0(x)$ is supposed to be continuous everywhere, from (3.4) we immediately obtain

$$[V_1]_{\{x=0\}} = 0, \tag{3.5}$$

which implies that $u = V_1$ is continuous for $t > 0$.

By using Lemma 2.2 in (3.4), we can obtain

Lemma 3.1 *(1) For $j = 4, 5$, the jump of V_j on Σ_j behaves as*

$$[V_j]_{\Sigma_j(t)} = [V_{0,j}]_{\{0\}} e^{-\frac{t}{2\tau}(1+O(\sqrt{\tau}))} \tag{3.6}$$

when all coefficient functions given in (1.1), $\alpha, \beta, \gamma, \delta, \kappa$ are bounded in (t,x), i.e. for any fixed $t > 0$, the jump of V_j on Σ_j decays exponentially as $\tau \to 0$.

(2) The jumps of V_2 and V_3 on Σ_2 and Σ_3 behave as

$$\lim_{\tau \to 0}([V_2]_{\Sigma_2(t)} - [V_{0,2}]_{\{0\}} e^{-\int_0^t (\frac{\beta\delta}{2\kappa\gamma} + \frac{\alpha_x}{2} - \frac{\alpha_t}{2\alpha})(s,\gamma_2(s))ds}) = 0 \tag{3.7}$$

and

$$\lim_{\tau \to 0}([V_3]_{\Sigma_3(t)} - [V_{0,3}]_{\{0\}} e^{-\int_0^t (\frac{\beta\delta}{2\kappa\gamma} - \frac{\alpha_x}{2} - \frac{\alpha_t}{2\alpha})(s,\gamma_3(s))ds}) = 0 \tag{3.8}$$

respectively.

From (3.7) and (3.8), it follows

Corollary 3.2 (1) *If there exists $\epsilon_0 > 0$ such that*

$$\frac{\beta\delta}{2\kappa\gamma} + \frac{\alpha_x}{2} - \frac{\alpha_t}{2\alpha} \ge \epsilon_0 \tag{3.9}$$

holds for all $t \ge t_0$ with some $t_0 > 0$, then we have

$$\lim_{\tau \to 0} |[V_2]_{\Sigma_2(t)}| \le C e^{-\epsilon_0 t} \tag{3.10}$$

for all $t \geq t_0 + 1$.

(2) *If there exists* $\epsilon_1 > 0$ *such that*

$$\frac{\beta\delta}{2\kappa\gamma} + \frac{\alpha_x}{2} - \frac{\alpha_t}{2\alpha} \leq -\epsilon_1 \tag{3.11}$$

holds for all $t \geq t_0$ *with some* $t_0 > 0$*, then we have*

$$\lim_{\tau\to 0} |[V_2]_{\Sigma_2(t)}| \geq Ce^{\epsilon_1 t} \tag{3.12}$$

for all $t \geq t_0 + 1$.

(3) *If the function* $\frac{\beta\delta}{2\kappa\gamma} - \frac{\alpha_x}{2} - \frac{\alpha_t}{2\alpha}$ *satisfies the same condition as in (3.9) or (3.11) for all* $t \geq t_0$*, then* $[V_3]_{\Sigma_3(t)}$ *has the same asymptotic property as in (3.10) or (3.12) when* $\tau \to 0$.

Remark 3.3 (1) For the problem (1.1), (3.1) with $f = g = 0$, when $\frac{\beta\delta}{\kappa\gamma}$ has a positive lower bound, and α is a positive constant for $t \geq t_0$, the assumption (3.9) holds obviously, so we deduce that the jumps of V_2 and V_3 on Σ_2 and Σ_3 respectively decay exponentially as $t \to +\infty$, more rapidly for smaller heat conduction $\kappa\gamma$, which is the same as in the constant coefficient case studied in [6].

(2) For the case that β, δ, κ and γ are constants, and α depends only on t, if when $t \geq t_0$, α satisfies $\partial_t(\ln\alpha) \geq C_\alpha$ for a constant C_α satisfying $C_\alpha > \frac{\beta\delta}{\kappa\gamma}$, then the jumps of V_2 and V_3 on Σ_2 and Σ_3 respectively increase exponentially as $t \to +\infty$, faster for larger C_α, which is a completely new phenomenon for variable coefficient problems.

Now, let us study the behavior of jumps of the unknown (u, θ, q) from the above conclusions. From (3.2), (2.4) and (2.5), we get

$$\begin{cases} u = V_1 \\ u_t + \alpha u_x = V_2 + O(\tau)V_3 + O(\sqrt{\tau})V_4 + O(\sqrt{\tau})V_5 \\ u_t - \alpha u_x = V_3 + O(\tau)V_2 + O(\sqrt{\tau})V_4 + O(\sqrt{\tau})V_5 \\ \theta = V_4 + V_5 + O(\tau)V_2 + O(\tau)V_3 \\ q = \sqrt{\frac{\kappa}{\gamma\tau}}(V_5 - V_4) - \frac{\delta}{2\gamma}(V_2 + V_3) + O(\tau)V_2 + O(\tau)V_3 + O(1)V_4 + O(1)V_5 \end{cases} \tag{3.13}$$

and

$$\begin{cases} V_{0,1} = u_0(x) \\ V_{0,2} = u_1(x) + \alpha(0,x)u_0'(x) + O(\tau) \\ V_{0,3} = u_1(x) - \alpha(0,x)u_0'(x) + O(\tau) \\ V_{0,j} = \frac{1}{2}\theta_0(x) + O(\sqrt{\tau}), \quad j = 4, 5 \end{cases} \tag{3.14}$$

By using the results given in (3.5) and Lemma 3.1, from (3.13) and (3.14) we conclude

Theorem 3.4 *When τ goes to zero, the solution to the problem (1.1)–(3.1) with $f = g = 0$ has the following properties:*

(1) $u(t,x)$ is continuous everywhere;

(2) The jump of $u_t + \alpha u_x$ behaves as

$$\lim_{\tau\to 0}\left\{[u_t+\alpha u_x]_{\Sigma_2(t)} - [u_1+\alpha u_0']_{\{0\}}e^{-\int_0^t(\frac{\beta\delta}{2\kappa\gamma}+\frac{\alpha_x}{2}-\frac{\alpha_t}{2\alpha})(s,\gamma_2(s))ds}\right\} = 0 \tag{3.15}$$

and

$$\lim_{\tau\to 0}[u_t+\alpha u_x]_{\Sigma_j(t)} = 0 \tag{3.16}$$

of order $O(\tau)$ for $j = 3$, and exponentially for $j = 4, 5$.

(3) The jump of $u_t - \alpha u_x$ behaves as

$$\lim_{\tau\to 0}\left\{[u_t-\alpha u_x]_{\Sigma_3(t)} - [u_1-\alpha u_0']_{\{0\}}e^{-\int_0^t(\frac{\beta\delta}{2\kappa\gamma}-\frac{\alpha_x}{2}-\frac{\alpha_t}{2\alpha})(s,\gamma_3(s))ds}\right\} = 0 \tag{3.17}$$

and

$$\lim_{\tau\to 0}[u_t-\alpha u_x]_{\Sigma_j(t)} = 0 \tag{3.18}$$

of order $O(\tau)$ for $j = 2$, and exponentially for $j = 4, 5$.

(4) The jump of temperature θ vanishes always, and it is described as

$$\lim_{\tau\to 0}[\theta]_{\Sigma_j(t)} = 0 \tag{3.19}$$

of order $O(\tau)$ for $j = 2, 3$, and exponentially for $j = 4, 5$.

(5) The jump of heat flux q behaves as

$$\begin{cases}\lim\limits_{\tau\to 0}\left\{[q]_{\Sigma_2(t)} + \frac{\delta}{2\gamma}[u_1+\alpha u_0']_{\{0\}}e^{-\int_0^t(\frac{\beta\delta}{2\kappa\gamma}+\frac{\alpha_x}{2}-\frac{\alpha_t}{2\alpha})(s,\gamma_2(s))ds}\right\} = 0\\ \lim\limits_{\tau\to 0}\left\{[q]_{\Sigma_3(t)} + \frac{\delta}{2\gamma}[u_1-\alpha u_0']_{\{0\}}e^{-\int_0^t(\frac{\beta\delta}{2\kappa\gamma}-\frac{\alpha_x}{2}-\frac{\alpha_t}{2\alpha})(s,\gamma_3(s))ds}\right\} = 0\end{cases} \tag{3.20}$$

and

$$\lim_{\tau\to 0}[q]_{\Sigma_j(t)} = 0 \tag{3.21}$$

exponentially for $j = 4, 5$.

4 Nonlinear problems

In this section, we are going to study the semilinear problem (1.1)–(3.1). First, for convenience we rewrite the nonlinear terms f and g depending on $(u, u_t + \alpha u_x, u_t - \alpha u_x, \theta)$ explicitly, i.e.

$$\begin{cases} f = f(u, u_t + \alpha u_x, u_t - \alpha u_x, \theta) \\ g = g(u, u_t + \alpha u_x, u_t - \alpha u_x, \theta) \end{cases} \tag{4.1}$$

and we always assume that

f and g are globally Lipschitz in their arguments

in the following discussion.

From the deduction given at the beginning of §2, we know that

$$V = L \cdot (u, u_t + \alpha u_x, u_t - \alpha u_x, \theta, q)^T$$

satisfy the diagonal equations (2.7) with

$$\tilde{F}(V) = L \cdot F(RV)$$

and $F = (0, f, f, g, 0)^T$.

By using the asymptotic behavior of the left eigenvectors $\{l_j\}_{j=1}^5$, we know that

$$\begin{cases} \tilde{F}_j(V) = f + O(\tau)f + O(\tau)g, \quad j = 2, 3 \\ \tilde{F}_j(V) = \frac{1}{2}g + O(\tau)g + O(\sqrt{\tau})f, \quad j = 4, 5 \end{cases} \tag{4.2}$$

From (2.9), one immediately obtains

$$[V_j]_{\Sigma_j(t)} = [V_{0,j}]_{\{0\}} e^{-\int_0^t b_{jj}(s,\gamma_j(s))ds} + \int_0^t [\tilde{F}_j(V)]_{\Sigma_j(t_1)} e^{-\int_{t_1}^t b_{jj}(s,\gamma_j(s))ds} dt_1 \tag{4.3}$$

for $2 \le j \le 5$.

Since V_j has jump only on Σ_j for any fixed $2 \le j \le 5$, it is obvious that

$$[\tilde{F}_j(V)]_{\Sigma_j(t_1)} = \partial_{V_j} \tilde{F}_j(\overline{V^j})[V_j]_{\Sigma_j(t_1)} \tag{4.4}$$

with the notation $\overline{V^j}$ representing $\overline{V^2} = (V_1, \overline{V_2}, V_3, V_4, V_5)^T$, for example, for certain $\overline{V_2}$ ranging between V_2^- and V_2^+.

From (4.2), it is easy to have

$$\begin{cases} \partial_{V_2} \tilde{F}_2(V) = (1 + O(\tau))\partial_2 f + O(\tau)(\partial_3 f + \partial_4 f) + O(\tau)(\partial_2 g + \partial_3 g + \partial_4 g) \\ \partial_{V_3} \tilde{F}_3(V) = (1 + O(\tau))\partial_3 f + O(\tau)(\partial_2 f + \partial_4 f) + O(\tau)(\partial_2 g + \partial_3 g + \partial_4 g) \\ \partial_{V_j} \tilde{F}_j(V) = (\frac{1}{2} + O(\tau))\partial_4 g + O(\sqrt{\tau})(\partial_2 f + \partial_3 f + \partial_4 f) \\ \qquad + O(\sqrt{\tau})(\partial_2 g + \partial_3 g), \quad j = 4, 5 \end{cases} \tag{4.5}$$

where the notation $\partial_k f$ means the partial derivative of f with respect to the $k-$th argument.

Substituting (4.4) and (4.5) into (4.3), it follows

$$\begin{aligned}[V_j]_{\Sigma_j(t)} &= [V_{0,j}]_{\{0\}}e^{-\int_0^t b_{jj}(s,\gamma_j(s))ds} \\ &\quad + \int_0^t \partial_j f(\overline{V}^j)[V_j]_{\Sigma_j(t_1)}e^{-\int_{t_1}^t b_{jj}(s,\gamma_j(s))ds}dt_1 + O(\tau)\end{aligned} \tag{4.6}$$

for $j = 2, 3$, and

$$\begin{aligned}[V_j]_{\Sigma_j(t)} &= [V_{0,j}]_{\{0\}}e^{-\int_0^t b_{jj}(s,\gamma_j(s))ds} \\ &\quad + \tfrac{1}{2}\int_0^t \partial_4 g(\overline{V}^j)[V_j]_{\Sigma_j(t_1)}e^{-\int_{t_1}^t b_{jj}(s,\gamma_j(s))ds}dt_1 + O(\sqrt{\tau})\end{aligned} \tag{4.7}$$

for $j = 4, 5$.

By applying the Gronwall inequality in the inequalities (4.6) and (4.7), and using Lemma 2.2, we obtain

$$|[V_j]_{\Sigma_j(t)}| \le (|[V_{0,j}]_{\{0\}}| + O(\tau))e^{L_{f,j}t-\int_0^t b_{jj}(s,\gamma_j(s))ds} \tag{4.8}$$

for $j = 2, 3$, and

$$|[V_j]_{\Sigma_j(t)}| \le (|[V_{0,j}]_{\{0\}}| + O(\sqrt{\tau}))e^{(L_{g,4}-\frac{1}{2\tau}+O(\tau^{-\frac{1}{2}}))t} \tag{4.9}$$

for $j = 4, 5$, where $L_{f,j}, L_{g,4}$ denote the Lipschitz constants of f and g in their j-*th* and 4-*th* arguments respectively.

Plugging (4.8) into the integrant of the second term on the right hand side of (4.6), we get

$$|[V_j]_{\Sigma_j(t)}| \ge |[V_{0,j}]_{\{0\}}|e^{-\int_0^t b_{jj}(s,\gamma_j(s))ds}(2 - e^{L_{f,j}t}) + O(\tau) \tag{4.10}$$

for $j = 2, 3$.

From (4.7) and (4.10), we conclude

Lemma 4.1 *Under the assumption that f and g are globally Lipschitz in their arguments, when $\tau \to 0$, the jumps of V_j on $\Sigma_j(t)$ decay exponentially for $j = 4, 5$, and are kept for $j = 2, 3$, and they are described by (4.10). Moreover, if there are $\epsilon_0, t_0 > 0$ such that*

$$\frac{\beta\delta}{2\kappa\gamma} + \frac{\alpha_x}{2} - \frac{\alpha_t}{2\alpha} - L_{f,2} \ge \epsilon_0 > 0 \tag{4.11}$$

$$\left(\frac{\beta\delta}{2\kappa\gamma} - \frac{\alpha_x}{2} - \frac{\alpha_t}{2\alpha} - L_{f,3} \ge \epsilon_0 > 0 \quad \text{resp.}\right) \tag{4.12}$$

holds for $t \ge t_0$, then the jump of V_2 (V_3 resp.) on $\Sigma_2(t)$ ($\Sigma_3(t)$ resp.) decays exponentially when t goes to infinity.

Similar to the discussion in §3, from Lemma 4.1 and (3.13)–(3.14), we can easily deduce

Theorem 4.2 *For the semilinear Cauchy problem (1.1)–(3.1) with the nonlinear terms f, g being the special form (4.1), suppose that f and g are globally Lipschitz in their arguments, u_0 is continuous, and $(u_0', u_1, \theta_0, q_0)$ are continuous away from the origin, and may have jumps at $\{x = 0\}$. When τ goes to zero, the solution to the problem (1.1)–(3.1) has the following properties:*

(1) $u(t, x)$ is continuous everywhere;

(2) The jump of $u_t + \alpha u_x$ behaves as

$$\begin{aligned} &|[u_1 + \alpha u_0']_{\{0\}}|e^{-\int_0^t(\frac{\beta\delta}{2\kappa\gamma}+\frac{\alpha_x}{2}-\frac{\alpha_t}{2\alpha})(s,\gamma_2(s))ds}(2 - e^{L_{f,2}t}) + O(\tau) \\ &\le |[u_t + \alpha u_x]_{\Sigma_2(t)}| \\ &\le |[u_1 + \alpha u_0']_{\{0\}}|e^{L_{f,2}t-\int_0^t(\frac{\beta\delta}{2\kappa\gamma}+\frac{\alpha_x}{2}-\frac{\alpha_t}{2\alpha})(s,\gamma_2(s))ds} + O(\tau) \end{aligned} \tag{4.13}$$

and

$$\lim_{\tau\to 0}[u_t + \alpha u_x]_{\Sigma_j(t)} = 0 \tag{4.14}$$

of order $O(\tau)$ for $j = 3$, and exponentially for $j = 4, 5$.

(3) The jump of $u_t - \alpha u_x$ behaves as

$$\begin{aligned} &|[u_1 - \alpha u_0']_{\{0\}}|e^{-\int_0^t(\frac{\beta\delta}{2\kappa\gamma}-\frac{\alpha_x}{2}-\frac{\alpha_t}{2\alpha})(s,\gamma_3(s))ds}(2 - e^{L_{f,3}t}) + O(\tau) \\ &\le |[u_t - \alpha u_x]_{\Sigma_3(t)}| \\ &\le |[u_1 - \alpha u_0']_{\{0\}}|e^{L_{f,3}t-\int_0^t(\frac{\beta\delta}{2\kappa\gamma}-\frac{\alpha_x}{2}-\frac{\alpha_t}{2\alpha})(s,\gamma_3(s))ds} + O(\tau) \end{aligned} \tag{4.15}$$

and

$$\lim_{\tau\to 0}[u_t - \alpha u_x]_{\Sigma_j(t)} = 0 \tag{4.16}$$

of order $O(\tau)$ for $j = 2$, and exponentially for $j = 4, 5$.

(4) The jump of temperature θ on Σ_j $(2 \le j \le 5)$ vanishes always, and it has the same decay rates as given in Theorem 3.4(4).

(5) The jump of heat flux q on Σ_2 (Σ_3 resp.) behaves in the same way as that of $-\frac{\delta}{2\gamma}V_2$ ($-\frac{\delta}{2\gamma}V_3$ resp.) given in (4.8)–(4.10), and the jumps of heat flux q on Σ_4, Σ_5 share the same decay properties as given in (3.21).

Remark 4.3 As in Lemma 4.1, when the coefficient functions in (1.1) satisfy the condition (4.11) ((4.12) resp.) for $t \ge t_0$, the jump of $u_t + \alpha u_x$ ($u_t - \alpha u_x$ resp.) on Σ_2 (Σ_3 resp.) decays exponentially when t goes to infinity. For example, it happens when β, δ, κ and γ are constants, and $\alpha = e^{Ct}$ with $C < \frac{\beta\delta}{\kappa\gamma} - 2\max\{L_{f,2}, L_{f,3}\}$.

References

[1] D. E. Carlson. Linear thermoelasticity. *Handbuch der Physik*, VIa/2 (1972), 297–346.

[2] D. S. Chandrasekharaiah. Hyperbolic thermoelasticity: A review of recent literature. *Appl. Mech. Rev.*, 51 (1998), 705–729.

[3] S. X. Chen and Y. G. Wang. Propagation of singularities of solutions to hyperbolic-parabolic coupled system. *Math. Nachr.* 242 (2002), 46–60.

[4] B. D. Coleman, W. J. Hrusa and D. R. Owen. Stability of equilibrium for a nonlinear hyperbolic system describing heat propagation by second sound in solids. *Arch. Ration. Mech. Anal.*, 94 (1986), 267–289.

[5] S. Jiang and R. Racke. *Evolution Equations in Thermoelasticity.* Π Monographys Surveys Pure and Applied Mathematics, Vol. 112, Chapman & Hall/CRC 2000.

[6] R. Racke and Y. G. Wang. Asymptotic behavior of discontinuous solutions to thermoelastic systems with second sound. *Z. Anal. Anwendungen*, 24 (2005), 117–135.

[7] R. Racke and Y. G. Wang. Asymptotic behavior of discontinuous solutions to thermoelastic systems with second sound in three space variables. To appear in *Quart. Appl. Math.*.

[8] R. Racke and Y. G. Wang. Nonlinear well-posedness and rates of decay in thermoelasticity with second sound. *J. Hyper. Diff. Equ.*, 5 (2008), 25–43.

[9] M. Reissig and Y. G. Wang. Cauchy problems for linear thermoelastic systems of type III in one space variable. *Math. Methods Appl. Sci.*, 28 (2005), 11, 1359–1381.

[10] M. A. Tarabek. On existence of smooth solutions in one-dimensional nonlinear thermoelasticity with second sound. *Quart. Appl. Math.*, 50 (1992), 727–742.

[11] Y. G. Wang. Microlocal analysis in nonlinear thermoelasticity. *Nonlinear Anal. TMA*, 54 (2003), 683–705.

[12] L. Yang and Y. G. Wang. Propagation of singularities in Cauchy problems for quasilinear thermoelastic systems in three space variables. *J. Math. Anal. Appl.*, 291 (2004), 638–652.

The Convergence of Euler-Poisson System to the Incompressible Euler Equations*

Shu Wang, Ke Wang, Jianwei Yang
College of Applied Sciences, Beijing University of Technology
Beijing 100124, China
Email: {wangshu, coco, yjw}@emails.bjut.edu.cn

Abstract

The quasineutral limit and the zero electron mass limit of Euler-Poisson system in plasma physics in the torus $\mathbb{T}^d, d \geq 1$ are studied. The convergence of Euler-Poisson system to the incompressible Euler equations is proven by performing quasineutral limit and/or zero electron mass limit. The proof of the result is based on a straightforward extension of the classical energy method, the iteration techniques and the standard compactness argument. One of the key points is to establish uniformly a priori estimate of the electric field with respect to the two singular parameters by dealing with the singular hyperbolic part and the singular Poisson part. Dealing with the former is just an application of Klainermann and Majda's singular limit theory. Another singular Poisson part caused by the coupling electric field is controlled by specially spatial dependent relation between the electric field and the density, which is established by using the Poisson equation and the mass conservation law carefully.

1 Introduction

The main objective of this paper is to study the asymptotic limits (quasineutral limit and/or zero electron mass limit) of Euler-Poisson

*The authors are partly supported by NSF of China (No. 10771009), Beijing NSF (No.1082001) and Funding Project for Academic Human Resources Development in Institutions of Higher Leading under the Jurisdiction of Beijing Municipality (PHR(IHLB))(N0.PHR200906103).

system

$$\partial_t n^{\epsilon,\lambda} + \mathrm{div}(n^{\epsilon,\lambda}\mathbf{u}^{\epsilon,\lambda}) = 0, \qquad x \in \mathbb{T}^d,\ t > 0, \tag{1.1}$$
$$\epsilon[\partial_t \mathbf{u}^{\epsilon,\lambda} + \mathbf{u}^{\epsilon,\lambda}\cdot\nabla\mathbf{u}^{\epsilon,\lambda}] + \nabla h(n^{\epsilon,\lambda}) = \nabla\phi^{\epsilon,\lambda}, \qquad x \in \mathbb{T}^d,\ t > 0, \tag{1.2}$$
$$\lambda^2\Delta\phi^{\epsilon,\lambda} = n^{\epsilon,\lambda} - 1, \qquad x \in \mathbb{T}^d,\ t > 0, \tag{1.3}$$
$$n^{\epsilon,\lambda}(t=0) = n_0^{\epsilon,\lambda}, \mathbf{u}^{\epsilon,\lambda}(t=0) = \mathbf{u}_0^{\epsilon,\lambda}, \qquad x \in \mathbb{T}^d, \tag{1.4}$$

for $x \in \mathbb{T}^d$, $t > 0$, where $\mathbb{T}^d$ is the d-dimensional torus, $d \geq 1$, ϵ is the ratio of the electron mass to the ion mass, λ is the (scaled) Debye length. Here $n^{\epsilon,\lambda}, \mathbf{u}^{\epsilon,\lambda}, \phi^{\lambda,\epsilon}$ denote the electron density, electron velocity and the electrostatic potential, respectively. The entropy function $h(s)$ satisfies $h'(s) = \frac{1}{s}P'(s)$ and $P(s)$ is the pressure-density function having the property that $s^2P'(s)$ is strictly increasing function from $[0, +\infty)$ onto $[0, +\infty)$.

The Euler-Poisson systems (1.1)–(1.4) are a simplified isentropic two-fluid model to describe the dynamics of a plasma, where the compressible electron fluid interacts with its own electric field against a constant charged ion background, see [7, 11]. The global existence of small amplitude smooth solution in R^d has been studied [11]. However, on the one hand, plasma physics is usually concerned with large scales structures with respect to the Debye length. For such scales, the plasma is electrically neutral, i.e. there is no charge separation or electric field. In this case, the formal limit system is the incompressible Euler equations of ideal fluid in the unknowns $(\mathbf{u}^{\epsilon,0}, p^{\epsilon,0})$ given by

$$\mathrm{div}\ \mathbf{u}^{\epsilon,0} = 0, \quad x \in \mathbb{T}^d, t > 0, \tag{1.5}$$
$$\epsilon[\mathbf{u}_t^{\epsilon,0} + \mathbf{u}^{\epsilon,0}\cdot\nabla\mathbf{u}^{\epsilon,0}] = \nabla p^{\epsilon,0}, \quad x \in \mathbb{T}^d, t > 0, \tag{1.6}$$
$$\mathbf{u}^{\epsilon,0}(t=0) = \mathbf{u}_0^{\epsilon,0}, \quad x \in \mathbb{T}^d \tag{1.7}$$

for any fixed $\varepsilon > 0$. Note that this limit is widely used in plasma physics.

On the other hand, usually the ion mass is much larger than the electron mass, so the zero electron mass limit $\epsilon \to 0$ makes sense, see [1]. If, set $\epsilon \to 0$ formally in (1.5)–(1.7), one gets the incompressible Euler equations uniformly on all $\epsilon > 0$

$$\mathrm{div}\ \mathbf{u}^{0,0} = 0, \quad x \in \mathbb{T}^d, t > 0, \tag{1.8}$$
$$\mathbf{u}_t^{0,0} + \mathbf{u}^{0,0}\cdot\nabla\mathbf{u}^{0,0} = \nabla p^{0,0}, \quad x \in \mathbb{T}^d, t > 0, \tag{1.9}$$
$$\mathbf{u}^{0,0}(t=0) = \mathbf{u}_0^{0,0}, \quad x \in \mathbb{T}^d. \tag{1.10}$$

Otherwise, if we change the turn of taking the limits, first let $\varepsilon \to 0$ and then let $\lambda \to 0$, then we also get the limit systems (1.8)–(1.10).

Moreover, if in the meantime let $\epsilon \to 0$ and $\lambda \to 0$, we can directly get the above formal limit systems (1.8)–(1.10) from the Euler-Poisson systems (1.1)–(1.4).

The aim of this work is to justify rigorously the above formal limits for the small time as well as large time and sufficiently smooth solutions.

Concerning the quasineutral limit $\lambda \to 0$, many results have been obtained recently. Particularly, the limit $\lambda \to 0$ has been performed in Vlasov-Poisson system by Brenier [2], Grenier [12, 13, 14] and Masmoudi [24], in Schrödinger-Poisson system by Puel [28] and Jüngel and Wang [17], in drift-diffusion-Poisson system by Gasser et al. [9, 10], Jüngel and Peng [16], Wang, Xin and Markowich [35], and in Euler-Poisson system by Cordier and Grenier [5, 6], Cordier et al. [4], Peng and Wang [27], Wang [33] and Wang and Jiang [34]. Recently, a zero-electron-mass limit of hydrodynamic model in the plasmas for any given $\lambda > 0$ is performed by Ali et al. [1].

From the point of view of the singular perturbation theory, the limit $\lambda \to 0$ in Euler-Poisson system is a problem of singular perturbation of an hyperbolic system by an operator of order -2, see [6]. For the theory of singular perturbations we refer to [20, 21], and, [20] in particular for the perturbation of hyperbolic systems by order one linear terms (the main example being the incompressible limit of weakly compressible fluids). However, combining quasineutral and zero-electron-mass limit problems is very different from the theory of singular low Mach number limit of symmetrizable hyperbolic system by Klainerman and Majda in [20, 21] because the extra singularity is caused by the coupling electric field, which can not be overcome by using the theory of singular perturbations formed by Klainerman, Majda et al., see [20, 21, 29].

One of the main difficulties in dealing with combining quasineutral and zero-electron-mass limits is the oscillatory behavior of the electric field with respect to ε and λ. Usually it is difficult to obtain the uniform estimates in the electric field with respect to the Debye length λ.

Our approach to construct local existence of smooth solutions in time uniformly with respect to ε and λ in this paper, motivated by [20, 21], is to introduce appropriate weighted energy norms with respect to the Debye length and then to use a straightforward extension of the classical energy method and iteration techniques. The proof of our asymptotic limit is based on uniformly a priori estimates and then on standard compactness arguments, whose key step of derivation is to establish the a priori λ-weighted Sobolev's norm estimates of the electric field by dealing with the estimate of such a crucial inner product like $\int D_x^\alpha \nabla V^{\varepsilon,\lambda} D_x^\alpha \mathbf{u}^{\varepsilon,\lambda}$, which can be controlled by using the specially spatial dependent relation, see below (3.1) and (3.2), between density and electric field, established by using the mass conservation equation and the Poisson equation carefully.

Let us summarize the main results of this paper as follows: Let $(n^{\epsilon,\lambda}, \mathbf{u}^{\epsilon,\lambda}, \phi^{\varepsilon,\lambda})$ be a solution of Euler-Poisson systems (1.1)–(1.4), then, under the assumption of well-prepared initial data, one has that $(n^{\epsilon,\lambda}$,

$\mathbf{u}^{\epsilon,\lambda}) \to (n, u)$ in the sense of Sobolev's norm for some fixed time interval $[0, T_0]$, where T_0 is independent of ε and λ, n is some positive constant and u is the classical smooth solution of the incompressible Euler equation of ideal fluid with some pressure function as $\sqrt{\varepsilon}\lambda \to 0$, and the convergence rate is also given(The precise statement will be given in Section 2). The limit $\sqrt{\varepsilon}\lambda \to 0$ includes all kinds of asymptotic limits of Euler-Poisson systems (1.1)–(1.4) for $\varepsilon \to 0$ and/or $\lambda \to 0$.

It should be pointed out that for the case of ill-prepared initial data, some further substantial techniques like the study of wave propagation used by Grenier [15], Masmoudi [25], Schochet [29], Ukai [32], et al. in the different directions to deal with the oscillations in time are required. But it is very likely to extend the present convergence results to the case of general initial data, even though convincing arguments require substantial technical efforts. This will be one of the topics of our further study in the future.

We also mention that many mathematicians have made contributions to the large time behavior and global existence of smooth or weak solutions to the related pure Euler model or the related Euler-Poisson model in semiconductor physics with momentum relaxation. See [3, 8, 18, 23, 31, 30] for more references on this subject.

Notations used throughout this paper. $H^s(\mathbb{T}^d)$ is the standard Sobolev's space in torus $\mathbb{T}^d$, which is defined by Fourier series, namely, $f \in H^s(\mathbb{T}^d)$ if and only if

$$\|f\|_s^2 = (2\pi)^d \sum_{k \in \mathcal{Z}^d} (1 + |k|^2)^s |(\mathcal{F}f)(k)|^2 < +\infty,$$

where $(\mathcal{F}f)(k) = \int_{\mathbb{T}^d} f(x) e^{-ikx} dx$ is the Fourier series of $f \in H^s(\mathbb{T}^d)$. Noting that if $\int_{\mathbb{T}^d} f(x)dx = 0$, then $\|f\|_{L^2(\mathbb{T}^d)} \leq \|\nabla f\|_{L^2(\mathbb{T}^d)}$. $\nabla = \nabla_x$ is the gradient. $\alpha = (\alpha_1, \cdots, \alpha_d), \beta$, etc. are the multi-index.

Also, we need the following basic Moser-type calculus inequalities [20, 21, 22]:

For $f, g, v \in H^s$, for any nonnegative multi index $\alpha, |\alpha| \leq s$,

$$(i) \quad \|D_x^\alpha (fg)\|_{L^2} \leq C_s(|f|_\infty \|D_x^s g\|_{L^2} + |g|_\infty \|D_x^s f\|_{L^2}), s \geq 0;$$

$$(ii) \quad \|D_x^\alpha (fg) - f D_x^\alpha g\|_{L^2} \leq C_s(|D_x f|_\infty \|D_x^{s-1} g\|_{L^2} + |g|_\infty \|D_x^s f\|_{L^2}), \quad s \geq 1;$$

$$(iii) \quad \|D_x^s A(v)\|_{L^2} \leq C_s \sum_{j=1}^{s} |D_v^j A(v)|_\infty (1 + |\nabla v|_\infty)^{s-1} \|D_x^s v\|_{L^2}, s \geq 1.$$

This paper is organized as follows: In Section 2, the precise convergence results are stated. Section 3 is devoted to the proofs of main convergence results of Euler-Poisson system.

2 Main result

In this section we state our main theorem. For this, we first recall the following classical result of the existence of sufficiently regular solutions of the incompressible Euler equations (see [19, 26]).

Proposition 2.1. *Let* $\mathbf{u}_0^{0,0}$ *satisfy* $\mathbf{u}_0^{0,0} \in H^s$ *(or* H^{s+1}*),* $s > \frac{d}{2}+2$*,* *div* $\mathbf{u}_0^{0,0} = 0$*. Then there exist* $0 < T_* \le \infty$ *(if* $d=2$*,* $T_* = \infty$*), the maximal existence time, and a unique smooth solution* $(\mathbf{u}^{0,0}, p^{0,0})$ *of the incompressible Euler equations on* $[0, T_*)$ *with initial datum* $\mathbf{u}_0^{0,0}$ *satisfying, for any* $T_0 < T_*$*,*

$$\sup_{0\le t\le T_0}\Big(\|\mathbf{u}^{0,0}\|_{H^s} + \|\partial_t \mathbf{u}^{0,0}\|_{H^{s-1}} + \|\nabla p^{0,0}\|_{H^s} + \|\partial_t \nabla p^{0,0}\|_{H^{s-1}}\Big) \le C(T_0). \quad (2.1)$$

Before stating our main results, it is convenient for us to rewrite the hyperbolic part of Euler-Poisson system in the symmetric form. First, using the entropy $h^{\epsilon,\lambda} = h(n^{\epsilon,\lambda})$ as new variable, one has

$$\begin{aligned}
&q(h^{\epsilon,\lambda})[\partial_t h^{\epsilon,\lambda} + \mathbf{u}^{\epsilon,\lambda}\cdot\nabla h^{\epsilon,\lambda}] + \mathrm{div}\mathbf{u}^{\epsilon,\lambda} = 0, \qquad x\in\mathbb{T}^d,\ t>0,\\
&\epsilon[\partial_t \mathbf{u}^{\epsilon,\lambda} + \mathbf{u}^{\epsilon,\lambda}\cdot\nabla\mathbf{u}^{\epsilon,\lambda}] + \nabla h^{\epsilon,\lambda} = \nabla\phi^{\epsilon,\lambda}, \qquad x\in\mathbb{T}^d,\ t>0,\\
&\lambda^2\Delta\phi^{\epsilon,\lambda} = n(h^{\epsilon,\lambda}) - 1, \qquad x\in\mathbb{T}^d,\ t>0,\\
&h^{\epsilon,\lambda}(t=0) = h_0^{\epsilon,\lambda}, \mathbf{u}^{\epsilon,\lambda}(t=0) = \mathbf{u}_0^{\epsilon,\lambda}, \qquad x\in\mathbb{T}^d,
\end{aligned}$$

where $q(s) = \frac{dn(s)}{ds}\frac{1}{n(s)}$, and $n^{\epsilon,\lambda} = n(h^{\epsilon,\lambda})$ is the inverse function of $h^{\epsilon,\lambda} = h(n^{\epsilon,\lambda})$. Then, by setting

$$h^{\epsilon,\lambda} = h^0 + \sqrt{\epsilon}\tilde{h}^{\epsilon,\lambda}, \phi^{\lambda,\epsilon} = \sqrt{\epsilon}V^{\epsilon,\lambda},$$

where h^0 is some constant (in fact, the Poisson equation in the torus implies that $n(h^0) = 1$), one has

$$q(h^0+\sqrt{\epsilon}\tilde{h}^{\epsilon,\lambda})[\partial_t\tilde{h}^{\epsilon,\lambda}+\mathbf{u}^{\epsilon,\lambda}\cdot\nabla\tilde{h}^{\epsilon,\lambda}]+\frac{\mathrm{div}\mathbf{u}^{\epsilon,\lambda}}{\sqrt{\epsilon}} = 0,\ x\in\mathbb{T}^d,\ t>0, \quad (2.2)$$

$$\partial_t\mathbf{u}^{\epsilon,\lambda} + \mathbf{u}^{\epsilon,\lambda}\cdot\nabla\mathbf{u}^{\epsilon,\lambda} + \frac{\nabla\tilde{h}^{\epsilon,\lambda}}{\sqrt{\epsilon}} = \frac{\nabla V^{\epsilon,\lambda}}{\sqrt{\epsilon}}, \qquad x\in\mathbb{T}^d,\ t>0, \quad (2.3)$$

$$\lambda^2\Delta V^{\epsilon,\lambda} = \frac{1}{\sqrt{\epsilon}}\big(n(h^0+\sqrt{\epsilon}\tilde{h}^{\epsilon,\lambda}) - 1\big), \qquad x\in\mathbb{T}^d,\ t>0, \quad (2.4)$$

$$\tilde{h}^{\epsilon,\lambda}(t=0) = \tilde{h}_0^{\epsilon,\lambda}, \mathbf{u}^{\epsilon,\lambda}(t=0) = \mathbf{u}_0^{\epsilon,\lambda}, \qquad x\in\mathbb{T}^d. \quad (2.5)$$

For simplicity, in this paper we assume that

$$n(h) = h \quad \text{or} \quad P(n) = \frac{1}{2}n^2. \quad (2.6)$$

Using the assumption (2.6), we know that $h^0 = 1$ and then we can rewrite (2.2)–(2.5) as

$$q(1+\sqrt{\epsilon}\tilde{h}^{\epsilon,\lambda})[\partial_t\tilde{h}^{\epsilon,\lambda}+\mathbf{u}^{\epsilon,\lambda}\cdot\nabla\tilde{h}^{\epsilon,\lambda}]+\frac{\mathrm{div}\mathbf{u}^{\epsilon,\lambda}}{\sqrt{\epsilon}}=0, x\in\mathbb{T}^d,\ t>0, \quad (2.7)$$

$$\partial_t\mathbf{u}^{\epsilon,\lambda}+\mathbf{u}^{\epsilon,\lambda}\cdot\nabla\mathbf{u}^{\epsilon,\lambda}+\frac{\nabla\tilde{h}^{\epsilon,\lambda}}{\sqrt{\epsilon}}=\frac{\nabla V^{\epsilon,\lambda}}{\sqrt{\epsilon}}, \qquad x\in\mathbb{T}^d,\ t>0, \quad (2.8)$$

$$\lambda^2\Delta V^{\epsilon,\lambda}=\tilde{h}^{\epsilon,\lambda}, \quad \int_{\mathbb{T}^d}V^{\epsilon,\lambda}dx=0, \qquad x\in\mathbb{T}^d,\ t>0, \quad (2.9)$$

$$\tilde{h}^{\epsilon,\lambda}(t=0)=\tilde{h}_0^{\epsilon,\lambda}, \mathbf{u}^{\epsilon,\lambda}(t=0)=\mathbf{u}_0^{\epsilon,\lambda}, \qquad x\in\mathbb{T}^d, \quad (2.10)$$

where $q(s)=\frac{1}{s}$.

Now we state our main results as follows.

Theorem 2.2 (Local convergence). *Assume that* $(\tilde{h}_0^{\epsilon,\lambda},\mathbf{u}_0^{\epsilon,\lambda})=(0,\mathbf{u}_0^{0,0})$ *satisfies* $\mathbf{u}_0^{0,0}\in H^s, s>\frac{d}{2}+3$, *div* $\mathbf{u}_0^{0,0}=0$. *Then there exists a fixed time interval* $[0,T]$ *with* $T>0$, *depending only on initial data but not on* λ *and* ϵ, *and a constant* ι_0, *depending only on initial data, such that Euler-Poisson systems (2.7)–(2.10) have a classical smooth solution* $(h^{\epsilon,\lambda},\mathbf{u}^{\epsilon,\lambda},V^{\epsilon,\lambda})$, *defined on* $[0,T]$, *satisfying*

$$\|(\tilde{h}^{\epsilon,\lambda},\mathbf{u}^{\epsilon,\lambda},\lambda\nabla V^{\epsilon,\lambda})(t,\cdot)\|_{H^s(\mathbb{T}^d)} +\|(\partial_t\tilde{h}^{\epsilon,\lambda},\partial_t\mathbf{u}^{\epsilon,\lambda},\lambda\partial_t\nabla V^{\epsilon,\lambda})(t,\cdot)\|_{H^{s-1}(\mathbb{T}^d)}\le 2M_0 \quad (2.11)$$

for all $0<\sqrt{\varepsilon}\lambda\le\iota_0$, $0<\epsilon\le 1$ *and* $0\le t\le T$, *where* $M_0=\|\mathbf{u}_0^{0,0}\|_{H^s}+\|\mathbf{u}_0^{0,0}\cdot\nabla\mathbf{u}_0^{0,0}\|_{H^{s-1}}$. *Particularly, as* $\sqrt{\varepsilon}\lambda\to 0$,

$$\begin{aligned}
&\tilde{h}^{\epsilon,\lambda}\to\tilde{h}^{0,0}\textit{strongly in } L^\infty([0,T],\\
&H^{s-1}(\mathbb{T}^d))\cap C([0,T],H^{s-\tau}(\mathbb{T}^d)) \textit{ for any } \tau>0,\\
&\partial_t\tilde{h}^{\epsilon,\lambda}\to\partial_t\tilde{h}^{0,0}\textit{strongly in} L^\infty([0,T],H^{s-2}(\mathbb{T}^d)),\\
&\mathbf{u}^{\epsilon,\lambda}\rightharpoonup\mathbf{u}^{0,0}\ \textit{weakly* in } L^\infty([0,T],H^s(\mathbb{T}^d)),\\
&\textit{div }\mathbf{u}^{\epsilon,\lambda}\to 0\ \textit{strongly in } L^\infty([0,T],H^{s-2}(\mathbb{T}^d)),\\
&\partial_t\mathbf{u}^{\epsilon,\lambda}\rightharpoonup\partial_t\mathbf{u}^{0,0}\ \textit{weakly* in } L^\infty([0,T],H^{s-1}(\mathbb{T}^d)),\\
&\mathbf{u}^{\epsilon,\lambda}\to\mathbf{u}^{0,0}\ \textit{uniformly in } C([0,T],H^{s-\tau}(\mathbb{T}^d)) \textit{ for any } \tau>0,\\
&\frac{\nabla(V^{\epsilon,\lambda}-\tilde{h}^{\epsilon,\lambda})}{\sqrt{\epsilon}}\rightharpoonup\nabla p^{0,0}\ \textit{weakly* in } L^\infty([0,T],H^{s-1}(\mathbb{T}^d)). \quad (2.12)
\end{aligned}$$

Moreover, $\tilde{h}^{0,0}=0$ *if* $\lambda\to 0$, *and* $(\mathbf{u}^{0,0},p^{0,0})$ *is a classical* $C^1([0,T]\times\mathbb{T}^d)$ *solution, defined on* $[0,T]$, *of the incompressible Euler equations with initial data* $\mathbf{u}_0^{0,0}$.

Remark 2.3. The same result holds for small well-prepared perturbations, avoiding the presence of the initial time layer, of the incompressible initial data $(0, \mathbf{u}_0^{0,0})$. We will discuss the case of ill-prepared initial data allowing the presence of initial layer in the future. Also, if $\tilde{h}_0^{\epsilon,\lambda} = 0, \mathbf{u}_0^{\epsilon,\lambda} \in C^\infty$, then a C^∞ result similar to that of Theorem 2.2 holds.

Remark 2.4. The condition $s > \frac{d}{2} + 3$ is required, see below (3.32), in the derivation of estimates of the Sobolev's norm of one order time derivative of the solutions, which yields necessary compactness in time in the limiting process as $\sqrt{\varepsilon}\lambda \to 0$.

Theorem 2.5 (Global convergence). *Assume that* $(\tilde{h}_0^{\epsilon,\lambda}, \mathbf{u}_0^{\epsilon,\lambda})$ *satisfies* $\|(\tilde{h}_0^{\epsilon,\lambda}, \mathbf{u}_0^{\epsilon,\lambda} - \mathbf{u}_0^{0,0}, \lambda\nabla(V^{\epsilon,\lambda} - \sqrt{\varepsilon}p^{0,0})(t = 0))\|_{H^s(\mathbb{T}^d)} \leq M_0\sqrt{\varepsilon}\lambda$ *with* $div\, \mathbf{u}_0^{0,0} = 0, \mathbf{u}_0^{0,0} \in H^{s+1}, s > \frac{d}{2} + 2$, *where* M_0 *is independent of* ε *and* λ, *and* $V^{\epsilon,\lambda}(t = 0)$ *solves the Poisson equation* $\lambda^2\Delta V^{\epsilon,\lambda}(t = 0) = \tilde{h}_0^{\epsilon,\lambda}$ *with* $\int_{\mathbb{T}^d} V^{\epsilon,\lambda}(t = 0)dx = 0$. *Let* T_*, $0 < T_* \leq \infty$ $(d = 2, T_* = \infty)$, *be the maximal existence time of smooth solution* $(\mathbf{u}^{0,0}, p^{0,0})$ *of the incompressible Euler equation with initial data* $\mathbf{u}_0^{0,0}$. *Then for any* $T_0 < T_*$, *there exist constants* $\iota_0(T_0)$ *and* $M(T_0)$, *depending only on* T_0 *and the initial data* M_0, *such that Euler-Poisson systems (2.7)–(2.10) have a classical smooth solution* $(\tilde{h}^{\epsilon,\lambda}, \mathbf{u}^{\epsilon,\lambda}, V^{\epsilon,\lambda})$, *defined on* $[0, T_0]$, *satisfying*

$$\|(\tilde{h}^{\epsilon,\lambda}, \mathbf{u}^{\epsilon,\lambda} - \mathbf{u}^{0,0}, \lambda\nabla(V^{\epsilon,\lambda} - \sqrt{\varepsilon}p^{0,0}))(t, \cdot)\|_{H^s(\mathbb{T}^d)} \leq M(T_0)\sqrt{\varepsilon}\lambda \quad (2.13)$$

for all $\sqrt{\varepsilon}\lambda \leq \iota_0$, $0 < \epsilon, \lambda \leq 1$ *and* $0 \leq t \leq T_0$.

Remark 2.6. If $\mathbf{u}_0^{0,0}, \tilde{h}_0^{\epsilon,\lambda}, \mathbf{u}_0^{\epsilon,\lambda} \in C^\infty$, then a C^∞ result similar to that of Theorem 2.5 holds.

Remark 2.7. Since, for $d = 2$, $T_* = \infty$, then the convergence holds globally in time.

Remark 2.8. Theorems 2.2 and 2.5 are complete, which include all kinds of asymptotic limits of Euler-Poisson systems (1.1)–(1.4) with respect to two small parameters ε and λ, and imply the convergence of Euler-Poisson systems (1.1)–(1.4) to the incompressible Euler equations of ideal fluid in any case($\varepsilon \to 0$ and/or $\lambda \to 0$) if coming back to the variables $(n^{\epsilon,\lambda}, \mathbf{u}^{\epsilon,\lambda}, \phi^{\varepsilon,\lambda})$.

Remark 2.9. The restriction $0 < \epsilon, \lambda \leq 1$ can be changed to $0 < \epsilon \leq \varepsilon_0, 0 \leq \lambda \leq \lambda_0$ for any given ε_0 and λ_0. Hence, the restriction $0 < \epsilon, \lambda \leq 1$ can be removed. Also, the assumption (2.6) is only a technical one. Similar results hold for general strictly increasing convex entropy $h = h(n)$. We omit this.

3 Euler-Poisson Systems (1.1)–(1.4)

In this section, we will prove our main theorem, Theorems 2.2 and 2.5, on Euler-Poisson systems (1.1)–(1.4) by a straightforward extension of the classical energy method, the iteration techniques and the standard compactness argument. Our key point is to deal with higher order Sobolev's energy estimates of the electric field term integrals caused by the Poisson part of Euler-Poisson system by using the idea formed by the author in [33].

3.1 Preliminary estimates on electric field

First we give the following Lemma, which is crucial for establishing the energy estimates of the electric fields.

Lemma 3.1. *If* $\mathbf{y}^{\epsilon,\lambda} = (y_0^{\varepsilon,\lambda}, \mathbf{y}_*^{\epsilon,\lambda})^T \in (H^{s_1})^{d+1}$ *with* $s_1 > \frac{d}{2}+2$, $\rho \in H^{s_1-1}, \mathbf{a} \in (H^{s_1-1})^d$, $\partial_t\rho, \nabla\rho, \nabla\mathbf{a} \in L^\infty$, $\nabla p^0 \in (H^{s_1+1})^d$, $\nabla p_t^0 \in (H^{s_1})^d$ *and* $f_0 \in H^{s_1-1}$. *Let* $(\mathbf{y}^{\epsilon,\lambda}, \widetilde{V}^{\epsilon,\lambda})$ *be the solution of*

$$\partial_t y_0^{\varepsilon,\lambda} + \mathbf{a}\cdot\nabla y_0^{\varepsilon,\lambda} + (1+\rho)\ div\ \mathbf{y}_*^{\epsilon,\lambda} = f_0, \tag{3.1}$$

$$\lambda^2\Delta\widetilde{V}^{\epsilon,\lambda} = \tau\lambda\Delta p^0 + y_0^{\varepsilon,\lambda}\ \ with \int_{\mathbb{T}^d}\widetilde{V}^{\epsilon,\lambda}dx = 0. \tag{3.2}$$

Then we have the following estimate

$$\begin{aligned}(D_x^\alpha\nabla\widetilde{V}^{\epsilon,\lambda}, D_x^\alpha\mathbf{y}_*^{\epsilon,\lambda}) \le& -\frac{\lambda^2}{2}\frac{d}{dt}\Big((\nabla\widetilde{V}^{\epsilon,\lambda},\nabla\widetilde{V}^{\epsilon,\lambda}) + (D_x^\beta\Delta\widetilde{V}^{\epsilon,\lambda}, D_x^\beta\Delta\widetilde{V}^{\epsilon,\lambda})\Big)\\ &+\lambda^2C_s\|\mathbf{a}\|_{s_1-1}\|\nabla\widetilde{V}^{\epsilon,\lambda}\|_{s_1}^2\\ &+\tau^2C_s\|\nabla p_t^0\|_{s_1}^2 + \tau^2C_s\|\mathbf{a}\|_{s_1-1}^2\|\nabla p^0\|_{s_1}^2\\ &+\frac{1}{\lambda^2}C_s\|\rho\|_{s_1-1}^2\|\mathbf{y}_*^{\epsilon,\lambda}\|_{s_1}^2 + \frac{1}{\lambda^2}C_s\|f_0\|_{s_1-1}^2,\end{aligned} \tag{3.3}$$

where α (β) *is a multi index of length* $\le s_1(s_1-1)$. *Here and in the following,* C_s *is a constant depending only on Sobolev's constant.*

Proof of Lemma 3.1 The proof is similar to Lemma 3.1 in [33]. We omit the details.

Remark 3.2. Singular two terms containing a multiplier $\frac{1}{\lambda^2}$ on the right hand side of estimate (3.3) of the electric potential are caused by the perturbation of hyperbolic equation of acoustics $\partial_t y_0^\lambda + \mathbf{a}\cdot\nabla y_0^\lambda +$ div $\mathbf{y}_*^{\epsilon,\lambda} = 0$. If $\rho = f_0 = 0$, then there is no singular term in (3.3). So in the following we must carefully estimate the perturbation terms resulting from nonlinear terms of Euler-Poisson system to establish uniform estimates with respect to ε and λ.

3.2 The proof of Theorem 2.2

In this subsection, we will give the proof of Theorem 2.2 by using Lemma 3.1, the carefully classical energy method and standard compactness arguments. The proof relies on the symmetrizable form of Euler part of Euler-Poisson systems (1.1)–(1.4), estimates of the λ-weighted high order Sobolev's norm and the specially space-dependent relation between the electric field and the density.

Denoting $\mathbf{v}^{\epsilon,\lambda} = (\tilde{h}^{\epsilon,\lambda}, \mathbf{u}^{\epsilon,\lambda})^T$, we can write (2.7)–(2.10) in the form

$$A_0^{\varepsilon}(1+\sqrt{\epsilon}\tilde{h}^{\epsilon,\lambda})\partial_t \mathbf{v}^{\epsilon,\lambda} + \sum_{j=1}^{d} A_j^{\varepsilon}(\mathbf{v}^{\epsilon,\lambda})\partial_j \mathbf{v}^{\epsilon,\lambda} = \frac{\widehat{\nabla V^{\epsilon,\lambda}}}{\sqrt{\varepsilon}}, x \in \mathbb{T}^d, t>0, \quad (3.4)$$

$$\lambda^2 \Delta V^{\epsilon,\lambda} = \tilde{h}^{\epsilon,\lambda}, \int_{\mathbb{T}^d} V^{\epsilon,\lambda} dx = 0, x \in \mathbb{T}^d, t>0, \quad (3.5)$$

$$\mathbf{v}^{\epsilon,\lambda}(t=0) = \mathbf{v}_0^{\epsilon,\lambda} = (0, \mathbf{u}_0^{0,0})^T, x \in \mathbb{T}^d. \quad (3.6)$$

Here and in the following we use $\widehat{\mathbf{q}}$ to denote $(0,\mathbf{q})^T$ for any $\mathbf{q} \in R^d$, $e_j^T = (\delta_{1j}, \cdots, \delta_{dj}), j = 1, \cdots, d$,

$$\delta_{ij} = \begin{cases} 1, \, i=j; \\ 0, \, i \neq j, \end{cases} \quad A_j^{\varepsilon}(\mathbf{v}^{\epsilon,\lambda}) = \begin{pmatrix} q(1+\sqrt{\epsilon}\tilde{h}^{\epsilon,\lambda})u_j^{\epsilon,\lambda} & \frac{e_j^T}{\sqrt{\epsilon}} \\ \frac{e_j}{\sqrt{\varepsilon}} & u_j^{\varepsilon,\lambda} I_d \end{pmatrix}$$

and

$$A_0^{\varepsilon}(1+\sqrt{\epsilon}\tilde{h}^{\epsilon,\lambda}) = \begin{pmatrix} q(1+\sqrt{\epsilon}\tilde{h}^{\epsilon,\lambda}) & \mathbf{O}^T \\ \mathbf{O} & I_d \end{pmatrix},$$

which is symmetric and positive since

$$q(1+\sqrt{\epsilon}\tilde{h}^{\epsilon,\lambda}) > 0$$

for all $\tilde{h}^{\epsilon,\lambda} : \sqrt{\varepsilon}|\tilde{h}^{\epsilon,\lambda}|_{L^\infty} \leq \frac{1}{2}$.

We next introduce the λ-weighted high order Sobolev's norm. For given $\mathbf{v}^{\epsilon,\lambda} = \begin{pmatrix} \tilde{h}^{\epsilon,\lambda} \\ \mathbf{u}^{\epsilon,\lambda} \end{pmatrix} \in L^\infty([0,T]; H^s) \cap C^{0,1}([0,T]; H^{s-1})$ with $s > \frac{d}{2}+3$, define the λ-weighted norm by

$$|||\mathbf{v}^{\epsilon,\lambda}|||_{\lambda,T} = \sup_{0\leq t\leq T} |||\mathbf{v}^{\epsilon,\lambda}|||_\lambda,$$

$$|||\mathbf{v}^{\epsilon,\lambda}|||_\lambda = \|\mathbf{v}^{\epsilon,\lambda}\|_s + \|\partial_t \mathbf{v}^{\epsilon,\lambda}\|_{s-1} + \lambda(\|\nabla V^{\epsilon,\lambda}\|_s + \|\partial_t \nabla V^{\epsilon,\lambda}\|_{s-1}),$$

where $\lambda^2 \Delta V^{\epsilon,\lambda} = \tilde{h}^{\epsilon,\lambda}$ in $\mathbb{T}^d$ with $\int_{\mathbb{T}^d} V^{\epsilon,\lambda}(t)dx = 0$ for any $t \in [0,T]$, which can be solved by

$$\nabla V^{\epsilon,\lambda} = \frac{1}{\lambda^2}\nabla(-\Delta)^{-1}(-\tilde{h}^{\epsilon,\lambda}) \quad \text{for any} \quad t \in [0,T].$$

Using Poisson equation, we have

$$\|\frac{\tilde{h}^{\epsilon,\lambda}}{\lambda}\|_{s-1} \leq \|\lambda\nabla V^{\epsilon,\lambda}\|_s, \quad \|\frac{\partial_t\tilde{h}^{\epsilon,\lambda}}{\lambda}\|_{s-2} \leq \|\lambda\partial_t\nabla V^{\epsilon,\lambda}\|_{s-1}. \tag{3.7}$$

Consider now the iteration scheme

$$(\mathbf{v}^{\epsilon,\lambda,0}, V^{\varepsilon,\lambda,0}) = (\mathbf{v}_0^{\epsilon,\lambda}, 0) = ((0, \mathbf{u}_0^{0,0})^T, 0), \tag{3.8}$$
$$(\mathbf{v}^{\varepsilon,\lambda,p+1}, V^{\varepsilon,\lambda,p+1}) = \Phi((\mathbf{v}^{\varepsilon,\lambda,p}, V^{\varepsilon,\lambda,p})), \tag{3.9}$$

where the generator Φ maps the vector $(\mathbf{v}^{\varepsilon,\lambda}, V^{\epsilon,\lambda}) = ((\tilde{h}^{\epsilon,\lambda}, \mathbf{u}^{\epsilon,\lambda})^T, V^{\epsilon,\lambda})$ into the solution $(\widetilde{\mathbf{v}}^{\lambda}, \widetilde{V}^{\epsilon,\lambda}) = ((\tilde{\tilde{h}}^{\epsilon,\lambda}, \widetilde{\mathbf{u}}^{\epsilon,\lambda})^T, \widetilde{V}^{\epsilon,\lambda})$ of the following linearized Euler-Poisson system

$$A_0^{\varepsilon}(1+\sqrt{\epsilon}\tilde{h}^{\epsilon,\lambda})\partial_t\widetilde{\mathbf{v}}^{\epsilon,\lambda} + \sum_{j=1}^{d} A_j^{\varepsilon}(\mathbf{v}^{\epsilon,\lambda})\partial_j\widetilde{\mathbf{v}}^{\epsilon,\lambda} = \frac{\widehat{\nabla\widetilde{V}^{\epsilon,\lambda}}}{\sqrt{\varepsilon}}, x \in \mathbb{T}^d, t > 0, \tag{3.10}$$

$$\lambda^2\Delta\widetilde{V}^{\epsilon,\lambda} = \tilde{\tilde{h}}^{\epsilon,\lambda} \quad \text{with} \quad \int_{\mathbb{T}^d} \widetilde{V}^{\epsilon,\lambda}dx = 0, x \in \mathbb{T}^d, t > 0, \tag{3.11}$$

$$\widetilde{\mathbf{v}}^{\epsilon,\lambda}(t=0) = \mathbf{v}_0^{\epsilon,\lambda} = (0, \mathbf{u}_0^{0,0})^T \text{ with } \quad \text{div } \mathbf{u}_0^{0,0} = 0, x \in \mathbb{T}^d. \tag{3.12}$$

We will prove the convergence of the approximating sequence $\{(\mathbf{v}^{\varepsilon,\lambda,p}, V^{\varepsilon,\lambda,p})\}_{p=0}^{\infty}$ via the uniform boundedness of this sequence in the above weighted high Sobolev's norm. This strategy is used in [20, 21].

To this end, we will establish the estimates of solutions of the linearized Euler-Poisson systems (3.10)–(3.12).

Lemma 3.3. *Assume that* $\mathbf{v}_0^{\epsilon,\lambda} \in H^s, s > \frac{d}{2} + 3,$ *div* $\mathbf{u}_0^{0,0} = 0$. *Then there exists* $\Theta = \Theta(C_s, M_0)$, *depending only on the initial data and Sobolev's constant* C_s, *such that, if*

$$|||\mathbf{v}^{\epsilon,\lambda}|||_{\lambda,T} \leq 2M_0, \; T: e^{\Theta T} = 4\min\{1, c_0\}, \; 0 < \sqrt{\varepsilon}\lambda \leq \iota_0, \varepsilon \leq 1, \tag{3.13}$$

where $\iota_0 = \min\{1, (4M_0C_s^*)^{-2}\}$, $0 < c_0 = \inf_{|s|\leq 1/2} q(1+s)$ *and* C_s^* *is Sobolev's constant, then the solution* $(\widetilde{\mathbf{v}}^{\epsilon,\lambda}, \widetilde{V}^{\epsilon,\lambda})$ *of (3.10)–(3.12) satisfies*

$$|||\widetilde{\mathbf{v}}^{\epsilon,\lambda}|||_{\lambda,T} \leq 2M_0, \; T: e^{\Theta T} = 4\min\{1, c_0\}, \; 0 < \sqrt{\varepsilon}\lambda \leq \iota_0, \varepsilon \leq 1. \tag{3.14}$$

Proof of Lemma 3.3 As usual in this framework, we determine the conditions on the constant $\Theta(M_0)$ in the estimate. As we will see, this constant depend only on the initial data and Sobolev's constant but does not depend on ε, λ and the choice of $\mathbf{v}^{\epsilon,\lambda}$. To show this, we will give the exact formulation of $\widetilde{C}$, see below (3.17), which determines the constant Θ. We divide the proof into several steps.

Step 1. The energy differential inequalities.

Introduce the energy norm $\|\cdot\|_E^2 = (A_0^\varepsilon\cdot,\cdot)$, and $(\cdot,\cdot)$ is the usual L^2 scalar product. We want to obtain the following energy differential inequalities with respect to $\|\cdot\|_E^2$.

$$\frac{d}{dt}Q(\widetilde{\mathbf{v}}^{\epsilon,\lambda}) \leq \widetilde{C}|||\widetilde{\mathbf{v}}^{\epsilon,\lambda}|||_\lambda^2, 0 < t \leq T, \tag{3.15}$$

where

$$\begin{aligned} Q(\widetilde{\mathbf{v}}^{\epsilon,\lambda}) &= \|(D_x^\alpha \widetilde{\mathbf{v}}^{\epsilon,\lambda}, D_x^\beta \partial_t \widetilde{\mathbf{v}}^{\epsilon,\lambda})\|_E^2 \\ &+ \|(\widehat{\lambda D_x^\beta \Delta \widetilde{V}^{\epsilon,\lambda}}, \widehat{\lambda \nabla \widetilde{V}^{\epsilon,\lambda}}, \widehat{\lambda D_x^\gamma \partial_t \Delta \widetilde{V}^{\epsilon,\lambda}}, \widehat{\lambda \partial_t \nabla \widetilde{V}^{\epsilon,\lambda}})\|_{L^2}^2, \end{aligned} \tag{3.16}$$

$$\begin{aligned} \widetilde{C} &= \widetilde{C}(C_s, \sum_{j=0}^d |D_v A_j^\varepsilon|_\infty, \sum_{j=1}^d \sum_{k=1}^s |D_v^k A_j^\varepsilon|_\infty, \sum_{k=0}^{s-1} |q^{(k)}|_\infty, |||\mathbf{v}^{\epsilon,\lambda}|||_\lambda) \\ &= C_s(|||\mathbf{v}^{\epsilon,\lambda}|||_\lambda + |||\mathbf{v}^{\epsilon,\lambda}|||_\lambda^2) + C_s(1 + \sum_{j=0}^d |D_v A_j^\varepsilon|_\infty)|||\mathbf{v}^{\epsilon,\lambda}|||_\lambda \\ &+ C_s \sum_{j=1}^d \sum_{k=1}^s |D_v^k A_j^\varepsilon(\mathbf{v}^{\epsilon,\lambda})|_\infty (1 + |||\mathbf{v}^{\epsilon,\lambda}|||_\lambda)^{s-1} |||\mathbf{v}^{\epsilon,\lambda}|||_\lambda \\ &+ C_s \sum_{j=1}^d \sum_{k=0}^{s-1} |D_v^k (D_v A_j^\varepsilon(\mathbf{v}^{\epsilon,\lambda}))|_\infty^2 (1 + |||\mathbf{v}^{\epsilon,\lambda}|||_\lambda)^{2(s-2)} |||\mathbf{v}^{\epsilon,\lambda}|||_\lambda^2 \\ &+ C_s \sum_{k=0}^{s-1} |q^{(k)}|_\infty^2 (|||\mathbf{v}^{\epsilon,\lambda}|||_\lambda^2 + (1 + |||\mathbf{v}^{\epsilon,\lambda}|||_\lambda)^{2(s-3)} \cdot \\ &\quad (|||\mathbf{v}^{\epsilon,\lambda}|||_\lambda^4 + |||\mathbf{v}^{\epsilon,\lambda}|||_\lambda^6)) \end{aligned} \tag{3.17}$$

and α, β, γ are the multi indexes of the length $\leq s, s-1, s-2$, respectively.

Taking the L^2 inner product of (3.10) with $\widetilde{\mathbf{v}}^{\epsilon,\lambda}$ and integration by parts, we have the basic energy equation of Friedrich

$$\frac{d}{dt}\|\widetilde{\mathbf{v}}^{\epsilon,\lambda}\|_E^2 = (DivA^\varepsilon(\mathbf{v}^{\epsilon,\lambda})\widetilde{\mathbf{v}}^{\epsilon,\lambda}, \widetilde{\mathbf{v}}^{\epsilon,\lambda}) + 2(\frac{\widehat{\nabla \widetilde{V}^{\epsilon,\lambda}}}{\sqrt{\varepsilon}}, \widetilde{\mathbf{v}}^{\epsilon,\lambda}), \tag{3.18}$$

where $DivA^\varepsilon(\mathbf{v}^{\epsilon,\lambda}) = \partial_t(A_0^\varepsilon(1+\sqrt{\varepsilon}\widetilde{h}^{\epsilon,\lambda})) + \sum_{j=1}^d \partial_j(A_j^\varepsilon(\mathbf{v}^{\epsilon,\lambda}))$.

Noting that $A_0^\varepsilon = A_0^\varepsilon(1+\sqrt{\varepsilon}\widetilde{h}^{\epsilon,\lambda})$ depends upon $\mathbf{v}^{\epsilon,\lambda}$ via $\sqrt{\varepsilon}\widetilde{h}^{\epsilon,\lambda}$ and the singular part $O(\frac{1}{\sqrt{\varepsilon}})$ of the matrix $A_j^\varepsilon, j = 1, \cdots, d$, is $\frac{1}{\sqrt{\varepsilon}}C_j$, where C_j is a constant symmetric matrix, independent of ε and λ, and C_j's diagonal elements are 0. One gets, for any $\sqrt{\varepsilon}\lambda \le 1$, that

$$
\begin{aligned}
&|DivA^{\varepsilon}(\mathbf{v}^{\epsilon,\lambda})|_{L^\infty}\\
&\le |D_vA_0^{\varepsilon}|_\infty|\sqrt{\varepsilon}\partial_t\tilde{h}^{\epsilon,\lambda}|_\infty+\sum_{j=1}^{d}|D_vA_j^{\varepsilon}|_\infty|\nabla\mathbf{v}^{\epsilon,\lambda}|_\infty\\
&\le C_s(|D_vA_0^{\varepsilon}|_\infty\|\sqrt{\varepsilon}\partial_t\tilde{h}^{\epsilon,\lambda}\|_{s-2}+\sum_{j=1}^{d}|D_vA_j^{\varepsilon}|_\infty\|\mathbf{v}^{\epsilon,\lambda}\|_s)\\
&\le C_s(|D_vA_0^{\varepsilon}|_\infty\sqrt{\varepsilon}\lambda\|\lambda\partial_t\nabla V^{\epsilon,\lambda}\|_{s-1}+\sum_{j=1}^{d}|D_vA_j^{\varepsilon}|_\infty\|\mathbf{v}^{\epsilon,\lambda}\|_s)\\
&\le C_s(|D_vA_0^{\varepsilon}|_\infty+\sum_{j=1}^{d}|D_vA_j^{\varepsilon}|_\infty)(\|\mathbf{v}^{\epsilon,\lambda}\|_s+\|\sqrt{\varepsilon}\partial_t\mathbf{v}^{\epsilon,\lambda}\|_{s-1})\\
&\le C_s(|D_vA_0^{\varepsilon}|_\infty+\sum_{j=1}^{d}|D_vA_j^{\varepsilon}|_\infty)|||\mathbf{v}^{\epsilon,\lambda}|||_\lambda,
\end{aligned}
$$

then

$$
\begin{aligned}
|(DivA^{\varepsilon}(\mathbf{v}^{\epsilon,\lambda})\widetilde{\mathbf{v}}^{\epsilon,\lambda},\widetilde{\mathbf{v}}^{\epsilon,\lambda})| &\le C_s(|D_vA_0^{\varepsilon}|_\infty\\
&+\sum_{j=1}^{d}|D_vA_j^{\varepsilon}|_\infty)|||\mathbf{v}^{\epsilon,\lambda}|||_\lambda\|\widetilde{\mathbf{v}}^{\epsilon,\lambda}\|_{L^2}^2. \qquad (3.19)
\end{aligned}
$$

Now we estimate another singular term, the electric potential term, which depends upon two parameters ε and λ. This singular term can be estimated by comparing (3.4)–(3.5) with (3.1)–(3.2) and then using Lemma 3.1. Taking $s_1 = s, y_0^{\varepsilon,\lambda} = \tilde{\tilde{h}}^{\varepsilon,\lambda}, \mathbf{a} = \mathbf{u}^{\epsilon,\lambda}, \rho = \sqrt{\varepsilon}\tilde{h}^{\epsilon,\lambda}, \mathbf{y}_*^{\epsilon,\lambda} = \frac{\widetilde{\mathbf{u}}^{\epsilon,\lambda}}{\sqrt{\varepsilon}}, \tau = 0, f_0 = 0$ in Lemma 3.1, we have, by (3.7),

$$
\begin{aligned}
&2(\widehat{\frac{\nabla\widetilde{V}^{\epsilon,\lambda}}{\sqrt{\varepsilon}}},\widetilde{\mathbf{v}}^{\epsilon,\lambda}) = 2(\frac{\nabla\widetilde{V}^{\epsilon,\lambda}}{\sqrt{\varepsilon}},\widetilde{\mathbf{u}}^{\epsilon,\lambda})\\
&\le -\lambda^2\frac{d}{dt}(\nabla\widetilde{V}^{\epsilon,\lambda},\nabla\widetilde{V}^{\epsilon,\lambda})+2\lambda^2(\nabla\widetilde{V}^{\epsilon,\lambda},\nabla\widetilde{V}^{\epsilon,\lambda})\\
&\quad+\frac{2}{\lambda^2}\|\sqrt{\varepsilon}\tilde{h}^{\epsilon,\lambda}\|_{s-1}^2\|\frac{\widetilde{\mathbf{u}}^{\epsilon,\lambda}}{\sqrt{\varepsilon}}\|_1^2+C_s\lambda^2\|\mathbf{u}^{\epsilon,\lambda}\|_{s-1}\|\nabla\widetilde{V}^{\epsilon,\lambda}\|_1^2\\
&\le -\lambda^2\frac{d}{dt}(\nabla\widetilde{V}^{\epsilon,\lambda},\nabla\widetilde{V}^{\epsilon,\lambda})+\lambda^2C_s(1+\|\mathbf{u}^{\epsilon,\lambda}\|_s)\|\nabla\widetilde{V}^{\epsilon,\lambda}\|_1^2\\
&\quad+\frac{2}{\lambda^2}\|\tilde{h}^{\epsilon,\lambda}\|_{s-1}^2\|\widetilde{\mathbf{v}}^{\epsilon,\lambda}\|_1^2
\end{aligned}
$$

$$\leq -\lambda^2 \frac{d}{dt}(\nabla\widetilde{V}^{\epsilon,\lambda}, \nabla\widetilde{V}^{\epsilon,\lambda}) + \lambda^2 C_s(1 + \|\mathbf{u}^{\epsilon,\lambda}\|_s)\|\nabla\widetilde{V}^{\epsilon,\lambda}\|_1^2$$
$$+C_s\|\lambda\nabla V^{\epsilon,\lambda}\|_s^2\|\widetilde{\mathbf{v}}^{\epsilon,\lambda}\|_1^2$$
$$\leq -\lambda^2 \frac{d}{dt}(\nabla\widetilde{V}^{\epsilon,\lambda}, \nabla\widetilde{V}^{\epsilon,\lambda}) + \lambda^2 C_s(1 + |||\mathbf{v}^{\epsilon,\lambda}|||_\lambda)\|\nabla\widetilde{V}^{\epsilon,\lambda}\|_1^2$$
$$+C_s|||\mathbf{v}^{\epsilon,\lambda}|||_\lambda^2\|\widetilde{\mathbf{v}}^{\epsilon,\lambda}\|_1^2. \tag{3.20}$$

In the last second inequality we use (3.7).

Thus, (3.18), together with (3.19) and (3.20), gives

$$\frac{d}{dt}\left(\|\widetilde{\mathbf{v}}^{\epsilon,\lambda}\|_E^2 + \lambda^2(\nabla\widetilde{V}^{\epsilon,\lambda}, \nabla\widetilde{V}^{\epsilon,\lambda})\right)$$
$$\leq C_s(|D_v A_0^\varepsilon|_\infty + \sum_{j=1}^d |D_v A_j^\varepsilon|_\infty)|||\mathbf{v}^{\epsilon,\lambda}|||_\lambda\|\widetilde{\mathbf{v}}^{\epsilon,\lambda}\|_{L^2}^2$$
$$+\lambda^2 C_s(1 + |||\mathbf{v}^{\epsilon,\lambda}|||_\lambda)\|\nabla\widetilde{V}^{\epsilon,\lambda}\|_1^2 + C_s|||\mathbf{v}^{\epsilon,\lambda}|||_\lambda^2\|\widetilde{\mathbf{v}}^{\epsilon,\lambda}\|_1^2. \tag{3.21}$$

Now we obtain higher order energy inequality. As in (3.18), by using the symmetry of the matrix A_j^ε and integration by parts, we have the basic energy equation of Friedrich

$$\frac{d}{dt}\|D_x^\alpha\widetilde{\mathbf{v}}^{\epsilon,\lambda}\|_E^2 = (DivA^\varepsilon(\mathbf{v}^{\epsilon,\lambda})D_x^\alpha\widetilde{\mathbf{v}}^{\epsilon,\lambda}, D_x^\alpha\widetilde{\mathbf{v}}^{\epsilon,\lambda})$$
$$+2(H_\alpha^{(2)}, D_x^\alpha\widetilde{\mathbf{v}}^{\epsilon,\lambda}) + 2(\frac{\widehat{D_x^\alpha\nabla\widetilde{V}^{\epsilon,\lambda}}}{\sqrt{\varepsilon}}, D_x^\alpha\widetilde{\mathbf{v}}^{\epsilon,\lambda}), \tag{3.22}$$

where $H_\alpha^{(2)}$ is a commutator defined by

$$H_\alpha^{(2)} = -\sum_{j=1}^d \left(D_x^\alpha(A_j^\varepsilon(\mathbf{v}^{\epsilon,\lambda})\partial_j\widetilde{\mathbf{v}}^{\epsilon,\lambda}) - A_j^\varepsilon(\mathbf{v}^{\epsilon,\lambda})\partial_j D_x^\alpha\widetilde{\mathbf{v}}^{\epsilon,\lambda}\right).$$

Similar to the above, the first term can be bounded by

$$|(DivA^\varepsilon(\mathbf{v}^{\epsilon,\lambda})D_x^\alpha\widetilde{\mathbf{v}}^{\epsilon,\lambda}, D_x^\alpha\widetilde{\mathbf{v}}^{\epsilon,\lambda})| \leq C_s(|D_v A_0^\varepsilon|_\infty$$
$$+\sum_{j=1}^d |D_v A_j^\varepsilon|_\infty)|||\mathbf{v}^{\epsilon,\lambda}|||_\lambda\|\widetilde{\mathbf{v}}^{\epsilon,\lambda}\|_s^2. \tag{3.23}$$

The standard commutator technique gives

$$2(H_\alpha^{(2)}, D_x^\alpha\widetilde{\mathbf{v}}^{\epsilon,\lambda}) \leq 2\|H_\alpha^{(2)}\|_{L^2}\|D_x^\alpha\widetilde{\mathbf{v}}^{\epsilon,\lambda}\|_{L^2}$$
$$\leq C_s\sum_{j=1}^d\sum_{k=1}^s |D_v^k A_j^\varepsilon|_\infty$$
$$(1 + |||\mathbf{v}^{\epsilon,\lambda}|||_\lambda)^{s-1}|||\mathbf{v}^{\epsilon,\lambda}|||_\lambda\|\widetilde{\mathbf{v}}^{\epsilon,\lambda}\|_s^2. \tag{3.24}$$

For electric field term, which is more difficult to deal with, we can control it as follows by using Lemma 3.1 and the Poisson equation. Taking $s_1 = s, y_0^{\varepsilon,\lambda} = \tilde{\tilde{h}}^{\varepsilon,\lambda}, \mathbf{a} = \mathbf{u}^{\epsilon,\lambda}, \rho = \sqrt{\varepsilon}\tilde{h}^{\epsilon,\lambda}, \mathbf{y}_*^{\epsilon,\lambda} = \frac{\widetilde{\mathbf{u}}^{\epsilon,\lambda}}{\sqrt{\varepsilon}}, \tau = 0, f_0 = 0$ in Lemma 3.1, we have

$$\begin{aligned}
&2\Big(\frac{\widehat{D_x^\alpha \nabla \widetilde{V}^{\epsilon,\lambda}}}{\sqrt{\varepsilon}}, D_x^\alpha \widetilde{\mathbf{v}}^{\epsilon,\lambda}\Big) = 2\Big(D_x^\alpha \nabla \widetilde{V}^{\epsilon,\lambda}, D_x^\alpha\Big(\frac{\widetilde{\mathbf{u}}^{\epsilon,\lambda}}{\sqrt{\varepsilon}}\Big)\Big) \\
&\le -\lambda^2 \frac{d}{dt}\Big((D_x^\beta \Delta \widetilde{V}^{\epsilon,\lambda}, D_x^\beta \Delta \widetilde{V}^{\epsilon,\lambda}) + (\nabla \widetilde{V}^{\epsilon,\lambda}, \nabla \widetilde{V}^{\epsilon,\lambda})\Big) \\
&\quad + \lambda^2 C_s \|\mathbf{u}^{\epsilon,\lambda}\|_{s-1} \|\nabla \widetilde{V}^{\epsilon,\lambda}\|_s^2 + \frac{1}{\lambda^2} C_s \|\sqrt{\varepsilon}\tilde{h}^{\epsilon,\lambda}\|_{s-1}^2 \Big\|\frac{\widetilde{\mathbf{u}}^{\epsilon,\lambda}}{\sqrt{\varepsilon}}\Big\|_s^2 \\
&= -\lambda^2 \frac{d}{dt}\Big((D_x^\beta \Delta \widetilde{V}^{\epsilon,\lambda}, D_x^\beta \Delta \widetilde{V}^{\epsilon,\lambda}) + (\nabla \widetilde{V}^{\epsilon,\lambda}, \nabla \widetilde{V}^{\epsilon,\lambda})\Big) \\
&\quad + \lambda^2 C_s \|\mathbf{u}^{\epsilon,\lambda}\|_{s-1} \|\nabla \widetilde{V}^{\epsilon,\lambda}\|_s^2 + \frac{1}{\lambda^2} C_s \|\tilde{h}^{\epsilon,\lambda}\|_{s-1}^2 \|\widetilde{\mathbf{u}}^{\epsilon,\lambda}\|_s^2 \\
&\le -\lambda^2 \frac{d}{dt}\Big((D_x^\beta \Delta \widetilde{V}^{\epsilon,\lambda}, D_x^\beta \Delta \widetilde{V}^{\epsilon,\lambda}) + (\nabla \widetilde{V}^{\epsilon,\lambda}, \nabla \widetilde{V}^{\epsilon,\lambda})\Big) \\
&\quad + \lambda^2 C_s \|\mathbf{u}^{\epsilon,\lambda}\|_{s-1} \|\nabla \widetilde{V}^{\epsilon,\lambda}\|_s^2 + C_s \|\lambda \nabla V^{\epsilon,\lambda}\|_s^2 \|\widetilde{\mathbf{u}}^{\epsilon,\lambda}\|_s^2 \\
&\le -\lambda^2 \frac{d}{dt}\Big((D_x^\beta \Delta \widetilde{V}^{\epsilon,\lambda}, D_x^\beta \Delta \widetilde{V}^{\epsilon,\lambda}) + (\nabla \widetilde{V}^{\epsilon,\lambda}, \nabla \widetilde{V}^{\epsilon,\lambda})\Big) \\
&\quad + (|||\mathbf{v}^{\epsilon,\lambda}|||_\lambda + |||\mathbf{v}^{\epsilon,\lambda}|||_\lambda^2) |||\widetilde{\mathbf{v}}^{\epsilon,\lambda}|||_\lambda^2,
\end{aligned} \tag{3.25}$$

where we have used

$$\Big\|\frac{\tilde{h}^{\epsilon,\lambda}}{\lambda}\Big\|_{s-1} \le \|\lambda \nabla V^{\epsilon,\lambda}\|_s.$$

Thus, (3.22), together with (3.23)-(3.25), gives

$$\begin{aligned}
&\frac{d}{dt}\Big(\|D_x^\alpha \widetilde{\mathbf{v}}^{\epsilon,\lambda}\|_E^2 + \|(\lambda \widehat{D_x^\beta \Delta \widetilde{V}}^{\epsilon,\lambda}, \lambda \widehat{\nabla \widetilde{V}}^{\epsilon,\lambda})\|_{L^2}^2\Big) \\
&\le C_s \sum_{j=0}^d |D_v A_j^\varepsilon|_\infty |||\mathbf{v}^{\epsilon,\lambda}|||_\lambda \|\widetilde{\mathbf{v}}^{\epsilon,\lambda}\|_s^2 + C_s(|||\mathbf{v}^{\epsilon,\lambda}||||_\lambda + |||\mathbf{v}^{\epsilon,\lambda}|||_\lambda^2) |||\widetilde{\mathbf{v}}^{\epsilon,\lambda}|||_\lambda^2 \\
&+ C_s \sum_{j=1}^d \sum_{k=1}^s |D_v^k A_j^\varepsilon|_\infty (1 + |||\mathbf{v}^{\epsilon,\lambda}|||_\lambda)^{s-1} |||\mathbf{v}^{\epsilon,\lambda}|||_\lambda \|\widetilde{\mathbf{v}}^{\epsilon,\lambda}\|_s^2.
\end{aligned} \tag{3.26}$$

To conclude (3.15), the rest is to establish the estimates of the first time derivatives.

Differentiating (3.10)–(3.12) with respect to t and denoting $\overline{\mathbf{v}^{\epsilon,\lambda}} =$

$\partial_t \widetilde{\mathbf{v}}^{\epsilon,\lambda} = (\overline{\tilde{h}^{\epsilon,\lambda}}, \overline{\mathbf{u}^{\epsilon,\lambda}})^T, \overline{V^{\epsilon,\lambda}} = \partial_t \widetilde{V}^{\epsilon,\lambda}$, one gets

$$\begin{aligned}
&A_0^\varepsilon \partial_t \overline{\mathbf{v}^{\epsilon,\lambda}} + \sum_{j=1}^d A_j(\mathbf{v}^{\epsilon,\lambda})\partial_j \overline{\mathbf{v}^{\epsilon,\lambda}} + \partial_t(A_0^\varepsilon)\overline{\mathbf{v}^{\epsilon,\lambda}} \\
&\quad = \mathbf{f} + \widehat{\nabla \overline{V^{\epsilon,\lambda}}}, x \in \mathbb{T}^d, t > 0; \qquad (3.27)\\
&\lambda^2 \Delta \overline{V^{\epsilon,\lambda}} = \overline{\tilde{h}^{\epsilon,\lambda}}, \int_{\mathbb{T}^d} \overline{V^{\epsilon,\lambda}} dx = 0, x \in \mathbb{T}^d, t > 0; \qquad (3.28)\\
&\overline{\mathbf{v}^{\epsilon,\lambda}}(t=0) = \partial_t \widetilde{\mathbf{v}}^{\epsilon,\lambda}(t=0)\\
&\quad = -\sum_{j=1}^d \big((A_0^\varepsilon)^{-1} A_j^\varepsilon(\mathbf{v}^{\epsilon,\lambda})\partial_j \widetilde{\mathbf{v}}^{\epsilon,\lambda}\big)(t=0) \in H^{s-1}, \qquad (3.29)
\end{aligned}$$

where $\mathbf{f} = -\sum_{j=1}^d \partial_t(A_j^\varepsilon(\mathbf{v}^{\epsilon,\lambda}))\partial_j \widetilde{\mathbf{v}}^{\epsilon,\lambda}$ and

$$\begin{aligned}
&\partial_t(A_j^\varepsilon(\mathbf{v}^{\epsilon,\lambda}))\\
&= \bar{A}_j^\varepsilon(\sqrt{\varepsilon}\tilde{h}^{\epsilon,\lambda}, \mathbf{u}^{\epsilon,\lambda}, \sqrt{\varepsilon}\tilde{h}_t^{\epsilon,\lambda}, \mathbf{u}_t^{\epsilon,\lambda})\\
&= \begin{pmatrix} q'(1+\sqrt{\varepsilon}\tilde{h}^{\epsilon,\lambda})\sqrt{\varepsilon}\tilde{h}_t^{\epsilon,\lambda} u_j^{\varepsilon,\lambda} + q(1+\sqrt{\varepsilon}\tilde{h}^{\epsilon,\lambda})\partial_t u_j^{\varepsilon,\lambda}) & 0 \\ 0 & \partial_t \mathbf{u}_j^{\epsilon,\lambda} I_d \end{pmatrix}
\end{aligned}$$

is a nonsingular matrix since the singular part of the matrix A_j^ε is a constant singular matrix.

It is clear that the systems (3.27)–(3.29) have the same structure as systems (3.10)–(3.12) except the additional non-homogenous source $\mathbf{f}$ (In fact, it yields the presence of the singular term). Therefore, proceeding as in the derivation of (3.26), we have, $|\beta| \le s-1$,

$$\begin{aligned}
\frac{d}{dt}\|D_x^\beta \overline{\mathbf{v}^{\epsilon,\lambda}}\|_E^2 &\le C_s \sum_{j=0}^d |D_v A_j^\varepsilon|_\infty |||\mathbf{v}^{\epsilon,\lambda}|||_\lambda \|\overline{\mathbf{v}^{\epsilon,\lambda}}\|_{s-1}^2\\
&\quad + C_s \sum_{j=1}^d \sum_{k=1}^s |D_v^k A_j^\varepsilon|_\infty (1+|||\mathbf{v}^{\epsilon,\lambda}|||_\lambda)^{s-1} |||\mathbf{v}^{\epsilon,\lambda}|||_\lambda \|\overline{\mathbf{v}^{\epsilon,\lambda}}\|_{s-1}^2\\
&\quad + \|\overline{\mathbf{v}^{\epsilon,\lambda}}\|_{s-1}^2 + \|D_x^\beta \mathbf{f}\|_{L^2}^2 + 2\Big(\frac{\widehat{D_x^\beta \nabla \overline{V^{\epsilon,\lambda}}}}{\sqrt{\varepsilon}}, D_x^\beta \overline{\mathbf{v}^{\epsilon,\lambda}}\Big). \qquad (3.30)
\end{aligned}$$

Now, we devote ourselves to the estimate of the final two terms on the right hand side of (3.30).

Using the definition of $\mathbf{f}$, with the aid of Moser-type calculus inequalities and Sobolev's lemma, one gets

$$\begin{aligned}
\|D_x^\beta \mathbf{f}\|_{L^2}^2 &= \| - D_x^\beta(\partial_t(A_j^\varepsilon(\mathbf{v}^{\epsilon,\lambda}))\partial_j \widetilde{\mathbf{v}}^{\epsilon,\lambda})\|_{L^2}^2 \\
&\le C_s(|\partial_t(A_j^\varepsilon(\mathbf{v}^{\epsilon,\lambda}))|_\infty^2 \|D_x^{s-1}\partial_j\widetilde{\mathbf{v}}^{\epsilon,\lambda}\|_{L^2}^2 \\
&\quad +|\partial_j\widetilde{\mathbf{v}}^{\epsilon,\lambda}|_\infty^2 \|D_x^{s-1}\partial_t(A_j^\varepsilon(\mathbf{v}^{\epsilon,\lambda}))\|_{L^2}^2) \\
&\le C_s\Big(|D_v A_j^\varepsilon(\mathbf{v}^{\epsilon,\lambda})|_\infty^2 |\partial_t \mathbf{v}^{\epsilon,\lambda}|_\infty^2 \|\widetilde{\mathbf{v}}^{\epsilon,\lambda}\|_s^2 \\
&\quad +|\partial_j\widetilde{\mathbf{v}}^{\epsilon,\lambda}|_\infty^2 \|D_x^{s-1}(D_v A_j^\varepsilon(\mathbf{v}^{\epsilon,\lambda})\partial_t \mathbf{v}^{\epsilon,\lambda})\|_{L^2}^2\Big) \\
&\le C_s\Big(|D_v A_j^\varepsilon(\mathbf{v}^{\epsilon,\lambda})|_\infty^2 |\partial_t \mathbf{v}^{\epsilon,\lambda}|_\infty^2 \|\widetilde{\mathbf{v}}^{\epsilon,\lambda}\|_s^2 \\
&\quad +|\partial_j\widetilde{\mathbf{v}}^{\epsilon,\lambda}|_\infty^2 \big(|D_v A_j^\varepsilon(\mathbf{v}^{\epsilon,\lambda})|_\infty^2 \|D_x^{s-1}\partial_t \mathbf{v}^{\epsilon,\lambda}\|_{L^2}^2 \\
&\quad +|\partial_t \mathbf{v}^{\epsilon,\lambda}|_\infty^2 \|D_x^{s-1} D_v A_j^\varepsilon(\mathbf{v}^{\epsilon,\lambda})\|_{L^2}^2\big)\Big) \\
&\le C_s\Big(|D_v A_j^\varepsilon(\mathbf{v}^{\epsilon,\lambda})|_\infty^2 |\partial_t \mathbf{v}^{\epsilon,\lambda}|_\infty^2 \|\widetilde{\mathbf{v}}^{\epsilon,\lambda}\|_s^2 \\
&\quad +|\partial_j\widetilde{\mathbf{v}}^{\epsilon,\lambda}|_\infty^2 \big(|D_v A_j^\varepsilon(\mathbf{v}^{\epsilon,\lambda})|_\infty^2 \|\partial_t \mathbf{v}^{\epsilon,\lambda}\|_{s-1}^2 \\
&\quad +|\partial_t \mathbf{v}^{\epsilon,\lambda}|_\infty^2 \sum_{k=0}^{s-1} |D_v^k(D_v A_j^\varepsilon(\mathbf{v}^{\epsilon,\lambda}))|_\infty^2 \\
&\quad (1+|\nabla \mathbf{v}^{\epsilon,\lambda}|_\infty)^{2(s-2)} \|D_x^{s-1}\mathbf{v}^{\epsilon,\lambda}\|_{L^2}^2\big)\Big) \\
&\le C_s\Big(|D_v A_j^\varepsilon(\mathbf{v}^{\epsilon,\lambda})|_\infty^2 \|\partial_t \mathbf{v}^{\epsilon,\lambda}\|_{s-1}^2 \|\widetilde{\mathbf{v}}^{\epsilon,\lambda}\|_s^2 \\
&\quad +\|\widetilde{\mathbf{v}}^{\epsilon,\lambda}\|_s^2 \big(|D_v A_j^\varepsilon(\mathbf{v}^{\epsilon,\lambda})|_\infty^2 \|\partial_t \mathbf{v}^{\epsilon,\lambda}\|_{s-1}^2 \\
&\quad +\|\partial_t \mathbf{v}^{\epsilon,\lambda}\|_{s-1}^2 \sum_{k=0}^{s-1} |D_v^k(D_v A_j^\varepsilon(\mathbf{v}^{\epsilon,\lambda}))|_\infty^2 \\
&\quad (1+\|\mathbf{v}^{\epsilon,\lambda}\|_s)^{2(s-2)} \|\mathbf{v}^{\epsilon,\lambda}\|_{s-1}^2\big)\Big) \\
&\le C_s \sum_{k=0}^{s-1} |D_v^k(D_v A_j^\varepsilon(\mathbf{v}^{\epsilon,\lambda}))|_\infty^2 \\
&\quad (1+|||\mathbf{v}^{\epsilon,\lambda}|||_\lambda)^{2(s-2)} |||\mathbf{v}^{\epsilon,\lambda}|||_\lambda^2 \|\widetilde{\mathbf{v}}^{\epsilon,\lambda}\|_s^2. \qquad (3.31)
\end{aligned}$$

In view of Lemma 3.1 with $s_1 = s-1 > \frac{d}{2}+2$ (Here we need $s > \frac{d}{2}+3$), as in (3.25), we have, by comparing the first equation in (3.27) with the equation (3.1), for $|\beta| \le s-1$, that

$$\begin{aligned}
2(\frac{\widehat{D_x^\beta \nabla \overline{V^{\epsilon,\lambda}}}}{\sqrt{\varepsilon}}, D_x^\beta \overline{\mathbf{v}^{\epsilon,\lambda}}) &= 2(D_x^\beta \nabla \overline{V^{\epsilon,\lambda}}, D_x^\beta(\frac{\overline{\mathbf{u}^{\epsilon,\lambda}}}{\sqrt{\varepsilon}})) \\
&\le -\lambda^2 \frac{d}{dt}((D_x^\gamma \Delta \overline{V^{\epsilon,\lambda}}, D_x^\gamma \Delta \overline{V^{\epsilon,\lambda}}) + (\nabla \overline{V^{\epsilon,\lambda}}, \nabla \overline{V^{\epsilon,\lambda}}))
\end{aligned}$$

$$
\begin{aligned}
&+\lambda^2 C_s\|\mathbf{u}^{\epsilon,\lambda}\|_{s-2}\|\nabla\overline{V^{\epsilon,\lambda}}\|_{s-1}^2+\frac{1}{\lambda^2}C_s\|\sqrt{\varepsilon}\tilde{h}^{\epsilon,\lambda}\|_{s-2}^2\|\frac{\overline{\mathbf{u}^{\epsilon,\lambda}}}{\sqrt{\varepsilon}}\|_{s-1}^2\\
&+\frac{1}{\lambda^2}C_s\|-[q'\sqrt{\varepsilon}\partial_t\tilde{h}^{\epsilon,\lambda}\overline{\tilde{h}^{\epsilon,\lambda}}+(q'\sqrt{\varepsilon}\partial_t\tilde{h}^{\epsilon,\lambda}\mathbf{u}^{\epsilon,\lambda}+q\partial_t\mathbf{u}^{\epsilon,\lambda})\cdot\nabla\tilde{\tilde{h}}^{\varepsilon,\lambda}]\|_{s-2}^2\\
\le&-\lambda^2\frac{d}{dt}((D_x^\gamma\Delta\overline{V^{\epsilon,\lambda}},D_x^\gamma\Delta\overline{V^{\epsilon,\lambda}})+(\nabla\overline{V^{\epsilon,\lambda}},\nabla\overline{V^{\epsilon,\lambda}}))\\
&+C_s\|\mathbf{u}^{\epsilon,\lambda}\|_{s-2}\|\lambda\nabla\overline{V^{\epsilon,\lambda}}\|_{s-1}^2+C_s\|\lambda\nabla V^{\epsilon,\lambda}\|_{s-1}^2\|\overline{\mathbf{u}^{\epsilon,\lambda}}\|_{s-1}^2\\
&+\frac{1}{\lambda^2}C_s\|-[q'\sqrt{\varepsilon}\partial_t\tilde{h}^{\epsilon,\lambda}\overline{\tilde{h}^{\epsilon,\lambda}}\\
&+(q'\sqrt{\varepsilon}\partial_t\tilde{h}^{\epsilon,\lambda}\mathbf{u}^{\epsilon,\lambda}+q\partial_t\mathbf{u}^{\epsilon,\lambda})\cdot\nabla\tilde{\tilde{h}}^{\varepsilon,\lambda}]\|_{s-2}^2.
\end{aligned}\tag{3.32}
$$

Now we estimate the final term in (3.32), which is singular since there is a multiplier $\frac{1}{\lambda^2}$.

$$
\begin{aligned}
&\frac{1}{\lambda^2}C_s\|-[q'\sqrt{\varepsilon}\partial_t\tilde{h}^{\epsilon,\lambda}\overline{\tilde{h}^{\epsilon,\lambda}}+(q'\sqrt{\varepsilon}\partial_t\tilde{h}^{\epsilon,\lambda}\mathbf{u}^{\epsilon,\lambda}+q\partial_t\mathbf{u}^{\epsilon,\lambda})\cdot\nabla\tilde{\tilde{h}}^{\varepsilon,\lambda}]\|_{s-2}^2\\
\le&\frac{1}{\lambda^2}C_s\Big(|q'\sqrt{\varepsilon}\partial_t\tilde{h}^{\epsilon,\lambda}|_\infty^2\|\overline{\tilde{h}^{\epsilon,\lambda}}\|_{s-2}^2+|\overline{\tilde{h}^{\epsilon,\lambda}}|_\infty^2\|q'\sqrt{\varepsilon}\partial_t\tilde{h}^{\epsilon,\lambda}\|_{s-2}^2\\
&+|q'\sqrt{\varepsilon}\partial_t\tilde{h}^{\epsilon,\lambda}\mathbf{u}^{\epsilon,\lambda}+q\partial_t\mathbf{u}^{\epsilon,\lambda}|_\infty^2\|\nabla\tilde{\tilde{h}}^{\varepsilon,\lambda}\|_{s-2}^2+|\nabla\tilde{\tilde{h}}^{\varepsilon,\lambda}|_\infty^2\|\\
&q'\sqrt{\varepsilon}\partial_t\tilde{h}^{\epsilon,\lambda}\mathbf{u}^{\epsilon,\lambda}+q\partial_t\mathbf{u}^{\epsilon,\lambda}\|_{s-2}^2\Big)\\
\le&\frac{1}{\lambda^2}C_s(|q'|_\infty^2\|\sqrt{\varepsilon}\partial_t\tilde{h}^{\epsilon,\lambda}\|_{s-2}^2\|\overline{\tilde{h}^{\epsilon,\lambda}}\|_{s-2}^2+\|\overline{\tilde{h}^{\epsilon,\lambda}}\|_{s-2}^2\\
&(|q'|_\infty^2\|\sqrt{\varepsilon}\partial_t\tilde{h}^{\epsilon,\lambda}\|_{s-2}+|\sqrt{\varepsilon}\partial_t\tilde{h}^{\epsilon,\lambda}|_\infty^2\|q'\|_{s-2}^2)\\
&+(|q'|_\infty^2\|\sqrt{\varepsilon}\partial_t\tilde{h}^{\epsilon,\lambda}\|_{s-2}^2\|\mathbf{u}^{\epsilon,\lambda}\|_{s-2}^2+|q|_\infty^2\|\partial_t\mathbf{u}^{\epsilon,\lambda}\|_{s-2}^2)\|\nabla\tilde{\tilde{h}}^{\varepsilon,\lambda}\|_{s-2}^2\\
&+\|\nabla\tilde{\tilde{h}}^{\varepsilon,\lambda}\|_{s-2}^2(\|q'\|_{s-2}^2\|\sqrt{\varepsilon}\partial_t\tilde{h}^{\epsilon,\lambda}\|_{s-2}^2\|\mathbf{u}^{\epsilon,\lambda}\|_{s-2}^2+\|q\|_{s-2}^2\|\partial_t\mathbf{u}^{\epsilon,\lambda}\|_{s-2}^2))\\
\le&C_s\Big(|q'|_\infty^2\|\frac{\sqrt{\varepsilon}\partial_t\tilde{h}^{\epsilon,\lambda}}{\lambda}\|_{s-2}^2\|\overline{\tilde{h}^{\epsilon,\lambda}}\|_{s-2}^2\\
&+\|\overline{\tilde{h}^{\epsilon,\lambda}}\|_{s-2}^2(|q'|_\infty^2\|\frac{\sqrt{\varepsilon}\partial_t\tilde{h}^{\epsilon,\lambda}}{\lambda}\|_{s-2}+|\sqrt{\varepsilon}\partial_t\tilde{h}^{\epsilon,\lambda}|_\infty^2\sum_{k=0}^{s-2}|q^{(k+1)}|_\infty^2\\
&(1+\sqrt{\varepsilon}|\nabla\tilde{h}^{\epsilon,\lambda}|_\infty)^{2(s-3)}\|\frac{\tilde{h}^{\epsilon,\lambda}}{\lambda}\|_{s-2}^2)\\
&+(|q'|_\infty^2\|\sqrt{\varepsilon}\partial_t\tilde{h}^{\epsilon,\lambda}\|_{s-2}^2\|\mathbf{u}^{\epsilon,\lambda}\|_{s-2}^2+|q|_\infty^2\|\partial_t\mathbf{u}^{\epsilon,\lambda}\|_{s-2}^2)\|\frac{\nabla\tilde{\tilde{h}}^{\varepsilon,\lambda}}{\lambda}\|_{s-2}^2\\
&+\|\frac{\nabla\tilde{\tilde{h}}^{\varepsilon,\lambda}}{\lambda}\|_{s-2}^2(\sum_{k=0}^{s-2}|q^{(k+1)}|_\infty^2\\
&(1+\sqrt{\varepsilon}|\nabla\tilde{h}^{\epsilon,\lambda}|_\infty)^{2(s-3)}\|\tilde{h}^{\epsilon,\lambda}\|_{s-2}^2\|\sqrt{\varepsilon}\partial_t\tilde{h}^{\epsilon,\lambda}\|_{s-2}^2\|\mathbf{u}^{\epsilon,\lambda}\|_{s-2}^2
\end{aligned}
$$

$$+\sum_{k=0}^{s-2}|q^{(k)}|_\infty^2(1+\sqrt{\varepsilon}|\nabla\tilde{h}^{\epsilon,\lambda}|_\infty)^{2(s-3)}\|\tilde{h}^{\epsilon,\lambda}\|_{s-2}^2\|\partial_t\mathbf{u}^{\epsilon,\lambda}\|_{s-2}^2)\Big)$$

$$\le C_s\sum_{k=0}^{s-1}|q^{(k)}|_\infty^2\Big(\varepsilon\big(\|\lambda\partial_t\nabla V^{\epsilon,\lambda}\|_{s-1}^2+(1+\sqrt{\varepsilon}\|\nabla\tilde{h}^{\epsilon,\lambda}\|_{s-2})^{2(s-3)}$$

$$|||\mathbf{v}^{\epsilon,\lambda}|||_\lambda^2\|\lambda\nabla V^{\epsilon,\lambda}\|_{s-1}^2\big)\|\overline{\mathbf{v}^{\epsilon,\lambda}}\|_{s-2}^2+\big(|||\mathbf{v}^{\epsilon,\lambda}|||_\lambda^2+\varepsilon|||\mathbf{v}^{\epsilon,\lambda}|||_\lambda^4$$

$$+(1+\sqrt{\varepsilon}\|\nabla\tilde{h}^{\epsilon,\lambda}\|_{s-2})^{2(s-3)}(|||\mathbf{v}^{\epsilon,\lambda}|||_\lambda^4+\varepsilon|||\mathbf{v}^{\epsilon,\lambda}|||_\lambda^6)\big)\|\lambda\nabla\widetilde{V}^{\epsilon,\lambda}\|_s^2\Big)$$

$$\le C_s\sum_{k=0}^{s-1}|q^{(k)}|_\infty^2\Big(\varepsilon\big(|||\mathbf{v}^{\epsilon,\lambda}|||_\lambda^2+(1+\sqrt{\varepsilon}|||\mathbf{v}^{\epsilon,\lambda}|||_\lambda)^{2(s-3)}|||\mathbf{v}^{\epsilon,\lambda}|||_\lambda^4\big)$$

$$\|\overline{\mathbf{v}^{\epsilon,\lambda}}\|_{s-2}^2+\big(|||\mathbf{v}^{\epsilon,\lambda}|||_\lambda^2+\varepsilon|||\mathbf{v}^{\epsilon,\lambda}|||_\lambda^4+(1+\sqrt{\varepsilon}|||\mathbf{v}^{\epsilon,\lambda}|||_\lambda)^{2(s-3)}$$

$$(|||\mathbf{v}^{\epsilon,\lambda}|||_\lambda^4+\varepsilon|||\mathbf{v}^{\epsilon,\lambda}|||_\lambda^6)\big)\|\lambda\nabla\widetilde{V}^{\epsilon,\lambda}\|_s^2\Big)$$

$$\le C_s\sum_{k=0}^{s-1}|q^{(k)}|_\infty^2\big(|||\mathbf{v}^{\epsilon,\lambda}|||_\lambda^2+(1+|||\mathbf{v}^{\epsilon,\lambda}|||_\lambda)^{2(s-3)}$$

$$(|||\mathbf{v}^{\epsilon,\lambda}|||_\lambda^4+|||\mathbf{v}^{\epsilon,\lambda}|||_\lambda^6)\big)\times\big(\|\overline{\mathbf{v}^{\epsilon,\lambda}}\|_{s-2}^2+\|\lambda\nabla\widetilde{V}^{\epsilon,\lambda}\|_s^2\big),\qquad(3.33)$$

where we have used $\varepsilon\le 1$, $\|\frac{\partial_t\tilde{h}^{\epsilon,\lambda}}{\lambda}\|_{s-2}\le\|\lambda\partial_t\nabla V^{\epsilon,\lambda}\|_{s-1}$, $\|\frac{\tilde{h}^{\epsilon,\lambda}}{\lambda}\|_{s-2}\le\|\lambda\nabla V^{\epsilon,\lambda}\|_{s-1}$ and $\|\frac{\nabla\tilde{\tilde{h}}^{\varepsilon,\lambda}}{\lambda}\|_{s-2}\le\|\lambda\nabla\widetilde{V}^{\epsilon,\lambda}\|_s$.

Thus, (3.32), together with (3.33), gives

$$2(D_x^\beta\widehat{\nabla\overline{V^{\epsilon,\lambda}}},D_x^\beta\overline{\mathbf{v}^{\epsilon,\lambda}})$$
$$\le-\lambda^2\frac{d}{dt}((D_x^\gamma\Delta\overline{V^{\epsilon,\lambda}},D_x^\gamma\Delta\overline{V^{\epsilon,\lambda}})+(\nabla\overline{V^{\epsilon,\lambda}},\nabla\overline{V^{\epsilon,\lambda}}))$$
$$+C_s|||\mathbf{v}^{\epsilon,\lambda}|||_\lambda\|\lambda\nabla\overline{V^{\epsilon,\lambda}}\|_{s-1}^2+C_s|||\mathbf{v}^{\epsilon,\lambda}|||_\lambda^2\|\overline{\mathbf{v}^{\epsilon,\lambda}}\|_{s-1}^2$$
$$+C_s\sum_{k=0}^{s-1}|q^{(k)}|_\infty^2\big(|||\mathbf{v}^{\epsilon,\lambda}|||_\lambda^2+(1+|||\mathbf{v}^{\epsilon,\lambda}|||_\lambda)^{2(s-3)}$$
$$(|||\mathbf{v}^{\epsilon,\lambda}|||_\lambda^4+|||\mathbf{v}^{\epsilon,\lambda}|||_\lambda^6)\big)\times\big(\|\overline{\mathbf{v}^{\epsilon,\lambda}}\|_{s-2}^2+\|\lambda\nabla\widetilde{V}^{\epsilon,\lambda}\|_s^2\big).\quad(3.34)$$

Thus, (3.30), together with (3.31) and (3.34), gives

$$\frac{d}{dt}\Big(\|D_x^\beta\overline{\mathbf{v}^{\epsilon,\lambda}}\|_E^2+\|(\lambda D_x^\gamma\widehat{\Delta\overline{V^{\epsilon,\lambda}}},\lambda\widehat{\nabla\overline{V^{\epsilon,\lambda}}})\|_{L^2}^2\Big)$$
$$\le C_s(1+\sum_{j=0}^{d}|D_vA_j^\varepsilon|_\infty)|||\mathbf{v}^{\epsilon,\lambda}|||_\lambda\|\overline{\mathbf{v}^{\epsilon,\lambda}}\|_{s-1}^2$$
$$+C_s\sum_{j=1}^{d}\sum_{k=1}^{s}|D_v^kA_j^\varepsilon|_\infty(1+|||\mathbf{v}^{\epsilon,\lambda}|||_\lambda)^{s-1}|||\mathbf{v}^{\epsilon,\lambda}|||_\lambda\|\overline{\mathbf{v}^{\epsilon,\lambda}}\|_{s-1}^2$$

$$+C_s\sum_{j=1}^{d}\sum_{k=0}^{s-1}|D_v^k(D_vA_j^\varepsilon(\mathbf{v}^{\epsilon,\lambda}))|_\infty^2(1+|||\mathbf{v}^{\epsilon,\lambda}|||_\lambda)^{2(s-2)}|||\mathbf{v}^{\epsilon,\lambda}|||_\lambda^2\|\widetilde{\mathbf{v}}^{\epsilon,\lambda}\|_s^2$$
$$+C_s|||\mathbf{v}^{\epsilon,\lambda}|||_\lambda\|\lambda\nabla\overline{V^{\epsilon,\lambda}}\|_{s-1}^2+C_s|||\mathbf{v}^{\epsilon,\lambda}|||_\lambda^2\|\overline{\mathbf{v}^{\epsilon,\lambda}}\|_{s-1}^2$$
$$+C_s\sum_{k=0}^{s-1}|q^{(k)}|_\infty^2\big(|||\mathbf{v}^{\epsilon,\lambda}|||_\lambda^2+(1+|||\mathbf{v}^{\epsilon,\lambda}|||_\lambda)^{2(s-3)}$$
$$\big(|||\mathbf{v}^{\epsilon,\lambda}|||_\lambda^4+|||\mathbf{v}^{\epsilon,\lambda}|||_\lambda^6\big)\big)\times\big(\|\overline{\mathbf{v}^{\epsilon,\lambda}}\|_{s-2}^2+\|\lambda\nabla\widetilde{V}^{\epsilon,\lambda}\|_s^2\big).\tag{3.35}$$

Combining (3.26) and (3.35), we have

$$\frac{d}{dt}Q(\widetilde{\mathbf{v}}^{\epsilon,\lambda})\le C_s(|||\mathbf{v}^{\epsilon,\lambda}|||_\lambda+|||\mathbf{v}^{\epsilon,\lambda}|||_\lambda^2)|||\widetilde{\mathbf{v}}^{\epsilon,\lambda}|||_\lambda^2$$
$$+C_s(1+\sum_{j=0}^{d}|D_vA_j^\varepsilon|_\infty)|||\mathbf{v}^{\epsilon,\lambda}|||_\lambda(\|\widetilde{\mathbf{v}}^{\epsilon,\lambda}\|_s^2+\|\overline{\mathbf{v}^{\epsilon,\lambda}}\|_{s-1}^2)$$
$$+C_s\sum_{j=1}^{d}\sum_{k=1}^{s}|D_v^kA_j^\varepsilon|_\infty(1+|||\mathbf{v}^{\epsilon,\lambda}|||_\lambda)^{s-1}|||\mathbf{v}^{\epsilon,\lambda}|||_\lambda$$
$$(\|\widetilde{\mathbf{v}}^{\epsilon,\lambda}\|_s^2+\|\overline{\mathbf{v}^{\epsilon,\lambda}}\|_{s-1}^2)+C_s\sum_{j=1}^{d}\sum_{k=0}^{s-1}|D_v^k$$
$$(D_vA_j^\varepsilon(\mathbf{v}^{\epsilon,\lambda}))|_\infty^2(1+|||\mathbf{v}^{\epsilon,\lambda}|||_\lambda)^{2(s-2)}|||\mathbf{v}^{\epsilon,\lambda}|||_\lambda^2\|\widetilde{\mathbf{v}}^{\epsilon,\lambda}\|_s^2$$
$$+C_s|||\mathbf{v}^{\epsilon,\lambda}|||_\lambda\|\lambda\nabla\overline{V^{\epsilon,\lambda}}\|_{s-1}^2+C_s|||\mathbf{v}^{\epsilon,\lambda}|||_\lambda^2\|\overline{\mathbf{v}^{\epsilon,\lambda}}\|_{s-1}^2$$
$$+C_s\sum_{k=0}^{s-1}|q^{(k)}|_\infty^2\big(|||\mathbf{v}^{\epsilon,\lambda}|||_\lambda^2+(1+|||\mathbf{v}^{\epsilon,\lambda}|||_\lambda)^{2(s-3)}$$
$$\big(|||\mathbf{v}^{\epsilon,\lambda}|||_\lambda^4+|||\mathbf{v}^{\epsilon,\lambda}|||_\lambda^6\big)\big)\times\big(\|\overline{\mathbf{v}^{\epsilon,\lambda}}\|_{s-2}^2+\|\lambda\nabla\widetilde{V}^{\epsilon,\lambda}\|_s^2\big)$$
$$\le\widetilde{C}|||\widetilde{\mathbf{v}}^{\epsilon,\lambda}|||_\lambda^2,\tag{3.36}$$

where $\widetilde{C}$ is given by (3.17). By (3.36) we get (3.15).

Step 2. The equivalences of the norm $Q(\widetilde{\mathbf{v}}^{\epsilon,\lambda})$ and the norm $|||\widetilde{\mathbf{v}}^{\epsilon,\lambda}|||_\lambda^2$.

First, from the definition of $|||\cdot|||_{\lambda,T}$ and Sobolev's lemma, it follows that

$$|\sqrt{\varepsilon}\tilde{h}^{\epsilon,\lambda}|_\infty\le\sqrt{\varepsilon}C_s^*\|\tilde{h}^{\epsilon,\lambda}\|_{s-1}$$
$$\le\sqrt{\varepsilon}\lambda C_s^*\|\lambda\nabla V^{\epsilon,\lambda}\|_s\le 2M_0\sqrt{\lambda}\varepsilon C_s^*\le\frac{1}{2}\tag{3.37}$$

for any $0<\sqrt{\varepsilon}\lambda\le\iota_0$ and $0<t\le T$ if $|||\mathbf{v}^{\epsilon,\lambda}|||_\lambda\le 2M_0$.

Thus, from (3.37) we know that the energy norm $\|\cdot\|_E^2$ and the norm $\|\cdot\|_{L^2}^2$ are equivalent.

Also, it follows from the fact

$$(D_x^\beta\Delta\cdot, D_x^\beta\Delta\cdot) = (D_x^\alpha\nabla\cdot, D_x^\alpha\nabla\cdot)$$

that the norm $(D_x^\beta\Delta\cdot, D_x^\beta\Delta\cdot)$ is equivalent to the norm $(D_x^\alpha\nabla\cdot, D_x^\alpha\nabla\cdot)$.

Therefore, the norm $Q(\widetilde{\mathbf{v}}^{\epsilon,\lambda})$ and the norm $|||\widetilde{\mathbf{v}}^{\epsilon,\lambda}|||_\lambda^2$ are equivalent.

Thus, we have

$$\frac{d}{dt}Q \le \widetilde{C}Q. \tag{3.38}$$

Step 3. The uniform bounds of the initial data and the coefficient $\widetilde{C}$.

Direct computation gives

$$Q(t=0) = |||\mathbf{v}^{\epsilon,\lambda}(t=0)|||_\lambda^2 = M_0^2. \tag{3.39}$$

Also, from the definition of the matrices $A_j^\varepsilon, j = 0, \cdots, d$, it follows that there exist constants c_1 and c_2, depending only on M_0, such that

$$0 < c_1 \le q^{(k)}(1+\sqrt{\varepsilon}\widetilde{h}^{\epsilon,\lambda}) \le c_2, \quad k = 0, \cdots, s, \tag{3.40}$$

$$\sum_{j=0}^{d}|D_vA_j^\varepsilon|_\infty \le \sum_{j=0}^{d} \sup_{\|\mathbf{v}^{\epsilon,\lambda}\|_s \le 2M_0} |D_vA_j^\varepsilon(\mathbf{v}^{\epsilon,\lambda})|_\infty \le c_2,$$

$$\sum_{j=1}^{d}\sum_{k=1}^{s}|D_v^kA_j^\varepsilon|_\infty \le \sum_{j=1}^{d} \sup_{\|\mathbf{v}^{\epsilon,\lambda}\|_s \le 2M_0} |\sum_{k=1}^{s} D_v^kA_j^\varepsilon(\mathbf{v}^{\epsilon,\lambda})|_\infty \le c_2 \tag{3.41}$$

for any $0 < \sqrt{\varepsilon}\lambda \le \iota_0$, $\varepsilon \le 1$ and $0 < t \le T$ if $|||\mathbf{v}^{\epsilon,\lambda}|||_\lambda \le 2M_0$.

Hence, if $|||\mathbf{v}^{\epsilon,\lambda}|||_\lambda \le 2M_0$, then there exists Θ, depending only on M_0, not on ε and λ, such that

$$\widetilde{C} \le \Theta. \tag{3.42}$$

Using (3.38), (3.39) and (3.42), Gronwall's inequality gives

$$\min\{1, c_0\}|||\widetilde{\mathbf{v}}^{\epsilon,\lambda}|||_\lambda^2 \le Q \le M_0^2 e^{\Theta(M_0)T}, \quad 0 < t \le T. \tag{3.43}$$

Taking T such that $e^{\Theta(M_0)T} = 4\min\{1, c_0\}$, we get our result.

This completes the proof of Lemma 3.3.

Noting that from the exact formulation of $\widetilde{C}$ we know that $\widetilde{C}$ depends only on the bound of $\mathbf{v}^{\epsilon,\lambda}$ other than on the choice of $\mathbf{v}^{\epsilon,\lambda}$, and so do the constant Θ and T. Thus, by Lemma 3.3, we have

Proposition 3.4. *There are $T > 0$ and an ι_0, depending only on the initial data, such that for any $0 < \sqrt{\varepsilon}\lambda \le \iota_0$, $0 < \varepsilon \le 1$ and $p = 0, 1, \cdots$,*

$$|||\mathbf{v}^{\epsilon,\lambda,p}|||_{\lambda,T} \le 2M_0,$$

where

$$M_0 = \|\mathbf{v}_0^{\epsilon,\lambda}\|_s + \|\partial_t \mathbf{v}^{\epsilon,\lambda}(t=0)\|_{s-1} = \|\mathbf{u}_0^{0,0}\|_s + \|\mathbf{u}_0^{0,0} \cdot \nabla \mathbf{u}_0^{0,0}\|_{s-1}.$$

Proof of Proposition 3.4 We shall prove this proposition by using Lemma 3.3.

First, since $(\mathbf{v}^{\varepsilon,\lambda,0}, V^{\varepsilon,\lambda,0}) = ((0, \mathbf{u}_0^{0,0})^T, 0)$, we have

$$|||\mathbf{v}^{\varepsilon,\lambda,0}|||_{\lambda,t} = M_0 < 2M_0 \tag{3.44}$$

for any $\varepsilon > 0, \lambda > 0$ and any $t > 0$. Particularly, take ι_0 and $T > 0$ given by Lemma 3.3, then

$$|||\mathbf{v}^{\varepsilon,\lambda,0}|||_{\lambda,t} = M_0 < 2M_0 \tag{3.45}$$

for any $0 < \sqrt{\varepsilon}\lambda \le \iota_0$, $0 < \varepsilon \le 1$ and any $0 < t \le T$. Thus, by using Lemma 3.3, we have

$$|||\mathbf{v}^{\varepsilon,\lambda,1}|||_{\lambda,T} \le 2M_0 \tag{3.46}$$

for any $0 < \sqrt{\varepsilon}\lambda \le \iota_0$ and $0 < \varepsilon \le 1$.

Now assume that

$$|||\mathbf{v}^{\varepsilon,\lambda,p}|||_{\lambda,T} \le 2M_0 \tag{3.47}$$

for any $0 < \sqrt{\varepsilon}\lambda \le \iota_0$ and $0 < \varepsilon \le 1$, we will show that $|||\mathbf{v}^{\varepsilon,\lambda,p+1}|||_{\lambda,T} \le 2M_0$ for the same ε, λ and T.

Because the constant Θ in Lemma 3.3 depends only on the initial data (more precisely, the bound of $\mathbf{v}^{\epsilon,\lambda}$) but does not depends on ε, λ and the choice of $\mathbf{v}^{\epsilon,\lambda}$ and the initial data on $\mathbf{v}^{\varepsilon,\lambda,p}$ are the same for any p, we can repeatedly use Lemma 3.3 with $\mathbf{v}^{\epsilon,\lambda} = \mathbf{v}^{\varepsilon,\lambda,p}$, $\widetilde{\mathbf{v}}^{\epsilon,\lambda} = \mathbf{v}^{\varepsilon,\lambda,p+1}$ and then get our result.

This completes the proof of Proposition 3.4.

The end of the proof of Theorem 2.2 It follows from Proposition 3.4 that approximating sequence $\{(\mathbf{v}^{\varepsilon,\lambda,p}, V^{\varepsilon,\lambda,p})\}$ satisfies $(\mathbf{v}^{\varepsilon,\lambda,p}, \nabla V^{\varepsilon,\lambda,p}) \in L^\infty([0,T]; H^s) \cap C^{0,1}([0,T]; H^{s-1})$ and the systems (3.10)–(3.12) with $\mathbf{v}^{\epsilon,\lambda} = \mathbf{v}^{\varepsilon,\lambda,p}$, $\widetilde{\mathbf{v}}^{\epsilon,\lambda} = \mathbf{v}^{\varepsilon,\lambda,p+1}$ and $\widetilde{V}^{\epsilon,\lambda} = V^{\varepsilon,\lambda,p+1}$ as well as the uniform estimates, for $0 < \sqrt{\varepsilon}\lambda \le \iota_0$ and $0 < \varepsilon \le 1$,

$$\begin{aligned} |||\mathbf{v}^{\varepsilon,\lambda,p}|||_{\lambda,T} = \sup_{0 \le t \le T} (&\|\mathbf{v}^{\varepsilon,\lambda,p}\|_s + \|\partial_t \mathbf{v}^{\varepsilon,\lambda,p}\|_{s-1} \\ &+ \|\lambda \nabla V^{\varepsilon,\lambda,p}\|_s + \|\lambda \partial_t \nabla V^{\varepsilon,\lambda,p}\|_{s-1}) \le 2M_0. \end{aligned} \tag{3.48}$$

It follows from the Arzela-Ascoli theorem that there exists

$$(\mathbf{v}^{\epsilon,\lambda}, \nabla V^{\epsilon,\lambda}) \in L^\infty([0,T]; H^s) \cap C^{0,1}([0,T]; H^{s-1}) \tag{3.49}$$

such that

$$\begin{aligned}(\mathbf{v}^{\varepsilon,\lambda,p}, \nabla V^{\varepsilon,\lambda,p}) &\to (\mathbf{v}^{\epsilon,\lambda}, \nabla V^{\epsilon,\lambda}) \\ &\text{strongly in } L^\infty([0,T]; H^{s-1}), p \to \infty\end{aligned} \tag{3.50}$$

and $(\mathbf{v}^{\epsilon,\lambda}, V^{\epsilon,\lambda})$ satisfies (3.4)–(3.6) as well as estimates

$$|||\mathbf{v}^{\epsilon,\lambda}|||_{\lambda,T} \leq 2M_0 \tag{3.51}$$

for $0 < \sqrt{\varepsilon}\lambda \leq \iota_0$ and $0 < \varepsilon \leq 1$, which gives (2.11).

Furthermore, from the standard Sobolev interpolation inequalities, it follows that

$$\begin{aligned}(\mathbf{v}^{\varepsilon,\lambda,p}, \nabla V^{\varepsilon,\lambda,p}) &\to (\mathbf{v}^{\epsilon,\lambda}, \nabla V^{\epsilon,\lambda}) \\ &\text{uniformly in } C([0,T]; H^{s-\tau}) \text{ for any } \tau > 0.\end{aligned} \tag{3.52}$$

Choosing τ such that $s - \tau > \frac{d}{2} + 1$, then we deduce

$$\sum_{j=1}^{d} A_j^\varepsilon(\mathbf{v}^{\varepsilon,\lambda,p})\partial_j \mathbf{v}^{\varepsilon,\lambda,p+1} + \nabla\widehat{V^{\varepsilon,\lambda,p+1}} \to \sum_{j=1}^{d} A_j^\varepsilon(\mathbf{v}^{\epsilon,\lambda})\partial_j \mathbf{v}^{\epsilon,\lambda} + \widehat{\nabla V^{\epsilon,\lambda}}$$

uniformly in $C([0,T]; H^{s-\tau-1})$ for any $\tau > 0$. Thus, we have $\partial_t \mathbf{v}^{\epsilon,\lambda} \in C([0,T]; H^{s-\tau-1})$ for any $\tau > 0$.

By Sobolev's lemma, we have $C([0,T]; H^{s-\tau}) \cap C^1([0,T]; H^{s-\tau-1}) \subset C^1([0,T] \times \mathbb{T}^d)$, and hence the constructed solutions $(\mathbf{v}^{\epsilon,\lambda}, V^{\epsilon,\lambda})$ are classical. Thus, we have proved the first part of Theorem 2.2.

Now we prove convergence of Euler-Poisson system to the incompressible Euler equations.

By (2.11) and the Poisson equation, we have

$$\|\frac{\tilde{h}^{\epsilon,\lambda}}{\lambda}\|_{s-1} \leq \|\lambda \nabla V^{\epsilon,\lambda}\|_s \leq 2M_0, \tag{3.53}$$

$$\|\frac{\partial_t \tilde{h}^{\epsilon,\lambda}}{\lambda}\|_{s-2} \leq \|\lambda \partial_t \nabla V^{\epsilon,\lambda}\|_{s-1} \leq 2M_0. \tag{3.54}$$

Therefore, as $\lambda \to 0$,

$$\tilde{h}^{\epsilon,\lambda} \to 0 \text{ strongly in } L^\infty([0,T]; H^{s-1}) \cap C([0,T]; H^{s-1-\tau}), \tag{3.55}$$

$$\partial_t \tilde{h}^{\epsilon,\lambda} \to 0 \text{ strongly in } L^\infty([0,T]; H^{s-2}). \tag{3.56}$$

Moreover since $\|(\tilde{h}^{\epsilon,\lambda}, \mathbf{u}^{\epsilon,\lambda})\|_s + \|\partial_t(\tilde{h}^{\epsilon,\lambda}, \mathbf{u}^{\epsilon,\lambda})\|_{s-1} \leq 2M_0$ for any $0 < \sqrt{\varepsilon}\lambda \leq \iota_0$ and $0 < \varepsilon \leq 1$, by Lions-Aubin lemma, we have that any subsequence $(\tilde{h}^{\epsilon,\lambda}, \mathbf{u}^{\epsilon,\lambda})$ has a subsequence (still denoted by $(\tilde{h}^{\epsilon,\lambda}, \mathbf{u}^{\epsilon,\lambda})$) with a limit $(\tilde{h}^{0,0}, \mathbf{u}^{0,0}) \in L^\infty([0,T]; H^s) \cap C([0,T]; H^{s-\tau}), (\partial_t \tilde{h}^{0,0}, \partial_t \mathbf{u}^{0,0}) \in L^\infty([0,T]; H^s)$ for any $\tau > 0$ satisfying, as $\sqrt{\varepsilon}\lambda \to 0$,

$$(\tilde{h}^{\epsilon,\lambda}, \mathbf{u}^{\epsilon,\lambda}) \to (\tilde{h}^{0,0}, \mathbf{u}^{0,0}) \text{ weakly* in } L^\infty([0,T]; H^s), \tag{3.57}$$

$$(\partial_t \tilde{h}^{\epsilon,\lambda}, \partial_t \mathbf{u}^{\epsilon,\lambda}) \to (\partial_t \tilde{h}^{0,0}, \partial_t \mathbf{u}^{0,0}) \text{ weakly* in } L^\infty([0,T]; H^{s-1}), \tag{3.58}$$

$$(\tilde{h}^{\epsilon,\lambda}, \mathbf{u}^{\epsilon,\lambda}) \to (\tilde{h}^{0,0}, \mathbf{u}^{0,0}) \text{uniformly in } C([0,T]; H^{s-\tau}) \text{for any} \tau > 0. \tag{3.59}$$

Let $\phi(x,t)$ be any smooth test function with div $\phi = 0$ and compact support in $t \in [0,T]$. Then,

$$\int_0^T \int_{\mathbb{T}^d} \phi(\partial_t \mathbf{u}^{\epsilon,\lambda} + \mathbf{u}^{\epsilon,\lambda} \cdot \nabla \mathbf{u}^{\epsilon,\lambda}) dx dt = \int_0^T \int_{\mathbb{T}^d} \phi \cdot \frac{\nabla(V^{\epsilon,\lambda} - \tilde{h}^{\epsilon,\lambda})}{\sqrt{\varepsilon}} dx dt$$
$$= -\int_0^T \int_{\mathbb{T}^d} \operatorname{div} \phi \frac{V^{\epsilon,\lambda} - \tilde{h}^{\epsilon,\lambda}}{\sqrt{\varepsilon}} dx dt = 0. \tag{3.60}$$

Furthermore, by (2.7), (2.11), (3.55) and (3.56)

$$\operatorname{div} \mathbf{u}^{\epsilon,\lambda} = -\sqrt{\varepsilon} q[\partial_t \tilde{h}^{\epsilon,\lambda} + \mathbf{u}^{\epsilon,\lambda} \cdot \nabla \tilde{h}^{\epsilon,\lambda}] \to 0 \tag{3.61}$$

strongly in $L^\infty([0,T]; H^{s-2})$ as $\sqrt{\varepsilon}\lambda \to 0$.

Thus, it follows from (3.57)–(3.61) that $\mathbf{u}^{0,0} \in C([0,T]; H^{s-\tau})$ satisfies

$$\partial_t \mathbf{u}^{0,0} + P(\mathbf{u}^{0,0} \cdot \nabla \mathbf{u}^{0,0}) = 0, \quad \operatorname{div} \mathbf{u}^{0,0} = 0, \quad \mathbf{u}^{0,0}(t=0) = \mathbf{u}_0^{0,0},$$

where P is the standard projection in the div zero vector fields. Since $\mathbf{u}^{0,0} \in C([0,T]; H^{s-\tau})$, $s - \tau > \frac{d}{2} + 1$, one infers that $\partial_t \mathbf{u}^{0,0} = -P(\mathbf{u}^{0,0} \cdot \nabla \mathbf{u}^{0,0}) \in C([0,T]; H^{s-1-\tau})$. Thus $\mathbf{u}^{0,0} \in C^1([0,T] \times \mathbb{T}^d)$ is a classical solution of the incompressible Euler equations (1.8)–(1.10) with some pressure function $p^{0,0}$. Because the classical solution is unique, it follows that the convergence is valid for $\mathbf{u}^{\epsilon,\lambda}$ as $\sqrt{\varepsilon}\lambda \to 0$ without passing to subsequences.

Now, we discuss the convergence of the pressure. From (1.2) and (3.57)–(3.59) we conclude easily that

$$\frac{\nabla(V^{\epsilon,\lambda} - \tilde{h}^{\epsilon,\lambda})}{\sqrt{\varepsilon}} = \partial_t \mathbf{u}^{\epsilon,\lambda} + \mathbf{u}^{\epsilon,\lambda} \cdot \nabla \mathbf{u}^{\epsilon,\lambda} \rightharpoonup \partial_t \mathbf{u}^{0,0} + \mathbf{u}^{0,0} \cdot \nabla \mathbf{u}^{0,0}$$
$$= \nabla p^{0,0} \tag{3.62}$$

weakly* in $L^\infty([0,T]; H^{s-1})$ as $\sqrt{\varepsilon}\lambda \to 0$.

Combining (3.55)–(3.59), (3.61) and (3.62), we get (2.12) and that $(\mathbf{u}^{0,0}, p^{0,0})$ is a solution of incompressible Euler equations (1.8)–(1.10) with the initial data $\mathbf{u}_0^{0,0}$.

The proof of Theorem 2.2 is complete.

Remark 3.5. Under much more assumptions on the initial data, we can obtain much stronger estimates. Namely, if assume that $\sum_{k=0}^{l}\|\partial_t^k(\mathbf{v}^{\epsilon,\lambda}+\lambda\nabla V^{\epsilon,\lambda})(t=0)\|_{s-k}\le C$ uniformly in ε,λ with $s>\frac{d}{2}+k+3$, then $\sum_{k=0}^{l}\|\partial_t^k(\mathbf{v}^{\epsilon,\lambda}+\lambda\nabla V^{\epsilon,\lambda})(t)\|_{s-k}\le C$. Particularly, if $\operatorname{div}\mathbf{u}_0^{0,0}=0$ and $\sum_{i,j=1}^{d}\partial_j u_{0i}^{0,0}\partial_i u_{0j}^{0,0}=0$, then $\sum_{k=0}^{2}\|\partial_t^k(\mathbf{v}^{\epsilon,\lambda}+\lambda\nabla V^{\epsilon,\lambda})(t)\|_{s-k}\le C$. In this case, we can directly prove that $\mathbf{u}^{0,0}\in C^1([0,T]\times\mathbb{T}^d)$. In fact, by Lions-Aubin Lemma, we have

$$\partial_t\mathbf{u}^{\epsilon,\lambda}\to\partial_t\mathbf{u}^{0,0} \text{ uniformly in } C([0,T],H^{s-\tau-1}(\mathbb{T}^d))\text{for any}\tau>0,$$

which, together with (3.59), gives $\mathbf{u}^{0,0}\in C([0,T],H^{s-\tau}(\mathbb{T}^d))\cap C^1([0,T],H^{s-\tau-1}(\mathbb{T}^d))$ for any $\tau>0$. Hence, $\mathbf{u}^0\in C^1([0,T]\times\mathbb{T}^d)$ by Sobolev's lemma and choosing τ such that $s-\tau-1>\frac{d}{2}+1$ due to $s>\frac{d}{2}+3$.

3.3 The proof of Theorem 2.5

In this subsection, we will prove Theorem 2.5 by using the asymptotic expansion of singular perturbation and the carefully classical energy method.

Let us start with the derivation of the 'error' equations by the asymptotic expansion of singular perturbation.

Set $\mathbf{u}^{\epsilon,\lambda}=\mathbf{u}^{0,0}+\mathbf{u}_1^{\epsilon,\lambda}$ and $V^{\epsilon,\lambda}=\sqrt{\varepsilon}p^{0,0}+V_1^{\epsilon,\lambda}$, then $(\tilde{h}^{\epsilon,\lambda},\mathbf{u}_1^{\epsilon,\lambda},V_1^{\epsilon,\lambda})$ satisfies

$$q(1+\sqrt{\epsilon}\tilde{h}^{\epsilon,\lambda})[\partial_t\tilde{h}^{\epsilon,\lambda}+(\mathbf{u}^{0,0}+\mathbf{u}_1^{\epsilon,\lambda})\cdot\nabla\tilde{h}^{\epsilon,\lambda}]+\frac{\operatorname{div}\mathbf{u}_1^{\epsilon,\lambda}}{\sqrt{\epsilon}}$$

$$=0,\quad x\in\mathbb{T}^d,\ t>0, \tag{3.63}$$

$$\partial_t\mathbf{u}_1^{\epsilon,\lambda}+(\mathbf{u}^{0,0}+\mathbf{u}_1^{\epsilon,\lambda})\cdot\nabla\mathbf{u}_1^{\epsilon,\lambda}+\frac{\nabla\tilde{h}^{\epsilon,\lambda}}{\sqrt{\epsilon}}+\mathbf{u}_1^{\epsilon,\lambda}\cdot\nabla\mathbf{u}^{0,0}$$

$$=\frac{\nabla V_1^{\epsilon,\lambda}}{\sqrt{\epsilon}},\quad x\in\mathbb{T}^d,\ t>0, \tag{3.64}$$

$$\lambda^2\Delta V_1^{\epsilon,\lambda}=-\sqrt{\varepsilon}\lambda^2\Delta p^{0,0}+\tilde{h}^{\epsilon,\lambda},\quad\int_{\mathbb{T}^d}V_1^{\epsilon,\lambda}dx=0,\ x\in\mathbb{T}^d,\ t>0, \tag{3.65}$$

$$\tilde{h}^{\epsilon,\lambda}(t=0)=\tilde{h}_0^{\epsilon,\lambda},\mathbf{u}_1^{\epsilon,\lambda}(t=0)=\mathbf{u}_{10}^{\epsilon,\lambda}=\mathbf{u}_0^{\epsilon,\lambda}-\mathbf{u}_0^{0,0},\ x\in\mathbb{T}^d. \tag{3.66}$$

Denoting $\mathbf{v}_1^{\epsilon,\lambda}=(\tilde{h}^{\epsilon,\lambda},\mathbf{u}_1^{\epsilon,\lambda})^T$, we can write (3.63)–(3.66) in the form

$$\mathcal{L}(t,x,\mathbf{u}^{0,0},\mathbf{v}_1^{\epsilon,\lambda},\varepsilon,\lambda)\mathbf{v}_1^{\epsilon,\lambda}$$

$$=A_0^{\varepsilon}\partial_t\mathbf{v}_1^{\epsilon,\lambda}+\sum_{j=1}^{d}\tilde{A}_j^{\varepsilon}(\mathbf{u}^{0,0},\mathbf{v}_1^{\epsilon,\lambda})\partial_j\mathbf{v}_1^{\epsilon,\lambda}+\widehat{\mathbf{u}_1^{\epsilon,\lambda}\cdot\nabla\mathbf{u}^{0,0}}=\widehat{\frac{\nabla V_1^{\epsilon,\lambda}}{\sqrt{\varepsilon}}}, \tag{3.67}$$

$$\lambda^2 \Delta V_1^{\epsilon,\lambda} = -\sqrt{\varepsilon}\lambda^2 \Delta p^{0,0} + \tilde{h}^{\epsilon,\lambda}, \int_{\mathbb{T}^d} V_1^{\epsilon,\lambda} dx = 0, \tag{3.68}$$

$$\mathbf{v}_1^{\epsilon,\lambda}(t=0) = \mathbf{v}_{10}^{\epsilon,\lambda} = (\tilde{h}_0^{\epsilon,\lambda}, \mathbf{u}_{10}^{\epsilon,\lambda})^T, \tag{3.69}$$

for $x \in \mathbb{T}^d, t > 0$, where A_0^ε is the same as before, $\tilde{A}_j^\varepsilon(\mathbf{u}^{0,0}, \mathbf{v}_1^{\epsilon,\lambda}) = \tilde{A}_j(\mathbf{u}^{0,0}, \mathbf{v}_1^{\epsilon,\lambda}) + \frac{1}{\sqrt{\varepsilon}} C_j$,

$$\tilde{A}_j(\mathbf{u}^{0,0}, \mathbf{v}_1^{\epsilon,\lambda}) = \begin{pmatrix} q(1+\sqrt{\varepsilon}\tilde{h}^{\epsilon,\lambda})(u_j^{0,0} + \mathbf{u}_{1,j}^{\epsilon,\lambda}) & O^T \\ O & (u_j^{0,0} + \mathbf{u}_{1,j}^{\epsilon,\lambda}) I_d \end{pmatrix},$$

$$C_j = \begin{pmatrix} 0 & e_j^T \\ e_j & O \end{pmatrix},$$

$\tilde{A}_j(\mathbf{u}^{0,0}, \mathbf{v}_1^{\epsilon,\lambda})$ is symmetric and C_j is a constant symmetric matrix.

Following the proof of Theorem 2.2 in Subsection 3.2, we introduce the set, S^λ, of functions to $L^\infty([0,T]; H^s) \cap C^{0,1}([0,T]; H^{s-1})$, $s > \frac{d}{2} + 2$, satisfying $\mathbf{v}_1^{\epsilon,\lambda}(t=0) = \mathbf{v}_{10}^{\epsilon,\lambda}$ and

$$\|\mathbf{v}_1^{\epsilon,\lambda}\|_s + \|\lambda \nabla V_1^{\epsilon,\lambda}\|_s \leq \tilde{M}(T_0)\sqrt{\varepsilon}\lambda, \tag{3.70}$$

$$\|\partial_t \mathbf{v}_1^{\epsilon,\lambda}\|_{s-1} \leq \tilde{M}(T_0), \tag{3.71}$$

where $\tilde{M}(T_0)$ is appropriate constant to be chosen later.

We want to prove that (3.67)–(3.69) have a smooth solution satisfying $(\mathbf{v}_1^{\epsilon,\lambda}, \nabla V_1^{\epsilon,\lambda}) \in S^\lambda$ for appropriate $\tilde{M}(T_0)$, ε and λ. This yields the desired estimates stated in Theorem 2.5.

As usual in this framework, we determine the conditions on the constant $\tilde{M}(T_0)$ in the estimate. As we will see, the constant $\tilde{M}(T_0)$ only depends on $(\mathbf{u}^{0,0}, p^{0,0})$. For a given choice for $\tilde{M}(T_0)$, we shall ultimately make a finite number of restriction for ε, λ. First, our first restriction for ε, λ will be the requirement that

$$\begin{aligned} \sqrt{\varepsilon}|\tilde{h}_0^{\epsilon,\lambda}|_\infty &\leq \sqrt{\varepsilon} C_s^* \|\tilde{h}_0^{\epsilon,\lambda}\|_s \\ &\leq \sqrt{\varepsilon} C_s^* M_0 \sqrt{\varepsilon}\lambda \leq C_s^* M_0 \sqrt{\varepsilon}\lambda \leq \frac{1}{2}, \end{aligned} \tag{3.72}$$

$$\begin{aligned} \sqrt{\varepsilon}|\tilde{h}^{\epsilon,\lambda}|_\infty &\leq \sqrt{\varepsilon} C_s^* \|\tilde{h}^{\epsilon,\lambda}\|_s \\ &\leq \sqrt{\varepsilon} C_s^* \tilde{M}(T_0) \sqrt{\varepsilon}\lambda \leq C_s^* \tilde{M}(T_0) \sqrt{\varepsilon}\lambda \leq \frac{1}{2}, \end{aligned} \tag{3.73}$$

which can be guaranteed provided that ε, λ satisfy $\varepsilon \leq 1$, $\sqrt{\varepsilon}\lambda C_s^* M_0 \leq \frac{1}{2}$ and $\sqrt{\varepsilon}\lambda C_s^* \tilde{M}(T_0) \leq \frac{1}{2}$.

Let $V_{10}^{\epsilon,\lambda} = V_1^{\epsilon,\lambda}(t=0)$ be a solution of

$$\lambda^2 \Delta V_{10}^{\epsilon,\lambda} = -\sqrt{\varepsilon}\lambda^2 \Delta p^{0,0}(t=0) + \tilde{h}_0^{\epsilon,\lambda}.$$

Consider now the iteration scheme

$$(\mathbf{v}_1^{\varepsilon,\lambda,0}, V_1^{\varepsilon,\lambda,0}) = (\mathbf{v}_{10}^{\epsilon,\lambda}, V_{10}^{\epsilon,\lambda}) = ((\tilde{h}_0^{\epsilon,\lambda}, \mathbf{u}_{10}^{\epsilon,\lambda}), V_{10}^{\epsilon,\lambda}), \tag{3.74}$$

$$(\mathbf{v}_1^{\varepsilon,\lambda,p+1}, V_1^{\varepsilon,\lambda,p+1}) = \Phi((\mathbf{v}_1^{\varepsilon,\lambda,p}, V_1^{\varepsilon,\lambda,p})), \tag{3.75}$$

where the generator Φ maps the vector $(\mathbf{v}_1^{\varepsilon,\lambda}, V_1^{\varepsilon,\lambda}) = ((\tilde{h}^{\epsilon,\lambda}, \mathbf{u}_1^{\varepsilon,\lambda})^T, V_1^{\epsilon,\lambda})$ into the solution $(\widetilde{\mathbf{v}}_1^{\varepsilon,\lambda}, \widetilde{V}_1^{\varepsilon,\lambda}) = ((\tilde{\tilde{h}}^{\varepsilon,\lambda}, \widetilde{\mathbf{u}}_1^{\epsilon,\lambda})^T, \widetilde{V}_1^{\epsilon,\lambda})$ of the following linearized Euler-Poisson system

$$A_0^\varepsilon \partial_t \widetilde{\mathbf{v}}_1^{\epsilon,\lambda} + \sum_{j=1}^d \tilde{A}_j^\varepsilon(\mathbf{u}^{0,0}, \mathbf{v}_1^{\epsilon,\lambda}) \partial_j \widetilde{\mathbf{v}}_1^{\epsilon,\lambda} + \widehat{\widetilde{\mathbf{u}}_1^{\epsilon,\lambda} \cdot \nabla \mathbf{u}^{0,0}} = \frac{\widehat{\nabla \widetilde{V}_1^{\epsilon,\lambda}}}{\sqrt{\varepsilon}}, x \in \mathbb{T}^d, t > 0, \tag{3.76}$$

$$\lambda^2 \Delta \widetilde{V}_1^{\epsilon,\lambda} = -\sqrt{\varepsilon}\lambda^2 \Delta p^{0,0} + \tilde{\tilde{h}}^{\varepsilon,\lambda} \quad \text{with} \quad \int_{\mathbb{T}^d} \widetilde{V}_1^{\epsilon,\lambda} dx = 0, x \in \mathbb{T}^d, t > 0, \tag{3.77}$$

$$\widetilde{\mathbf{v}}_1^{\epsilon,\lambda}(t=0) = \mathbf{v}_{10}^{\epsilon,\lambda} = (\tilde{h}_0^{\epsilon,\lambda}, \mathbf{u}_{10}^{\epsilon,\lambda})^T, x \in \mathbb{T}^d. \tag{3.78}$$

As before, we will prove the convergence of the approximating sequence $\{(\mathbf{v}_1^{\varepsilon,\lambda,p}, V_1^{\varepsilon,\lambda,p})\}_{p=0}^\infty$ via the uniform boundedness of this sequence in some weighted high Sobolev's norm.

Now we will establish the estimates of the sequence $\{(\mathbf{v}_1^{\varepsilon,\lambda,p}, V_1^{\varepsilon,\lambda,p})\}_{p=0}^\infty$.

Lemma 3.6. *For any $T_0 < T_*$, there exist a constant $\tilde{M}(T_0) > 0$ and a constant $\iota_0(T_0) > 0$ such that*

$$\|\mathbf{v}_1^{\varepsilon,\lambda,p}\|_s + \|\lambda \nabla V_1^{\varepsilon,\lambda,p}\|_s + \|\sqrt{\varepsilon}\partial_t \tilde{h}^{\varepsilon,\lambda,p}\|_{s-1} \le \tilde{M}(T_0)\sqrt{\varepsilon}\lambda, \|\partial_t \mathbf{v}_1^{\varepsilon,\lambda,p}\|_{s-1} \le \tilde{M}(T_0) \tag{3.79}$$

for $0 < \sqrt{\varepsilon}\lambda \le \iota_0$, $0 < \varepsilon, \lambda \le 1$ and any $p = 0, 1, \cdots$.

Proof of Lemma 3.6 First, since $(\mathbf{v}_1^{\epsilon,\lambda,0}, V_1^{\varepsilon,\lambda,0}) = ((\tilde{h}_0^{\varepsilon,\lambda}, \mathbf{u}_{10}^{\epsilon,\lambda}), V_{10}^{\epsilon,\lambda})$, it follows from the assumption on the initial data, (3.72) and the system (3.63)–(3.66) that

$$\|\mathbf{v}_1^{\varepsilon,\lambda,0}\|_s + \|\lambda \nabla V_1^{\varepsilon,\lambda,0}\|_s + \|\sqrt{\varepsilon}\partial_t \tilde{h}^{\varepsilon,\lambda,0}\|_{s-1} \le \tilde{M}(T_0)\sqrt{\varepsilon}\lambda, \|\partial_t \mathbf{v}_1^{\varepsilon,\lambda,0}\|_{s-1} \le \tilde{M}(T_0) \tag{3.80}$$

for any $\tilde{M}(T_0) \ge \max\{M_0, C_0(M_0)\}$ and any $0 < \varepsilon \le 1$ and $0 < \lambda \le 1$, where $C_0(M_0) = \sup_{0<\varepsilon\le 1, 0<\lambda\le 1} \|\partial_t \mathbf{v}_1^{\varepsilon,\lambda,0}|_{t=0}\|_{s-1}$ can be exactly given.

So, our second restriction for ε, λ will be the requirement that $0 < \varepsilon \le 1$ and $0 < \lambda \le 1$.

Now assume that there exist an $\tilde{M}(T_0)$ and an $\iota_0(T_0)$ such that

$$\|\mathbf{v}_1^{\varepsilon,\lambda,p}\|_s + \|\lambda\nabla V_1^{\varepsilon,\lambda,p}\|_s + \|\sqrt{\varepsilon}\partial_t\tilde{h}^{\varepsilon,\lambda,p}\|_{s-1} \le \tilde{M}(T_0)\sqrt{\varepsilon}\lambda,$$
$$\|\partial_t\mathbf{v}_1^{\varepsilon,\lambda,p}\|_{s-1} \le \tilde{M}(T_0), 0 \le t \le T_0 \quad (3.81)$$

for all $0 < \sqrt{\varepsilon}\lambda \le \iota_0$, $0 < \varepsilon, \lambda \le 1$, and we shall show

$$\|\mathbf{v}_1^{\varepsilon,\lambda,p+1}\|_s + \|\lambda\nabla V_1^{\varepsilon,\lambda,p+1}\|_s + \|\sqrt{\varepsilon}\partial_t\tilde{h}^{\varepsilon,\lambda,p+1}\|_{s-1} \le \tilde{M}(T_0)\sqrt{\varepsilon}\lambda,$$
$$\|\partial_t\mathbf{v}_1^{\varepsilon,\lambda,p+1}\|_{s-1} \le \tilde{M}(T_0), 0 \le t \le T_0 \quad (3.82)$$

for all $0 < \sqrt{\varepsilon}\lambda \le \iota_0$, $0 < \varepsilon, \lambda \le 1$.

Denote $\mathbf{v}_1^{\varepsilon,\lambda,p} = \mathbf{v}_1^{\epsilon,\lambda}, \mathbf{v}_1^{\varepsilon,\lambda,p+1} = \widetilde{\mathbf{v}}_1^{\epsilon,\lambda}, V_1^{\varepsilon,\lambda,p} = V_1^{\epsilon,\lambda}, V_1^{\varepsilon,\lambda,p+1} = \widetilde{V}_1^{\epsilon,\lambda}$, then $(\mathbf{v}_1^{\epsilon,\lambda}, \widetilde{\mathbf{v}}_1^{\epsilon,\lambda}, V_1^{\epsilon,\lambda}, \widetilde{V}_1^{\epsilon,\lambda})$ satisfies (3.76)–(3.78) and $\tilde{h}^{\epsilon,\lambda}$ satisfies (3.73) by assumption (3.81).

Obviously, if $\|(\mathbf{v}_1^{\varepsilon,\lambda,p+1}, \lambda\nabla V_1^{\varepsilon,\lambda,p+1})\|_s \le \tilde{M}(T_0)\sqrt{\varepsilon}\lambda$, then from (3.76), by the calculus inequality, it easily follows that there exists a constant $C_1(T_0)$ such that

$$\|\sqrt{\varepsilon}\partial_t\tilde{h}^{\varepsilon,\lambda,p+1}\|_{s-1} \le \sqrt{\varepsilon}\lambda C_1(T_0), \|\partial_t\mathbf{v}_1^{\varepsilon,\lambda,p+1}\|_{s-1}$$
$$\le C_1(T_0), \quad 0 \le t \le T_0 \quad (3.83)$$

for any $0 < \sqrt{\varepsilon}\lambda \le \iota_0$, $0 < \varepsilon, \lambda \le 1$. Thus, we only need to prove

$$\|\mathbf{v}_1^{\varepsilon,\lambda,p+1}\|_s + \|\lambda\nabla V_1^{\varepsilon,\lambda,p+1}\|_s \le \tilde{M}(T_0)\sqrt{\varepsilon}\lambda. \quad (3.84)$$

Also, from the Poisson equation, we have

$$\lambda^2\|\partial_t\nabla V_1^{\varepsilon,\lambda,p+1}\|_{s-1} \le 3\|\partial_t\tilde{h}_0^{\varepsilon,\lambda,p+1}\|_{s-1} + 3\sqrt{\varepsilon}\lambda^2\|\partial_t\nabla p^{0,0}\|_{s-1}$$
$$\le 4C_1(T_0). \quad (3.85)$$

Now we determine $\tilde{M}(T_0)$ and $\iota_0(T_0)$.

As in (3.22), the standard higher-order energy estimates of Friedrich imply that

$$\frac{d}{dt}\|D_x^\alpha\widetilde{\mathbf{v}}_1^{\epsilon,\lambda}\|_E^2 = (Div\tilde{A}^\varepsilon(\mathbf{u}^{0,0}, \mathbf{v}_1^{\epsilon,\lambda})D_x^\alpha\widetilde{\mathbf{v}}_1^{\epsilon,\lambda}, D_x^\alpha\widetilde{\mathbf{v}}_1^{\epsilon,\lambda}) + 2(H_\alpha^{(3)}, D_x^\alpha\widetilde{\mathbf{v}}_1^{\epsilon,\lambda})$$
$$-2(\widehat{D_x^\alpha\widetilde{\mathbf{v}}_1^{\epsilon,\lambda}\cdot\nabla\mathbf{u}^{0,0}}, D_x^\alpha\widetilde{\mathbf{v}}_1^{\epsilon,\lambda}) - 2(H_\alpha^{(4)}, D_x^\alpha\widetilde{\mathbf{v}}_1^{\epsilon,\lambda})$$
$$+2(\frac{\widehat{D_x^\alpha\widetilde{V}_1^{\epsilon,\lambda}}}{\sqrt{\varepsilon}}, D_x^\alpha\widetilde{\mathbf{v}}_1^{\epsilon,\lambda}), \quad (3.86)$$

where $H_\alpha^{(k)}, k = 3, 4$ are commutators defined by

$$H_\alpha^{(3)} = -\sum_{j=1}^{d}\left(D_x^\alpha(\tilde{A}_j^\varepsilon(\mathbf{u}^{0,0}, \mathbf{v}_1^{\epsilon,\lambda})\partial_j\widetilde{\mathbf{v}}_1^{\epsilon,\lambda}) - A_j^\varepsilon(\mathbf{u}^{0,0}, \mathbf{v}_1^{\epsilon,\lambda})\partial_j D_x^\alpha\widetilde{\mathbf{v}}_1^{\epsilon,\lambda}\right)$$

and

$$H_\alpha^{(4)} = D_x^\alpha(\widehat{\widetilde{\mathbf{v}}_1^{\epsilon,\lambda} \cdot \nabla\mathbf{u}^{0,0}}) - D_x^\alpha \widehat{\widetilde{\mathbf{v}}_1^{\epsilon,\lambda} \cdot \nabla\mathbf{u}^{0,0}}.$$

The coefficient $Div\tilde{A}^\varepsilon = \partial_t A_0^\varepsilon + \partial_j \tilde{A}_j^\varepsilon$ in the first term on the right hand side of (3.86) can be bounded by

$$\begin{aligned}
|Div\tilde{A}^\varepsilon(\mathbf{v}^{\epsilon,\lambda})|_{L^\infty} &\le |D_v A_0^\varepsilon|_\infty |\sqrt{\varepsilon}\partial_t \tilde{h}^{\epsilon,\lambda}|_\infty + \sum_{j=1}^d |D_v A_j^\varepsilon|_\infty |\nabla \mathbf{v}^{\epsilon,\lambda}|_\infty \\
&\le C_s(|D_v A_0^\varepsilon|_\infty \|\sqrt{\varepsilon}\partial_t \tilde{h}^{\epsilon,\lambda}\|_{s-1} + \sum_{j=1}^d |D_v A_j^\varepsilon|_\infty \|\mathbf{v}^{\epsilon,\lambda}\|_s) \\
&\le C_s(\sup_{\|\mathbf{v}_1^{\epsilon,\lambda}\|_s \le 1} |D_v A_0^\varepsilon|_\infty \sqrt{\varepsilon} \|\partial_t \tilde{h}^{\epsilon,\lambda}\|_{s-1} \\
&\quad + \sum_{j=1}^d \sup_{\|\mathbf{v}_1^{\epsilon,\lambda}\|_s \le 1} |D_v A_j^\varepsilon|_\infty \|\mathbf{v}^{\epsilon,\lambda}\|_s) \\
&\le C_s \Bigg(\sup_{\|\mathbf{v}_1^{\epsilon,\lambda}\|_s \le 1} |D_v A_0^\varepsilon|_\infty \sqrt{\varepsilon}\lambda \tilde{M}(T_0) \\
&\quad + \sum_{j=1}^d \sup_{\|\mathbf{v}_1^{\epsilon,\lambda}\|_s \le 1} |D_v A_j^\varepsilon|_\infty \sqrt{\varepsilon}\lambda \tilde{M}(T_0) \Bigg) \\
&\le C_s \Bigg(\sup_{\|\mathbf{v}_1^{\epsilon,\lambda}\|_s \le 1} |D_v A_0^\varepsilon|_\infty + \sum_{j=1}^d \sup_{\|\mathbf{v}_1^{\epsilon,\lambda}\|_s \le 1} |D_v A_j^\varepsilon|_\infty \Bigg),
\end{aligned}$$

which yields the third restriction $\sqrt{\varepsilon}\lambda \tilde{M}(T_0) \le 1$.

Thus, by using (3.73), $\varepsilon \le 1$, $\|\mathbf{v}_1^{\epsilon,\lambda}\|_s \le \sqrt{\varepsilon}\lambda\tilde{M}(T_0) \le 1$ and Sobolev's lemma, we have

$$|Div\tilde{A}^\varepsilon|_\infty \le C_2(T_0).$$

Here and in the following $C_2(T_0)$ denotes a constant that may take different values during the same proof and depends only on $\mathbf{u}^{0,0}$ but does not depend on $\tilde{M}(T_0)$, ε and λ.

Hence, we have

$$(Div\tilde{A}^\varepsilon(\mathbf{u}^{0,0}, \mathbf{v}_1^{\epsilon,\lambda}) D_x^\alpha \widetilde{\mathbf{v}}_1^{\epsilon,\lambda}, D_x^\alpha \widetilde{\mathbf{v}}_1^{\epsilon,\lambda}) \le C_2(T_0) \|\widetilde{\mathbf{v}}_1^{\epsilon,\lambda}\|_s^2. \tag{3.87}$$

The usual estimates on commutators lead to

$$\begin{aligned}
2(H_\alpha^{(3)}, D_x^\alpha \widetilde{\mathbf{v}}_1^{\epsilon,\lambda}) &\le C_2(\|\mathbf{u}^{0,0}\|_s, \|\mathbf{v}_1^{\epsilon,\lambda}\|_s) \|\widetilde{\mathbf{v}}_1^{\epsilon,\lambda}\|_s^2 \\
&\le C_2(\|\mathbf{u}^{0,0}\|_s, 1) \|\widetilde{\mathbf{v}}_1^{\epsilon,\lambda}\|_s^2
\end{aligned} \tag{3.88}$$

and

$$-2(H_\alpha^{(4)}, D_x^\alpha \widetilde{\mathbf{v}}_1^{\epsilon,\lambda}) \leq C_2(\|\mathbf{u}^{0,0}\|_{s+1})\|\widetilde{\mathbf{v}}_1^{\epsilon,\lambda}\|_s^2. \tag{3.89}$$

Here we have used $s > \frac{d}{2}+2$ and $\|\mathbf{v}_1^{\epsilon,\lambda}\|_s \leq 1$ and the bound of $\|\mathbf{u}^{0,0}\|_{s+1}$.

Similarly, we have

$$\begin{aligned}
-2(\widehat{D_x^\alpha \widetilde{\mathbf{v}}_1^{\epsilon,\lambda} \cdot \nabla \mathbf{u}^{0,0}}, D_x^\alpha \widetilde{\mathbf{v}}_1^{\epsilon,\lambda}) &\leq 2|\nabla \mathbf{u}^{0,0}|_\infty \|\widetilde{\mathbf{v}}_1^{\epsilon,\lambda}\|_s^2 \\
&\leq 2C_s^* \|\mathbf{u}^{0,0}\|_s \|\widetilde{\mathbf{v}}_1^{\epsilon,\lambda}\|_s^2 \\
&\leq C_2(T_0)\|\widetilde{\mathbf{v}}_1^{\epsilon,\lambda}\|_s^2. \qquad (3.90)
\end{aligned}$$

Now we estimate the most singular term in (3.86) by using Lemma 3.1.

Taking $\mathbf{a} = \mathbf{u}^{0,0} + \mathbf{u}_1^{\epsilon,\lambda}, \rho = \sqrt{\varepsilon}\tilde{h}^{\epsilon,\lambda}, y_0^{\varepsilon,\lambda} = \tilde{\tilde{h}}^{\varepsilon,\lambda}, \mathbf{y}^{\epsilon,\lambda} = \frac{\widetilde{\mathbf{u}}_1^{\epsilon,\lambda}}{\sqrt{\varepsilon}}, p^0 = -p^{0,0}, \tau = \sqrt{\varepsilon}\lambda, f_0 = 0$ in Lemma 3.1, we have

$$\begin{aligned}
&2(\frac{\widehat{D_x^\alpha \nabla \widetilde{V}_1^{\epsilon,\lambda}}}{\sqrt{\varepsilon}}, D_x^\alpha \widetilde{\mathbf{v}}_1^{\epsilon,\lambda}) = 2(D_x^\alpha \nabla \widetilde{V}_1^{\epsilon,\lambda}, D_x^\alpha (\frac{\widetilde{\mathbf{u}}_1^{\epsilon,\lambda}}{\sqrt{\varepsilon}})) \\
&\leq -\lambda^2 \frac{d}{dt}\Big((D_x^\beta \Delta \widetilde{V}_1^{\epsilon,\lambda}, D_x^\beta \Delta \widetilde{V}_1^{\epsilon,\lambda}) + (\nabla \widetilde{V}_1^{\epsilon,\lambda}, \nabla \widetilde{V}_1^{\epsilon,\lambda})\Big) \\
&\quad +\lambda^2 C_s \|\mathbf{u}^{0,0} + \mathbf{u}_1^{\epsilon,\lambda}\|_{s-1} \|\nabla \widetilde{V}_1^{\epsilon,\lambda}\|_s^2 + \frac{1}{\lambda^2} C_s \|\sqrt{\varepsilon}\tilde{h}^{\epsilon,\lambda}\|_{s-1}^2 \|\frac{\widetilde{\mathbf{u}}_1^{\epsilon,\lambda}}{\sqrt{\varepsilon}}\|_s^2 \\
&\quad +\varepsilon\lambda^2 C_s \|\nabla \partial_t p^{0,0}\|_s^2 + \varepsilon\lambda^2 C_s \|\mathbf{u}^{0,0} + \mathbf{u}_1^{\epsilon,\lambda}\|_{s-1}^2 \|\nabla p^{0,0}\|_s^2 \\
&\leq -\lambda^2 \frac{d}{dt}\Big((D_x^\beta \Delta \widetilde{V}_1^{\epsilon,\lambda}, D_x^\beta \Delta \widetilde{V}_1^{\epsilon,\lambda}) + (\nabla \widetilde{V}_1^{\epsilon,\lambda}, \nabla \widetilde{V}_1^{\epsilon,\lambda})\Big) \\
&\quad +C_s(\|\mathbf{u}^{0,0}\|_{s-1} + \|\mathbf{u}_1^{\epsilon,\lambda}\|_{s-1})\|\lambda \nabla \widetilde{V}_1^{\epsilon,\lambda}\|_s^2 + \varepsilon\lambda^2 C_s \|\nabla \partial_t p^{0,0}\|_s^2 \\
&\quad +\varepsilon\lambda^2 C_s (\|\mathbf{u}^{0,0}\|_{s-1}^2 + \|\mathbf{u}_1^{\epsilon,\lambda}\|_{s-1}^2)\|\nabla p^{0,0}\|_s^2 \\
&\quad +\frac{1}{\lambda^2} C_s \|\tilde{h}^{\epsilon,\lambda}\|_{s-1}^2 \|\widetilde{\mathbf{u}}_1^{\epsilon,\lambda}\|_s^2. \qquad (3.91)
\end{aligned}$$

It follows from the Poisson equation that

$$\frac{\tilde{h}^{\epsilon,\lambda}}{\lambda} = \lambda \Delta V_1^{\epsilon,\lambda} + \sqrt{\varepsilon}\lambda \Delta p^{0,0}$$

and hence

$$\begin{aligned}
\|\frac{\tilde{h}^{\epsilon,\lambda}}{\lambda}\|_{s-1}^2 &\leq (\|\lambda \Delta V_1^{\epsilon,\lambda}\|_{s-1} + \sqrt{\varepsilon}\lambda \|\Delta p^{0,0}\|_{s-1})^2 \\
&\leq (1 + \sqrt{\varepsilon}\lambda \|\nabla p^{0,0}\|_s)^2 \leq C_2(T_0). \qquad (3.92)
\end{aligned}$$

Combining (3.91) and (3.92), one have

$$
\begin{aligned}
&2(\frac{\widehat{D_x^\alpha \nabla \widetilde{V}_1^{\epsilon,\lambda}}}{\sqrt{\varepsilon}}, D_x^\alpha \widetilde{\mathbf{v}}_1^{\epsilon,\lambda}) = 2(D_x^\alpha \nabla \widetilde{V}_1^{\epsilon,\lambda}, D_x^\alpha (\frac{\widetilde{\mathbf{u}}_1^{\epsilon,\lambda}}{\sqrt{\varepsilon}}))\\
&\leq -\lambda^2 \frac{d}{dt}\Big((D_x^\beta \Delta \widetilde{V}_1^{\epsilon,\lambda}, D_x^\beta \Delta \widetilde{V}_1^{\epsilon,\lambda}) + (\nabla \widetilde{V}_1^{\epsilon,\lambda}, \nabla \widetilde{V}_1^{\epsilon,\lambda})\Big)\\
&\quad + C_s(\|\mathbf{u}^{0,0}\|_{s-1} + \|\mathbf{u}_1^{\epsilon,\lambda}\|_{s-1})\|\lambda \nabla \widetilde{V}_1^{\epsilon,\lambda}\|_s^2 + \varepsilon\lambda^2 C_s \|\nabla \partial_t p^{0,0}\|_s^2\\
&\quad + \varepsilon\lambda^2 C_s(\|\mathbf{u}^{0,0}\|_{s-1}^2 + \|\mathbf{u}_1^{\epsilon,\lambda}\|_{s-1}^2)\|\nabla p^{0,0}\|_s^2 + C_s C_2(T_0)\|\widetilde{\mathbf{u}}_1^{\epsilon,\lambda}\|_s^2\\
&\leq -\lambda^2 \frac{d}{dt}\Big((D_x^\beta \Delta \widetilde{V}_1^{\epsilon,\lambda}, D_x^\beta \Delta \widetilde{V}_1^{\epsilon,\lambda}) + (\nabla \widetilde{V}_1^{\epsilon,\lambda}, \nabla \widetilde{V}_1^{\epsilon,\lambda})\Big)\\
&\quad + C_s(\|\mathbf{u}^{0,0}\|_{s-1} + 1)\|\lambda \nabla \widetilde{V}_1^{\epsilon,\lambda}\|_s^2 + \varepsilon\lambda^2 C_s \|\nabla \partial_t p^{0,0}\|_s^2\\
&\quad + \varepsilon\lambda^2 C_s(\|\mathbf{u}^{0,0}\|_{s-1}^2 + 1)\|\nabla p^{0,0}\|_s^2 + C_s C_2(T_0)\|\widetilde{\mathbf{u}}_1^{\epsilon,\lambda}\|_s^2\\
&\leq -\lambda^2 \frac{d}{dt}\Big((D_x^\beta \Delta \widetilde{V}_1^{\epsilon,\lambda}, D_x^\beta \Delta \widetilde{V}_1^{\epsilon,\lambda}) + (\nabla \widetilde{V}_1^{\epsilon,\lambda}, \nabla \widetilde{V}_1^{\epsilon,\lambda})\Big)\\
&\quad + C_s C_2(T_0)(\|\lambda \nabla \widetilde{V}_1^{\epsilon,\lambda}\|_s^2 + \|\widetilde{\mathbf{u}}_1^{\epsilon,\lambda}\|_s^2) + \varepsilon\lambda^2 C_s C_2(T_0).
\end{aligned} \tag{3.93}
$$

Hence, by (3.86), together with (3.87)–(3.90), and (3.93) we have

$$
\begin{aligned}
&\frac{d}{dt}(\|D_x^\alpha \widetilde{\mathbf{v}}_1^{\epsilon,\lambda}\|_E^2 + \|(\lambda \widehat{D_x^\beta \Delta \widetilde{V}_1^{\epsilon,\lambda}}, \lambda \widehat{\nabla \widetilde{V}_1^{\epsilon,\lambda}})\|_{L^2}^2)\\
&\leq C_2(T_0)\|(\widetilde{\mathbf{v}}_1^{\epsilon,\lambda}, \lambda \nabla \widetilde{V}_1^{\epsilon,\lambda})\|_s^2 + C_2(T_0)\varepsilon\lambda^2.
\end{aligned} \tag{3.94}
$$

Similar to Step 2 of Lemma 3.3, we know that there exist constants $c_k, k=3,4$, for example, $c_3=\min\{\min_{|s|\leq\frac{1}{2}} q(1+s), 1\}$, $c_4=\max\{\max_{|s|\leq\frac{1}{2}} q(1+s), 1\}$ such that

$$
\begin{aligned}
&c_3(\|D_x^\alpha \widetilde{\mathbf{v}}_1^{\epsilon,\lambda}\|_E^2 + \|(\lambda \widehat{D_x^\beta \Delta \widetilde{V}_1^{\epsilon,\lambda}}, \lambda \widehat{\nabla \widetilde{V}_1^{\epsilon,\lambda}})\|_{L^2}^2)\\
&\leq \|(\widetilde{\mathbf{v}}_1^{\epsilon,\lambda}, \lambda \nabla \widetilde{V}_1^{\epsilon,\lambda})\|_s^2\\
&\leq c_4(\|D_x^\alpha \widetilde{\mathbf{v}}_1^{\epsilon,\lambda}\|_E^2 + \|(\lambda \widehat{D_x^\beta \Delta \widetilde{V}_1^{\epsilon,\lambda}}, \lambda \widehat{\nabla \widetilde{V}_1^{\epsilon,\lambda}})\|_{L^2}^2).
\end{aligned} \tag{3.95}
$$

Since $\|(\widetilde{\mathbf{v}}_1^{\epsilon,\lambda}, \lambda \nabla \widetilde{V}_1^{\epsilon,\lambda})(t=0)\|_s \leq M_0\sqrt{\varepsilon}\lambda$, by Gronwall's lemma, (3.94), together with (3.95), implies

$$
\|(\widetilde{\mathbf{v}}_1^{\epsilon,\lambda}, \lambda \nabla \widetilde{V}_1^{\epsilon,\lambda})(t)\|_s^2 \leq \varepsilon\lambda^2 c_4(\frac{M_0^2}{c_3} + C_2(T_0)T_0)e^{C_2(T_0)c_4T_0} \tag{3.96}
$$

for any $0<t\leq T_0$. Taking $\tilde{M}(T_0)=\max\{(c_4(\frac{M_0^2}{c_3}+C_2(T_0)T_0)e^{C_2(T_0)c_4T_0})^{\frac{1}{2}}$, $\max\{M_0, C_0(M_0)\}$, $C_1(T_0)\}$, we get (3.84) from (3.96). Then for this fixed choice of $\tilde{M}(T_0)$, we can take $\iota_0(T_0) = \min\{(\tilde{M}(T_0))^{-1}, (2C_s^*\tilde{M}(T_0))^{-1}\}$ by using the above three restrictions.

This completes the proof of Lemma 3.6.

The end of the proof of Theorem 2.5 As in the proof of Theorem 2.2, from Lemma 3.6 and (3.83)–(3.85) it follows that for $\sqrt{\varepsilon}\lambda \le \iota_0$ and $0 < \varepsilon, \lambda \le 1$, there exists $(\mathbf{v}_1^{\epsilon,\lambda}, V_1^{\epsilon,\lambda})$ such that $\mathbf{v}_1^{\epsilon,\lambda} \in C([0,T], H^{s-\tau}(\mathbb{T}^d)) \cap C^1([0,T], H^{s-\tau-1}(\mathbb{T}^d))$, $\nabla V_1^{\epsilon,\lambda} \in C([0,T], H^{s-\tau}(\mathbb{T}^d))$, the subsequence (still denoted by) $(\mathbf{v}_1^{\varepsilon,\lambda,p}, \nabla V_1^{\varepsilon,\lambda,p})$ converges to $(\mathbf{v}_1^{\epsilon,\lambda}, \nabla V_1^{\epsilon,\lambda})$ strongly in $L^\infty([0,T], H^{s-1}(\mathbb{T}^d))$ and $(\mathbf{v}_1^{\epsilon,\lambda}, V_1^{\epsilon,\lambda})$ satisfies (3.67)–(3.69) as well as the uniform estimates

$$\|(\mathbf{v}_1^{\epsilon,\lambda}, \lambda\nabla V_1^{\epsilon,\lambda})\|_s \le \tilde{M}(T_0)\sqrt{\varepsilon}\lambda.$$

The proof of Theorem 2.5 is complete.

References

[1] G. Ali, L. Chen, A. Jungel and Y. J. Peng, The zero-electron-mass limit in the hydrodynamic model for plasmas. Preprint, 2005.

[2] Y. Brenier, Converence of the Vlasov-Poisson system to the incompressible Euler equations, *Commun. Partial Diff. Eqs.*, **25** (2000), 737–754.

[3] G. Q. Chen and J. Glimm, Global solution to the compressible Euler equations with geometrical structure, *Comm. Math. Phys.*, **180** (1996), 153–193.

[4] S. Cordier, P. Degond, P. Markowich and C. Schmeiser, A travelling wave analysis of a hydrodynamical model of quasineutral plasma, *Asymptotic Anal.*, **11** (1995), 209–224.

[5] S. Cordier and E. Grenier, Quasineutral limit of an Euler-Poisson system arising from plasma physics, *Comm. PDE*, **23** (2000), 1099–1113.

[6] S. Cordier and E. Grenier, Quasineutral limit of two species Euler-Poisson systems, *Proceedings of the Workshop "Recent Progress in the Mathematical Theory on Vlasov-Maxwell Equations" (Paris)*, Preprint, 1997.

[7] P. Degond, Mathematical modelling of microelectronics semiconductor derices. Some current topics on nonlinear conservation laws, 77–110. AMS/IP Stud. Adv. Math. 15, Amer. Math. Soc., Providence, RI, 2000.

[8] P. Degond and P. A. Markowich, A steady-state potential flow model for semiconductors, *Ann. Mat. Pura Appl.*,**IV** (1993), 87–98.

[9] I. Gasser, L. Hsiao, P. Markowich and S. Wang, Quasineutral limit of a nonlinear drift-diffusion model for semiconductor models, *J. Math. Anal. Appl.*, **268** (2002), 184–199.

[10] I. Gasser, C. D. Levermore, P. Markowich and C. Schmeiser, The initial time layer problem and the quasineutral limit in the semiconductor drift-diffusion model, *Europ. J. Appl. Math.*, **12** (2001), 497–512.

[11] Y. Guo, Smooth irrotational flows in the large to the Euler-Poisson system in R^{3+1}, *Comm. Math. Phys.*, **195** (1998), 249–265.

[12] E. Grenier, Oscillations in quasineutral plasmas, *Commun. Partial Diff. Eqs.*, **21** (1996), 363–394.

[13] E. Grenier, Defect measures of the Vlasov-Poisson system, *Commun. Partial Diff. Eqs.*, **20** (1995), 1189–1215.

[14] E. Grenier, Pseudodifferential estimates of singular perturbations, *Commun. Pure Appl. Math.*, **50** (1997), 821–865.

[15] E. Grenier, Oscillartory perturbations of the Navier-Stokes equations, *Journal de Maths Pures et appl.(9)*, **76** (1997), 477–498.

[16] A. Jüngel and Y.-J. Peng, A hierarchy of hydrodynamic models for plasmas: quasi-neutral limits in the drift-diffusion equations, *Asympt. Anal.*, **28** (2001), 49–73.

[17] A. Jüngel and S. Wang, Convergence of nonlinear Schrödinger-Poisson systems to the compressible Euler equations, *Comm. PDE*, **28** (2003), 1005–1022.

[18] L. Hsiao, P. A. Markowich and S. Wang, The asymptotic behavior of globally smooth solutions of the multidimensional isentropic hydrodynamic model for semiconductors, *J. Diff. Eqns.*, **192** (2003), 111–133.

[19] T. Kato, Nonstationary flow of viscous and ideal fluids in R^3, *J. Funct. Anal.,* **9** (1972), 296–305.

[20] S. Klainerman and A. Majda, Singular limits of quasilinear hyperbolic systems with large parameters and the incompressible limit of compressible fluids, *Comm. Pure and Appli. Math.,* Vol. XXXIV (1981), 481–524.

[21] S. Klainerman and A. Majda, Compressible and incompressible fluids, *Comm. Pure and Appli. Math.,* Vol. XXXV (1982), 629–651.

[22] A. Majda, *Compressible Fluid Flow and Systems of Conservation laws in Several Space Variables*, Springer-Verlag, Berlin, 1984.

[23] P. A. Marcati and R. Natalini, Weak solutions to hydrodynamic model for semiconductors and relaxation to the drift-diffusion equation, *Arch. Rational Mech. Anal.*, **129** (1995), 129–145.

[24] N. Masmoudi, From Vlasov-Poisson system to the incompressible Euler system, *Comm. Partial Differential Equations*, **26** (2001), 1913–1928.

[25] N. Masmoudi, Ekman layers of rotating fluids, the case of general initial data, *Commun. Pure and Appl. Math.*, **LIII** (2000), 0432–0483.

[26] F. J. McGrath, Nonstationary plane flow of viscous and ideal fluds, *Arch. Rational Mech. Anal.*, **27** (1968), 229–348.

[27] Y. J. Peng and Y. G. Wang, Convergence of compressible Euler-Poisson equations to incompressible type Euler equations. *Asymptotic Anal.*, 41 (2005), 141–160.

[28] M. Puel, *Etudes Variationnelles et Asymptotiques des Problèmes de la Mécanique des Fluides et des Plasmas*, PhD thesis, Université Paris 6, 2000.

[29] S. Schochet, Fast singular limits of hyperbolic PDEs, *J. Diff. Eqns.*, **114** (1994), 476–512.

[30] T. Sideris, The lifespan of smooth solutions to the three-dimensional compressible Euler equations and the incompressible limit, *Indiana Univ. Math. J.*, **40** (1991), 536–550.

[31] P. A. Markowich, C. Ringhofer and C. Schmeiser, *Semiconductors Equations*, Springer-Verlag, Vienna, New York, 1990.

[32] S. Ukai, The incompressible limit and the initial layer of the compressible Euler equation, *J. Math. Kyoto Univ.*, **26** (1986), 323–331.

[33] S. Wang, Quasineutral limit of Euler-Poisson system with and without viscosity, *Commun. P. D. E.*, **29** (2004), 419–456.

[34] S. Wang and S. Jiang, The convergence of the Navier-Stokes-Poisson system to the incompressible Euler equations. *Comm. PDE*, **31** (2006), 571–591.

[35] S. Wang, Z. P. Xin and P. A. Markowich, Quasineutral limit of the Drift Diffusion models for semiconductors: The case of general sign-changing doping profile. *SIAM J. Math. Anal.*, **37** (2006), 1854–1889.

On the Relaxation-time Limits in Bipolar Hydrodynamic Models for Semiconductors

Jiang Xu*
Department of Mathematics
Nanjing University of Aeronautics and Astronautics
Nanjing 211106, China
Email: jiangxu_79@nuaa.edu.cn

Wen-An Yong
Zhou Pei-Yuan Center for Appl. Math.
Tsinghua University
Beijing 100084, China
Email: wayong@tsinghua.edu.cn

Abstract

This work is concerned with bipolar hydrodynamic models for semiconductors with short momentum relaxation-time and energy relaxation-time. Inspired by the Maxwell iteration, we construct a new approximation solution and show that the periodic initial-value problems of certain scaled hydrodynamic model have unique smooth solutions in a time interval independent of the two relaxation-times. Furthermore, as the relaxation-times both tend to zero, the smooth solutions converge to the smooth solution of the corresponding bipolar drift-diffusion model.

1 Introduction

With the increasing demand of semiconductor devices, the mathematical theory about various device models has become an active area in applied mathematics. It ranges from kinetic transport equations for charged carriers (electrons and holes) to fluid-dynamical models. The Boltzmann equation is an accurate kinetic model which describes the transport of charged carriers. However, it needs much computing power in practical

*This work is supported by NUAA's Scientific Fund for the Introduction of Qualified Personnel.

applications. Based on the moment method, one derives some simpler fluid-dynamical equations for macroscopic quantities like density, velocity and energy, which can represent a reasonable compromise between physical accuracy and the reduction of computational cost. For details, see [19].

Bipolar drift-diffusion models are the most popular fluid-dynamical equations for simulations in semiconductor devices. These models work very well in the case of low carrier densities and small electric fields. By contrast, bipolar hydrodynamic models are usually considered to describe high field phenomena or submicron devices. The main aim of this paper is to discuss the relation between the hydrodynamic models and drift-diffusion models.

Consider a typical semiconductor device (*e.g.*, P–N diodes or bipolar transistor), where the current flow is generated by electrons with charge $q_e = -1$ and holes with charge $q_i = +1$. We denote by $n_e = n_e(t,x), \mathbf{u}_e = \mathbf{u}_e(t,x)$ ($n_i, \mathbf{u}_i$, respectively) the density and velocity of electrons (holes, respectively). $\Phi = \Phi(t,x)$ represents the electrostatic potential generated by the Coulomb force from the particles. After a re-scaling of the time variable, these variables satisfy the bipolar hydrodynamic model:

$$\begin{cases} \partial_t n_a + \frac{1}{\tau}\mathrm{div}(n_a\mathbf{u}_a) = 0, \\ \partial_t(n_a\mathbf{u}_a) + \frac{1}{\tau}\mathrm{div}(n_a\mathbf{u}_a \otimes \mathbf{u}_a) + \frac{1}{\tau}\nabla P_a \\ = -\frac{q_a}{\tau} n_a \nabla\Phi - \frac{1}{\tau^2} n_a\mathbf{u}_a, \\ \partial_t(\frac{n_a|\mathbf{u}_a|^2}{2} + \frac{P_a}{\gamma-1}) + \frac{1}{\tau}\mathrm{div}((\frac{n_a|\mathbf{u}_a|^2}{2} + \frac{\gamma P_a}{\gamma-1})\mathbf{u}_a) \\ = -\frac{q_a}{\tau} n_a\mathbf{u}_a\nabla\Phi - \frac{1}{\tau\sigma}(\frac{n_a|\mathbf{u}_a|^2}{2} + \frac{P_a - n_a T_L}{\gamma-1}), \\ \Delta\Phi = n_e - n_i - b, \end{cases} \tag{1.1}$$

where $a = e, i$ and $(t,x) \in [0,+\infty) \times \mathbb{R}^d (d \geq 2)$. The dimensionless parameters $\tau, \sigma > 0$ are the momentum relaxation-time and energy relaxation-time of electrons and holes, respectively. The pressure function P_a satisfies the state equation $P_a = (\gamma - 1)n_a e_a$ (the adiabatic exponent $\gamma > 1$), in which e_a is the specific internal energy. For simplicity, we only handle the polytropic gas case. Furthermore, we may set $P_a = n_a T_a$ and $e_a = \frac{1}{\gamma-1}T_a$, where $T_a (a = e, i)$ are the temperatures of electrons and holes respectively. $T_L = T_L(x) > 0$ is the given lattice temperature of semiconductor device, and the given function $b = b(x) > 0$ stands for the density of fixed, positively charged background ions (doping profile).

With variables $(n_a, \mathbf{u}_a, T_a)(a = e, i)$, the system (1.1) for classical solutions is equivalent to

$$
\begin{cases}
\partial_t n_a + \frac{1}{\tau}\mathrm{div}(n_a \mathbf{u}_a) = 0, \\
\partial_t (n_a \mathbf{u}_a) + \frac{1}{\tau}\mathrm{div}(n_a \mathbf{u}_a \otimes \mathbf{u}_a) + \frac{1}{\tau}\nabla(n_a T_a) \\
= -\frac{q_a}{\tau} n_a \nabla \Phi - \frac{1}{\tau^2} n_a \mathbf{u}_a, \\
\partial_t T_a + \frac{1}{\tau}\mathbf{u}_a \cdot \nabla T_a + \frac{\gamma-1}{\tau} T_a \mathrm{div}\mathbf{u}_a \\
= (\gamma-1)(\frac{1}{\tau^2} - \frac{1}{2\tau\sigma})|\mathbf{u}_a|^2 - \frac{1}{\tau\sigma}(T_a - T_L), \\
\Delta \Phi = n_e - n_i - b.
\end{cases}
\tag{1.2}
$$

In what follows, we focus mainly on the system (1.2). Note that the scaling

$$t = \tau \tilde{t}$$

converts (1.2) back into the original bipolar hydrodynamic model in [19] with $\tilde{t}$ as its time variable.

The scaled-time variable t was first introduced in [17] to establish the relation between the unipolar hydrodynamic model and drift-diffusion model, via the zero-relaxation-time limit. Since [17], this kind of limit problem for the bipolar isentropic model has been investigated by many authors in the compensated compactness framework for weak entropy solutions [7, 8, 9, 10, 12, 18, 20, 22, 27], and in the Aubin-Lions [21] compactness framework for small smooth solutions [1, 13]. Recently, Y. P. Li [15] considered the relaxation limit in the bipolar isentropic hydrodynamic model by generalizing the analysis in [23] for the unipolar model (one carrier type). The main idea of [15, 23] is essentially different from the previous ideas for weak entropy solutions or small smooth solutions. However, only one small parameter τ is considered in [15].

In this paper, we consider the genuine two-parameter singular limit problems where both relaxation-times τ, σ are much smaller than 1 and $\tau = O(\sigma)$. These seemingly contain all the physically relevant cases, since Monte Carlo simulations on the bipolar Boltzmann-Poisson equations show that the momentum relaxation-time τ is much smaller than the energy relaxation-time σ [2].

To show our approach, we rewrite the momentum and temperature equations in (1.2) as ($a = e, i$)

$$
\begin{cases}
n_a \mathbf{u}_a = -q_a \tau n_a \nabla \Phi - \tau \nabla(n_a T_a) - \tau \mathrm{div}(n_a \mathbf{u}_a \otimes \mathbf{u}_a) - \tau^2 \partial_t (n_a \mathbf{u}_a), \\
T_a = T_L + \tau\sigma(\gamma - 1)\Big(\frac{1}{\tau^2} - \frac{1}{2\tau\sigma}\Big)|\mathbf{u}_a|^2 \\
\qquad -\tau\sigma\Big(\partial_t T_a + \frac{1}{\tau}\mathbf{u}_a \cdot \nabla T_a + \frac{\gamma-1}{\tau} T_a \mathrm{div}\mathbf{u}_a\Big).
\end{cases}
$$

For $\tau, \sigma \ll 1$ and $\tau = O(\sigma)$, these equations show that $\mathbf{u}_a = O(\tau)$ and $T_a = T_L + O(\tau\sigma)$ formally. With these, we iterate the momentum equation once to obtain

$$n_a\mathbf{u}_a = -q_a\tau n_a\nabla\Phi - \tau\nabla(n_aT_L) + O(\tau^2).$$

Substituting the truncation $n_a\mathbf{u}_a = -q_a\tau n_a\nabla\Phi - \tau\nabla(n_aT_L)$ into the mass equations in (1.2), we immediately obtain the bipolar drift-diffusion model:

$$\begin{cases} \partial_t n_e = \Delta(n_eT_L) - \operatorname{div}(n_e\nabla\Phi), \\ \partial_t n_i = \Delta(n_iT_L) + \operatorname{div}(n_i\nabla\Phi), \\ \Delta\Phi = n_e - n_i - b, \end{cases} \tag{1.3}$$

which is a semilinear parabolic-elliptic system, for $T_L(x) > 0$.

The main aim of this paper is to justify the above formal procedure. Let (n_e, n_i, Φ) solve the drift-diffusion model (1.3). Inspired by the above (Maxwell) iteration, we construct ($\epsilon := (\tau, \sigma)$)

$$\begin{cases} n_{e\epsilon} = n_e, \\ \mathbf{u}_{e\epsilon} = \tau\nabla\Phi - \tau\frac{\nabla(n_eT_L)}{n_e}, \\ T_{e\epsilon} = T_L + (\gamma-1)(\tau\sigma - \frac{1}{2}\tau^2)|\nabla\Phi - \frac{\nabla(n_eT_L)}{n_e}|^2 \\ \quad +\tau\sigma(\frac{\nabla(n_eT_L)}{n_e} - \nabla\Phi)\nabla T_L - (\gamma-1)\tau\sigma T_L\operatorname{div}(\nabla\Phi - \frac{\nabla(n_eT_L)}{n_e}), \\ n_{i\epsilon} = n_i, \\ \mathbf{u}_{i\epsilon} = -\tau\nabla\Phi - \tau\frac{\nabla(n_iT_L)}{n_i}, \\ T_{i\epsilon} = T_L + (\gamma-1)(\tau\sigma - \frac{1}{2}\tau^2)|\nabla\Phi + \frac{\nabla(n_iT_L)}{n_i}|^2 \\ \quad +\tau\sigma(\frac{\nabla(n_iT_L)}{n_i} + \nabla\Phi)\nabla T_L + (\gamma-1)\tau\sigma T_L\operatorname{div}(\nabla\Phi + \frac{\nabla(n_iT_L)}{n_i}), \\ \Phi_\epsilon = \Phi = \Delta^{-1}(n_e - n_i - b) \end{cases} \tag{1.4}$$

as an approximation for the solution $(n_e^\epsilon, \mathbf{u}_e^\epsilon, T_e^\epsilon; n_i^\epsilon, \mathbf{u}_i^\epsilon, T_i^\epsilon, \Phi^\epsilon)$ to the system (1.2) with initial data

$$(n_e^\epsilon, \mathbf{u}_e^\epsilon, T_e^\epsilon, n_i^\epsilon, \mathbf{u}_i^\epsilon, T_i^\epsilon)(0,x) = (n_{e\epsilon}, \mathbf{u}_{e\epsilon}, T_{e\epsilon}, n_{i\epsilon}, \mathbf{u}_{i\epsilon}, T_{i\epsilon})(0,x). \tag{1.5}$$

Obviously, these initial data are in equilibrium. Then, by using the energy method, we can prove the validity of the approximation (1.4) and establish the following main result.

Theorem 1.1. *Let $s > 1 + d/2$ be an integer. Suppose that $T_L = T_L(x), b = b(x)$ satisfy conditions*

$$b \in H^{s+1}(\mathbb{T}^d), \quad T_L \in H^{s+3}(\mathbb{T}^d), \quad \textit{and} \quad T_L(x), b(x) \ge C_0 > 0, \tag{1.6}$$

and the semilinear drift-diffusion equations (1.3) have a solution

$$(n_e, n_i) \in C([0, \mathbb{T}_*], H^{s+3}(\mathbb{T}^d)) \cap C^1([0, \mathbb{T}_*], H^{s+2}(\mathbb{T}^d))$$

with positive lower bounds. Then, for sufficiently small τ *and* σ *with* $\tau = O(\sigma)$*, there is an* ϵ*-independent positive number* $\mathbb{T}_{**} \le \mathbb{T}_*$ *such that the system (1.2) with periodic initial data (1.5) has a unique solution* $(n_e^\epsilon, \boldsymbol{u}_e^\epsilon, T_e^\epsilon, n_i^\epsilon, \boldsymbol{u}_i^\epsilon, T_i^\epsilon)$ *satisfying*

$$(n_e^\epsilon, \boldsymbol{u}_e^\epsilon, T_e^\epsilon, n_i^\epsilon, \boldsymbol{u}_i^\epsilon, T_i^\epsilon) \in C([0, \mathbb{T}_{**}], H^s(\mathbb{T}^d)).$$

Moreover, there exists an ϵ*-independent constant* $K > 0$ *such that for all* $t \in [0, \mathbb{T}_{**}]$*,*

$$\|(n_{e\epsilon} - n_e^\epsilon, \boldsymbol{u}_{e\epsilon} - \boldsymbol{u}_e^\epsilon, T_{e\epsilon} - T_e^\epsilon, n_{i\epsilon} - n_i^\epsilon, \boldsymbol{u}_{i\epsilon} - \boldsymbol{u}_i^\epsilon, T_{i\epsilon} - T_i^\epsilon)(t)\|_{H^s(\mathbb{T}^d)} \le K\tau\sigma. \tag{1.7}$$

Remark 1.2. From (1.7) and Lemma 2.1 in the next section it simply follows that

$$\sup_{t\in[0,\mathbb{T}_{**}]} \|\varPhi^\epsilon - \varPhi_\epsilon\|_{H^s(\mathbb{T}^d)} \le K'\tau\sigma, \tag{1.8}$$

where $K' > 0$ is a constant independent of ϵ.

Remark 1.3. From (1.7)–(1.8) and (1.4) we see that the exact solution $(n_e^\epsilon, \mathbf{u}_e^\epsilon, T_e^\epsilon, n_i^\epsilon, \mathbf{u}_i^\epsilon, T_i^\epsilon, \Phi^\epsilon)$ of the system (1.2) has the following asymptotic expression

$$\begin{cases} n_e^\epsilon(t,x) = n_e(t,x) + O(\tau\sigma), \\ \mathbf{u}_e^\epsilon(t,x) = \tau\nabla\varPhi - \tau\frac{\nabla(n_e T_L)}{n_e} + O(\tau\sigma), \\ T_e^\epsilon(t,x) = T_L(x) + O(\tau\sigma), \\ n_i^\epsilon(t,x) = n_i(t,x) + O(\tau\sigma), \\ \mathbf{u}_i^\epsilon(t,x) = -\tau\nabla\varPhi - \tau\frac{\nabla(n_i T_L)}{n_i} + O(\tau\sigma), \\ T_i^\epsilon(t,x) = T_L(x) + O(\tau\sigma), \\ \varPhi^\epsilon(t,x) = \varPhi(t,x) + O(\tau\sigma) \end{cases}$$

for $(t,x) \in [0, \mathbb{T}_{**}] \times \mathbb{T}^d$. Therefore, Theorem 1.1 characterizes the limiting behavior more precisely than previous results in [1, 13], where the convergence was proven but no convergence rates were given. Moreover, let us mention that no smallness conditions are required by Theorem 1.1.

Remark 1.4. This paper deals with the case where the initial data are in equilibrium. For more general periodic initial data, the initial-layers will occur and the similar results of form (1.7) may still be verified by using the matched expansion methods, see [24].

The paper is arranged as follows. In Section 2, we review the continuation principle developed in [25, 3] for singular limit problems. The approximation solution (1.4) is discussed in Section 3. Section 4 is devoted to the proof of Theorem 1.1.

Notations. The symbol C is a generic positive constant independent of $\epsilon = (\tau, \sigma)$. $|U|$ denotes the standard Euclidean norm of a vector or matrix U. $L^2 = L^2(\mathbb{T}^d)$ is the space of square integrable (vector- or matrix-valued) functions on the d-dimensional unit $\mathbb{T}^d = (0,1]^d$. $H^s(\mathbb{T}^d)$ is the usual Sobolev space on the d-dimensional unit $\mathbb{T}^d$ whose distribution derivatives of order $\leq s$ are all in $L^2(\mathbb{T}^d)$. We use notations $\|U\|_s$ and $\|U\|$ as the space norms respectively. We also label $\|(a,b,c,d)\|_s^2 = \|a\|_s^2 + \|b\|_s^2 + \|c\|_s^2 + \|d\|_s^2$, where $a,b,c,d \in H^s(\mathbb{T}^d)$. Finally, we denote by $C([0,T],\mathbf{X})$ (resp., $C^1([0,T],\mathbf{X})$) the space of continuous (resp., continuously differentiable) functions on $[0,T]$ with values in a Banach space $\mathbf{X}$.

At the end of introduction, we mention many other efforts made for the bipolar hydrodynamic model on the whole position space or spatial bounded domain, such as well-posedness of steady-state solutions, global existence of classical or entropy weak solutions and large time behavior of solutions. The interested readers may refer to [1, 4, 5, 6, 8, 11, 18, 22, 27, 28] and the literatures quoted therein.

2 Preliminaries

In this section, we first rewrite the scaled bipolar hydrodynamical model (1.1) as a symmetrizable hyperbolic system. Then we review a continuation principle for singular limit problems [25, 3]. To begin with, we recall an elementary fact from [23].

Lemma 2.1. $\nabla\Delta^{-1}$ *is a bounded linear operator on* $L^2(\mathbb{T}^d)$.

This proposition can be easily proved by using the Fourier series. It is this proposition that requires the initial data to be periodic.

Having this proposition, we see that $(1.2)(a = e, i)$ for classical solutions is equivalent to

$$\begin{cases} \partial_t n_a + \frac{1}{\tau}\mathrm{div}(n_a \mathbf{u}_a) = 0, \\ \partial_t \mathbf{u}_a + \frac{1}{\tau}(\mathbf{u}_a \cdot \nabla)\mathbf{u}_a + \frac{1}{\tau}\left(\frac{T_a}{n_a}\nabla n_a + \nabla T_a\right) \\ = -\frac{q_a}{\tau}\nabla\Delta^{-1}(n_e - n_i - b) - \frac{1}{\tau^2}\mathbf{u}_a, \\ \partial_t T_a + \frac{1}{\tau}\mathbf{u}_a \cdot \nabla T_a + \frac{\gamma-1}{\tau} T_a \mathrm{div}\mathbf{u}_a \\ = (\gamma-1)(\frac{1}{\tau^2} - \frac{1}{2\tau\sigma})|\mathbf{u}_a|^2 - \frac{1}{\tau\sigma}(T_a - T_L) \end{cases} \tag{2.1}$$

and $\Phi = \Delta^{-1}(n_e - n_i - b)$. Obviously, (2.1) is a symmetrizable hyperbolic system, since $\nabla\Delta^{-1}(n_e - n_i - b)$ is a zero-order (but non-local) term.

Next, we review the continuation principle in [25, 3] for general singular limit problems of quasi-linear symmetrizable hyperbolic systems depending (singularly) on parameters

$$U_t + \sum_{j=1}^{d} A_j(U, \epsilon)U_{x_j} = Q(U, \epsilon) \tag{2.2}$$

for $x \in \Omega = \mathbb{R}^d$ or $\mathbb{T}^d$. Here ϵ represents a parameter in a *topological* space (in this paper $\epsilon = (\tau, \sigma)$ is a vector), $A_j(U, \epsilon)$ and $Q(U, \epsilon)$ are (matrix- or vector-valued) smooth functions of $U \in G \subset \mathbb{R}^d$ for each ϵ different from a certain singular point, say $\mathbf{0}$.

For each fixed $\epsilon(\neq \mathbf{0})$, consider the initial-value problem of (2.2) with initial data $\bar{U}(x, \epsilon)$. Assume $\bar{U}(x, \epsilon) \in G_0 \subset G$ for all $x \in \Omega$ and $\bar{U}(\cdot, \epsilon) \in H^s$ with $s > 1 + d/2$. Let G_1 be a subset of the state space satisfying $G_0 \subset\subset G_1$. According to the local-in-time existence theory for the initial-value problem of symmetrizable hyperbolic systems (see Theorem 2.1 in [16]), there exists $T > 0$ such that (2.2) with initial data $\bar{U}(x, \epsilon)$ has a classical solution

$$U^\epsilon \in C([0, T], H^s(\Omega)) \quad \text{and} \quad U^\epsilon(t, x) \in G_1 \quad \text{for } (t, x) \in [0, T] \times \Omega.$$

Define

$$\mathbb{T}_\epsilon = \sup\{T > 0 : U^\epsilon \in C([0, T], H^s(\Omega)) \\ \text{and} \quad U^\epsilon(t, x) \in G_1 \quad \text{for } (t, x) \in [0, T] \times \Omega\}.$$

Obviously, $[0, \mathbb{T}_\epsilon)$ is the maximal time interval for the existence of H^s-solutions with values in G_1. Note that $\mathbb{T}_\epsilon = \mathbb{T}_\epsilon(G_1)$ depends on G_1 and may tend to zero as ϵ approaches to the singular point $\mathbf{0}$.

In order to show that $\underline{\lim}_{\epsilon \to 0} \mathbb{T}_\epsilon > 0$, we make the following assumption.

Convergence assumption: there exist $\mathbb{T}_* > 0$ and

$$U_\epsilon(t,x) \in L^\infty([0,\mathbb{T}_*], H^s(\Omega))$$

for each $\epsilon(\neq \mathbf{0})$, satisfying

$$\bigcup_{t,x,\epsilon} \{U_\epsilon(t,x)\} \subset\subset G \quad \text{and} \quad \sup_{t\in[0,\mathbb{T}_*]} \|U_\epsilon(\cdot,t)\|_s < \infty,$$

such that for $t \in [0, \min\{\mathbb{T}_*, \mathbb{T}_\epsilon\})$,

$$\sup_{t,x} |U^\epsilon(t,x) - U_\epsilon(t,x)| = o(1),$$

$$\sup_t \|U^\epsilon(t,\cdot) - U_\epsilon(t,\cdot)\|_s = O(1),$$

as ϵ goes to the singular point $\mathbf{0}$.

With such an assumption, we have the following fact established in [25, 3].

Lemma 2.2. *Suppose $\bar{U}(x,\epsilon) \in G_0 \subset G$ for all $x \in \Omega$ and $\epsilon(\neq \mathbf{0})$, $\bar{U}(\cdot,\epsilon) \in H^s$ with $s > 1+d/2$ an integer, and the convergence assumption holds. Then, for each G_1 satisfying*

$$G_0 \bigcup_{t,x,\epsilon} \{U_\epsilon(t,x)\} \subset\subset G_1 \subset G,$$

there is a neighborhood of the singular point such that

$$\mathbb{T}_\epsilon(G_1) > \mathbb{T}_*$$

for all ϵ in the neighborhood of the singular point $\mathbf{0}$.

Thanks to Lemma 2.2, our prime task is reduced to find an approximation solution $U_\epsilon(t,x)$ such that the above convergence-assumption holds.

3 Construction of approximation solutions

Let (n_e, n_i) solve the semilinear and nonlocal parabolic equations (1.3):

$$\begin{cases} \partial_t n_e = \Delta(n_e T_L) - \mathrm{div}(n_e \nabla \Delta^{-1}(n_e - n_i - b)), \\ \partial_t n_i = \Delta(n_i T_L) + \mathrm{div}(n_i \nabla \Delta^{-1}(n_e - n_i - b)). \end{cases} \tag{3.1}$$

Inspired by the Maxwell iteration described in Introduction, we construct a formal approximation solution as in (1.4):

$$\begin{cases} n_{e\epsilon} = n_e, \\ \mathbf{u}_{e\epsilon} = \tau\nabla\Phi - \tau\frac{\nabla(n_eT_L)}{n_e}, \\ T_{e\epsilon} = T_L + (\gamma-1)(\tau\sigma - \frac{1}{2}\tau^2)|\nabla\Phi - \frac{\nabla(n_eT_L)}{n_e}|^2 \\ \qquad +\tau\sigma(\frac{\nabla(n_eT_L)}{n_e} - \nabla\Phi)\nabla T_L - (\gamma-1)\tau\sigma T_L\mathrm{div}(\nabla\Phi - \frac{\nabla(n_eT_L)}{n_e}), \\ n_{i\epsilon} = n_i, \\ \mathbf{u}_{i\epsilon} = -\tau\nabla\Phi - \tau\frac{\nabla(n_iT_L)}{n_i}, \\ T_{i\epsilon} = T_L + (\gamma-1)(\tau\sigma - \frac{1}{2}\tau^2)|\nabla\Phi + \frac{\nabla(n_iT_L)}{n_i}|^2 \\ \qquad +\tau\sigma(\frac{\nabla(n_iT_L)}{n_i} + \nabla\Phi)\nabla T_L + (\gamma-1)\tau\sigma T_L\mathrm{div}(\nabla\Phi + \frac{\nabla(n_iT_L)}{n_i}), \\ \Phi_\epsilon = \Phi = \Delta^{-1}(n_e - n_i - b). \end{cases}$$

It is easy to show that this approximation solution satisfies the following equations

$$\begin{cases} \partial_t n_{e\epsilon} + \frac{1}{\tau}\mathrm{div}(n_{e\epsilon}\mathbf{u}_{e\epsilon}) = 0, \\ \partial_t\mathbf{u}_{e\epsilon} + \frac{1}{\tau}(\mathbf{u}_{e\epsilon}\cdot\nabla)\mathbf{u}_{e\epsilon} + \frac{1}{\tau}\frac{\nabla(n_{e\epsilon}T_{e\epsilon})}{n_{e\epsilon}} \\ = \frac{1}{\tau}\nabla\Phi_\epsilon - \frac{1}{\tau^2}\mathbf{u}_{e\epsilon} + \tau\mathcal{R}_{e1} + \sigma\mathcal{R}_{e2}, \\ \partial_t T_{e\epsilon} + \frac{1}{\tau}\mathbf{u}_{e\epsilon}\cdot\nabla T_{e\epsilon} + \frac{\gamma-1}{\tau}T_{e\epsilon}\mathrm{div}\mathbf{u}_{e\epsilon} \\ = (\gamma-1)(\frac{1}{\tau^2} - \frac{1}{2\tau\sigma})|\mathbf{u}_{e\epsilon}|^2 - \frac{1}{\tau\sigma}(T_{e\epsilon} - T_L) + \tau\sigma\mathcal{R}_{e3}, \\ \partial_t n_{i\epsilon} + \frac{1}{\tau}\mathrm{div}(n_{i\epsilon}\mathbf{u}_{i\epsilon}) = 0, \\ \partial_t\mathbf{u}_{e\epsilon} + \frac{1}{\tau}(\mathbf{u}_{i\epsilon}\cdot\nabla)\mathbf{u}_{i\epsilon} + \frac{1}{\tau}\frac{\nabla(n_{i\epsilon}T_{i\epsilon})}{n_{i\epsilon}} \\ = -\frac{1}{\tau}\nabla\Phi_\epsilon - \frac{1}{\tau^2}\mathbf{u}_{i\epsilon} + \tau\mathcal{R}_{i1} + \sigma\mathcal{R}_{i2}, \\ \partial_t T_{i\epsilon} + \frac{1}{\tau}\mathbf{u}_{i\epsilon}\cdot\nabla T_{i\epsilon} + \frac{\gamma-1}{\tau}T_{i\epsilon}\mathrm{div}\mathbf{u}_{i\epsilon} \\ = (\gamma-1)(\frac{1}{\tau^2} - \frac{1}{2\tau\sigma})|\mathbf{u}_{i\epsilon}|^2 - \frac{1}{\tau\sigma}(T_{i\epsilon} - T_L) + \tau\sigma\mathcal{R}_{i3}, \\ \Phi_\epsilon = \Delta^{-1}(n_{e\epsilon} - n_{i\epsilon} - b(x)). \end{cases} \tag{3.2}$$

Here $(a = i, e)$

$$\begin{aligned}
\mathcal{R}_{a1} &= \frac{\partial_t \mathbf{u}_{a\epsilon} + (\mathbf{u}_{a\epsilon}\cdot\nabla)\mathbf{u}_{a\epsilon}/\tau}{\tau} \\
&= \partial_t\Big(- q_a\nabla\Delta^{-1}(n_{e\epsilon} - n_{i\epsilon} - b) - \frac{\nabla(n_aT_L)}{n_a}\Big) \\
&\quad +\Big(- q_a\nabla\Delta^{-1}(n_{e\epsilon} - n_{i\epsilon} - b) - \frac{\nabla(n_aT_L)}{n_a}\Big) \\
&\quad \cdot\nabla\Big(- q_a\nabla\Delta^{-1}(n_{e\epsilon} - n_{i\epsilon} - b) - \frac{\nabla(n_aT_L)}{n_a}\Big), \\
\mathcal{R}_{a2} &= \frac{\nabla(n_{a\epsilon}(T_{a\epsilon} - T_L))}{\tau\sigma n_{a\epsilon}} \\
&= \nabla\Big\{n_a\Big[(\gamma-1)\Big(1-\frac{\tau}{2\sigma}\Big)\Big| - q_a\nabla\Delta^{-1}(n_{e\epsilon} - n_{i\epsilon} - b) - \frac{\nabla(nT_L)}{n}\Big|^2 \\
&\quad +\Big(\frac{\nabla(nT_L)}{n} + q_a\nabla\Delta^{-1}(n_{e\epsilon} - n_{i\epsilon} - b)\Big)\nabla T_L \\
&\quad -(\gamma-1)T_L\mathrm{div}\Big(- q_a\nabla\Delta^{-1}(n_{e\epsilon} - n_{i\epsilon} - b) - \frac{\nabla(n_aT_L)}{n_a}\Big)\Big]\Big\}\Big/ n_a, \\
\mathcal{R}_{a3} &= \frac{1}{\tau\sigma}\Big[\partial_t T_{a\epsilon} + \frac{1}{\tau}\mathbf{u}_{a\epsilon}\cdot\nabla T_{a\epsilon} + \frac{\gamma-1}{\tau}T_{a\epsilon}\mathrm{div}\mathbf{u}_{a\epsilon} \\
&\quad -(\gamma-1)\Big(\frac{1}{\tau^2} - \frac{1}{2\tau\sigma}\Big)|\mathbf{u}_{a\epsilon}|^2 + \frac{1}{\tau\sigma}\Big(T_{a\epsilon} - T_L\Big)\Big] \\
&= \partial_t\Big\{(\gamma-1)\Big(1-\frac{\tau}{2\sigma}\Big)\Big|q_a\nabla\Delta^{-1}(n_e - n_i - b) + \frac{\nabla(n_aT_L)}{n_a}\Big|^2 \\
&\quad +\Big(\frac{\nabla(n_aT_L)}{n_a} + q_a\nabla\Delta^{-1}(n_e - n_i - b)\Big)\nabla T_L \\
&\quad +(\gamma-1)T_L\mathrm{div}\Big(q_a\nabla\Delta^{-1}(n_e - n_i - b) + \frac{\nabla(n_aT_L)}{n_a}\Big)\Big\} \\
&\quad -\Big(q_a\nabla\Delta^{-1}(n_e - n_i - b) + \frac{\nabla(n_aT_L)}{n_a}\Big) \\
&\quad \cdot\nabla\Big[\Big(\frac{2\sigma-\tau}{2\sigma}\Big)(\gamma-1)\Big|q_a\nabla\Delta^{-1}(n_e - n_i - b) + \frac{\nabla(n_aT_L)}{n_a}\Big|^2 \\
&\quad +\Big(\frac{\nabla(n_aT_L)}{n_a} + q_a\nabla\Delta^{-1}(n_e - n_i - b)\Big)\nabla T_L \\
&\quad +(\gamma-1)T_L\mathrm{div}\Big(q_a\nabla\Delta^{-1}(n_e - n_i - b) + \frac{\nabla(n_aT_L)}{n_a}\Big)\Big] \\
&\quad -(\gamma-1)\Big[(\gamma-1)\Big(1-\frac{\tau}{2\sigma}\Big)\Big|q_a\nabla\Delta^{-1}(n_e - n_i - b) + \frac{\nabla(n_aT_L)}{n_a}\Big|^2 \\
&\quad +\Big(\frac{\nabla(n_aT_L)}{n_a} + q_a\nabla\Delta^{-1}(n_e - n_i - b)\Big)\nabla T_L \\
&\quad +(\gamma-1)T_L\mathrm{div}\Big(q_a\nabla\Delta^{-1}(n_e - n_i - b) + \frac{\nabla(n_aT_L)}{n_a}\Big)\Big] \\
&\quad \times\mathrm{div}\Big(q_a\nabla\Delta^{-1}(n_e - n_i - b) + \frac{\nabla(nT_L)}{n}\Big).
\end{aligned}$$

Observe that the residues $\mathcal{R}_{a1}, \mathcal{R}_{a2}$ and $\mathcal{R}_{a3} (a = i, e)$ are bounded with respect to $\epsilon = (\tau, \sigma)$ under certain regularity assumptions on T_L, b, n_e and n_i, while (n_e, n_i) solves the semilinear and non-local parabolic equation (3.1). This observation is crucial for subsequent analysis. Moreover, the regularity of the approximation solution also depends on the four quantities. To clarify these, we formulate the following lemma.

Lemma 3.1. *Let $s > d/2 + 1$ be an integer. Assume $T_L(x)$ and $b(x)$ satisfy (1.6). If $(n_e, n_i) \in C([0, \mathbb{T}_*], H^{s+3}) \cap C^1([0, \mathbb{T}_*], H^{s+2})$ has positive lower bounds, then $\boldsymbol{u}_{e\epsilon}, T_{e\epsilon}, \boldsymbol{u}_{i\epsilon}, T_{i\epsilon} \in C([0, \mathbb{T}_*], H^{s+1})$ and $\mathcal{R}_{e1}, \mathcal{R}_{e2}, \mathcal{R}_{e3}, \mathcal{R}_{i1}, \mathcal{R}_{i2}, \mathcal{R}_{i3} \in C([0, \mathbb{T}_*], H^s)$ for $\tau, \sigma \ll 1$ and $\tau = O(\sigma)$.*

The proof of this lemma needs some calculus inequalities in Sobolev spaces, whose proofs can be found in [26]:

Lemma 3.2. *(Moser-type calculus inequalities)*

(I). *Let $A, V \in H^s$ with $s \geq [d/2] + 1$. Then, for any multi-index α with $|\alpha| \leq s$, it holds that*

$$\|\partial^\alpha (AV)\| \leq C_s \|A\|_s \|V\|_s.$$

(II). *For integer $s \geq [d/2] + 2$ and multi-index α with $|\alpha| \leq s$,*

$$\|\partial^\alpha (AV) - A\partial^\alpha V\| \leq C_s \|\partial A\|_{s-1} \|V\|_{s-1}.$$

(III). *Let $A = A(x, V)$ be a smooth function satisfying $A(x, 0) \equiv 0$ and $V \in H^s$ with $s \geq [d/2] + 1$. For multi-index α with $|\alpha| \leq s$, it holds that*

$$\|\partial^\alpha A(x, V)\| \leq C_s |A|_{C^{s+1}} (1 + \|V\|_s^{s-1}) \|V\|_s.$$

Here $C_s > 0$ is a generic constant depending only on s and d.

4 Proof of the main result

In this section, we prove Theorem 1.1.

Since (n_e, n_i) has positive lower bounds, there are two positive constants a and b such that $n_{e\epsilon}(0, x), n_{i\epsilon}(0, x), T_{e\epsilon}(0, x), T_{i\epsilon}(0, x) \in (2a, b)$, $|\mathbf{u}_{i\epsilon}(0, x)| \leq b$ and $|\mathbf{u}_{e\epsilon}(0, x)| \leq b$ for all x. Denote by $[0, \mathbb{T}_\epsilon)$ the maximal time interval where the system (2.1) with initial data (1.5) has a unique H^s-solution $(n_a^\epsilon, \mathbf{u}_a^\epsilon, T_a^\epsilon)(a = i, e)$ with values in $(a, 2b) \times (-2b, 2b)^d \times (a, 2b) \equiv G_1$. Thanks to Lemma 2.2, it suffices to prove the error estimate in (1.7) for $t \in [0, \min\{\mathbb{T}_{**}, \mathbb{T}_\epsilon\})$ with $\mathbb{T}_{**} \leq \mathbb{T}_*$ independent of ϵ and to be determined.

To this end, we set

$$\begin{pmatrix} N_e \\ U_e \\ \Theta_e \\ N_i \\ U_i \\ \Theta_i \end{pmatrix} := \begin{pmatrix} n_{e\epsilon} - n_e^\epsilon \\ \mathbf{u}_{e\epsilon} - \mathbf{u}_e^\epsilon \\ T_{e\epsilon} - T_e^\epsilon \\ n_{i\epsilon} - n_i^\epsilon \\ \mathbf{u}_{i\epsilon} - \mathbf{u}_i^\epsilon \\ T_{i\epsilon} - T_i^\epsilon \end{pmatrix}.$$

From the equations in (2.1) and (3.2), it follows that the error $(N_a, U_a, \Theta_a)(a = e, i)$ satisfies

$$\begin{cases} \partial_t N_a + \frac{1}{\tau}(\mathbf{u}_a^\epsilon \cdot \nabla N_a + n_a^\epsilon \mathrm{div} U_a) \\ = -\frac{1}{\tau}(U_a \cdot \nabla n_{a\epsilon} + N_a \mathrm{div}\mathbf{u}_{a\epsilon}), \\ \partial_t U_a + \frac{1}{\tau}(\mathbf{u}_a^\epsilon \cdot \nabla) U_a + \frac{1}{\tau}(\nabla \Theta_a + \frac{T_a^\epsilon}{n_a^\epsilon} \nabla N_a) \\ = -\frac{1}{\tau}(U_a \cdot \nabla)\mathbf{u}_{a\epsilon} - \frac{1}{\tau}\left(\frac{T_{a\epsilon}}{n_{a\epsilon}} - \frac{T_a^\epsilon}{n_a^\epsilon}\right)\nabla n_\epsilon \\ -\frac{q_a}{\tau}\nabla\Delta^{-1}(N_e - N_i) - \frac{1}{\tau^2}U_a + \tau\mathcal{R}_{a1} + \sigma\mathcal{R}_{a2}, \\ \partial_t \Theta_a + \frac{1}{\tau}\mathbf{u}_a^\epsilon \cdot \nabla\Theta_a + \frac{\gamma-1}{\tau}T_a^\epsilon \mathrm{div} U_a \\ = -\frac{1}{\tau}U_a \cdot \nabla T_{a\epsilon} - \frac{(\gamma-1)}{\tau}\Theta \mathrm{div}\mathbf{u}_\epsilon \\ +(\gamma-1)(\frac{1}{\tau^2} - \frac{1}{2\tau\sigma})(|\mathbf{u}_{a\epsilon}|^2 - |\mathbf{u}_a^\epsilon|^2) - \frac{1}{\tau\sigma}\Theta_a + \tau\sigma\mathcal{R}_{a3}. \end{cases} \tag{4.1}$$

Then we differentiate (4.1) with ∂_x^α for a multi-index α satisfying $|\alpha| \le s$ to get

$$\begin{cases} \partial_t\partial_x^\alpha N_a + \frac{1}{\tau}(\mathbf{u}_a^\epsilon \cdot \nabla\partial_x^\alpha N_a + n_a^\epsilon \mathrm{div}\partial_x^\alpha U_a) = \mathcal{F}_a^1, \\ \partial_t\partial_x^\alpha U_a + \frac{1}{\tau}(\mathbf{u}_a^\epsilon \cdot \nabla)\partial_x^\alpha U_a + \frac{1}{\tau}(\nabla\partial_x^\alpha\Theta_a + \frac{T_a^\epsilon}{n_a^\epsilon}\nabla\partial_x^\alpha N_a) \\ = -\frac{1}{\tau^2}\partial_x^\alpha U_a + \mathcal{F}_a^2, \\ \partial_t\partial_x^\alpha\Theta_a + \frac{1}{\tau}\mathbf{u}_a^\epsilon \cdot \nabla\partial_x^\alpha\Theta_a + \frac{\gamma-1}{\tau}T_a^\epsilon \mathrm{div}\partial_x^\alpha U_a \\ = -\frac{1}{\tau\sigma}\partial_x^\alpha\Theta_a + \mathcal{F}_a^3. \end{cases} \tag{4.2}$$

Here $a = e, i$ and

$$\begin{aligned} \mathcal{F}_a^1 =& -\frac{1}{\tau}\partial_x^\alpha(U_a \cdot \nabla n_{a\epsilon} + N_a \mathrm{div}\mathbf{u}_{a\epsilon}) \\ &- \frac{1}{\tau}\Big(\partial_x^\alpha(\mathbf{u}_a^\epsilon \cdot \nabla N_a + n_a^\epsilon \mathrm{div} U_a) - (\mathbf{u}_a^\epsilon \cdot \nabla\partial_x^\alpha N_a + n_a^\epsilon \mathrm{div}\partial_x^\alpha U_a)\Big) \\ :=& f_a^{11} + f_a^{12}, \end{aligned}$$

$$
\begin{aligned}
\mathcal{F}_a^2 =& \frac{1}{\tau}\partial_x^\alpha((-q_a)\nabla\Delta^{-1}(N_e - N_i) + \tau^2\mathcal{R}_{a1} + \tau\sigma\mathcal{R}_{a2}) \\
&- \frac{1}{\tau}\partial_x^\alpha\Big((U_a\cdot\nabla)\mathbf{u}_{a\epsilon} + \Big(\frac{T_{a\epsilon}}{n_{a\epsilon}} - \frac{T^{a\epsilon}}{n^{a\epsilon}}\Big)\nabla n_{a\epsilon}\Big) \\
&- \frac{1}{\tau}\Big\{\partial_x^\alpha\Big((\mathbf{u}_a^\epsilon\cdot\nabla)U_a + \frac{T_a^\epsilon}{n_a^\epsilon}\nabla N_a\Big) - \Big((\mathbf{u}_a^\epsilon\cdot\nabla)\partial_x^\alpha U_a + \frac{T_a^\epsilon}{n_a^\epsilon}\nabla\partial_x^\alpha N_a\Big)\Big\} \\
:=& f_a^{21} + f_a^{22} + f_a^{23},
\end{aligned}
$$

$$
\begin{aligned}
\mathcal{F}_a^3 =& \tau\sigma\partial_x^\alpha\mathcal{R}_{a3} - \frac{1}{\tau}\partial_x^\alpha\Big[U_a\cdot\nabla T_{a\epsilon} - (\gamma-1)\Theta_a \mathrm{div}\mathbf{u}_{a\epsilon} \\
&+ (\gamma-1)\Big(\frac{1}{\tau} - \frac{1}{2\sigma}\Big)(|\mathbf{u}_{a\epsilon}|^2 - |\mathbf{u}_a^\epsilon|^2)\Big] \\
&- \frac{1}{\tau}\Big\{\partial_x^\alpha\Big(\mathbf{u}_a^\epsilon\cdot\nabla\Theta_a + (\gamma-1)T_a^\epsilon \mathrm{div} U_a\Big) \\
&- \Big(\mathbf{u}_a^\epsilon\cdot\nabla\partial_x^\alpha\Theta_a + (\gamma-1)T_a^\epsilon \mathrm{div}\partial_x^\alpha U_a\Big)\Big\} \\
:=& f_a^{31} + f_a^{32} + f_a^{33}.
\end{aligned}
$$

For the sake of clarity, we divide the following arguments into lemmas.

Lemma 4.1. *Under the assumptions of Theorem 1.1, we have*

$$
\begin{aligned}
&\frac{d}{dt}\int\Big\{\Big(\frac{T_e^\epsilon}{n_e^\epsilon}\Big)^2|\partial_x^\alpha N_e|^2 + \Big(\frac{T_i^\epsilon}{n_i^\epsilon}\Big)^2|\partial_x^\alpha N_i|^2 + T_e^\epsilon|\partial_x^\alpha U_e|^2 + T_i^\epsilon|\partial_x^\alpha U_i|^2 \\
&+ (\gamma-1)\Big(|\partial_x^\alpha\Theta_e|^2 + |\partial_x^\alpha\Theta_i|^2\Big)\Big\}dx \\
&+ \frac{1}{\tau^2}\|(\partial_x^\alpha U_e, \partial_x^\alpha U_i)\|^2 + \frac{1}{\tau\sigma}\|(\partial_x^\alpha\Theta_e, \partial_x^\alpha\Theta_i)\|^2 \\
\le& \frac{C}{\tau}\int\Big[\mathcal{J}_e(|\partial_x^\alpha N_e|^2 + |\partial_x^\alpha U_e|^2 + |\partial_x^\alpha\Theta_e|^2) \qquad (4.3)\\
&\quad + \mathcal{J}_i(|\partial_x^\alpha N_i|^2 + |\partial_x^\alpha U_i|^2 + |\partial_x^\alpha\Theta_i|^2)\Big]dx \\
&+ C_{e1}\|\mathcal{F}_e^1\|\|\partial_x^\alpha N_e\| + C_{e2}\|\mathcal{F}_e^2\|\|\partial_x^\alpha U_e\| + C_{e3}\|\mathcal{F}_e^3\|\|\partial_x^\alpha\Theta_e\| \\
&+ C_{i1}\|\mathcal{F}_i^1\|\|\partial_x^\alpha N_i\| + C_{i2}\|\mathcal{F}_i^2\|\|\partial_x^\alpha U_i\| + C_{i3}\|\mathcal{F}_i^3\|\|\partial_x^\alpha\Theta_i\|,
\end{aligned}
$$

where $C, C_{e1}, C_{e2}, C_{e3}, C_{i1}C_{i2}$ and C_{i3} are all generic constants depend-

ing only on the range $[a, 2b]$ *of* $n_e^\epsilon, n_i^\epsilon, T_e^\epsilon, T_i^\epsilon$ *but independent of* ϵ, *and*

$$\mathcal{J}_a = |\mathrm{div}\boldsymbol{u}_a^\epsilon| + |\boldsymbol{u}_a^\epsilon \cdot \nabla n_a^\epsilon| + |\boldsymbol{u}_a^\epsilon \cdot \nabla T_a^\epsilon| + \frac{1}{\tau}|\boldsymbol{u}_a^\epsilon|^2 +$$

$$\frac{1}{\sigma}|T_a^\epsilon - T_L| + \tau|\nabla n_a^\epsilon|^2 + \tau|\nabla T_a^\epsilon|^2 \qquad (a = e, i).$$

Proof. By multiplying (4.2) by $\left(\frac{T_a^\epsilon}{n_a^\epsilon}\right)^2 \partial_x^\alpha N_a$, $T_a^\epsilon \partial_x^\alpha U_a$, $(\gamma - 1)\partial_x^\alpha \Theta_a (a = e, i)$, respectively, and integrating them with respect to x over $\mathbb{T}^d$, after tedious but straight calculations we can obtain (4.3). See also [14] for similar details. □

For the right-hand side of (4.3), we have the following lemma.

Lemma 4.2. *Set*

$$D = D(t) = \frac{\|(N_e, U_e, \Theta_e/\sigma, N_i, U_i, \Theta_i/\sigma)(\cdot, t)\|_s}{\tau}.$$

For $\tau, \sigma \ll 1$ *and* $\tau = O(\sigma)$, *the following estimates hold:*

$$\mathcal{J}_e \leq C\tau(1 + D^2), \tag{4.4}$$

$$\mathcal{J}_i \leq C\tau(1 + D^2), \tag{4.5}$$

$$C_{e1}\|\mathcal{F}_e^1\|\|\partial_x^\alpha N_e\| \leq \kappa_e \frac{\|U_e\|_s^2}{\tau^2} + C(1 + D^2)\|N_e\|_s^2, \tag{4.6}$$

$$C_{i1}\|\mathcal{F}_i^1\|\|\partial_x^\alpha N_i\| \leq \kappa_i \frac{\|U_i\|_s^2}{\tau^2} + C(1 + D^2)\|N_i\|_s^2, \tag{4.7}$$

$$C_{e2}\|\mathcal{F}_e^2\|\|\partial_x^\alpha U_e\| \leq \frac{3\|\partial_x^\alpha U_e\|^2}{4\tau^2} + C\tau^2\sigma^2 + C\|N_i\|_s^2$$

$$+ C(1 + D)\|U_e\|_s^2 + C(1 + D^{2s})(\|N_e\|_s^2 + \|\Theta_e\|_s^2), \tag{4.8}$$

$$C_{i2}\|\mathcal{F}_i^2\|\|\partial_x^\alpha U_i\| \leq \frac{3\|\partial_x^\alpha U_i\|^2}{4\tau^2} + C\tau^2\sigma^2 + C\|N_e\|_s^2$$

$$+ C(1 + D)\|U_i\|_s^2 + C(1 + D^{2s})(\|N_i\|_s^2 + \|\Theta_i\|_s^2), \tag{4.9}$$

$$C_{e3}\|\mathcal{F}_e^3\|\|\partial_x^\alpha \Theta_e\| \leq \frac{\|\partial_x^\alpha \Theta_e\|^2}{4\tau\sigma} + \kappa_e \frac{\|U_e\|_s^2}{\tau^2}$$

$$+ C\tau^3\sigma^3 + C(1 + D^2)\|\Theta_e\|_s^2, \tag{4.10}$$

$$C_{i3}\|\mathcal{F}_i^3\|\|\partial_x^\alpha\Theta_i\| \le \frac{\|\partial_x^\alpha\Theta_i\|^2}{4\tau\sigma} + \kappa_i\frac{\|U_i\|_s^2}{\tau^2}$$

$$+C\tau^3\sigma^3 + C(1+D^2)\|\Theta_i\|_s^2, \qquad (4.11)$$

where $C > 0$ is a generic constant (independent of ϵ) depending only on the range $[a, 2b]$ of $n_e^\epsilon, n_i^\epsilon, T_e^\epsilon, T_i^\epsilon$ and κ_e, κ_i are two positive constants (independent of ϵ) to be determined below, see (4.19).

Proof. Note that

$$\mathbf{u}_{e\epsilon} = -\tau\frac{\nabla(n_eT_L)}{n_e} + \tau\nabla\Delta^{-1}(n_e - n_i - b).$$

Thus, for $s > 1 + d/2$, we use the well-known embedding inequality in Sobolev spaces to get

$$|\mathrm{div}\mathbf{u}_e^\epsilon| \le |\mathrm{div}U_e| + |\mathrm{div}\mathbf{u}_{e\epsilon}| \le C\|U_e\|_s + C\tau \le C\tau(1+D),$$

$$|\mathbf{u}_e^\epsilon\cdot\nabla n_e^\epsilon| \le (|\nabla N_e| + |\nabla n_{e\epsilon}|)|\mathbf{u}_e^\epsilon| \le C\tau(1+\tau D)(1+D),$$

$$|\mathbf{u}_e^\epsilon\cdot\nabla T_e^\epsilon| \le (|\nabla\Theta_e| + |\nabla T_{e\epsilon}|)|\mathbf{u}_e^\epsilon| \le C\tau(1+\tau\sigma D)(1+D),$$

$$\frac{1}{\tau}|\mathbf{u}_e^\epsilon|^2 \le \frac{2}{\tau}(|U_e|^2 + |\mathbf{u}_{e\epsilon}|^2) \le \frac{C}{\tau}(\tau^2D^2 + \tau^2) = C\tau(1+D^2),$$

$$\frac{1}{\sigma}|T_e^\epsilon - T_L| \le \frac{1}{\sigma}(|\Theta_e| + |T_{e\epsilon} - T_L|) \le \frac{C}{\sigma}(\tau\sigma D + \tau\sigma) = C\tau(1+D),$$

$$\tau|\nabla n_e^\epsilon|^2 \le 2\tau(|\nabla N_e|^2 + |\nabla n_{e\epsilon}|^2) \le C\tau(1+\tau^2D^2),$$

and

$$\tau|\nabla T_e^\epsilon|^2 \le 2\tau(|\nabla\Theta_e|^2 + |\nabla T_{e\epsilon}|^2) \le C\tau(1+\tau^2\sigma^2D^2).$$

Putting all above inequalities together gives the estimate (4.4) for $\mathcal{J}_e$ defined in Lemma 4.1. Through a similar process, we can deduce the estimate (4.5).

Next we turn to estimate $C_{e1}\|\mathcal{F}_e^1\|\|\partial_x^\alpha N_e\|$, where $\mathcal{F}_e^1 = f_e^{11} + f_e^{12}$. For f_e^{11}, we use Lemma 3.2 and the boundedness of $\|(n_{e\epsilon}, \mathbf{u}_{e\epsilon}, T_{e\epsilon})\|_{s+1}$ indicated in Lemma 3.1 to obtain

$$\begin{aligned}\tau\|f_e^{11}\| &= \|\partial_x^\alpha(U_e\cdot\nabla n_{e\epsilon} + N_e\mathrm{div}\mathbf{u}_{e\epsilon})\| \\ &\le C(\|\nabla n_{e\epsilon}\|_s\|U_e\|_s + \|\mathrm{div}\mathbf{u}_{e\epsilon}\|_s\|N_e\|_s) \\ &\le C(\|U_e\|_s + \tau\|N_e\|_s)\end{aligned}$$

and the second term f_e^{12} can be estimated as

$$\begin{aligned}\tau\|f_e^{12}\| &= \|\partial_x^\alpha(\mathbf{u}_e^\epsilon\cdot\nabla N_e)-\mathbf{u}_e^\epsilon\cdot\nabla\partial_x^\alpha N_e+\partial_x^\alpha(n_e^\epsilon\mathrm{div}U_e)-n_e^\epsilon\mathrm{div}\partial_x^\alpha U_e\|\\ &\le C\|\partial\mathbf{u}_e^\epsilon\|_{s-1}\|\nabla N_e\|_{s-1}+C\|\partial n_e^\epsilon\|_{s-1}\|\mathrm{div}U_e\|_{s-1}\\ &\le C[(\|U_e\|_s+\|\mathbf{u}_{e\epsilon}\|_s)\|N_e\|_s+(\|N_e\|_s+\|n_{e\epsilon}\|_s)\|U_e\|_s]\\ &\le C[\tau(1+D)\|N_e\|_s+(1+\tau D)\|U_e\|_s].\end{aligned}$$

Thus, we have

$$C_{e1}\|\mathcal{F}_e^1\|\|\partial_x^\alpha N_e\|\le\kappa_e\frac{\|U_e\|_s^2}{\tau^2}+C(1+D^2)\|N_e\|_s^2,$$

where $\kappa_e>0$ is a constant to be determined in (4.19). This is just the inequality (4.6). In a similar way, we also can arrive at (4.7) and $\kappa_i>0$ is a constant to be determined in (4.19), too.

For $C_{e2}\|\mathcal{F}_e^2\|\|\partial_x^\alpha U_e\|$, we recall $\mathcal{F}_e^2=f_e^{21}+f_e^{22}+f_e^{23}$ and estimate them as follows:

$$\begin{aligned}&C_{e2}\|f_e^{21}\|\|\partial_x^\alpha U_e\|\\ &=\frac{C_{e2}}{\tau}\|\partial_x^\alpha(\nabla\Delta^{-1}(N_e-N_i)+\tau^2\mathcal{R}_{e1}+\tau\sigma\mathcal{R}_{e2})\|\|\partial_x^\alpha U_e\|\\ &\le C\frac{\|\partial_x^\alpha U_e\|}{\tau}(\|\partial_x^\alpha N_e\|+\|\partial_x^\alpha N_i\|+\tau^2\|\partial_x^\alpha\mathcal{R}_{e1}\|+\tau\sigma\|\partial_x^\alpha\mathcal{R}_{e2}\|)\\ &\le C\frac{\|\partial_x^\alpha U_e\|}{\tau}(\tau^2+\tau\sigma+\|N_e\|_s+\|N_i\|_s)\\ &\le\frac{\|\partial_x^\alpha U_e\|^2}{4\tau^2}+C(\tau^4+\tau^2\sigma^2+\|N_e\|_s^2+\|N_i\|_s^2),\end{aligned}\tag{4.12}$$

$$\begin{aligned}&C_{e2}\|f_e^{22}\|\|\partial_x^\alpha U_e\|\\ &=\frac{C_{e2}}{\tau}\left\|\partial_x^\alpha\Big((U_e\cdot\nabla)\mathbf{u}_{e\epsilon}+\Big(\frac{T_{e\epsilon}}{n_{e\epsilon}}-\frac{T_e^\epsilon}{n_e^\epsilon}\Big)\nabla n_{e\epsilon}\Big)\right\|\|\partial_x^\alpha U_e\|\\ &\le C\frac{\|\partial_x^\alpha U_e\|}{\tau}\Big(\|U_e\|_s\|\nabla\mathbf{u}_{e\epsilon}\|_s+\left\|\frac{T_{e\epsilon}}{n_{e\epsilon}}-\frac{T_e^\epsilon}{n_e^\epsilon}\right\|_s\|\nabla n_{e\epsilon}\|_s\Big)\\ &\le C\frac{\|\partial_x^\alpha U_e\|}{\tau}\Big(\tau\|U_e\|_s+(1+D^{s-1})(\|N_e\|_s+\|\Theta_e\|_s)\Big)\\ &\le\frac{\|\partial_x^\alpha U_e\|^2}{4\tau^2}+C\|U_e\|_s^2+C(1+D^{2(s-1)})(\|N_e\|_s^2+\|\Theta_e\|_s^2),\end{aligned}\tag{4.13}$$

where (**III**) of Lemma 3.2 is used to estimate

$$A(x,N_e,\Theta_e):=\frac{T_{e\epsilon}}{n_{e\epsilon}}-\frac{T_e^\epsilon}{n_e^\epsilon}=\frac{T_{e\epsilon}(x)}{n_{e\epsilon}(x)}-\frac{T_{e\epsilon}(x)-\Theta_e}{n_{e\epsilon}(x)-N_e}$$

as

$$\begin{aligned}&\|A(x,N_e,\Theta_e)\|_s\\ &\le C_s|A|_{C^{s+1}}(1+(\|N_e\|_s+\|\Theta_e\|_s)^{s-1})(\|N_e\|_s+\|\Theta_e\|_s)\\ &\le C(1+D^{s-1})(\|N_e\|_s+\|\Theta_e\|_s),\end{aligned}$$

and

$$\begin{aligned}&C_{e2}\|f_e^{23}\|\|\partial_x^\alpha U_e\|\\ &=C_{e2}\frac{\|\partial_x^\alpha U_e\|}{\tau}\Big\|\partial_x^\alpha\Big((\mathbf{u}_e^\epsilon\cdot\nabla)U_e+\frac{T_e^\epsilon}{n_e^\epsilon}\nabla N_e\Big)\\ &\quad-\Big((\mathbf{u}_e^\epsilon\cdot\nabla)\partial_x^\alpha U_e+\frac{T_e^\epsilon}{n_e^\epsilon}\nabla\partial_x^\alpha N_e\Big)\Big\|\\ &\le C\frac{\|\partial_x^\alpha U_e\|}{\tau}\Big(\|\partial\mathbf{u}_e^\epsilon\|_{s-1}\|\nabla U_e\|_{s-1}+\|\partial(T_e^\epsilon/n_e^\epsilon)\|_{s-1}\|\nabla N_e\|_{s-1}\Big)\\ &\le C\frac{\|\partial_x^\alpha U_e\|}{\tau}\Big(\tau(1+D)\|U_e\|_s+(\|T_{e\epsilon}/n_{e\epsilon}\|_s+\|A\|_s)\|N_e\|_s\Big)\\ &\le C\frac{\|\partial_x^\alpha U_e\|}{\tau}\Big(\tau(1+D)\|U_e\|_s+(1+D^s)\|N_e\|_s\Big)\\ &\le\frac{\|\partial_x^\alpha U_e\|^2}{4\tau^2}+C(1+D^{2s})\|N_e\|_s^2+C(1+D)\|U_e\|_s^2.\end{aligned}\tag{4.14}$$

Adding (4.12)–(4.14) immediately gives the inequality (4.8). In the similar spirit, we can achieve (4.9) without extra troubles.

Finally, we estimate $C_{e3}\|\mathcal{F}_e^3\|\|\partial_x^\alpha\Theta_e\|$ as follows:

$$\begin{aligned}C_{e3}\|f_e^{31}\|\|\partial_x^\alpha\Theta_e\|&=C_{e3}\|\tau\sigma\partial_x^\alpha\mathcal{R}_{e3}\|\|\partial_x^\alpha\Theta_e\|\\ &\le\frac{\|\partial_x^\alpha\Theta_e\|^2}{4\tau\sigma}+C\tau^3\sigma^3,\end{aligned}\tag{4.15}$$

$$\tau\|f_e^{32}\| = \Big\|\partial_x^\alpha\Big(U_e\cdot\nabla T_{e\epsilon} - (\gamma-1)\Theta_e \mathrm{div}\mathbf{u}_{e\epsilon} + (\gamma-1)\Big(\frac{1}{\tau}-\frac{1}{2\sigma}\Big)(|\mathbf{u}_{e\epsilon}|^2 - |\mathbf{u}_e^\epsilon|^2)\Big)\Big\|$$

$$\le C\Big(\|\nabla T_{e\epsilon}\|_s\|U_e\|_s + \|\mathrm{div}\mathbf{u}_{e\epsilon}\|_s\|\Theta_e\|_s + \frac{1}{\tau}(2\|\mathbf{u}_{e\epsilon}\|_s + \|U_e\|_s)\|U_e\|_s\Big)$$

$$\le C(1+D)\|U_e\|_s + C\tau\|\Theta_e\|_s \tag{4.16}$$

and

$$\tau\|f_e^{33}| = \Big\|\partial_x^\alpha\Big(\mathbf{u}_e^\epsilon\cdot\nabla\Theta_e + (\gamma-1)T_e^\epsilon \mathrm{div}U_e\Big) - \Big(\mathbf{u}_e^\epsilon\cdot\nabla\partial_x^\alpha\Theta_e + (\gamma-1)T_e^\epsilon\mathrm{div}\partial_x^\alpha U_e\Big)\Big\|$$

$$\le C(\|\partial\mathbf{u}^\epsilon\|_{s-1}\|\nabla\Theta_e\|_{s-1} + \|\partial T^\epsilon\|_{s-1}\|\mathrm{div}U_e\|_{s-1})$$

$$\le C\tau(1+D)\|\Theta_e\|_s + C(1+\tau\sigma D)\|U_e\|_s. \tag{4.17}$$

From (4.15)–(4.17) we easily deduce that

$$C_{e3}\|\mathcal{F}_e^3\|\|\partial_x^\alpha\Theta_e\| \le \frac{\|\partial_x^\alpha\Theta_e\|^2}{4\tau\sigma} + \kappa_e\frac{\|U_e\|_s^2}{\tau^2} + C\tau^3\sigma^3 + C(1+D^2)\|\Theta_e\|_s^2.$$

This is the inequality (4.10). Similarly, it is not difficult to get (4.11).

The proof of Lemma 4.2 is complete. □

Substituting the estimates in (4.4)–(4.11) into (4.3) gives

$$\frac{d}{dt}\int\Big\{\Big(\frac{T_e^\epsilon}{n_e^\epsilon}\Big)^2|\partial_x^\alpha N_e|^2 + \Big(\frac{T_i^\epsilon}{n_i^\epsilon}\Big)^2|\partial_x^\alpha N_i|^2 + T_e^\epsilon|\partial_x^\alpha U_e|^2 + T_i^\epsilon|\partial_x^\alpha U_i|^2 + (\gamma-1)\Big(|\partial_x^\alpha\Theta_e|^2 + |\partial_x^\alpha\Theta_i|^2\Big)\Big\}dx$$

$$+\frac{1}{4\tau^2}\|(\partial_x^\alpha U_e, \partial_x^\alpha U_i)\|^2 + \frac{3}{4\tau\sigma}\|(\partial_x^\alpha\Theta_e, \partial_x^\alpha\Theta_i)\|^2$$

$$\le C\tau^2\sigma^2 + 2\kappa_e\frac{\|U_e\|_s^2}{\tau^2} + 2\kappa_i\frac{\|U_i\|_s^2}{\tau^2} + C(1+D^{2s})\|(N_e, U_e, \Theta_e, N_i, U_i, \Theta_i)\|_s^2$$

By integrating this inequality from 0 to $t \le \min\{\mathbb{T}_\epsilon, \mathbb{T}_*\}$, we get

$$\int \Big\{ \Big(\frac{T_e^\epsilon}{n_e^\epsilon}\Big)^2 |\partial_x^\alpha N_e|^2 + \Big(\frac{T_i^\epsilon}{n_i^\epsilon}\Big)^2 |\partial_x^\alpha N_i|^2 + T_e^\epsilon |\partial_x^\alpha U_e|^2 + T_i^\epsilon |\partial_x^\alpha U_i|^2$$

$$+(\gamma-1)\Big(|\partial_x^\alpha \Theta_e|^2 + |\partial_x^\alpha \Theta_i|^2\Big)\Big\} dx + \frac{1}{4\tau^2}\int_0^t \|(\partial_x^\alpha U_e, \partial_x^\alpha U_i)(t',\cdot)\|^2 dt'$$

$$\le Ct\tau^2\sigma^2 + \frac{2\kappa_e}{\tau^2}\int_0^t \|U_e(t',\cdot)\|_s^2 dt' + \frac{2\kappa_i}{\tau^2}\int_0^t \|U_i(t',\cdot)\|_s^2 dt'$$

$$+C\int_0^t (1+D^{2s})\|(N_e, U_e, \Theta_e, N_i, U_i, \Theta_i)(t',\cdot)\|_s^2 dt', \tag{4.18}$$

where we have used the fact that the initial data are in equilibrium (1.5). It is easy to show

$$C^{-1}\|(\partial_x^\alpha N_e, \partial_x^\alpha U_e, \partial_x^\alpha \Theta_e, \partial_x^\alpha N_i, \partial_x^\alpha U_i, \partial_x^\alpha \Theta_i)\|^2$$

$$\le \int \Big\{ \Big(\frac{T_e^\epsilon}{n_e^\epsilon}\Big)^2 |\partial_x^\alpha N_e|^2 + \Big(\frac{T_i^\epsilon}{n_i^\epsilon}\Big)^2 |\partial_x^\alpha N_i|^2 + T_e^\epsilon |\partial_x^\alpha U_e|^2 + T_i^\epsilon |\partial_x^\alpha U_i|^2$$

$$+(\gamma-1)\Big(|\partial_x^\alpha \Theta_e|^2 + |\partial_x^\alpha \Theta_i|^2\Big)\Big\} dx$$

$$\le C\|(\partial_x^\alpha N_e, \partial_x^\alpha U_e, \partial_x^\alpha \Theta_e, \partial_x^\alpha N_i, \partial_x^\alpha U_i, \partial_x^\alpha \Theta_i)\|^2,$$

since n^ϵ and T^ϵ are bounded from the below and the above for $t \le \min\{\mathbb{T}_\epsilon, \mathbb{T}_*\}$. Now we take κ_e, κ_i to be so small that

$$8\kappa_e \sum_{|\alpha|\le s} 1 \le 1, \quad 8\kappa_i \sum_{|\alpha|\le s} 1 \le 1 \tag{4.19}$$

respectively and sum up (4.18) over all α satisfying $|\alpha| \le s$ to obtain

$$\|(N_e, U_e, \Theta_e, N_i, U_i, \Theta_i)(t,\cdot)\|_s^2 \le C\mathbb{T}_{**}\tau^2\sigma^2$$

$$+C\int_0^t (1+D^{2s})\|(N_e, U_e, \Theta_e, N_i, U_i, \Theta_i)(t',\cdot)\|_s^2 dt' \tag{4.20}$$

for $t \le \mathbb{T}_{**} \le \mathbb{T}_*$ with $\mathbb{T}_{**}$ to be determined. Then we apply Gronwall's lemma to (4.20) and get

$$\|(N_e, U_e, \Theta_e, N_i, U_i, \Theta_i)(t,\cdot)\|_s^2$$

$$\le C\mathbb{T}_{**}\tau^2\sigma^2 \exp\Big[C\int_0^t (1+D^{2s})dt'\Big]. \tag{4.21}$$

Recalling that $\|(N_e, U_e, \Theta_e/\sigma, N_i, U_i, \Theta_i/\sigma)\|_s = \tau D$ and $0 < \tau << 1$, it follows from (4.21) that

$$D(t)^2 \leq C\mathbb{T}_{**} \exp\left[C \int_0^t (1 + D^{2s})dt'\right] \equiv \Phi(t), \qquad (4.22)$$

thus

$$\Phi'(t) = C(1 + D^{2s})\Phi(t) \leq C\Phi(t) + C\Phi^{s+1}(t).$$

With the help of the nonlinear Gronwall-type inequality in [24], we obtain

$$\Phi(t) \leq e^{C\mathbb{T}_{**}}$$

for $t \in [0, \min\{\mathbb{T}_\epsilon, \mathbb{T}_{**}\})$, if we choose $\mathbb{T}_{**} > 0$ (independent of ϵ!) to be so small that

$$\Phi(0) = C\mathbb{T}_{**} \leq e^{-C\mathbb{T}_{**}}.$$

Then, in view of (4.22), there exists a constant c, independent of ϵ, such that

$$D(t) \leq c \qquad (4.23)$$

for $t \in [0, \min\{\mathbb{T}_\epsilon, \mathbb{T}_{**}\})$. Finally, according to the inequalities (4.21) and (4.23) we conclude the proof of Theorem 1.1.

References

[1] G. Alì and A. Jüngel, Global smooth solutions to the multi-dimensional hydrodynamic model for two-carrier plasmas, *J. Differential Equations*, **190** (2003), 663–685.

[2] A. M. Anile, An extended thermodynamic framework for the hydrodynamic modeling of semiconductors. In: P. Marcati, et al, eds., *Mathematial Problem in Semiconductor Physics.* Pitman Research Notes In Mathematical Sciences **340**. Longman, (1995), 3–41.

[3] Y. Brenier and W.-A. Yong, Derivation of particle, string, and membrane motions from the Born-Infeld electromagnetism, *J. Math. Phys.*, **46** (2005), 062305.

[4] G. Q. Chen, J. Jerome, C. W. Shu and D. H. Wang, Two carrier semiconductor devices models with geometric structure and symmetry properties. In: J. Jerome eds., *Modelling and Computation for Applications in Mathematics, Science, and Engineering*, Oxford University Press, Oxford, 1998, 103–140.

[5] W. F. Fang and K. Ito, Weak solutions to a one-dimensional hydrodynamic model of two carrier types for semiconductors, *Nonlinear Anal.*, **28** (1997), 947–963.

[6] I. Gasser, L. Hsiao and H. L. Li, Large time behavior of solutions of the bipolar hydrodynamic model for semiconductors, *J. Differential Equations* **192** (2003), 326–359.

[7] L. Hsiao, N. J. Mauser and K. J. Zhang, The relaxation limits to the bipolar hydrodynamic model for semiconductors, *Appl. Math. Letters*, **17** (2004), 95–100.

[8] L. Hsiao and K. J. Zhang, Global weak solution and relaxation limits of the initial-boundary value problem to the bipolar hydrodynamic model for semiconductors, *Math. Model Methods Appl. Sci.*, **10** (2000), 1333–1361.

[9] A. Jüngel and Y. J. Peng, A hierarchy of hydrodynamic models for plasmas: Zero-relaxation-time limits, *Comm. Partial Differential Equations*, **24** (1999), 1007–1033.

[10] A. Jüngel and Y. J. Peng, Zero-relaxation-time limits in hydrodynamic model for plasmas revisited, *Z. Angew. Math. Phys.*, **51** (2000), 385–396.

[11] Q. C. Ju, Global smooth solutions to the multidimensional hydrodynamic model for plasmas with insulating boundary conditions, *J. Math. Anal. Appl.*, **336** (2007), 888–904.

[12] C. Lattanzio, On the 3-D bipolar isentropic Euler-Poisson model for semiconductors and the drift-diffusion limit, *Math. Models Methods Appl. Sci.*, **10** (2000), 351–360.

[13] Y. Li, Relaxation time limit problem for hydrodynamic models in semiconductor science, *Acta Mathematica Scientia* **27B** (2007), 437–448.

[14] Y. P. Li, Diffusion relaxation limit of a nonisentropic hydrodynamic model for semiconductors, *Math. Methods Appl. Sci.* **30** (2007), 2247–2261.

[15] Y. P. Li, Diffusion relaxation limit of a bipolar hydrodynamic model for semiconductor, *J. Math. Anal. Appl.*, **336** (2007), 1341–1356.

[16] A. Majda, *Compressible Fluid Flow and Conservation Laws in Several Space Variables*, Berlin, New York: Springer-Verlag, 1984.

[17] P. Marcati and R. Natalini, Weak solutions to a hydrodynamic model for semiconductors and relaxation to the drift-diffusion equations, *Arch. Ration. Mech. Anal.*, **129** (1995), 129–145.

[18] N. J. Mauser, Y. C. Qiu and K. J. Zhang, Global existence and Asymptotic limits of weak solutions of the bipolar hydrodynamic model for semiconductor, *Monatsh. Math.*, **140** (2003), 285–313.

[19] P. A. Markowich, C. Ringhofer and C. Schmeiser, *Semiconductor Equations*, Vienna, Springer-Verlag, 1990.

[20] R. Natalini, The bipolar hydrodynamic model for semiconductors and the Drift-Diffusion equations, *J. Math. Anal. Appl.*, **198** (1996), 262–281.

[21] J. Simon, Compact sets in the space $L^p(0,T;B)$, *Ann. Math. Pura Appl.*, **146** (1987), 65–96.

[22] D. H. Wang, Global solutions to the Euler-Poisson equations of two-carrier types in one dimension, *Z. Angew. Math. Phys.*, **48** (1997), 680–693.

[23] W.-A. Yong, Diffusive relaxation limit of multi-dimensional isentropic hydrodynamic models for semiconductor, *SIAM J. Appl. Math.* **64** (2004), 1737–1748.

[24] W.-A. Yong, Singular pertubations of first-order hyperbolic systems with stiff source terms, *J. Differential Equations* **155** (1999), 89–132.

[25] W.-A. Yong, *Basic aspects of hyperbolic relaxation systems*, In: Advances in the Theory of Shock Waves, Freistühler, H., Szepessy, A. eds., Progress in Nonlinear Differential Equations and Their Applications, Vol. 47, Birkhäuser, Boston, 2001, 259–305.

[26] W.-A. Yong, Singular Pertubations of First-Order Hyperbolic Systems, Ph.D. thesis, Universität Heidelberg, 1992.

[27] K. J. Zhang, On the initial-boundary value problem for the bipolar hydrodynamic model for semiconductors, *J. Differential Equations*, **171** (2001), 251–293.

[28] C. Zhu and H. Hattori, Stability of stead state solutions for an isentropic hydrodynamic model of semiconductors of two species, *J. Differential Equations* **166** (2000), 1–32.

Part II

Exact Controllability and Observability for Quasilinear Hyperbolic Systems and Applications

Observability in Arbitrary Small Time for Discrete Approximations of Conservative Systems*

Sylvain Ervedoza†
Laboratoire de Mathématiques de Versailles
Université de Versailles-Saint-Quentin
45, avenue des États-Unis
78035 Versailles Cedex, France
Email: sylvain.ervedoza@math.uvsq.fr

Abstract

The goal of this article is to analyze observability results of arbitrary small time for discrete approximations of conservative systems. In previous works, under the assumption that the continuous conservative system is admissible and exactly observable, observability results of the corresponding discrete approximation schemes have been proved within the class of conveniently filtered solutions using resolvent estimates. However, in several situations and for Schrödinger equations in particular when the Geometric Control Condition is satisfied, the exact observability property holds in any arbitrary small time. We prove that in several cases, namely under a stronger resolvent condition, the time-discrete approximations of conservative systems also enjoy uniform observability properties in arbitrary small time, still within the class of conveniently filtered solutions. Particularly, our methodology applies to space semi-discrete and fully discrete approximation schemes of Schrödinger equations for which the Geometric Control Condition is satisfied. Our approach is based on the resolvent characterization of the exact observability property and a constructive argument by Haraux in [14].

*This work was done during the French-Chinese Summer Institute on Applied Mathematics at the Institut Sino-Français de Mathématiques Appliquées, Fudan University, Shanghaï. The author acknowledges the institute for its hospitality.

†The author was partially supported by the "Agence Nationale de la Recherche" (ANR), Project C-QUID, number BLAN-3-139579.

1 Introduction

Let X be a Hilbert space endowed with the norm $\|\cdot\|_X$ and let $A : \mathcal{D}(A) \to X$ be a skew-adjoint operator with compact resolvent. Let us consider the following abstract system:

$$\dot{z}(t) = Az(t), \quad z(0) = z_0. \tag{1.1}$$

Here and henceforth, a dot ($\dot{}$) denotes differentiation with respect to the time t. The element $z_0 \in X$ is called the *initial state*, and $z = z(t)$ is the *state* of the system. Note that since A is skew-adjoint, solutions of (1.1) have constant energy: $\forall t \in \mathbb{R}$, $\|z(t)\|_X = \|z_0\|_X$. Particularly, (1.1) can be solved for all time $t \in \mathbb{R}$.

Such systems are often used as models of vibrating systems (e.g., the wave equation), electromagnetic phenomena (Maxwell equations) or in quantum mechanics (Schrödinger equation).

Assume that Y is another Hilbert space equipped with the norm $\|\cdot\|_Y$. We denote by $\mathfrak{L}(X, Y)$ the space of bounded linear operators from X to Y, endowed with the classical operator norm. Let $B \in \mathfrak{L}(\mathcal{D}(A), Y)$ be an observation operator and define the output function

$$y(t) = Bz(t). \tag{1.2}$$

In order to give a sense to (1.2), we make the assumption that B is an admissible observation operator in the following sense (see [20]):

Definition 1.1. The operator B is an admissible observation operator for systems (1.1)–(1.2) if for every $T > 0$ there exists a constant $K_T > 0$ such that

$$\int_0^T \|Bz(t)\|_Y^2 \, dt \leq K_T \|z_0\|_X^2, \qquad \forall z_0 \in \mathcal{D}(A). \tag{1.3}$$

Note that if B is *bounded* on X, i.e. if it can be extended such that $B \in \mathfrak{L}(X, Y)$, then B is obviously an admissible observation operator. However, in applications, this is often not the case, and the admissibility condition is then a consequence of a suitable "hidden regularity" property of the solutions of the evolution equation (1.1).

The exact observability property of systems (1.1)–(1.2) can be formulated as follows:

Definition 1.2. Systems (1.1)–(1.2) are exactly observable in time T if there exists $k_T > 0$ such that

$$k_T \|z_0\|_X^2 \leq \int_0^T \|Bz(t)\|_Y^2 \, dt, \qquad \forall z_0 \in \mathcal{D}(A). \tag{1.4}$$

Moreover, (1.1)–(1.2) are said to be exactly observable if they are exactly observable in some time $T > 0$.

Note that observability issues arise naturally when dealing with controllability and stabilization properties of linear systems (see for instance [20]). Indeed, controllability and observability are dual notions, and therefore each statement concerning observability has its counterpart in controllability. In the sequel, we focus on the observability properties of (1.1)–(1.2).

It was proved in [21] that systems (1.1)–(1.2) are exactly observable if and only if the following assertion holds:

Condition 1. There exist positive constants $M, m > 0$ such that

$$M^2 \left\|(A - i\omega I)z\right\|_X^2 + m^2 \left\|Bz\right\|_Y^2 \geq \left\|z\right\|_X^2, \quad \forall \omega \in \mathbb{R},\ \forall z \in \mathcal{D}(A). \tag{1.5}$$

This spectral condition can be viewed as a Hautus-type test, and generalizes the classical Kalman rank condition (see for instance [26]). To be more precise, if Condition 1 holds, then systems (1.1)–(1.2) are exactly observable in any time $T > T_0 = \pi M$ (see [21]).

The following stronger resolvent condition is more interesting for our purpose:

Condition 2. There exist a positive constant $m > 0$ and a function $M = M(\omega)$ of $\omega \in \mathbb{R}$, bounded on $\mathbb{R}$, satisfying

$$\lim_{|\omega| \to \infty} M(\omega) = 0, \tag{1.6}$$

such that for all $\omega \in \mathbb{R}$,

$$M(\omega)^2 \left\|(A - i\omega I)z\right\|_X^2 + m^2 \left\|Bz\right\|_Y^2 \geq \left\|z\right\|_X^2, \qquad \forall z \in \mathcal{D}(A). \tag{1.7}$$

This condition appears naturally when considering Schrödinger equations for which the Geometric Control Condition is satisfied (see [21] and Section 4 below).

Theorem 1.3 ([5, 21]). *When Condition* 2 *is fulfilled, systems* (1.1)–(1.2) *are observable in any time* $T^* > 0$.

The proof of Theorem 1.3 in [21] is based on a decoupling argument of high- and low-frequency components. Given $T^* > 0$, take $M > 0$ such that $\pi M < T^*$, and choose a frequency cut $\Omega = \Omega_0 + 1/M$ where Ω_0 satisfies $\sup_{|\omega| \geq \Omega_0}\{M(\omega)\} \leq M$. Then the frequencies higher than Ω are exactly observable in any time $\tilde{T} \in (\pi M, T^*)$. The low-frequency components then correspond to a finite dimensional observability problem and can be handled in any time $T > 0$. Finally these two partial observability properties are combined together using a compactness argument (or simultaneous exact controllability results as in [26]).

But these methods are not constructive and are then not sufficient to obtain observability results of family of operators satisfying (1.7) uniformly. It is then of particular interest to design a constructive proof of Theorem 1.3 when dealing for instance with discrete approximation schemes of (1.1)–(1.2).

In the sequel, we will then propose a constructive proof of Theorem 1.3, based on an explicit method proposed by Haraux in [14]. This allows us to deal with families of operators satisfying Condition 2 uniformly. We then explain how our method applies to time semi-discretizations of (1.1)–(1.2).

Particularly, our method implies that when the Geometric Control Condition is satisfied, time continuous and time semi-discrete Schrödinger equations are exactly observable in arbitrary small time, as we will see in Section 4. In this case, based on the abstract approach developed in [8, 9], we can also deal with space semi-discrete and fully discrete approximation schemes. Particularly, we will prove uniform (with respect to the discretization parameters) observability results in arbitrary small time for discrete approximations of Schrödinger equations satisfying the Geometric Control Condition, within the class of conveniently filtered solutions.

Let us now briefly comment the literature. This article follows the works [10, 8, 9] on observability properties for discrete approximation schemes of abstract conservative systems which, in the continuous setting, are exactly observable. The main underlying idea there is to use spectral criteria such as (1.5) which yield explicit dependence on the parameters for the constants entering in the exact observability property (1.4). Indeed, one can then use the following diagram to prove uniform observability results of discrete approximations of (1.1)–(1.2):

Exact observability property for the continuous system		Uniform observability for discretizations of (1.1)–(1.2)
$\Downarrow$		$\Uparrow$
Spectral criterion for the continuous system	$\Longrightarrow$	Spectral criterion for the discrete systems

The spectral criteria used in [10, 8, 9] for the exact observability property are due to [5, 21, 24, 26] in particular. As already noticed in [26], if the operators A and B satisfy estimate (1.7) for a function $M(\omega)$ satisfying $\lim_{|\omega|\to\infty} M(\omega) = M$ (M may be different from 0), then systems (1.1)–(1.2) is exactly observable in any time $T > \pi M$.

Let us mention that one has to look for *uniform* observability properties for the discrete approximation schemes of (1.1)–(1.2) to guarantee the convergence of the controls computed in the discrete setting to one of the continuous systems (1.1)–(1.2). However, as already noticed in

[11, 12, 13], observability properties for discrete versions of (1.1)–(1.2) do not hold uniformly with respect to the discretization parameters due to spurious high-frequencies. We thus need to restrict ourselves to prove uniform observability properties within the class of conveniently filtered solutions.

There are of course several other techniques to study observability properties for discrete versions of (1.1)–(1.2), such as Ingham's Lemma [16], whose use is essentially limited to the 1d cases (see [15, 6, 7]), and discrete multiplier methods in [22, 23]. For extensive references and the state of the art for the observability properties of discrete approximations of the wave equation, we refer to [27].

The paper is organized as follows:

In Section 2, we give a constructive proof of Theorem 1.3. In Section 3, we explain how this can yield uniform observability results in arbitrary small time for time semi-discrete versions of (1.1)–(1.2). in Section 4, we present an application of these techniques to discrete Schrödinger equations, including the fully discrete case, when the Geometric Control Condition is satisfied. We finally end up with some further comments.

Acknowledgments. The author acknowledges Vilmos Komornik for pointing out the constructive argument of Haraux.

2 A constructive proof of Theorem 1.3

Before going into the proof, we introduce some notations.

For a function $f \in L^2(\mathbb{R}; X)$ depending on time $t \in \mathbb{R}$, we define its time Fourier transform $\hat{f} \in L^2(\mathbb{R}, X)$ by

$$\hat{f}(\omega) = \frac{1}{\sqrt{2\pi}} \int_{\mathbb{R}} f(t) e^{-i\omega t} \, dt. \tag{2.1}$$

The Parseval identity then reads:

$$\int_{\mathbb{R}} \|f(t)\|_X^2 \, dt = \int_{\mathbb{R}} \left\| \hat{f}(\omega) \right\|_X^2 \, d\omega, \quad \forall f \in L^2(\mathbb{R}; X). \tag{2.2}$$

It is convenient to introduce the spectrum of the operator A. Since A is skew-adjoint with compact resolvent, its spectrum is discrete and $\sigma(A) = \{i\mu_j : j \in \mathbb{Z}\}$, where $(\mu_j)_{j\in\mathbb{Z}}$ is an increasing sequence of real numbers. Set $(\Phi_j)_{j\in\mathbb{N}}$ an orthonormal basis of eigenvectors of A associated with the eigenvalues $(i\mu_j)_{j\in\mathbb{Z}}$:

$$A\Phi_j = i\mu_j \Phi_j. \tag{2.3}$$

Moreover, we define the filtered class

$$\mathcal{C}(s) = \text{span}\{\Phi_j : \text{the corresponding } i\mu_j \text{ satisfying } |\mu_j| < s\}, \tag{2.4}$$

and its orthogonal $\mathcal{C}(s)^\perp$ in X, which coincides with

$$\operatorname{span}\{\Phi_j : \text{the corresponding } i\mu_j \text{ satisfying } |\mu_j| \geq s\}.$$

We can now focus on the proof of Theorem 1.3, which is decomposed in two main steps, which will be explained in the following subsections:

1. We prove an observability inequality in arbitrary small time for the high-frequency solutions of (1.1).
2. We use the constructive argument in [14] to obtain an observability inequality for any solutions of (1.1).

2.1 High frequency components

We first prove a high-frequency resolvent estimate:

Lemma 2.1. *For all $M > 0$ there exists a constant Ω such that*

$$M^2 \|(A - i\omega I)z\|_X^2 + m^2 \|Bz\|_Y^2 \geq \|z\|_X^2, \qquad \forall \omega \in \mathbb{R},\ \forall z \in \mathcal{D}(A) \cap \mathcal{C}(\Omega)^\perp. \tag{2.5}$$

Proof of Lemma 2.1. Fix $M > 0$. Then there exists Ω_0 such that

$$\forall \omega \geq \Omega_0, \quad |M(\omega)| \leq M.$$

This implies the following version of (2.5):

$$M^2 \|(A - i\omega I)z\|_X^2 + m^2 \|Bz\|_Y^2 \geq \|z\|_X^2, \qquad \forall \omega \text{ such that } |\omega| \geq \Omega_0,\ \forall z \in \mathcal{D}(A).$$

Particularly, this implies (2.5) for all ω such that $|\omega| \geq \Omega_0$. We thus only need to prove (2.5) for ω such that $|\omega| \leq \Omega_0$. This can be done using the following remark: If $\Omega \geq \Omega_0$,

$$\forall \omega \text{ such that } |\omega| \leq \Omega_0,\ \forall z \in \mathcal{D}(A) \cap \mathcal{C}(\Omega)^\perp, \quad \|(A - i\omega)z\|_X^2 \geq (\Omega - \Omega_0)^2 \|z\|_X^2.$$

Then, with the choice $\Omega = \Omega_0 + 1/M$, (2.5) holds. □

We now prove the following lemma:

Lemma 2.2. *If* (2.5) *holds for given M and Ω, the observability inequality* (1.4) *holds in any time $T > \pi M$ for solutions of* (1.1) *with initial data lying in $\mathcal{C}(\Omega)^\perp$. Besides the corresponding constant $k_T > 0$ of observability in* (1.4) *can be chosen as*

$$k_T = \frac{1}{2m^2T^2}(T^2 - \pi^2 M^2).$$

This lemma can actually be found in [5, 21]. We provide the proof for completeness.

Proof of Lemma 2.2. Given $z_0 \in \mathcal{C}(\Omega)^\perp \cap \mathcal{D}(A)$, let $z(t)$ be the corresponding solution of (1.1) and define, for $\chi \in C_0^\infty(\mathbb{R})$,

$$g(t) = \chi(t)z(t), \quad f(t) = g'(t) - Ag(t) = \chi'(t)z(t).$$

Then $\hat{f}(\omega) = (i\omega - A)\hat{g}(\omega)$. Besides, $\hat{g}$ belongs to $L^2(\mathbb{R}; \mathcal{D}(A) \cap \mathcal{C}(\Omega)^\perp)$. We can thus apply the resolvent estimate (2.5) to $\hat{g}(\omega)$:

$$\forall \omega \in \mathbb{R}, \quad \|\hat{g}(\omega)\|_X^2 \leq m^2 \left\|\widehat{Bg}(\omega)\right\|_Y^2 + M^2 \left\|\hat{f}(\omega)\right\|_X^2.$$

Integrating with ω and using the Parseval identity, we obtain

$$\Big(\int_{\mathbb{R}} \chi(t)^2 \, dt - M^2 \int_{\mathbb{R}} \chi'(t)^2 \, dt\Big) \|z_0\|_X^2 \leq m^2 \int_{\mathbb{R}} \chi(t)^2 \|Bz(t)\|_Y^2 \, dt,$$

where we see that the energy of solutions of (1.1), given by $\|z(t)\|_X^2$, is constant.

We then look for a function χ which makes the left hand side positive. This can be achieved by taking $\chi(t) = \sin(\pi t/T)$ in $(0, T)$ and vanishing anywhere else. This is not in $C_0^\infty(\mathbb{R})$ but in $H^1(\mathbb{R})$ with compact support, which is sufficient for the proof developed above.

This gives the desired estimate for any initial data $z_0 \in \mathcal{D}(A) \cap \mathcal{C}(\Omega)^\perp$ and we conclude by density. □

2.2 Haraux's constructive argument

To present the construction precisely, remark that since A has compact resolvent, there is only a finite number of eigenvalues for which $|\mu_j| < \Omega$. For convenience, we introduce the finite sequence $(m_j)_{1 \leq j \leq N}$ of strictly increasing real numbers such that $\{m_j\} = \{\mu_j$ such that $|\mu_j| < \Omega\}$. For $j \in \{1, \cdots, N\}$, we then denote by X_j the finite-dimensional vector space spanned by the eigenvectors corresponding to eigenvalues $i\mu_j$ with $\mu_j = m_j$. Note that these notations are not needed when the eigenvalues are simple.

Lemma 2.3 ([14]). *Let B be an admissible operator for* (1.1)–(1.2). *Assume that there exist positive constants $\tilde{k} > 0$ and $\tilde{T}$ such that any solution of* (1.1) *with initial data $z_0 \in \mathcal{C}(\Omega)^\perp$ satisfies*

$$\tilde{k} \|z_0\|_X^2 \leq \int_0^{\tilde{T}} \|Bz(t)\|_Y^2 \, dt. \tag{2.6}$$

Also assume that there exists a strictly positive number β such that

$$\forall j \in \{1, \cdots, N\}, \ \forall z \in X_j, \quad \|Bz\|_Y \geq \beta \|z\|_X. \tag{2.7}$$

Then the observability inequality (1.4) holds in any time $T > \tilde{T}$, with a strictly positive observability constant $k_T > 0$ depending explicitly on the parameters β, $T - \tilde{T}$, $\tilde{k}$, the number N of low frequencies and the low frequency gap

$$\gamma = \inf_{j \in \{0, \cdots, N\}} \{m_{j+1} - m_j\}, \text{ where } m_0 = -\Omega \text{ and } m_{N+1} = +\Omega. \tag{2.8}$$

Note that γ in (2.8) is strictly positive as an infimum of a finite number of strictly positive quantities.

We give the proof of this lemma below, since it will later be generalized to more complex situations. Note that this proof can also be found in [18].

Proof of Lemma 2.3. The argument is inductive. We then just need to describe the first step, for the others are similar. We then focus on the observability inequality (1.4) for initial data in $X_N + \mathcal{C}(\Omega)^\perp$.

Set $z_0 \in X_N + \mathcal{C}(\Omega)^\perp$, and expand it as $z_{0,N} + z_{0,hf}$ with $z_{0,N} \in X_N$ and $z_{0,hf} \in \mathcal{C}(\Omega)^\perp$.

Let $z(t)$ be the solution of (1.1) corresponding to the initial data z_0, and define, for $\delta > 0$,

$$v(t) = z(t) - \frac{1}{2\delta} \int_{-\delta}^{\delta} e^{i m_N s} z(t-s)\, ds. \tag{2.9}$$

Writing $z_0 = \sum a_j \Phi_j$, the solution $z(t)$ of (1.1) can be explicitly written as $\sum a_j \Phi_j \exp(i\mu_j t)$. Particularly,

$$\begin{aligned} v(t) &= \sum_j a_j \Phi_j \exp(i\mu_j t)\Big(1 - \operatorname{sinc}(\delta(m_N - \mu_j))\Big) \\ &= \sum_{j \text{ with } |\mu_j| \geq \Omega} a_j \Phi_j \exp(i\mu_j t)\Big(1 - \operatorname{sinc}(\delta(m_N - \mu_j))\Big) \end{aligned} \tag{2.10}$$

Note particularly that (2.10) implies that the norms of $v_0 = v(0)$ and z_0 satisfy

$$\|z_{0,hf}\|_X^2 \leq \frac{1}{(1 - \operatorname{sinc}(\delta\gamma))^2} \|v_0\|_X^2. \tag{2.11}$$

Besides, (2.10) also implies that v is a solution of (1.1) with initial data in $\mathcal{C}(\Omega)^\perp$. Hence, it shall also satisfy the observability inequality (2.6):

$$\|v_0\|_X^2 \leq \frac{1}{\tilde{k}} \int_0^{\tilde{T}} \|Bv(t)\|_Y^2 \, dt. \tag{2.12}$$

We then have to estimate the right hand side of (2.12). From (2.9), we get:

$$\begin{aligned}\int_0^{\tilde{T}} \|Bv(t)\|_Y^2 \, dt \le 2\int_0^{\tilde{T}} \|Bz(t)\|_Y^2 \, dt & \\ +2\int_0^{\tilde{T}} \left\| B\Big(\frac{1}{2\delta}\int_{-\delta}^{\delta} e^{im_N s} z(t-s)\, ds\Big)\right\|_Y^2 & \, dt \\ \le 2\int_0^{\tilde{T}} \|Bz(t)\|_Y^2 \, dt + 2\int_{-\delta}^{\tilde{T}+\delta} \|Bz(t)\|_Y^2 \, dt & \\ \le 4\int_{-\delta}^{\tilde{T}+\delta} \|Bz(t)\|_Y^2 \, dt. & \qquad (2.13)\end{aligned}$$

Combined with (2.11) and (2.12), this yields

$$\|z_{0,hf}\|_X^2 \le \frac{4}{\tilde{k}(1-\operatorname{sinc}(\gamma\delta))^2}\int_{-\delta}^{\tilde{T}+\delta} \|Bz(t)\|_Y^2 \, dt. \qquad (2.14)$$

We then focus on the component of the solution in X_N. Obviously, denoting by z_N, z_{hf} the solutions of (1.1) with initial data $z_{0,N}, z_{0,hf}$ respectively, and applying (2.7) we obtain

$$\begin{aligned}\|z_{0,N}\|_X^2 &\le \frac{1}{\tilde{T}\beta^2}\int_0^{\tilde{T}} \|Bz_N(t)\|_Y^2 \, dt \\ &\le \frac{1}{\tilde{T}\beta^2}\Big(2\int_0^{\tilde{T}} \|Bz(t)\|_Y^2 \, dt + 2\int_0^{\tilde{T}} \|Bz_{hf}(t)\|_Y^2 \, dt\Big).\end{aligned}$$

Using the admissibility of B for systems (1.1)–(1.2), we obtain

$$\|z_{0,N}\|_X^2 \le \frac{2}{\tilde{T}\beta^2}\int_0^{\tilde{T}} \|Bz(t)\|_Y^2 \, dt + \frac{2K_{\tilde{T}}}{\tilde{T}\beta^2}\|z_{0,hf}\|_X^2 . \qquad (2.15)$$

Using (2.14) and the orthogonality of X_N and $\mathcal{C}(\Omega)^\perp$, we conclude

$$\|z_0\|_X^2 \le \Big[\frac{2}{\tilde{T}\beta^2} + \Big(\frac{2K_{\tilde{T}}}{\tilde{T}\beta^2}+1\Big)\frac{4}{\tilde{k}(1-\operatorname{sinc}(\gamma\delta))^2}\Big]\int_{-\delta}^{\tilde{T}+\delta} \|Bz(t)\|_Y^2 \, dt, \qquad (2.16)$$

or, using the conservation of the energy for solutions of (1.1) and the semi-group property,

$$\|z_0\|_X^2 \le \Big[\frac{2}{\tilde{T}\beta^2} + \Big(\frac{2K_{\tilde{T}}}{\tilde{T}\beta^2}+1\Big)\frac{4}{\tilde{k}(1-\operatorname{sinc}(\gamma\delta))^2}\Big]\int_0^{\tilde{T}+2\delta} \|Bz(t)\|_Y^2 \, dt. \qquad (2.17)$$

Since $\delta > 0$ is arbitrarily small, we have proved the observability inequality (2.17) in any time $T_N > \tilde{T}$ for any solution of (1.1) with initial data in $X_N + \mathcal{C}(\Omega)^\perp$.

The induction argument is then left to the reader. □

2.3 End of the proof of Theorem 1.3

Set $T^* > 0$. Choose $M > 0$ such that $\pi M = T^*/4$. From Lemma 2.1, one can choose Ω such that (2.5) holds for $z \in \mathcal{D}(A) \cap \mathcal{C}(\Omega)^\perp$. From Lemma 2.2, this implies that any solution of (1.1) with initial data in $\mathcal{C}(\Omega)^\perp$ satisfies (2.6) in time $\tilde{T} = T^*/2$.

Since A has compact resolvent, there is only a finite number of eigenvalues μ_j such that $|\mu_j| < \Omega$ and then the low frequency gap γ defined in (2.8) is strictly positive.

We only have to check that estimate (2.7) indeed holds. This is actually obvious, since for $j \in \{1, \cdots, N\}$ and $z \in X_j$, taking $\omega = m_j$ in (1.7), we obtain:

$$m^2 \|Bz\|_Y^2 \geq \|z\|_X^2 .$$

The proof is then completed by applying Lemma 2.3. □

3 Applications to time-discrete approximations of (1.1)–(1.2)

This section aims at describing how the previous result can be adapted to time-discrete approximations of systems (1.1)–(1.2) satisfying Condition 2. Particularly, we shall prove that in that case, time semi-discrete approximations of (1.1)–(1.2) indeed are exactly observable in arbitrary small time within the class of conveniently filtered solutions, uniformly with respect to the time discretization parameter.

3.1 Time discrete approximations of (1.1)–(1.2)

To simplify the presentation, we will focus on the following natural approximation of (1.1)–(1.2), the so-called midpoint scheme. For $\Delta t > 0$, consider

$$\begin{cases} \dfrac{z^{k+1} - z^k}{\Delta t} = A\Big(\dfrac{z^k + z^{k+1}}{2}\Big), & \text{in } X, \quad k \in \mathbb{Z}, \\ z^0 = z_0 \text{ given.} \end{cases} \tag{3.1}$$

Here, z^k denotes the approximation of the solution z of (1.1) at time $t_k = k\Delta t$. Note that the discrete system (3.1) is conservative, in the sense that $k \mapsto \|z^k\|_X^2$ is constant.

The output function is now given by the discrete sample

$$y^k = Bz^k. \tag{3.2}$$

The admissibility and observability properties for (3.1)–(3.2) have been studied in [10] using spectral criteria such as Condition 1 for the observability properties. Particularly, [10] states the following result:

Theorem 3.1 ([10]). *Assume that the continuous systems* (1.1)–(1.2) *are admissible and exactly observable in some time $T > 0$. Then for all $\delta > 0$:*

- *For all $T > 0$, there exists a constant $K_{\delta,T}$ such that, for all $\Delta t > 0$, any solution z^k of* (3.1) *with initial data $z_0 \in \mathcal{C}(\delta/\Delta t)$ satisfies*

$$\Delta t \sum_{k\Delta t \in (0,T)} \left\| Bz^k \right\|_Y^2 \leq K_{\delta,T} \left\| z_0 \right\|_X^2 . \tag{3.3}$$

- *There exist a time T_δ and a positive constant $k_\delta > 0$ such that, for all $\Delta t > 0$ small enough, any solution z^k of* (3.1) *with initial data $z_0 \in \mathcal{C}(\delta/\Delta t)$ satisfies*

$$k_\delta \left\| z_0 \right\|_X^2 \leq \Delta t \sum_{k\Delta t \in (0,T_\delta)} \left\| Bz^k \right\|_Y^2 . \tag{3.4}$$

Note that the observability property (3.4) requires the time to be large enough. Actually, a precise estimate is given in [10] in terms of the resolvent parameters in (1.5) and the scaling parameter δ, but this is not completely satisfactory since, to our knowledge, even in the continuous setting, we are not able in general to recover the optimal time of controllability from (1.5).

But, as explained in Introduction, Condition 2 is sufficient to prove observability of the continuous systems (1.1)–(1.2) in any positive time. We thus ask whether or not it is also possible to prove discrete observability properties for (3.1)–(3.2) in arbitrary small time when Condition 2 is satisfied.

Theorem 3.2. *Assume that Condition 2 is satisfied.*
- *If $B \in \mathfrak{L}(\mathcal{D}(A^\kappa), Y)$ with $\kappa < 1$. Then for any $\delta > 0$, for any time $T^* > 0$, there exists a positive constant $k_{\delta,T^*} > 0$ such that, for all $\Delta t > 0$ small enough, any solution z^k of* (3.1) *with initial data $z_0 \in \mathcal{C}(\delta/\Delta t)$ satisfies*

$$k_{\delta,T^*} \left\| z_0 \right\|_X^2 \leq \Delta t \sum_{k\Delta t \in (0,T^*)} \left\| Bz^k \right\|_Y^2 . \tag{3.5}$$

- *If B simply belongs to $\mathfrak{L}(\mathcal{D}(A), Y)$. Then for any time $T^* > 0$, there exist two positive constants $\delta > 0$ and $k_{\delta,T^*} > 0$ such that, for all $\Delta t > 0$*

small enough, any solution z^k *of* (3.1) *with initial data* $z_0 \in \mathcal{C}(\delta/\Delta t)$ *satisfies* (3.5).

Theorem 3.2 is the exact counterpart in the discrete setting of Theorem 1.3, and we will only indicate the modifications needed in its proof to derive Theorem 3.2.

Proof of Theorem 3.2. As we said, the proof of Theorem 3.2 closely follows the one of Theorem 1.3, and we thus only sketch it briefly.

We first deal with the high-frequency components. Lemma 2.1 still holds, since it is by nature independent of time, no matter whether this time is continuous or not. However, Lemma 2.2 has to be modified and replaced by the following

Lemma 3.3. *Assume that* (2.5) *holds for given* m, M *and* Ω. *For any* $\delta > 0$, *set*

$$T_{M,\delta} = \begin{cases} \pi M\left(1+\dfrac{\delta^2}{4}\right), & \text{if } B \in \mathfrak{L}(\mathcal{D}(A^\kappa), Y) \text{ with } \kappa < 1, \\ \pi\left[M^2\left(1+\dfrac{\delta^2}{4}\right)^2 + m^2\,\|B\|^2_{\mathfrak{L}(\mathcal{D}(A),Y)}\dfrac{\delta^4}{16}\right]^{1/2} & \\ \qquad\qquad\qquad\qquad \text{if } B \in \mathfrak{L}(\mathcal{D}(A), Y). & \end{cases} \tag{3.6}$$

Then the observability inequality (3.4) *holds in any time* $T > T_M$ *for some positive constant* $k_T > 0$ *for any solution of* (3.1) *with initial data lying in* $\mathcal{C}(\Omega)^\perp \cap \mathcal{C}(\delta/\Delta t)$. *Besides,* k_T *can be chosen explicitly as a function of* T, m, M *and the norm of* B *in* $\mathfrak{L}(\mathcal{D}(A), Y)$.

Proof of Lemma 3.3. In the case $B \in \mathfrak{L}(\mathcal{D}(A), Y)$, this lemma corresponds exactly to Theorem 1.3 in [10], and might be seen as an extension of Lemma 2.2 to the time discrete case, involving discrete Fourier transforms in particular instead of continuous ones.

In the case $B \in \mathfrak{L}(\mathcal{D}(A^\kappa), Y)$ with $\kappa < 1$, the proof of Lemma 3.3 can be adapted immediately from the one of Theorem 1.3 in [10] by modifying estimate (2.19) in [10] using

$$\begin{aligned} \left\| B\left(\frac{z^{k+1}-z^k}{\Delta t}\right)\right\|_Y &\leq \|B\|_{\mathcal{L}(\mathcal{D}(A^\kappa),Y)}\left\|A^{1+\kappa}\left(\frac{z^k+z^{k+1}}{2}\right)\right\|_X \\ &\leq \left(\frac{\delta}{\Delta t}\right)^{1+\kappa}\|B\|_{\mathcal{L}(\mathcal{D}(A^\kappa),Y)}\left\|\frac{z^0+z^1}{2}\right\|_X, \end{aligned}$$

and the following ones accordingly. Particularly, with the notations of

[10], a_2 in Lemma 2.4 shall be replaced by

$$a_2 = M^2\left(1+\frac{\delta^2}{4}\right) + \frac{\delta^2(\Delta t)^2}{16}(\beta-1) + m^2\,\|B\|^2_{\mathfrak{L}(\mathcal{D}(A^\kappa),Y)}\,\frac{\delta^{2+2\kappa}(\Delta t)^{2-2\kappa}}{16}\left(1+\frac{1}{\alpha}\right).$$

Details are then left to the reader. □

We then deal with the low-frequency components. This is done as in the continuous case:

Lemma 3.4. *Let B be an admissible operator for* (1.1). *Assume that there exist positive constants $\delta > 0$, $\tilde{k} > 0$ and $\tilde{T}$ such that, for all $\Delta t > 0$ small enough, any solution of* (3.1) *with initial data $z_0 \in \mathcal{C}(\Omega)^{\perp} \cap \mathcal{C}(\delta/\Delta t)$ satisfies*

$$\tilde{k}\,\|z_0\|_X^2 \le \Delta t \sum_{k\Delta t\in(0,\tilde{T})} \left\|Bz^k\right\|_Y^2. \tag{3.7}$$

Also assume that there exists a strictly positive number β such that (2.7) *holds. Then, for any $\Delta t > 0$ small enough, for any solutions of* (3.1) *with initial data $z_0 \in \mathcal{C}(\delta/\Delta t)$, the observability inequality* (3.4) *holds in any time $T > \tilde{T}$, with a strictly positive observability constant $k_T > 0$ depending explicitly on the parameters β, $T-\tilde{T}$, $\tilde{k}$, the number N of low frequencies and the low frequency gap* (2.8).

Proof of Lemma 3.4. The proof of Lemma 3.4 closely follows the one of Lemma 2.3.

Fix $\Delta t > 0$. First remark that solutions of (3.1) can be written as

$$z^k = \sum_j a_j \Phi_j \exp(\lambda_{j,\Delta t} k\Delta t), \qquad \text{with } \lambda_{j,\Delta t} = \frac{1}{2\Delta t}\tan\left(\frac{\mu_j \Delta t}{2}\right).$$

Then define, similarly as in (2.8),

$$m_{j,\Delta t} = \frac{1}{2\Delta t}\tan\left(\frac{m_j\Delta t}{2}\right) \text{ and } \gamma_{\Delta t} = \inf_{j\in\{0,\cdots,N\}}\{m_{j+1,\Delta t} - m_{j,\Delta t}\}.$$

For simplicity, choose $\delta_{\Delta t}$ such that $\delta/\Delta t$ is an integer. Introduce, similarly as in (2.9),

$$v^k = z^k - \frac{\Delta t}{2\delta}\sum_{\ell\Delta t\in(-\delta,\delta)} \exp(im_{N,\Delta t}\ell\Delta t)\, z^{k-l}.$$

The proof of Lemma 3.4 then follows all along the line the one of Lemma 2.3, replacing all the continuous integrals in time by discrete summations and using the admissibility result (3.3) in the class $\mathcal{C}(\delta/\Delta t)$. This yields, similarly as in (2.17), that any solution of (3.1) with initial data $z_0 \in X_N + \mathcal{C}(\Omega)^\perp \cap \mathcal{C}(\delta/\Delta t)$ satisfies

$$\|z_0\|_X^2 \leq \Big[\frac{2}{\tilde{T}\beta^2} + \Big(\frac{2K_{\tilde{T}}}{\tilde{T}\beta^2} + 1\Big)\frac{4}{\tilde{k}(1-\operatorname{sinc}(\gamma_{\Delta t}\delta_{\Delta t}))^2}\Big]\Delta t \sum_{k\Delta t \in (0, \tilde{T}+2\delta_{\Delta t})} \|Bz^k\|_Y^2 .$$

Particularly, when Δt goes to zero, one can choose $(\delta_{\Delta t})$ converging to δ. Besides, when $\Delta t \to 0$, $(\gamma_{\Delta t})$ obviously converges to γ. We thus obtain, for $\Delta t > 0$ small enough, that any solution of (3.1) with initial data $z_0 \in X_N + \mathcal{C}(\Omega)^\perp \cap \mathcal{C}(\delta/\Delta t)$ satisfies

$$\|z_0\|_X^2 \leq \Big[\frac{2}{\tilde{T}\beta^2} + \Big(\frac{2K_{\tilde{T}}}{\tilde{T}\beta^2} + 1\Big)\frac{4}{\tilde{k}(1-\operatorname{sinc}(\gamma\delta))^2}\Big]\Delta t \sum_{k\Delta t \in (0, \tilde{T}+2\delta_{\Delta t})} \|Bz^k\|_Y^2 .$$

The inductive argument then again works and allows one to conclude Lemma 3.4. □

We now finish the proof of Theorem 3.2. Set $T^* > 0$.
• If $B \in \mathfrak{L}(\mathcal{D}(A^\kappa), Y)$ with $\kappa < 1$. Set $\delta > 0$. Choose M such that $\pi M(1 + \delta^2/4) = T^*/4$.
• If $B \in \mathfrak{L}(\mathcal{D}(A), Y)$. Set $\delta < \delta_0$, where δ_0 is

$$\pi m \|B\|_{\mathfrak{L}(\mathcal{D}(A),Y)} \frac{\delta_0^2}{2} = T^*/8.$$

Choose $M > 0$ such that

$$\pi\left[M^2\Big(1+\frac{\delta^2}{4}\Big)^2 + m^2\|B\|^2_{\mathfrak{L}(\mathcal{D}(A),Y)}\frac{\delta^4}{16}\right]^{1/2} = T^*/4.$$

Applying successively Lemmas 2.2 and 3.3, we prove uniform observability properties (3.7) for any solution of (3.1) with initial data $z_0 \in \mathcal{C}(\Omega)^\perp \cap \mathcal{C}(\delta/\Delta t)$ in time $T^*/2$. We then conclude as in the continuous case by Lemma 3.4 and estimate (2.7). □

Remark 3.5. Note that the approach developed in this section can also be applied to more general time discrete approximation schemes. We refer to [10] for the precise assumptions needed on the time discrete numerical schemes. Roughly speaking, any time discrete scheme which preserves the eigenvectors and for which the energy is constant enters into our setting. This includes, for instance, the fourth order Gauss method, or the Newmark method for the wave equation.

4 Schrödinger equations

In this section, we present an application to the above results of Schrödinger equations. Condition 2 is indeed typically satisfied by Schrödinger equations, and can be guaranteed when the corresponding wave equation is observable.

4.1 The continuous case

Let Ω be a smooth bounded domain of $\mathbb{R}^N$, and ω a subdomain of Ω.

Let us consider the following Schrödinger equation:

$$\begin{cases} i\dot{z} + \Delta z = 0, & \text{in } \Omega \times (0, \infty), \\ z = 0, & \text{on } \partial\Omega \times (0, \infty), \\ z(0) = z_0 \in L^2(\Omega), & \end{cases} \tag{4.1}$$

observed through $y(t) = \chi_\omega\, z(t)$, where $\chi_\omega = \chi_\omega(x)$ denotes the characteristic function of the set ω.

We thus consider the following observability property: for $T^* > 0$, find a strictly positive constant k_* such that any solution of (4.1) satisfies

$$k_* \left\| z_0 \right\|_{L^2(\Omega)}^2 \leq \int_0^{T^*} \left\| z(t) \right\|_{L^2(\omega)}^2 \, dt. \tag{4.2}$$

Note that this fits the abstract setting presented above: $X = L^2(\Omega)$, $A = i\Delta$ with Dirichlet boundary conditions, the domain of the operator A is $\mathcal{D}(A) = H^2 \cap H_0^1(\Omega)$ and B simply is the multiplication operator by χ_ω, which is continuous from $L^2(\Omega)$ to $L^2(\omega)$ (and therefore admissible).

For Schrödinger equations, due to the infinite velocity of propagation of rays, there are many cases in which the observability inequality (4.2) holds in any time $T^* > 0$, for instance, when the Geometric Control Condition (GCC) is satisfied in some time T.

The GCC in time T can be, roughly speaking, formulated as follows (see [2] for the precise setting): The subdomain ω of Ω is said to satisfy the GCC in time T if all rays of Geometric Optics that propagate inside the domain Ω at velocity one reach the set ω in time less than T.

Note that this is not a necessary condition. For instance, in [17], it has been proved that when the domain Ω is a square, for any non-empty bounded open subset ω, the observability inequality (4.2) holds for system (4.1). Other geometry has also been dealt with: we refer to [18, 19, 3, 1].

Similarly, Condition 2 is not guaranteed in general: it is indeed not clear whether the observability property in arbitrary small time for Schrödinger equation (4.1) implies Condition 2.

But there are several cases in which Condition 2 is satisfied: when it has been proven directly to prove observability in arbitrary small time (see for instance [5]), which has not been fully developed in the literature, or when the Geometric Control Condition is satisfied.

Theorem 4.1 ([21]). *Assume that the Geometric Control Condition holds. Then Condition 2 is satisfied for system* (4.1) *observed through* $y(t) = \chi_\omega\, z(t)$.

Before going into the proof, we recall that the Geometric Control Condition in time $T > 0$ is equivalent [4] to the exact observability property in time T of the corresponding wave equation

$$\begin{cases} \ddot{u} - \Delta u = 0, & \text{in } \Omega \times (0,\infty), \\ u = 0, & \text{on } \partial\Omega \times (0,\infty), \\ (u,\dot{u})(0) = (u_0,u_1) \in H_0^1(\Omega) \times L^2(\Omega), & \end{cases} \tag{4.3}$$

observed by $y(t) = \chi_\omega \dot{u}(t)$. In this case, the observability inequality reads as the existence of a strictly positive constant $c_T > 0$ such that solutions of (4.3) satisfy

$$c_T \left\|(u_0,u_1)\right\|^2_{H_0^1(\Omega)\times L^2(\Omega)} \leq \int_0^T \left\|\dot{u}(t)\right\|^2_{L^2(\omega)}\, dt. \tag{4.4}$$

It is then convenient to introduce an abstract setting, which generalizes Theorem 4.1.

Theorem 4.2 ([21]). *Let A_0 be a positive definite operator on X, and let B be a continuous operator from $\mathcal{D}(A_0^{1/2})$ to Y. Assume that the wave like equation*

$$\ddot{u} + A_0 u = 0, \quad t \geq 0, \qquad (u(0),\dot{u}(0)) = (u_0,u_1) \in \mathcal{D}(A_0^{1/2}) \times X \tag{4.5}$$

observed through

$$y(t) = B\dot{u}(t), \tag{4.6}$$

is admissible and exactly observable, meaning that there exist a time T and positive constants $c_T, K_T > 0$ such that solutions of (4.5) *satisfy*

$$c_T \left\|(u_0,u_1)\right\|^2_{\mathcal{D}(A_0^{1/2})\times X} \leq \int_0^T \left\|B\dot{u}(t)\right\|^2_Y\, dt \leq K_T \left\|(u_0,u_1)\right\|^2_{\mathcal{D}(A_0^{1/2})\times X}. \tag{4.7}$$

Then the operators $A = -iA_0$ and B satisfy Condition 2.

Particularly, the Schrödinger like equation

$$i\dot{z} = A_0 z, \quad t \geq 0, \qquad z(0) = z_0 \in X, \tag{4.8}$$

observed by

$$y(t) = Bz(t) \tag{4.9}$$

satisfies Condition 2 *and is therefore observable in arbitrary small time: for all* $T^* > 0$*, there exists a positive constant* $k_* > 0$ *such that any solution of* (4.8) *satisfies*

$$k_* \|z_0\|_X^2 \leq \int_0^{T^*} \|Bz(t)\|_Y^2 \, dt \tag{4.10}$$

Proof. Assume that (4.5)–(4.6) are exactly observable. Remark that, setting $\mathfrak{X} = \mathcal{D}(A_0^{1/2}) \times X$, and

$$\mathcal{A} = \begin{pmatrix} 0 & Id \\ -A_0 & 0 \end{pmatrix}, \quad \mathcal{B} = (\ 0\ ,\ B), \tag{4.11}$$

equation (4.5) fits the abstract setting given above. Particularly, the domain of $\mathcal{A}$ simply is $\mathcal{D}(A_0) \times \mathcal{D}(A_0^{1/2})$ and then the conditions $B \in \mathfrak{L}(\mathcal{D}(A_0^{1/2}), Y)$ and $\mathcal{B} \in \mathfrak{L}(\mathcal{D}(\mathcal{A}), Y)$ are equivalent.

The admissibility and observability properties (4.7) then imply (see [21]) Condition 1: There exist positive constants $M, m > 0$ such that

$$M^2 \left\| (\mathcal{A} - i\omega I) \begin{pmatrix} u \\ v \end{pmatrix} \right\|_{\mathfrak{X}}^2 + m^2 \left\| \mathcal{B} \begin{pmatrix} u \\ v \end{pmatrix} \right\|_Y^2 \geq \left\| \begin{pmatrix} u \\ v \end{pmatrix} \right\|_{\mathfrak{X}}^2, \qquad \forall \omega \in \mathbb{R}, \ \forall \begin{pmatrix} u \\ v \end{pmatrix} \in \mathcal{D}(\mathcal{A}). \tag{4.12}$$

Particularly, for all $\omega \in \mathbb{R}$ and $u \in \mathcal{D}(A_0)$, taking $v = i\omega u$ yields

$$\begin{aligned} M^2 \|(A_0 - \omega^2 I)u\|_X^2 + m^2\omega^2 \|Bu\|_Y^2 &\geq \left\| A_0^{1/2} u \right\|_X^2 + \omega^2 \|u\|_X^2 \\ &\geq \omega^2 \|u\|_X^2 . \end{aligned}$$

Hence

$$\frac{M^2}{\omega^2} \|(A_0 - \omega^2 I)u\|_X^2 + m^2 \|Bu\|_Y^2 \geq \|u\|_X^2 , \ \forall \omega \in \mathbb{R}, \forall u \in \mathcal{D}(A_0),$$

or, equivalently,

$$\frac{M^2}{\omega} \|(A_0 - \omega I)u\|_X^2 + m^2 \|Bu\|_Y^2 \geq \|u\|_X^2 , \ \forall \omega \in \mathbb{R}_+, \forall u \in \mathcal{D}(A_0). \tag{4.13}$$

Of course, this estimate does not hold for $\omega < 0$ and is not interesting for small values of ω. But this actually corresponds to the easy case.

Indeed, if $\omega < \lambda_1(A_0)$, where $\lambda_1(A_0)$ is the first eigenvalue of A_0 (which is strictly positive since A_0 is positive definite),

$$\|(A_0 - \omega I)u\|_X^2 \geq (\lambda_1(A_0) - \omega)^2 \|u\|_X^2, \quad \forall u \in \mathcal{D}(A).$$

Combining with (4.13), by taking

$$M(\omega) = \begin{cases} \dfrac{M}{\sqrt{\omega}} & \text{for } \omega > \dfrac{\lambda_1(A_0)}{2}, \\ \dfrac{1}{\lambda_1(A_0) - \omega} & \text{for } \omega \leq \dfrac{\lambda_1(A_0)}{2}, \end{cases}$$

we then obtain

$$M(\omega)^2 \|(A_0 - \omega I)u\|_X^2 + m^2 \|Bu\|_Y^2 \geq \|u\|_X^2, \ \forall \omega \in \mathbb{R}, \forall u \in \mathcal{D}(A_0). \tag{4.14}$$

This completes the proof of Theorem 4.2 and, as a particular instance of it, of Theorem 4.1. □

The interest of this approach is that it also applies to space semi-discrete, as well as fully discrete approximation schemes of (4.8) and (4.1) in particular.

4.2 Space semi-discrete approximation schemes

Let us now introduce the finite element method to (4.8).

Let $(V_h)_{h>0}$ be a sequence of vector spaces of finite dimension n_h which embed into X via a linear injective map $\pi_h : V_h \to X$. For each $h > 0$, the inner product $< \cdot, \cdot >_X$ in X induces a structure of Hilbert space for V_h endowed with the scalar product $< \cdot, \cdot >_h = < \pi_h \cdot, \pi_h \cdot >_X$.

We assume that for each $h > 0$, the vector space $\pi_h(V_h)$ is a subspace of $\mathcal{D}(A_0^{1/2})$. We thus define the linear operator $A_{0h} : V_h \to V_h$ by

$$< A_{0h}\phi_h, \psi_h >_h = < A_0^{1/2}\pi_h\phi_h, A_0^{1/2}\pi_h\psi_h >_X, \quad \forall (\phi_h, \psi_h) \in V_h^2. \tag{4.15}$$

The operator A_{0h} defined in (4.15) obviously is self-adjoint and positively definite. If we introduce the adjoint π_h^* of π_h, definition (4.15) reads:

$$A_{0h} = \pi_h^* A_0 \pi_h. \tag{4.16}$$

This operator A_{0h} corresponds to the finite element discretization of the operator A_0. We thus consider the following space semi-discretization of (4.8):

$$i\dot{z}_h = A_{0h} z_h, \quad t \in \mathbb{R}, \qquad z_h(0) = z_{0h} \in V_h. \tag{4.17}$$

In this context, for all $h > 0$, the observation then naturally becomes

$$y_h(t) = B\pi_h z_h = B_h z_h. \tag{4.18}$$

Note that we shall impose $B \in \mathfrak{L}(\mathcal{D}(A_0^{1/2}), Y)$ for this definition to make sense.

We now make precisely the assumptions we have, usually, on π_h, which will be needed in our analysis. One easily checks that $\pi_h^* \pi_h = Id_h$. The injection π_h describes the finite element approximation we have chosen. Particularly, the vector space $\pi_h(V_h)$ approximates, in the sense given hereafter, the space $\mathcal{D}(A_0^{1/2})$: There exist $\theta > 0$ and $C_0 > 0$, such that for all $h > 0$,

$$\begin{cases} \left\| A_0^{1/2}(\pi_h \pi_h^* - I)\phi \right\|_X \leq C_0 \left\| A_0^{1/2} \phi \right\|_X, & \forall \phi \in \mathcal{D}(A_0^{1/2}), \\ \left\| A_0^{1/2}(\pi_h \pi_h^* - I)\phi \right\|_X \leq C_0 h^\theta \left\| A_0 \phi \right\|_X, & \forall \phi \in \mathcal{D}(A_0). \end{cases} \tag{4.19}$$

When considering finite element discretizations of the Schrödinger equation (4.1), which, as we said, corresponds to taking A_0 as the Laplace operator with Dirichlet boundary conditions, estimates (4.19) are satisfied [25] for $\theta = 1$ when using P1 finite elements on a regular mesh (in the sense of finite elements).

We will not discuss convergence results of the numerical approximation schemes presented here, which are classical under assumption (4.19), and can be found for instance in [25].

In [8, 9], we proved uniform observability properties for (4.17)–(4.18) in classes of conveniently filtered initial data. In the sequel, our goal is to obtain uniform observability properties for (4.17) similar to (4.10), but in arbitrary small time, also for conveniently filtered initial data.

Therefore, we shall introduce the filtered classes of data. For all $h > 0$, since A_{0h} is a self-adjoint positive definite operator, the spectrum of A_{0h} is given by a sequence of positive eigenvalues

$$0 < \lambda_1^h \leq \lambda_2^h \leq \cdots \leq \lambda_{n_h}^h, \tag{4.20}$$

and normalized (in V_h) eigenvectors $(\Phi_j^h)_{1 \leq j \leq n_h}$. For any $s > 0$, we can now define, for any $h > 0$, the filtered space

$$\mathcal{C}_h(s) = \operatorname{span}\left\{ \Phi_j^h \text{ with the corresponding eigenvalue satisfying } |\lambda_j^h| \leq s \right\}.$$

We have then proved in Theorem 1.3 in [8]:

Theorem 4.3. *Let A_0 be a self-adjoint positive definite operator with compact resolvent, and $B \in \mathfrak{L}(\mathcal{D}(A_0^\kappa), Y)$, with $\kappa < 1/2$. Assume that the maps $(\pi_h)_{h>0}$ satisfy property* (4.19). *Set*

$$\sigma = \theta \min\left\{ 2(1 - 2\kappa), \frac{2}{5} \right\}. \tag{4.21}$$

Assume that systems (4.8)–(4.9) *are admissible and exactly observable. Then there exist* $\varepsilon > 0$, *a time* T^* *and positive constants* $k_*, K_* > 0$ *such that, for any* $h > 0$, *any solution of* (4.17) *with initial data*

$$z_{0h} \in \mathcal{C}_h(\varepsilon/h^\sigma) \tag{4.22}$$

satisfies

$$k_* \|z_{0h}\|_h^2 \leq \int_0^{T^*} \|B_h z_h(t)\|_Y^2 \, dt \leq K_* \|z_{0h}\|_h^2 . \tag{4.23}$$

In this result, based on spectral criteria for the admissibility and admissibility of Schrödinger operators, the time of observability T^* cannot be made as small as desired.

When the Geometric Control Condition is satisfied, the following has been proved in Theorem 8.3 in [9] as a by product on our analysis of the abstract wave like equation (4.5):

Theorem 4.4. *Let* A_0 *be a positive definite unbounded operator with compact resolvent and* $B \in \mathfrak{L}(\mathcal{D}(A_0^\kappa), Y)$, *with* $\kappa < 1/2$. *Assume that the approximations* $(\pi_h)_{h>0}$ *satisfy property* (4.19). *Set*

$$\varsigma = \theta \min\left\{2(1-2\kappa), \frac{2}{3}\right\}. \tag{4.24}$$

Assume that systems (4.5)–(4.6) *is admissible and exactly observable. Then there exist* $\varepsilon > 0$, *a time* T^* *and positive constants* $k_*, K_* > 0$ *such that, for any* $h > 0$, *any solution of* (4.17) *with initial data in*

$$z_{0h} \in \mathcal{C}_h(\varepsilon/h^\varsigma) \tag{4.25}$$

satisfies (4.23).

Theorem 4.4 indeed improves Theorem 4.3 since $\varsigma \geq \sigma$. This is expected since the assumptions of admissibility and observability for the abstract wave systems (4.5)–(4.6) are stronger than the admissibility and observability of Schrödinger equations (4.8)–(4.9).

The proof of Theorem 4.4 is made in [9]. However, Theorem 4.4 requires the time of observability to be large enough. We shall prove below that it can actually be chosen to be arbitrarily small.

Theorem 4.5. *Under the assumptions of Theorem 4.4, assume that systems* (4.5)–(4.6) *are admissible and exactly observable. Then there exists* $\varepsilon > 0$ *such that for all* $T^* > 0$, *there exist positive constants* $k_*, K_* > 0$ *such that, for any* $h > 0$, *any solution of* (4.17) *with initial data in* (4.25) *satisfies* (4.23).

Proof. The admissibility result of (4.23) follows from the one in Theorem 4.4 since, when the admissibility inequality holds for some time $T > 0$, it holds for any time. We shall thus not deal further with that question.

Assume that systems (4.5)–(4.6) are admissible and exactly observable. Then we can use Theorem 1.1 in [9], which states that, under the assumptions of Theorem 4.4, the space semi-discrete wave systems

$$\ddot{u}_h + A_{0h} u_h = 0, \quad t \geq 0, \qquad y_h(t) = B_h \dot{u}_h,$$

are

- uniformly (with respect to $h > 0$) admissible for any initial data $(u_{0h}, u_{1h}) \in \mathcal{C}_h(\eta\, h^{-\varsigma})^2$, whatever $\eta > 0$ is.
- uniformly (with respect to $h > 0$) observable in some time $T > 0$ for initial data $(u_{0h}, u_{1h}) \in \mathcal{C}_h(\varepsilon\, h^{-\varsigma})^2$, provided ε is small enough.

These uniform admissibility and observability properties imply, as proved in [21], that the resolvent condition (4.12) for the operators A_{0h} and B_h holds uniformly with respect to h for data $z_h \in \mathcal{C}_h(\varepsilon/h^{\varsigma})$. The proof of Theorem 4.2 then gives that, uniformly with respect to $h > 0$, we can find positive constants $M,\ m > 0$ such that

$$\frac{M^2}{\omega} \left\| (A_{0h} - \omega I_h) u_h \right\|_h^2 + m^2 \left\| B_h u_h \right\|_Y^2 \geq \left\| u_h \right\|_h^2, \qquad \forall \omega \in \mathbb{R}_+, \forall u_h \in \mathcal{C}_h(\varepsilon/h^{\varsigma}).$$

To conclude that Condition 2 is uniformly satisfied, following the proof of Theorem 4.2, we only need to check that the first eigenvalue λ_1^h corresponding to the operator A_{0h} stays away from 0. But, writing the Rayleigh coefficient which characterizes λ_1^h and $\lambda_1(A_0)$, one instantaneously checks that $\lambda_1^h \geq \lambda_1(A_0) > 0$ for all $h > 0$.

In other words, we have proved that there exists a bounded positive function $M = M(\omega)$ satisfying $\lim_{|\omega| \to \infty} M(\omega) = 0$ and a positive constant $m > 0$ such that for all $h > 0$

$$M(\omega)^2 \left\| (A_{0h} - \omega I_h) u_h \right\|_h^2 + m^2 \left\| B_h u_h \right\|_Y^2 \geq \left\| u_h \right\|_h^2, \qquad \forall \omega \in \mathbb{R}, \forall u_h \in \mathcal{C}_h(\varepsilon/h^{\varsigma}). \quad (4.26)$$

Now, we use our constructive proof of Theorem 1.3 to deduce uniform observability properties in any time T^*. However, though this might seem at first a direct consequence of Theorem 1.3, one needs to be cautious.

Following the proof of Theorem 1.3, we see that the high-frequency components can be dealt with uniformly without modification. Particularly, for all $\tilde{T} > 0$, there exists Ω such that, for all $h > 0$, any solution of (4.17) with initial data $z_{0h} \in \mathcal{C}_h(\Omega)^\perp \cap \mathcal{C}_h(\varepsilon/h^\varsigma)$ satisfies

$$\tilde{k} \|z_{0h}\|_h^2 \leq \int_0^{\tilde{T}} \|B_h z_h(t)\|_Y^2 \, dt, \tag{4.27}$$

for some positive constant $\tilde{k} > 0$ independent of $h > 0$.

Besides the systems (4.17)–(4.18) are uniformly admissible because of Theorem 4.4.

But the low-frequency components require an estimate on the low-frequency gap for each $h > 0$. The constant Ω is independent of $h > 0$ and setting $(m_j^h)_{j\in\{1,\cdots,N_h\}}$ for the increasing sequence of the values taken by the eigenvalues of A_{0h} which are smaller than Ω, we shall estimate

$$\gamma_h = \inf_{j\in\{0,\cdots,N_h\}} \{m_{j+1}^h - m_j^h\} \text{ where } m_0^h = -\Omega \text{ and } m_{N_h+1}^h = \Omega. \tag{4.28}$$

Note particularly that N_h might depend on h. However, since all these correspond to the discrete spectrum of A_{0h}, it shall converge to the spectrum of A_0.

Case 1: Each eigenvalue of the spectrum of A_0 is simple. Then the convergence of the discrete spectrum of A_{0h} in the band of eigenvalues smaller than the constant Ω is guaranteed [25]. Particularly, N_h is constant for $h > 0$ small enough and the sequence (γ_h) then simply converges to γ.

Case 2: The general case. When the spectrum of A_0 is not simple, this is harder since a multiple eigenvalue of the continuous operator may yield different but close eigenvalues, making γ_h dangerously small for our argument. The idea then is to refine Haraux' argument, and to think directly about this convergence property of the spectrum.

For each positive $\alpha > 0$ smaller than $\gamma/4$ (γ being the continuous low frequency gap defined in (2.8)), there exists $h_\alpha > 0$ such that for $h \in (0, h_\alpha)$, the spectrum of the operator A_{0h} satisfies

$$\{\lambda_\ell^h \text{ such that } \lambda_\ell^h < \Omega\} \subset \bigcup_{j\in\{1,\cdots,N\}} [m_j - \alpha, m_j + \alpha]. \tag{4.29}$$

Define then the sets $X_j^{h,\alpha} = \text{span}\{\Phi_\ell^h \text{ such that } |\lambda_\ell^h - m_j| \leq \alpha\}$. Since the discrete operators satisfy (4.26), further assuming that α is smaller than $1/(2\sup M(\omega))$, choosing for instance $\beta = 1/(2m)$, we obtain

$$\forall j \in \{1,\cdots,N\},\ \forall z \in X_j^{h,\alpha}, \quad \|B_h z_h\|_Y \geq \beta \|z\|_h. \tag{4.30}$$

Once we have seen (4.29)–(4.30), the inductive argument developed in Lemma 2.3 works as before, except some small error terms. Let us present it briefly below at the first step.

To write it properly, we shall introduce the orthogonal projections $\mathbb{P}_N^{h,\alpha}$ on $X_N^{h,\alpha}$ and $\mathbb{P}_{hf}^h$ on $\mathcal{C}_h(\Omega)^\perp$, respectively.

Set then $z_{0h} \in X_N^{h,\alpha} + \mathcal{C}_h(\Omega)^\perp \cap \mathcal{C}_h(\varepsilon/h^\varsigma)$, and decompose it into $z_{0h,N} = \mathbb{P}_N^{h,\alpha} z_{0h}$ and $z_{0h,hf} = \mathbb{P}_{hf}^h z_{0h}$. Let $z_h(t)$ be the solution of (4.17) with intial data z_{0h} and, for $\delta > 0$, define v_h as

$$v_h(t) = z_h(t) - \frac{1}{2\delta}\int_{-\delta}^{\delta} e^{im_N s} z_h(t-s)\, ds.$$

Expanding z_{0h} on the basis of Φ_j^h, similarly as in (2.11), we obtain

$$\begin{cases} \left\|\mathbb{P}_{hf}^h z_{0h}\right\|_h^2 \leq \dfrac{1}{(1-\operatorname{sinc}(\delta\gamma))^2}\left\|v_h(0)\right\|_X^2 . \\ \left\|\mathbb{P}_N^{h,\alpha} v_h(0)\right\|_h^2 \leq (1-\operatorname{sinc}(\alpha\delta))^2 \left\|\mathbb{P}_N^{h,\alpha} z_{0h}\right\|_h^2 . \end{cases} \tag{4.31}$$

Besides, v_h is a solution of (4.17), so is $\mathbb{P}_{hf}^h v_h$. But $\mathbb{P}_{hf}^h v_h$ lies in $\mathcal{C}_h(\Omega)^\perp \cap \mathcal{C}_h(\varepsilon/h^\varsigma)$, and then one can use (4.27):

$$\begin{aligned} \left\|\mathbb{P}_{hf}^h v_h(0)\right\|_X^2 &\leq \frac{1}{\tilde{k}}\int_0^{\tilde{T}} \left\|B_h \mathbb{P}_{hf}^h v_h(t)\right\|_Y^2 \, dt \\ &\leq \frac{2}{\tilde{k}}\int_0^{\tilde{T}} \left\|B_h v_h(t)\right\|_Y^2 \, dt + \frac{2K_{\tilde{T}}}{\tilde{k}}\left\|\mathbb{P}_N^{h,\alpha} v_h(0)\right\|_h^2 \\ &\leq \frac{2}{\tilde{k}}\int_0^{\tilde{T}} \left\|B_h v_h(t)\right\|_Y^2 \, dt + \frac{2K_{\tilde{T}}}{\tilde{k}}(1-\operatorname{sinc}(\alpha\delta))^2 \left\|\mathbb{P}_N^{h,\alpha} z_h(0)\right\|_h^2 \end{aligned} \tag{4.32}$$

Using the same estimates as in (2.13), combining with (2.11), we get

$$\begin{aligned} \left\|\mathbb{P}_{hf}^h z_{0h}\right\|_h^2 \leq{}& \frac{4}{\tilde{k}(1-\operatorname{sinc}(\gamma\delta))^2}\int_{-\delta}^{\tilde{T}+\delta} \left\|B_h z_h(t)\right\|_Y^2 \, dt \\ &+ \frac{2K_{\tilde{T}}}{\tilde{k}}\left(\frac{1-\operatorname{sinc}(\alpha\delta)}{1-\operatorname{sinc}(\gamma\delta)}\right)^2 \left\|\mathbb{P}_N^{h,\alpha} z_h(0)\right\|_h^2 . \end{aligned} \tag{4.33}$$

We then focus on the component of the solution in $X_N^{h,\alpha}$. Arguing as in (2.15) and using (4.30), we obtain

$$\left\|\mathbb{P}_N^{h,\alpha} z_{0h}\right\|_h^2 \leq \frac{2}{\tilde{T}\beta^2}\int_0^{\tilde{T}} \left\|B_h z_h(t)\right\|_Y^2 \, dt + \frac{2K_{\tilde{T}}}{\tilde{T}\beta^2}\left\|\mathbb{P}_{hf}^h z_{0h}\right\|_h^2 . \tag{4.34}$$

Equations (4.33) and (4.34), together, give

$$\left\|\mathbb{P}_N^{h,\alpha} z_{0h}\right\|_h^2 \left(1 - \frac{4K_{\tilde{T}}^2}{\tilde{T}\beta^2\tilde{k}}\left(\frac{1-\operatorname{sinc}(\alpha\delta)}{1-\operatorname{sinc}(\gamma\delta)}\right)^2\right)$$
$$\leq \left(\frac{2}{\tilde{T}\beta^2} + \frac{8K_{\tilde{T}}}{\tilde{T}\beta^2\tilde{k}(1-\operatorname{sinc}(\gamma\delta))^2}\right)\int_{-\delta}^{\tilde{T}+\delta} \|B_h z_h(t)\|_Y^2 \, dt. \quad (4.35)$$

Particularly, if one can guarantee that the left hand side is positive, which can be done simply by choosing $\alpha > 0$ small enough and

$$(1-\operatorname{sinc}(\alpha\delta))^2 \leq \frac{\tilde{T}\beta^2\tilde{k}}{16K_{\tilde{T}}^2}(1-\operatorname{sinc}(\gamma\delta))^2, \quad (4.36)$$

we deduce

$$\left\|\mathbb{P}_N^{h,\alpha} z_{0h}\right\|_h^2 \leq \left(\frac{4}{\tilde{T}\beta^2} + \frac{16K_{\tilde{T}}}{\tilde{T}\beta^2\tilde{k}(1-\operatorname{sinc}(\gamma\delta))^2}\right)\int_{-\delta}^{\tilde{T}+\delta} \|B_h z_h(t)\|_Y^2 \, dt.$$

From (4.33), we obtain an estimate for $\left\|\mathbb{P}_{hf}^h z_{0h}\right\|_h^2$. Using the orthogonality of $X_N^{h,\alpha}$ and $\mathcal{C}_h(\Omega)^\perp \cap \mathcal{C}_h(\varepsilon/h^\varsigma)$, this proves that the observability inequality holds in any time $T > \tilde{T}$, uniformly with respect to $h \in (0, h_\alpha)$, for solutions of (4.17) with initial data in $X_N^{h,\alpha} + \mathcal{C}_h(\Omega)^\perp \cap \mathcal{C}_h(\varepsilon/h^\varsigma)$.

Note that (4.36) does not depend on $h > 0$. Thus, once α is chosen according to (4.36), the above proof stands for any $h \in (0, h_\alpha)$.

This concludes the inductive argument, and this slightly generalized Haraux's technique can be applied to concluding the proof of Theorem 4.5. □

4.3 Fully discrete approximation schemes

We can also prove observability properties for fully discrete approximations of (4.8)–(4.9), uniformly with both discretization parameters $\Delta t > 0$ and $h > 0$, in arbitrary small time.

To be more precise, we consider, for $h > 0$ and $\Delta t > 0$, the following system:

$$\begin{cases} i\left(\dfrac{z_h^{k+1} - z_h^k}{\Delta t}\right) = A_{0h}\Big(\dfrac{z_h^k + z_h^{k+1}}{2}\Big), & \text{in } V_h, \quad k \in \mathbb{Z}, \\ z_h^0 = z_{0h}, \end{cases} \quad (4.37)$$

observed by

$$y_h^k = B_h z_h^k. \quad (4.38)$$

For these systems, admissibility and observability results have been derived in [8] using [10] in the class $\mathcal{C}_h(\delta/\Delta t) \cap \mathcal{C}_h(\varepsilon h^{-\sigma})$, with σ in (4.21), but the observability results in [8] need the time T to be large enough. Later in [9], these admissibility and observability results have been improved by using the Geometric Control Condition, yielding the filtering class $\mathcal{C}_h(\delta/\Delta t) \cap \mathcal{C}_h(\varepsilon/h^{\varsigma})$ with ς in (4.24), but the observability time is again required to be large enough.

However, using [9] and the techniques developed above, we can prove that the discrete systems (4.37)–(4.38) actually are observable in arbitrary small time.

Theorem 4.6. *Under the assumptions of Theorem 4.4. Assume that systems* (4.5)–(4.6) *are admissible and exactly observable. Then, for any time $T^* > 0$, for any $\delta > 0$, there exist two positive constants $\varepsilon > 0$ and $k_{\delta,T^*} > 0$ such that, for all $h, \Delta t > 0$ small enough, any solution of* (4.37) *with initial data*

$$z_{0h} \in \mathcal{C}_h\Big(\inf\Big\{ \frac{\delta}{\Delta t}, \frac{\varepsilon}{h^{\varsigma}} \Big\} \Big),$$

where ς is given by (4.24), *satisfies*

$$k_{\delta,T^*} \left\| z_{0h} \right\|_h^2 \leq \Delta t \sum_{k\Delta t \in (0,T^*)} \left\| B_h z_h^k \right\|_Y^2 . \tag{4.39}$$

The proof of Theorem 4.6, which can be adapted easily from the previous theorems, is left to the reader. The keynote is the convergence of the low components of the spectrum and the fact that all the above proofs are explicit and shortcut any compacity argument.

5 Further comments

This work is based on the resolvent estimate given in Condition 2. Under Condition 2, observability properties hold in arbitrary small time. However, there might be systems fitting the abstract setting (1.1)–(1.2) which are observable in arbitrary small time but for which Condition 2 does not hold. In this sense, we did not completely solve the problem.

This is actually part of a more general question: can we read on the operators A and B and their spectral properties the critical time of observability ? To our knowledge, this is still not clear if the resolvent estimates keep precisely track of this information, which is of primary importance in applications, for instance, when dealing with waves. It would then be interesting to try to design an efficient spectral characterization of the time of observability.

References

[1] B. Allibert. Contrôle analytique de l'équation des ondes et de l'équation de Schrödinger sur des surfaces de revolution. *Comm. Partial Differential Equations*, 23(9-10): 1493–1556, 1998.

[2] C. Bardos, G. Lebeau and J. Rauch. Sharp sufficient conditions for the observation, control and stabilization of waves from the boundary. *SIAM J. Control and Optim.*, 30(5): 1024–1065, 1992.

[3] N. Burq. Contrôle de l'équation des plaques en présence d'obstacles strictement convexes. *Mém. Soc. Math. Fr. (N.S.)*, (55): 126, 1993.

[4] N. Burq and P. Gérard. Condition nécessaire et suffisante pour la contrôlabilité exacte des ondes. *C. R. Acad. Sci. Paris Sér. I Math.*, 325(7): 749–752, 1997.

[5] N. Burq and M. Zworski. Geometric control in the presence of a black box. *J. Amer. Math. Soc.*, 17(2): 443–471 (electronic), 2004.

[6] C. Castro and S. Micu. Boundary controllability of a linear semi-discrete 1-d wave equation derived from a mixed finite element method. *Numer. Math.*, 102(3): 413–462, 2006.

[7] S. Ervedoza. Observability of the mixed finite element method for the 1d wave equation on nonuniform meshes. *To appear in ESAIM Control Optim. Calc. Var.*

[8] S. Ervedoza. Admissibility and observability for Schrödinger systems: Applications to finite element approximation schemes. *Preprint*, 2008.

[9] S. Ervedoza. Admissibility and observability for Wave systems: Applications to finite element approximation schemes. *Preprint*, 2008.

[10] S. Ervedoza, C. Zheng and E. Zuazua. On the observability of time-discrete conservative linear systems. *J. Funct. Anal.*, 254(12): 3037–3078, 2008.

[11] R. Glowinski. Ensuring well-posedness by analogy: Stokes problem and boundary control for the wave equation. *J. Comput. Phys.*, 103(2): 189–221, 1992.

[12] R. Glowinski, C. H. Li and J.-L. Lions. A numerical approach to the exact boundary controllability of the wave equation. I. Dirichlet controls: description of the numerical methods. *Japan J. Appl. Math.*, 7(1): 1–76, 1990.

[13] R. Glowinski, J.-L. Lions and J. He. *Exact and approximate controllability for distributed parameter systems*, volume 117 of Encyclopedia of Mathematics and its Applications. Cambridge University Press, Cambridge, 2008.

[14] A. Haraux. Séries lacunaires et contrôle semi-interne des vibrations d'une plaque rectangulaire. *J. Math. Pures Appl. (9)*, 68(4): 457–465 (1990), 1989.

[15] J.A. Infante and E. Zuazua. Boundary observability for the space semi discretizations of the 1-d wave equation. *Math. Model. Num. Ann.*, 33: 407–438, 1999.

[16] A. E. Ingham. Some trigonometrical inequalities with applications to the theory of series. *Math. Z.*, 41(1): 367–379, 1936.

[17] S. Jaffard. Contrôle interne exact des vibrations d'une plaque rectangulaire. *Port. Math.*, 47(4): 423–429, 1990.

[18] V. Komornik. A new method of exact controllability in short time and applications. *Ann. Fac. Sci. Toulouse Math. (5)*, 10(3): 415–464, 1989.

[19] V. Komornik. On the exact internal controllability of a Petrowsky system. *J. Math. Pures Appl. (9)*, 71(4): 331–342, 1992.

[20] J.-L. Lions. *Contrôlabilité exacte, Stabilisation et Perturbations de Systèmes Distribués. Tome 1. Contrôlabilité exacte*, volume RMA 8. Masson, 1988.

[21] L. Miller. Controllability cost of conservative systems: resolvent condition and transmutation. *J. Funct. Anal.*, 218(2): 425–444, 2005.

[22] A. Münch and A. F. Pazoto. Uniform stabilization of a viscous numerical approximation for a locally damped wave equation. *ESAIM Control Optim. Calc. Var.*, 13(2): 265–293 (electronic), 2007.

[23] M. Negreanu and E. Zuazua. Convergence of a multigrid method for the controllability of a 1-d wave equation. *C. R. Math. Acad. Sci. Paris*, 338(5): 413–418, 2004.

[24] K. Ramdani, T. Takahashi, G. Tenenbaum and M. Tucsnak. A spectral approach for the exact observability of infinite-dimensional systems with skew-adjoint generator. *J. Funct. Anal.*, 226(1): 193–229, 2005.

[25] P.-A. Raviart and J.-M. Thomas. *Introduction à l'analyse numérique des équations aux dérivées partielles.* Collection Mathématiques Appliquées pour la Maitrise. [Collection of Applied Mathematics for the Master's Degree]. Masson, Paris, 1983.

[26] M. Tucsnak and G. Weiss. Observation and control for operator semigroups, 2008.

[27] E. Zuazua. Propagation, observation, and control of waves approximated by finite difference methods. *SIAM Rev.*, 47(2): 197–243 (electronic), 2005.

Logarithmic Decay of Hyperbolic Equations with Arbitrary Small Boundary Damping*

Xiaoyu Fu
School of Mathematics, Sichuan University
Chengdu 610064, China
Email: rj_xy@163.com

Abstract

This paper is addressed to an analysis of the longtime behavior of the hyperbolic equations with a partially boundary damping, under sharp regularity assumptions on the coefficients appearing in the equation. Based on a global Carleman estimate, we establish an estimate on the underlying resolvent operator of the equation, via which we show the logarithmic decay rate for solutions of the hyperbolic equations without any geometric assumption on the subboundary in which the damping is effective.

1 Introduction and main results

Let $\Omega \subset \mathbb{R}^n$ ($n \in \mathbb{N}$) be a bounded domain with boundary $\partial\Omega$ of class C^2. Denote by $\nu = (\nu_1, \cdots, \nu_n)$ the unit outward normal field along the boundary $\partial\Omega$, and $\overline{\Omega}$ the closure of Ω. For simplicity, we use the notation $u_j = \frac{\partial u}{\partial x_j}$, where x_j is the j-th coordinate of a point $x = (x_1, \cdots, x_n)$ in $\mathbb{R}^n$. In a similar manner, we use the notations w_j, v_j, etc. for the partial derivatives of w and v with respect to x_j. By $\overline{c}$ we denote the complex conjugate of $c \in \mathbb{C}$. Throughout this paper, we will use C to denote a positive constant which may vary from line to line (unless otherwise stated).

Let $a^{jk}(\cdot) \in C^2(\overline{\Omega}; \mathbb{R})$ be fixed functions satisfying

$$a^{jk}(x) = a^{kj}(x), \qquad \forall\, x \in \overline{\Omega},\ j,k = 1,2,\cdots,n, \tag{1.1}$$

and for some constant $s_0 > 0$,

$$\sum_{j,k=1}^{n} a^{jk}(x)\xi^j\overline{\xi}^k \geq s_0|\xi|^2, \qquad \forall\, (x,\xi) \in \overline{\Omega} \times \mathbb{C}^n, \tag{1.2}$$

*The author was partially supported by the NSF of China (No. 10525105, No. 10771149 and No. 10831007).

where $\xi = (\xi^1, \cdots, \xi^n)$.

Next, we fix a function $a(\cdot) \in L^\infty(\partial\Omega;\ \mathbb{R}^+)$ satisfying

$$\Gamma_0 \stackrel{\triangle}{=} \{x \in \partial\Omega;\ a(x) > 0\} \neq \emptyset. \tag{1.3}$$

The main purpose of this paper is to study the longtime behavior of solutions of the following hyperbolic equation with a boundary damping term $a(x)u_t$:

$$\begin{cases} u_{tt} - \sum_{j,k=1}^{n} (a^{jk}u_j)_k = 0 & \text{in } \mathbb{R}^+ \times \Omega, \\ \sum_{j,k=1}^{n} a^{jk}u_j\nu_k + a(x)u_t = 0 & \text{on } \mathbb{R}^+ \times \partial\Omega, \\ (u(0), u_t(0)) = (u^0, u^1) & \text{in } \Omega. \end{cases} \tag{1.4}$$

Put

$$H \stackrel{\triangle}{=} \left\{ (f, g) \in H^1(\Omega) \times L^2(\Omega) \;\middle|\; \int_\Omega f dx = 0 \right\},$$

which is a Hilbert space, whose norm is given by

$$||(f, g)||_H = \sqrt{\int_\Omega \Big[\sum_{j,k=1}^{n} a^{jk} f_j \overline{f_k} + |g|^2 \Big] dx}, \qquad \forall\, (f, g) \in H.$$

Define an unbounded operator $\mathcal{A} : H \to H$ by (recalling that $u_j^0 = \frac{\partial u^0}{\partial x_j}$)

$$\begin{cases} \mathcal{A} \stackrel{\triangle}{=} \begin{pmatrix} 0 & I \\ \sum_{j,k=1}^{n} \partial_k(a^{jk}\partial_j) & 0 \end{pmatrix}, \\ D(\mathcal{A}) \stackrel{\triangle}{=} \Big\{ u = (u^0, u^1) \in H; \mathcal{A}u \in H; \big(\sum_{j,k=1}^{n} a^{jk}u_j^0\nu_k + au^1 \big)\Big|_{\partial\Omega} = 0 \Big\}. \end{cases} \tag{1.5}$$

It is easy to show that $\mathcal{A}$ generates a C_0-semigroup $\{e^{t\mathcal{A}}\}_{t\in\mathbb{R}}$ on H. Therefore, system (1.4) is well posed in H. Clearly, H is the *finite energy space* of system (1.4). By the classical energy method, it is easy to check that

$$\frac{d}{dt}||(u, u_t)||_H^2 = -2 \int_{\Gamma_0} a(x)|u_t|^2 d\Gamma_0. \tag{1.6}$$

Formula (1.6) shows that the only dissipative mechanism acting on the system is through the sub-boundary Γ_0.

According to the energy dissipation law (1.6) and the well-known unique continuation property for solutions of the wave equation, it is easy to show that there are no nonzero solutions of (1.4) which conserve energy. Hence, using LaSalle's invariance principle ([12, p.18]), we may conclude that the energy of every solution of (1.4) tends to zero as $t \to \infty$, without any geometric conditions on the domain Ω. This paper is devoted to analyzing further the decay rate of solutions of system (1.4) tending to zero as $t \to \infty$. In this respect, very interesting logarithmic decay result was given in [14] for the above system under the regularity assumption that the coefficients $a^{jk}(\cdot)$ and $a(\cdot)$, and the boundary $\partial\Omega$ are C^∞-smooth. Note that since the sub-boundary Γ_0 in which the damping $a(x)u_t$ is effective may be very "small" with respect to the whole boundary $\partial\Omega$, the "geometric optics condition" introduced in [1] is not guaranteed for system (1.4), and therefore, in general, one can not expect exponential stability of this system. On the other hand, as pointed in [14], for some special case of system (1.4), logarithmic stability is the best decay rate.

The main results of this paper are stated as follows:

Theorem 1.1. Let $a^{jk}(\cdot) \in C^2(\overline{\Omega};\ \mathbb{R})$ satisfy (1.1)–(1.2), and $a(\cdot) \in L^\infty(\partial\Omega;\ \mathbb{R}^+)$ satisfy (1.3). Then solutions $e^{t\mathcal{A}}(u^0, u^1) \equiv (u, u_t) \in C(\mathbb{R};\ D(\mathcal{A})) \cap C^1(\mathbb{R};\ H)$ of system (1.4) satisfy

$$\begin{aligned} ||e^{t\mathcal{A}}(u^0,u^1)||_H \le \frac{C}{\ln(2+t)}||(u^0,u^1)||_{D(\mathcal{A})}, \\ \forall\,(u^0,u^1) \in D(\mathcal{A}),\ \forall\, t>0. \end{aligned} \tag{1.7}$$

Following [3] (see also [5]), Theorem 1.1 is a consequence of the following resolvent estimate for operator $\mathcal{A}$:

Theorem 1.2. There exists a constant $C > 0$ such that for any

$$\operatorname{Re}\lambda \in \left[-\frac{e^{-C|\operatorname{Im}\lambda|}}{C}, 0\right],$$

we have

$$||(\mathcal{A}-\lambda I)^{-1}||_{\mathcal{L}(H)} \le Ce^{C|\operatorname{Im}\lambda|}, \qquad \text{for } |\lambda| > 1.$$

We shall develop an approach based on global Carleman estimate to prove Theorem 1.2, which is the main novelty of this paper. Our approach, stimulated by [13] (see also [7, 9, 21]), is different from that in [14], which instead employed the classical local Carleman estimate and therefore needs C^∞-regularity for the data.

It would be quite interesting to establish better decay rate (than logarithmic one) for system (1.4) under further conditions (without geometric optics condition). There are some impressive results in this respect, say [4, 5, 15, 16, 17], for polynomial decay of system (1.4) with special geometry. However, to the best of the author's knowledge, the full picture of this problem is still unclear. We refer to [6, 19, 22] for some related work.

The rest of this paper is organized as follows. In Section 2, we collect some useful preliminary results which will be useful later. Another key result, an interpolation inequality for an elliptic equation with an inhomogeneous boundary condition, is brought about in Section 3. Sections 4–5 are devoted to the proof of our main results.

2 Some preliminaries

In this section, we collect some preliminaries which will be useful in the sequel.

First of all, we recall the following result which is an easy consequence of known result in [11, 20], for example.

Lemma 2.1. Let Γ_0 be given by (1.3). Then there exists a real-valued function $\hat{\psi} \in C^2(\overline{\Omega})$ such that

$$\begin{cases} \hat{\psi} > 0 & \text{in } \Omega, \\ |\nabla \hat{\psi}| > 0 & \text{in } \overline{\Omega}, \\ \hat{\psi} = 0 & \text{on } \partial\Omega \setminus \Gamma_0, \\ \displaystyle\sum_{j,k=1}^{n} a^{jk} \hat{\psi}_j \nu_k \leq 0 & \text{on } \partial\Omega \setminus \Gamma_0. \end{cases} \tag{2.1}$$

Next, to establish the desired interpolation inequality via global Carleman estimate, we need the following point-wise estimate for second-order differential operators with symmetric coefficients, which is a consequence of [9, Theorem 2.1] (see also [8, Theorem 1.1]).

Lemma 2.2. Let $b^{jk} \in C^2(\mathbb{R}^n;\ \mathbb{R})$ satisfy $b^{jk} = b^{kj}$ $(j, k = 1, \cdots, n)$. Assume that $w \in C^2(\mathbb{R}^{1+n}; \mathbb{C})$ and $\ell \in C^2(\mathbb{R}^{1+n}; \mathbb{R})$. Set

$$\theta = e^{\ell}, \quad v = \theta w, \quad \Psi = -2\ell_{ss} - 2\sum_{j,k=1}^{n} (b^{jk}\ell_j)_k. \tag{2.2}$$

Then

$$
\begin{aligned}
&\theta^2\Big|w_{ss}+\sum_{j,k=1}^n (b^{jk}w_j)_k\Big|^2+M_s+\operatorname{div}V\\
&\geq 2\Big(3\ell_{ss}+\sum_{j,k=1}^n (b^{jk}\ell_j)_k\Big)|v_s|^2+4\sum_{j,k=1}^n b^{jk}\ell_{js}(v_k\overline{v}_s+\overline{v}_k v_s) \qquad (2.3)\\
&\quad+\sum_{j,k=1}^n c^{jk}(v_k\overline{v}_j+\overline{v}_k v_j)+B|v|^2,
\end{aligned}
$$

where

$$
\left\{
\begin{aligned}
&A=\ell_s^2+\sum_{j,k=1}^n b^{jk}\ell_j\ell_k-\ell_{ss}-\sum_{j,k=1}^n (b^{jk}\ell_j)_k-\Psi,\\
&M=2\ell_s(|v_s|^2-\sum_{j,k=1}^n b^{jk}\overline{v}_j v_k)+2\sum_{j=1}^n b^{jk}\ell_j(\overline{v}_s v_j+v_s\overline{v}_j)\\
&\qquad -\Psi(\overline{v}_s v+v_s\overline{v})+(2A\ell_s+\Psi_s)|v|^2,\\
&V=[V^1,\cdots,V^k,\cdots,V^n],\\
&V^k=\sum_{j,j',k'=1}^n\Big\{-2b^{jk}\ell_j|v_s|^2+2b^{jk}\ell_s(\overline{v}_j v_s+v_j\overline{v}_s)\\
&\qquad -\Psi b^{jk}(v_j\overline{v}+\overline{v}_j v)+b^{jk}(2A\ell_j+\Psi_j)|v|^2\\
&\qquad +\Big(2b^{jk'}b^{j'k}-b^{jk}b^{j'k'}\Big)\ell_j(v_{j'}\overline{v}_{k'}+\overline{v}_{j'}v_{k'})\Big\},
\end{aligned}
\right. \qquad (2.4)
$$

and,

$$
\left\{
\begin{aligned}
&c^{jk}=\sum_{j',k'=1}^n\Big[2(b^{j'k}\ell_{j'})_{k'}b^{jk'}-b^{jk}_{k'}b^{j'k'}\ell_{j'}+b^{jk}(b^{j'k'}\ell_{j'})_{k'}\Big]+b^{jk}\ell_{ss},\\
&B=\sum_{j,k=1}^n (b^{jk}\Psi_k)_j+\Psi_{ss}+2(A\ell_s)_s+2\sum_{j,k=1}^n (Ab^{jk}\ell_j)_k+2A\Psi.
\end{aligned}
\right.
$$

(2.5)

Remark 2.1. Since θ is a weight function, (2.3) can be viewed as a weighted inequality. Although this inequality looks very complicated, its proof is considerably simple and elementary.

Remark 2.2. For any function $\tilde{\ell}\in C^2(\mathbb{R}^{1+n};\ \mathbb{R})$, set

$$
\tilde{\theta}=e^{\tilde{\ell}},\quad \tilde{v}=\tilde{\theta}w,\quad \tilde{\Psi}=-2\tilde{\ell}_{ss}-2\sum_{j,k=1}^n (b^{jk}\tilde{\ell}_j)_k. \qquad (2.6)
$$

Then, by Lemma 2.2, one obtains a similar inequality as (2.3) with M, V and c^{jk} replaced by $\tilde{M}$, $\tilde{V}$ and $\tilde{c}^{jk}$, respectively.

Proof of Lemma 2.2. Using Theorem 2.1 in [9] with $m = 1+n$, and

$$\alpha=\beta=0, \quad t=s, \quad (a^{jk}(t,x))_{m\times m}=\begin{pmatrix}1 & 0\\ 0 & (b^{jk}(x))_{n\times n}\end{pmatrix}.$$

By a direct calculation, we obtain (2.3). □

Finally, proceeding exactly as [21, Lemma 3.3] and [10, Lemma 3.2], we obtain the following identity.

Lemma 2.3. Let $b^{jk}\in C^2(\mathbb{R}^n;\ \mathbb{R})$ satisfy $b^{jk}=b^{kj}\,(j,k=1,\cdots,n)$, and $g \triangleq (g^1,\cdots,g^n):\mathbb{R}_s\times\mathbb{R}^n_x\to\mathbb{R}^n$ be a vector field of class C^1. Then for any $w\in C^2(\mathbb{R}_s\times\mathbb{R}^n_x;\mathbb{C})$, it holds

$$\begin{aligned}
&-\sum_{k=1}^n\Big[(g\cdot\nabla\overline{w})\sum_{j=1}^n b^{jk}w_j+(g\cdot\nabla w)\sum_{j=1}^n b^{jk}\overline{w}_j\\
&\qquad\qquad -g^k\Big(|w_s|^2+\sum_{i,l=1}^n b^{jl}w_j\overline{w}_l\Big)\Big]_k\\
&=-\Big[w_{ss}+\sum_{j,k=1}^n(b^{jk}w_j)_k\Big]g\cdot\nabla\overline{w}-\Big[\overline{w}_{ss}+\sum_{j,k=1}^n(b^{jk}\overline{w}_j)_k\Big]g\cdot\nabla w\\
&\quad+(w_sg\cdot\nabla\overline{w}+\overline{w}_sg\cdot\nabla w)_s-(w_sg_s\cdot\nabla\overline{w}+\overline{w}_sg\cdot\nabla w)\\
&\quad+(\nabla\cdot g)|w_s|^2-2\sum_{j,k,l=1}^n b^{jk}w_j\overline{w}_l\frac{\partial g^l}{\partial x_k}+\sum_{j,k=1}^n w_j\overline{w}_k\nabla\cdot(b^{jk}g).
\end{aligned} \tag{2.7}$$

3 Interpolation inequality for an elliptic equation with an inhomogeneous boundary condition

In this section, by means of the global Carleman estimate, we shall derive an interpolation inequality for an elliptic equation with a nonhomogeneous and complex Neumann-like boundary condition.

Denote

$$X=(-2,2)\times\Omega,\ \Sigma=(-2,2)\times\partial\Omega,\ Y=(-1,1)\times\Omega,\ Z=(-2,2)\times\Gamma_0.$$

Let us consider the following elliptic equation:

$$\begin{cases} z_{ss} + \sum_{j,k=1}^{n} \left(a^{jk} z_j\right)_k = z^0 & \text{in } (-2,2) \times \Omega, \\ \sum_{j,k=1}^{n} a^{jk} z_j \nu_k - ia(x) z_s = a(x) z^1 & \text{on } (-2,2) \times \partial\Omega, \end{cases} \tag{3.1}$$

where $z^0 \in L^2(X)$ and $z^1 \in L^2(\Sigma)$.

The desired interpolation inequality is stated as follows:

Theorem 3.1. Under the assumptions in Theorem 1.1, there exists a constant $C > 0$ such that for any $\varepsilon > 0$, any solution z of system (3.1) satisfies

$$\begin{aligned} ||z||_{H^1(Y)} \leq Ce^{C/\varepsilon}\Big[||z^0||_{L^2(X)} + ||z^1||_{L^2(\Sigma)} + ||z||_{L^2(Z)} + ||z_s||_{L^2(Z)}\Big] \\ + Ce^{-2/\varepsilon}||z||_{H^1(X)}. \end{aligned} \tag{3.2}$$

Proof. The proof is based on the point-wise estimate presented in Section 2. The point is to estimate the "energy-terms" (on the right hand side of (2.3)) and the "divergence-terms" (M_s and $\operatorname{div} V$ on the left hand side of (2.3)). Note however that we need to consider the problem with nonhomogeneous Neumann-like boundary in this paper. Hence, the treatment on the corresponding boundary terms becomes much more complicated than the usual case with homogeneous Dirichlet boundary condition. Note also that in the present case, we shall choose two weight functions such that many boundary terms vanish on $\partial\Omega \setminus \Gamma_0$. The proof is long, and therefore, we divided it into several steps.

Step 1. **Choosing of the weight functions.**

We borrow some ideas from [18]. For any $\mu > \ln 2$, put

$$b \triangleq \sqrt{1 + \frac{1}{\mu}\ln(2 + e^{\mu})}, \qquad b_0 \triangleq \sqrt{b^2 - \frac{1}{\mu}\ln\left(\frac{1 + e^{\mu}}{e^{\mu}}\right)}. \tag{3.3}$$

It is easy to check that

$$1 < b_0 < b \leq 2. \tag{3.4}$$

Note however that there is no boundary condition for z at $s = \pm 2$. Therefore, we need to introduce a cut-off function $\varphi = \varphi(s) \in C_0^{\infty}(-b, b) \subset C_0^{\infty}(\mathbb{R})$ such that

$$\begin{cases} 0 \leq \varphi(s) \leq 1, & |s| < b, \\ \varphi(s) = 1, & |s| \leq b_0. \end{cases} \tag{3.5}$$

Put

$$\hat{z} = \varphi z. \tag{3.6}$$

Then, noting that φ does not depend on x, by (3.1), it follows

$$\begin{cases} \hat{z}_{ss} + \sum_{j,k=1}^{n} \left(a^{jk}\hat{z}_j\right)_k = \varphi_{ss} z + 2\varphi_s z_s + \varphi z^0 & \text{in } (-2,2)\times\Omega, \\ \sum_{j,k=1}^{n} a^{jk}\hat{z}_j \nu_k - ia(x)\hat{z}_s = -ia(x)\varphi_s z + a(x)\varphi z^1 & \text{on } (-2,2)\times\partial\Omega. \end{cases} \tag{3.7}$$

Now, we choose the desired weight functions as follows.

$$\begin{cases} \psi = \psi(s,x) \stackrel{\triangle}{=} \dfrac{\hat{\psi}(x)}{||\hat{\psi}||_{L^\infty(\Omega)}} + b^2 - s^2, \quad \phi = e^{\mu\psi}, \quad \theta = e^{\ell} = e^{\lambda\phi}, \\ \tilde{\psi} = \psi(s,x) \stackrel{\triangle}{=} -\dfrac{\hat{\psi}(x)}{||\hat{\psi}||_{L^\infty(\Omega)}} + b^2 - s^2, \tilde{\phi} = e^{\mu\tilde{\psi}}, \quad \tilde{\theta} = e^{\tilde{\ell}} = e^{\lambda\tilde{\phi}}, \end{cases} \tag{3.8}$$

where $\hat{\psi} \in C^2(\overline{\Omega})$ is given by Lemma 2.1. Thus, by Lemma 2.1 and (3.8), we find

$$h \stackrel{\triangle}{=} |\nabla\psi| = |\nabla\tilde{\psi}| = \frac{1}{||\hat{\psi}||_{L^\infty(\Omega)}}|\nabla\hat{\psi}(x)| > 0, \qquad \text{in } \overline{\Omega}, \tag{3.9}$$

and

$$\begin{cases} \phi(s,\cdot) \geq 2 + e^{\mu}, & \text{for any } s \text{ satisfying } |s| \leq 1, \\ \phi(s,\cdot) \leq 1 + e^{\mu}, & \text{for any } s \text{ satisfying } b_0 \leq |s| \leq b. \end{cases} \tag{3.10}$$

Next, it is easy to check that

$$\tilde{\psi} \leq \psi, \quad 0 < \tilde{\phi} \leq \phi, \quad 0 < \tilde{\theta} \leq \theta. \tag{3.11}$$

Finally, by (3.8), it is easy to see that

$$\begin{cases} \ell_s = \lambda\mu\phi\psi_s, \qquad \ell_j = \lambda\mu\phi\psi_j, \quad \ell_{js} = \lambda\mu^2\phi\psi_s\psi_j \\ \ell_{ss} = \lambda\mu^2\phi\psi_s^2 + \lambda\mu\phi\psi_{ss}, \quad \ell_{jk} = \lambda\mu^2\phi\psi_j\psi_k + \lambda\mu\phi\psi_{jk} \end{cases} \tag{3.12}$$

and

$$\begin{cases} \tilde{\ell}_s = \lambda\mu\tilde{\phi}\tilde{\psi}_s, \qquad \tilde{\ell}_j = \lambda\mu\tilde{\phi}\tilde{\psi}_j, \quad \tilde{\ell}_{js} = \lambda\mu^2\tilde{\phi}\tilde{\psi}_s\tilde{\psi}_j \\ \tilde{\ell}_{ss} = \lambda\mu^2\tilde{\phi}\tilde{\psi}_s^2 + \lambda\mu\tilde{\phi}\tilde{\psi}_{ss}, \quad \tilde{\ell}_{jk} = \lambda\mu^2\tilde{\phi}\tilde{\psi}_j\tilde{\psi}_k + \lambda\mu\tilde{\phi}\tilde{\psi}_{jk}. \end{cases} \tag{3.13}$$

In what follows, for $n \in \mathbb{N}$, we denote by $O(\mu^n)$ a function of order μ^n for large μ (which is independent of λ); by $O_\mu(\lambda^n)$ a function of order λ^n for fixed μ and for large λ.

Step 2. **Estimates for the energy terms.**

First, recalling the definition of c^{jk} in (2.5), by (3.12) and with b^{jk} replaced by a^{jk} in Lemma 2.2, note that $a^{jk} = a^{kj}$, we have

$$\begin{aligned}
\sum_{j,k=1}^{n} c^{jk}(v_k\overline{v}_j + \overline{v}_k v_j) &= 2\sum_{j,k=1}^{n} c^{jk} v_k \overline{v}_j \\
&= 2\sum_{j,k,j',k'=1}^{n} \Big\{\Big[2(a^{j'k}\ell_{j'})_{k'}a^{jk'} - a^{jk}_{k'}a^{j'k'}\ell_{j'} + a^{jk}(a^{j'k'}\ell_{j'})_{k'}\Big] \\
&\qquad\qquad + a^{jk}\ell_{ss}\Big\} v_k\overline{v}_j \qquad (3.14) \\
&= 4\lambda\mu^2\Big|\sum_{j,k=1}^{n} a^{jk}\psi_j\overline{v}_k\Big|^2 \\
&\quad + 2\Big\{\lambda\mu^2\phi\Big[\sum_{j,k=1}^{n} a^{jk}\psi_j\psi_k + |\psi_s|^2\Big] + \lambda\phi O(\mu)\Big\}\sum_{j,k=1}^{n} a^{jk}v_k\overline{v}_j.
\end{aligned}$$

Hence, by (3.14), we have the following estimate.

$$\begin{aligned}
&2\Big(3\ell_{ss} + \sum_{j,k=1}^{n}(a^{jk}\ell_j)_k\Big)|v_s|^2 \\
&\quad + 4\sum_{j,k=1}^{n} a^{jk}\ell_{js}(v_k\overline{v}_s + \overline{v}_k v_s) + \sum_{j,k=1}^{n} c^{jk}(v_k\overline{v}_j + \overline{v}_k v_j) \\
&= 2\Big\{\lambda\mu^2\phi\Big[3|\psi_s|^2 + \sum_{j,k=1}^{n} a^{jk}\psi_j\psi_k\Big] + \lambda\phi O(\mu)\Big\}|v_s|^2 \\
&\quad + 8\lambda\mu^2\phi\sum_{j,k=1}^{n} a^{jk}\psi_j\psi_s v_k\overline{v}_s + 4\lambda\mu^2\Big|\sum_{j,k=1}^{n} a^{jk}\psi_j\overline{v}_k\Big|^2 \qquad (3.15) \\
&\quad + 2\Big\{\lambda\mu^2\phi\Big[\sum_{j,k=1}^{n} a^{jk}\psi_j\psi_k + |\psi_s|^2\Big] + \lambda\phi O(\mu)\Big\}\sum_{j,k=1}^{n} a^{jk}v_k\overline{v}_j \\
&= 4\lambda\mu^2\phi\Big|\psi_s v_s + \sum_{j,k=1}^{n} a^{jk}\psi_j v_k\Big|^2 + 2\Big\{\lambda\mu^2\phi\Big[\sum_{j,k=1}^{n} a^{jk}\psi_j\psi_k \\
&\qquad + |\psi_s|^2\Big] + \lambda\phi O(\mu)\Big\}\Big(|v_s|^2 + \sum_{j,k=1}^{n} a^{jk}v_j\overline{v}_k\Big) \\
&\geq 2\Big[\lambda\mu^2\phi\sum_{j,k=1}^{n} a^{jk}\psi_j\psi_k + \lambda\phi O(\mu)\Big]\Big(|v_s|^2 + \sum_{j,k=1}^{n} a^{jk}v_j\overline{v}_k\Big).
\end{aligned}$$

Further, by (3.12) and recalling A and Ψ in (2.4) and (2.2), respectively, we have

$$\begin{cases} \Psi = -2\lambda\mu^2\phi\Big[|\psi_s|^2 + \sum\limits_{j,k=1}^{n} a^{jk}\psi_j\psi_k\Big] + \lambda\phi O(\mu), \\ A = (\lambda^2\mu^2\phi^2 + \lambda\mu^2\phi)\Big[|\psi_s|^2 + \sum\limits_{j,k=1}^{n} a^{jk}\psi_j\psi_k\Big] + \lambda\phi O(\mu). \end{cases} \tag{3.16}$$

Therefore, by (2.4), (2.5) and (2.2), we have

$$\begin{aligned} B &= \sum_{j,k=1}^{n} (a^{jk}\Psi_k)_j + \Psi_{ss} + 2(A\ell_s)_s + 2\sum_{j,k=1}^{n} (Aa^{jk}\ell_j)_k + 2A\Psi \\ &= 2A_s\ell_s + 2\sum_{j,k=1}^{n} a^{jk}\ell_j A_k + A\Psi + \sum_{j,k=1}^{n} (a^{jk}\Psi_k)_j + \Psi_{ss} \\ &= 2\lambda^3\mu^4\phi^3|\psi_s|^4 + 2\lambda^3\mu^4\phi^3\Big|\sum_{j,k=1}^{n} a^{jk}\psi_j\psi_k\Big|^2 + \lambda^3\phi^3 O(\mu^3) + O_\mu(\lambda^2) \\ &\geq 2\lambda^3\mu^4\phi^3\Big|\sum_{j,k=1}^{n} a^{jk}\psi_j\psi_k\Big|^2 + \lambda^3\phi^3 O(\mu^3) + O_\mu(\lambda^2). \end{aligned} \tag{3.17}$$

Combining (2.3), (3.15) and (3.17), by using (1.2) and (3.9), we conclude that there exists a $\mu_0 > 1$, such that for any $\mu \geq \mu_0$, there exists $\lambda_0(\mu) > 1$ such that for any $\lambda \geq \lambda_1$, it holds (recall that $v = \theta w$)

$$\begin{aligned} &\theta^2\Big|w_{ss} + \sum_{j,k=1}^{n} (a^{jk}w_j)_k\Big|^2 + M_s + \operatorname{div} V \\ &\geq 2\Big[\lambda\mu^2\phi\sum_{j,k=1}^{n} a^{jk}\psi_j\psi_k + \lambda\phi O(\mu)\Big]\Big(|v_s|^2 + \sum_{j,k=1}^{n} a^{jk}v_j\overline{v}_k\Big) \\ &\quad +2\Big[\lambda^3\mu^4\phi^3\Big|\sum_{j,k=1}^{n} a^{jk}\psi_j\psi_k\Big|^2 + \lambda^3\phi^3 O(\mu^3) + O_\mu(\lambda^2)\Big]|v|^2 \\ &\geq 2[s_0h^2\lambda\mu^2\phi + \lambda\phi O(\mu)](|v_s|^2 + s_0|\nabla v|^2) \\ &\quad +2[s_0^2h^4\lambda^3\mu^4\phi^3 + \lambda^3\phi^3 O(\mu^3) + O_\mu(\lambda^2)]|v|^2 \\ &\geq C\Big[\lambda\mu^2\phi(|v_s|^2 + |\nabla v|^2) + \lambda^3\mu^4\phi^3|v|^2\Big]. \end{aligned} \tag{3.18}$$

Similar to (3.18), by (1.2), (3.8), (3.13) and Remark 2.2, we have (recall that $\tilde{v} = \tilde{\theta} w$)

$$\begin{aligned}
&\tilde{\theta}^2 \Big| w_{ss} + \sum_{j,k=1}^{n} (a^{jk} w_j)_k \Big|^2 + \tilde{M}_s + \operatorname{div} \tilde{V} \\
&\geq C\Big[\lambda\mu^2\tilde{\phi}(|\tilde{v}_s|^2 + |\nabla\tilde{v}|^2) + \lambda^3\mu^4\tilde{\phi}^3|\tilde{v}|^2\Big].
\end{aligned} \tag{3.19}$$

Combining (3.18) and (3.19), we have

$$\begin{aligned}
&(\theta^2 + \tilde{\theta}^2)\Big| w_{ss} + \sum_{j,k=1}^{n} (a^{jk} w_j)_k \Big|^2 + (M + \tilde{M})_s + \operatorname{div} (V + \tilde{V}) \\
&\geq C\Big[\lambda\mu^2\phi(|v_s|^2 + |\nabla v|^2) + \lambda^3\mu^4\phi^3|v|^2\Big] \\
&\quad + C\Big[\lambda\mu^2\tilde{\phi}(|\tilde{v}_s|^2 + |\nabla\tilde{v}|^2) + \lambda^3\mu^4\tilde{\phi}^3|\tilde{v}|^2\Big].
\end{aligned} \tag{3.20}$$

Now, integrating inequality (3.20) (with w replaced by $\hat{z}$) in $(-b, b) \times \Omega$, recalling that φ vanishes near $s = \pm b$, by (3.7) and (3.11), one arrives at

$$\begin{aligned}
&\lambda\mu^2 \int_{-b}^{b} \int_\Omega \phi(|\nabla v|^2 + |v_s|^2) dx ds + \lambda^3\mu^4 \int_{-b}^{b} \int_\Omega \phi^3 |v|^2 dx ds \\
&\leq C\Big[\int_{-b}^{b} \int_\Omega \theta^2 |\varphi_{ss} z + 2\varphi_s z_s + \varphi z^0|^2 dx ds \\
&\qquad + \int_{-b}^{b} \int_{\partial\Omega} (V + \tilde{V}) \cdot \nu dx ds\Big].
\end{aligned} \tag{3.21}$$

Recalling that $v = \theta\hat{z}$, by (3.12), we get

$$\begin{aligned}
\frac{1}{C}\theta^2(|\nabla\hat{z}|^2 + \lambda^2\mu^2\phi^2|\hat{z}|^2) &\leq |\nabla v|^2 + \lambda^2\mu^2\phi^2|v|^2 \\
&\leq C\theta^2(|\nabla\hat{z}|^2 + \lambda^2\mu^2\phi^2|\hat{z}|^2).
\end{aligned} \tag{3.22}$$

Therefore, by (3.21) and (3.22), we end up with

$$\begin{aligned}
&\lambda\mu^2 \int_{-b}^{b} \int_\Omega \theta^2\phi(|\nabla\hat{z}|^2 + |\hat{z}_s|^2) dx ds + \lambda^3\mu^4 \int_{-b}^{b} \int_\Omega \theta^2\phi^3 |\hat{z}|^2 dx ds \\
&\leq C\Big[\int_{-b}^{b} \int_\Omega \theta^2 |\varphi_{ss} z + 2\varphi_s z_s + \varphi z^0|^2 dx ds \\
&\qquad + \int_{-b}^{b} \int_{\partial\Omega} (V + \tilde{V}) \cdot \nu dx ds\Big].
\end{aligned} \tag{3.23}$$

Step 3. **Estimate for the boundary term** $\displaystyle\int_{-b}^{b}\int_{\partial\Omega}(V+\tilde{V})\cdot\nu dxds$.

By recalling (2.4) for V and Remark 2.2 for $\tilde{V}$, we have (with b^{jk} replaced by a^{jk})

$$
\begin{aligned}
&\int_{-b}^{b}\int_{\partial\Omega}(V+\tilde{V})\cdot\nu dxds\\
&=\int_{-b}^{b}\int_{\partial\Omega}\sum_{k=1}^{n}(V^k+\tilde{V}^k)\nu_k dxds\\
&=-2\int_{-b}^{b}\int_{\partial\Omega}\sum_{j,k=1}^{n}\Big[a^{jk}\ell_j\nu_k|v_s|^2+a^{jk}\tilde{\ell}_j\nu_k|\tilde{v}_s|^2\Big]dxds\\
&\quad+2\int_{-b}^{b}\int_{\partial\Omega}\sum_{j,k=1}^{n}a^{jk}[\ell_s(\overline{v}_jv_s+v_j\overline{v}_s)+\tilde{\ell}_s(\overline{\tilde{v}}_j\tilde{v}_s+\tilde{v}_j\overline{\tilde{v}}_s)]\nu_k dxds\\
&\quad-\int_{-b}^{b}\int_{\partial\Omega}\sum_{j,k=1}^{n}a^{jk}[\Psi(v_j\overline{v}+\overline{v}_jv)+\tilde{\Psi}(\tilde{v}_j\overline{\tilde{v}}+\overline{\tilde{v}}_j\tilde{v})]\nu_k dxds\\
&\quad+\int_{-b}^{b}\int_{\partial\Omega}\sum_{j,k=1}^{n}a^{jk}[(2A\ell_j+\Psi_j)|v|^2+(2\tilde{A}\tilde{\ell}_j+\tilde{\Psi}_j)|\tilde{v}|^2]\nu_k dxds\\
&\quad+2\int_{-b}^{b}\int_{\partial\Omega}\sum_{j,k,j',k'=1}^{n}a^{jk'}a^{j'k}\Big[\ell_j(v_{j'}\overline{v}_{k'}+\overline{v}_{j'}v_{k'})\nu_k\\
&\qquad\qquad+\tilde{\ell}_j(\tilde{v}_{j'}\overline{\tilde{v}}_{k'}+\overline{\tilde{v}}_{j'}\tilde{v}_{k'})\nu_k\Big]dxds\\
&\quad-\int_{-b}^{b}\int_{\partial\Omega}\sum_{j,k,j',k'=1}^{n}a^{jk}a^{j'k'}\Big[\ell_j(v_{j'}\overline{v}_{k'}+\overline{v}_{j'}v_{k'})\nu_k\\
&\qquad\qquad+\tilde{\ell}_j(\tilde{v}_{j'}\overline{\tilde{v}}_{k'}+\overline{\tilde{v}}_{j'}\tilde{v}_{k'})\nu_k\Big]dxds.
\end{aligned}
\tag{3.24}
$$

We will estimate the above six terms on the right left side of (3.24) one by one. Before doing this, we note that by (2.1) , (3.8) and (3.12)–(3.13), it holds

$$
\begin{cases}
\phi=\tilde{\phi},\ \ell=\tilde{\ell},\ \theta=\tilde{\theta},\ \ell_s=\tilde{\ell}_s, & \text{on } \partial\Omega\setminus\Gamma_0,\\
\displaystyle\sum_{j,k=1}^{n}a^{jk}\ell_j\nu_k=\frac{\lambda\mu\phi}{||\hat{\psi}||_{L^\infty(\Omega)}}\sum_{j,k=1}^{n}a^{jk}\hat{\psi}_j\nu_k\le 0, & \text{on } \partial\Omega\setminus\Gamma_0,\\
\displaystyle\sum_{j,k=1}^{n}a^{jk}\tilde{\ell}_j\nu_k=-\frac{\lambda\mu\tilde{\phi}}{||\hat{\psi}||_{L^\infty(\Omega)}}\sum_{j,k=1}^{n}a^{jk}\hat{\psi}_j\nu_k\ge 0, & \text{on } \partial\Omega\setminus\Gamma_0.
\end{cases}
\tag{3.25}
$$

Hence, by (3.6) and (3.25), we have (recalling that $v=\theta\hat{z},\ \tilde{v}=\tilde{\theta}\hat{z}$)

$$\begin{aligned}
&-2\int_{-b}^{b}\int_{\partial\Omega}\sum_{j,k=1}^{n}\Big[a^{jk}\ell_j\nu_k|v_s|^2+a^{jk}\tilde{\ell}_j\nu_k|\tilde{v}_s|^2\Big]dxds\\
&=-2\int_{-b}^{b}\int_{\Gamma_0}\sum_{j,k=1}^{n}\Big[a^{jk}\ell_j\nu_k|v_s|^2+a^{jk}\tilde{\ell}_j\nu_k|\tilde{v}_s|^2\Big]dxds \qquad (3.26)\\
&\le Ce^{C\lambda}\int_{-b}^{b}\int_{\Gamma_0}(|z_s|^2+|z|^2)dxds.
\end{aligned}$$

Next, by (1.3), (3.6)–(3.7) and (3.25), we have (recalling that $v=\theta\hat{z},\ \tilde{v}=\tilde{\theta}\hat{z}$)

$$\begin{aligned}
&2\int_{-b}^{b}\int_{\partial\Omega}\sum_{j,k=1}^{n}a^{jk}[\ell_s(\overline{v}_jv_s+v_j\overline{v}_s)+\tilde{\ell}_s(\overline{\tilde{v}}_j\tilde{v}_s+\tilde{v}_j\overline{\tilde{v}}_s)]\nu_kdxds\\
&=2\int_{-b}^{b}\int_{\partial\Omega}(\theta^2\ell_s+\tilde{\theta}^2\tilde{\ell}_s)\sum_{j,k=1}^{n}a^{jk}\nu_k(\overline{\hat{z}_j}\hat{z}_s+\hat{z}_j\overline{\hat{z}_s})dxds\\
&\quad+2\int_{-b}^{b}\int_{\partial\Omega}(\theta^2\ell_s^2+\tilde{\theta}^2\tilde{\ell}_s^2)\sum_{j,k=1}^{n}a^{jk}\nu_k(\overline{\hat{z}_j}\hat{z}+\hat{z}_j\overline{\hat{z}})dxds\\
&\quad+2\int_{-b}^{b}\int_{\partial\Omega}\sum_{j,k=1}^{n}(\theta^2\ell_sa^{jk}\ell_j\nu_k+\tilde{\theta}^2\tilde{\ell}_sa^{jk}\tilde{\ell}_j\nu_k)(\hat{z}\overline{\hat{z}_s}+\overline{\hat{z}}\hat{z}_s)dxds \qquad (3.27)\\
&\quad+4\int_{-b}^{b}\int_{\partial\Omega}\sum_{j,k=1}^{n}(\theta^2\ell_s^2a^{jk}\ell_j\nu_k+\tilde{\theta}^2\tilde{\ell}_s^2a^{jk}\tilde{\ell}_j\nu_k)|\hat{z}|^2dxds\\
&\le Ce^{C\lambda}\int_{-b}^{b}\int_{\Gamma_0}(|z_s|^2+|z|^2+|z^1|^2)dxds.
\end{aligned}$$

Similarly, by (1.3), (2.2), (2.6), (3.6)–(3.7) and (3.25), we get (recalling that $v=\theta\hat{z},\ \tilde{v}=\tilde{\theta}\hat{z}$)

$$\begin{aligned}
&-\int_{-b}^{b}\int_{\partial\Omega}\sum_{j,k=1}^{n}a^{jk}[\Psi(v_j\overline{v}+\overline{v}_jv)+\tilde{\Psi}(\tilde{v}_j\overline{\tilde{v}}+\overline{\tilde{v}}_j\tilde{v})]\nu_kdxds\\
&+\int_{-b}^{b}\int_{\partial\Omega}\sum_{j,k=1}^{n}a^{jk}[(2A\ell_j+\Psi_j)|v|^2+(2\tilde{A}\tilde{\ell}_j+\tilde{\Psi}_j)|\tilde{v}|^2]\nu_kdxds \qquad (3.28)\\
&\le Ce^{C\lambda}\int_{-b}^{b}\int_{\Gamma_0}(|z_s|^2+|z|^2+|z^1|^2)dxds
\end{aligned}$$

Further, by (3.25), and noting that $v = \theta\hat{z},\ \tilde{v} = \tilde{\theta}\hat{z}$, we get

$$\begin{aligned}
&2\int_{-b}^{b}\int_{\partial\Omega}\sum_{j,k,j',k'=1}^{n} a^{jk'}a^{j'k}[\ell_j(v_{j'}\overline{v}_{k'}+\overline{v}_{j'}v_{k'}) \\
&\qquad\qquad +\tilde{\ell}_j(\tilde{v}_{j'}\overline{\tilde{v}}_{k'}+\overline{\tilde{v}}_{j'}\tilde{v}_{k'})]\nu_k dxds \\
&=2\int_{-b}^{b}\int_{\Gamma_0}\sum_{j,k,j',k'=1}^{n} a^{jk'}a^{j'k}[\ell_j(v_{j'}\overline{v}_{k'}+\overline{v}_{j'}v_{k'}) \\
&\qquad\qquad +\tilde{\ell}_j(\tilde{v}_{j'}\overline{\tilde{v}}_{k'}+\overline{\tilde{v}}_{j'}\tilde{v}_{k'})]\nu_k dxds \\
&=X_1+X_2+X_3
\end{aligned} \tag{3.29}$$

where $X_j (j = 1, 2, 3)$ will be given below. First, by (3.6)–(3.7) and (3.12), we have

$$\begin{aligned}
&X_1 \\
&=2\int_{-b}^{b}\int_{\Gamma_0}\theta^2\sum_{j,k,j',k'=1}^{n} a^{jk'}a^{j'k}\ell_j[\overline{\hat{z}}_{k'}\hat{z}_{j'}+\hat{z}_{k'}\overline{\hat{z}}_{j'}]\nu_k dxds \\
&=2\int_{-b}^{b}\int_{\Gamma_0}\sum_{j,k'=1}^{n}(a^{jk'}\psi_j\overline{\hat{z}}_{k'})[a(x)\lambda\mu\theta^2\phi(i\hat{z}_s-i\varphi_s z+\varphi z^1)]dxds \\
&\quad+2\int_{-b}^{b}\int_{\Gamma_0}\sum_{j,k'=1}^{n}(a^{jk'}\psi_j\hat{z}_{k'})[a(x)\lambda\mu\theta^2\phi(-i\overline{\hat{z}}_s+i\varphi_s\overline{z}+\varphi\overline{z}^1)]dxds \\
&\le c\int_{-b}^{b}\int_{\Gamma_0}|\nabla z|^2dxds+Ce^{C\lambda}\int_{-b}^{b}\int_{\Gamma_0}(|z_s|^2+|z|^2+|z^1|^2)dxds,
\end{aligned} \tag{3.30}$$

where $c > 0$ independent of $\lambda,\ \mu$.

Similarly, by (3.6)–(3.7) and (3.12)–(3.13), we have

$$\begin{aligned}
&X_2 \\
&=2\int_{-b}^{b}\int_{\Gamma_0}\tilde{\theta}^2\sum_{j,k,j',k'=1}^{n} a^{jk'}a^{j'k}\tilde{\ell}_j[\overline{\hat{z}}_{k'}\hat{z}_{j'}+\hat{z}_{k'}\overline{\hat{z}}_{j'}]\nu_k dxds \\
&\quad+2\int_{-b}^{b}\int_{\Gamma_0}\theta^2\sum_{j,k,j',k'=1}^{n} a^{j'k}\ell_{j'}\nu_k a^{jk'}\ell_j[\overline{\hat{z}}_{k'}\hat{z}+\hat{z}_{k'}\overline{\hat{z}}]dxds \\
&\quad+2\int_{-b}^{b}\int_{\Gamma_0}\tilde{\theta}^2\sum_{j,k,j',k'=1}^{n} a^{j'k}\tilde{\ell}_{j'}\nu_k a^{jk'}\tilde{\ell}_j[\overline{\hat{z}}_{k'}\hat{z}+\hat{z}_{k'}\overline{\hat{z}}]dxds \\
&\le c\int_{-b}^{b}\int_{\Gamma_0}|\nabla z|^2dxds+Ce^{C\lambda}\int_{-b}^{b}\int_{\Gamma_0}(|z_s|^2+|z|^2+|z^1|^2)dxds.
\end{aligned} \tag{3.31}$$

Further, by (3.6)–(3.7), we obtain

$$
\begin{aligned}
&X_3 \\
&= 2\int_{-b}^{b}\int_{\Gamma_0}\sum_{j,k,j',k'=1}^{n} a^{jk'}[\theta^2\ell_j\ell_{k'} + \tilde{\theta}^2\tilde{\ell}_j\tilde{\ell}_{k'}]a^{j'k}(\hat{z}_{j'}\overline{\hat{z}} + \overline{\hat{z}}_{j'}\hat{z})\nu_k dxds \\
&\quad +4\int_{-b}^{b}\int_{\Gamma_0}\sum_{j,k,j'k'=1}^{n} a^{jk'}a^{j'k}[\theta^2\ell_j\ell_{k'}\ell_{j'} + \tilde{\theta}^2\tilde{\ell}_j\tilde{\ell}_{k'}\tilde{\ell}_{j'}]\nu_k|\hat{z}|^2 dxds \\
&\le Ce^{C\lambda}\int_{-b}^{b}\int_{\Gamma_0}(|z_s|^2 + |z|^2 + |z^1|^2)dxds.
\end{aligned}
\tag{3.32}
$$

Combining (3.33) and (3.30)–(3.32), we have

$$
\begin{aligned}
&2\int_{-b}^{b}\int_{\partial\Omega}\sum_{j,k,j',k'=1}^{n} a^{jk'}a^{j'k}[\ell_j(v_{j'}\overline{v}_{k'} + \overline{v}_{j'}v_{k'}) \\
&\qquad\qquad + \tilde{\ell}_j(\tilde{v}_{j'}\overline{\tilde{v}}_{k'} + \overline{\tilde{v}}_{j'}\tilde{v}_{k'})]\nu_k dxds \\
&\le c\int_{-b}^{b}\int_{\Gamma_0}|\nabla z|^2 dxds + Ce^{C\lambda}\int_{-b}^{b}\int_{\Gamma_0}(|z_s|^2 + |z|^2 + |z^1|^2)dxds.
\end{aligned}
\tag{3.33}
$$

Finally, by (1.1), (3.6)–(3.7), (3.11) and (3.25), and noting that $v = \theta\hat{z}$, $\tilde{v} = \tilde{\theta}\hat{z}$, we get

$$
\begin{aligned}
&-\int_{-b}^{b}\int_{\partial\Omega}\sum_{j,k,j',k'=1}^{n} a^{jk}a^{j'k'}\Big[\ell_j(v_{j'}\overline{v}_{k'} + \overline{v}_{j'}v_{k'})\nu_k \\
&\qquad\qquad + \tilde{\ell}_j(\tilde{v}_{j'}\overline{\tilde{v}}_{k'} + \overline{\tilde{v}}_{j'}\tilde{v}_{k'})\nu_k\Big]dxds \\
&= -2\int_{-b}^{b}\int_{\Gamma_0}\sum_{j,k,j',k'=1}^{n} a^{jk}(\theta^2\ell_j + \tilde{\theta}^2\tilde{\ell}_j)\nu_k a^{j'k'}\hat{z}_{j'}\overline{\hat{z}}_{k'}dxds \\
&\quad -2\int_{-b}^{b}\int_{\Gamma_0}\sum_{j,k,j',k'=1}^{n}\Big[\theta^2(a^{jk}\ell_j\nu_k)(a^{j'k'}\ell_{k'}\ell_{j'}) \\
&\qquad\qquad + \tilde{\theta}^2(a^{jk}\tilde{\ell}_j\nu_k)(a^{j'k'}\tilde{\ell}_{k'}\tilde{\ell}_{j'})\Big]|\hat{z}|^2 dxds \\
&\le -c\lambda\mu\int_{-b}^{b}\int_{\Gamma_0}(\theta^2\phi - \tilde{\theta}^2\tilde{\phi})\sum_{j,k,j',k'=1}^{n}(a^{jk}\hat{\psi}_j\nu_k)(a^{j'k'}\hat{z}_{j'}\overline{\hat{z}}_{k'})dxds \\
&\quad +Ce^{C\lambda}\int_{-b}^{b}\int_{\Gamma_0}|z|^2 dxds \le Ce^{C\lambda}\int_{-b}^{b}\int_{\Gamma_0}|z|^2 dxds.
\end{aligned}
\tag{3.34}
$$

Thus, Combining (3.24), (3.26)–(3.28) and (3.33)– (3.34), we end up with

$$\begin{aligned}&\int_{-b}^{b}\int_{\partial\Omega}(V+\tilde{V})\cdot\nu dxds\\&\leq c\int_{-b}^{b}\int_{\Gamma_0}|\nabla z|^2dxds\\&\quad+Ce^{C\lambda}\int_{-b}^{b}\int_{\Gamma_0}(|z_s|^2+|z|^2+|z^1|^2)dxds.\end{aligned}\tag{3.35}$$

Step 4. **Estimate for** $\int_{-b}^{b}\int_{\Gamma_0}|\nabla z|^2dxds$**.**

First, we choose a function $g\in C^1(\overline{\Omega};\mathbb{R})$ such that $g=\nu$ on $\partial\Omega$ in Lemma 2.3. Integrating (2.7) in $(-b,b)\times\Omega$, with w replaced by $\hat{z}$ and b^{jk} replaced by a^{jk}, using integration by parts, and noting $\hat{z}(-b)=\hat{z}(b)=0$, by (3.7), we have

$$\begin{aligned}&\int_{-b}^{b}\int_{\partial\Omega}\Big(|\hat{z}_s|^2+\sum_{j,l=1}^{n}a^{jl}\hat{z}_j\overline{\hat{z}}_l\Big)dxds\\&-\int_{-b}^{b}\int_{\partial\Omega}\sum_{j,k=1}^{n}[(g\cdot\nabla\overline{\hat{z}})a^{jk}\hat{z}_j\nu_k+(g\cdot\nabla\hat{z})a^{jk}\overline{\hat{z}}_j\nu_k]dxds\\&=-\int_{-b}^{b}\int_{\Omega}[\hat{z}_{ss}+\sum_{j,k=1}^{n}(a^{jk}\hat{z}_j)_k]g\cdot\nabla\overline{\hat{z}}dxds\\&-\int_{-b}^{b}\int_{\Omega}[\overline{\hat{z}}_{ss}+\sum_{j,k=1}^{n}(a^{jk}\overline{\hat{z}}_j)_k]g\cdot\nabla\hat{z}dxds\\&-\int_{-b}^{b}\int_{\Omega}(\hat{z}_sg_s\cdot\nabla\overline{\hat{z}}+\overline{\hat{z}}_sg\cdot\nabla\hat{z})dxds\\&+\int_{-b}^{b}\int_{\Omega}\Big[(\nabla\cdot g)|\hat{z}_s|^2-2\sum_{j,k,l=1}^{n}a^{jk}\hat{z}_j\overline{\hat{z}}_l\frac{\partial g^l}{\partial x_k}\\&\qquad\qquad+\sum_{j,k=1}^{n}\hat{z}_j\overline{\hat{z}}_k\nabla\cdot(a^{jk}g)\Big]dxds\\&\leq C\int_{-b}^{b}\int_{\Omega}\Big[|\varphi_{ss}z+2\varphi_sz_s+\varphi z^0|^2+(|\hat{z}_s|^2+|\nabla\hat{z}|^2)\Big]dxds.\end{aligned}\tag{3.36}$$

Next, by (1.2), (1.3), the boundary condition in (3.7), and by (3.36), we have

$$\int_{-b}^{b}\int_{\partial\Omega}(|\hat{z}_s|^2+s_0|\nabla\hat{z}|^2)dxds \le \int_{-b}^{b}\int_{\partial\Omega}\Big(|\hat{z}_s|^2+\sum_{j,l=1}^{n}a^{jl}\hat{z}_j\bar{\hat{z}}_l\Big)dxds$$
$$\le C\int_{-b}^{b}\int_{\Omega}\Big[|\varphi_{ss}z+2\varphi_s z_s+\varphi z^0|^2+(|\hat{z}_s|^2+|\nabla\hat{z}|^2)\Big]dxds$$
$$+\delta\int_{-b}^{b}\int_{\Gamma_0}|\nabla\hat{z}|^2dxds+C(\delta)\int_{-b}^{b}\int_{\Gamma_0}(|z|^2+|\hat{z}_s|^2+|z^1|^2)dxds, \tag{3.37}$$

where $0<\delta<s_0$ is small.

Then, by (3.6) and (3.37), we deduce that

$$\int_{-b}^{b}\int_{\Gamma_0}|\nabla z|^2dxds$$
$$\le C\int_{-b}^{b}\int_{\Omega}\Big[|\varphi_{ss}z+2\varphi_s z_s+\varphi z^0|^2+(|\hat{z}_s|^2+|\nabla\hat{z}|^2)\Big]dxds \tag{3.38}$$
$$+C\int_{-b}^{b}\int_{\Gamma_0}(|z|^2+|z_s|^2+|z^1|^2)dxds.$$

Step 5. **End of the proof.**

Combining (3.23), (3.35) and (3.38), we end up with

$$\lambda\mu^2\int_{-b}^{b}\int_{\Omega}\theta^2\phi(|\nabla z|^2+|z_s|^2)dxds+\lambda^3\mu^4\int_{-b}^{b}\int_{\Omega}\theta^2\phi^3|z|^2dxds$$
$$\le C\int_{-b}^{b}\int_{\Omega}e^{2\lambda\phi}|\varphi_{ss}z+2\varphi_s z_s+\varphi z^0|^2dxds \tag{3.39}$$
$$+Ce^{C\lambda}\int_{-b}^{b}\int_{\Gamma_0}(|z|^2+|z_s|^2+|z^1|^2)dxds.$$

Denote $c_0=2+e^{\mu}>1$, and recall (3.3) for $b_0\in(1,b)$. Fixing the parameter μ in (3.39), using (3.5) and (3.10), one finds

$$\lambda e^{2\lambda c_0}\int_{-1}^{1}\int_{\Omega}(|\nabla z|^2+|z_s|^2+|z|^2)dxds$$
$$\le Ce^{C\lambda}\Big\{\int_{-2}^{2}\int_{\Omega}|z^0|^2dxds+\int_{-2}^{2}\int_{\partial\Omega}|z^1|^2dxds$$
$$+\int_{-2}^{2}\int_{\Gamma_0}(|z|^2+|z_s|^2)dxds\Big\} \tag{3.40}$$
$$+Ce^{2\lambda(c_0-1)}\int_{(-b,-b_0)\bigcup(b_0,b)}\int_{\Omega}(|z|^2+|z_s|^2)dxds.$$

From (3.40), one concludes that there exists an $\varepsilon_2 > 0$ such that the desired inequality (3.2) holds for $\varepsilon \in (0, \varepsilon_2]$, which, in turn, implies that it holds for any $\varepsilon > 0$. This completes the proof of Theorem 3.1. □

4 Proof of Theorem 1.2

In this section, we will prove the existence and the estimate of the norm of the resolvent $(\mathcal{A} - \lambda I)^{-1}$ when $\operatorname{Re}\lambda \in \Big[- e^{-C|\operatorname{Im}\lambda|}/C, 0\Big]$, stated in Theorem 1.2.

Proof of Theorem 1.2. We divide the proof into two steps.

Step 1. First, fix $f = (f^0, f^1) \in H$ and $u = (u^0, u^1) \in D(\mathcal{A})$ satisfying the boundary condition $\Big(\sum_{j,k=1}^{n} a^{jk} u_j^0 \nu_k + a u^1 \Big)\Big|_{\partial\Omega} = 0$. It is easy to see that the following equation

$$(\mathcal{A} - \lambda I)u = f \tag{4.1}$$

is equivalent to

$$\begin{cases} -\lambda u^0 + u^1 = f^0, \\ \sum_{j,k=1}^{n} (a^{jk} u_j^0)_k - \lambda u^1 = f^1. \end{cases} \tag{4.2}$$

Substituting u^1 by u^0 in the second equation of (4.2) and noting the boundary condition, we conclude that

$$\begin{cases} \sum_{j,k=1}^{n} (a^{jk} u_j^0)_k - \lambda^2 u^0 = \lambda f^0 + f^1 & \text{in } \Omega, \\ \sum_{j,k=1}^{n} a^{jk} u_j^0 \nu_k + a\lambda u^0 = -a f^0 & \text{on } \partial\Omega, \\ u^1 = f^0 + \lambda u^0 & \text{in } \Omega. \end{cases} \tag{4.3}$$

Put

$$v = e^{i\lambda s} u^0. \tag{4.4}$$

It is easy to check that v satisfies the following equation:

$$\begin{cases} v_{ss} + \sum_{j,k=1}^{n} (a^{jk} v_j)_k = (\lambda f^0 + f^1) e^{i\lambda s} & \text{in } \mathbb{R} \times \Omega, \\ \sum_{j,k=1}^{n} a^{jk} v_j \nu_k - i a v_s = -a f^0 e^{i\lambda s} & \text{on } \mathbb{R} \times \partial\Omega. \end{cases} \tag{4.5}$$

Step 2. By (4.4), we have the following estimates.

$$\begin{cases} |u^0|_{H^1(\Omega)} \le Ce^{C|\mathrm{Im}\,\lambda|}|v|_{H^1(Y)}, \\ |v|_{H^1(X)} \le C(|\lambda|+1)e^{C|\mathrm{Im}\,\lambda|}|u^0|_{H^1(\Omega)}, \\ |v|_{L^2(Z)} \le Ce^{C|\mathrm{Im}\,\lambda|}|u^0|_{L^2(\Gamma_0)}, \\ |v_s|_{L^2(Z)} \le C|\lambda|e^{C|\mathrm{Im}\,\lambda|}|u^0|_{L^2(\Gamma_0)}. \end{cases} \tag{4.6}$$

Now, applying Theorem 3.1 to v, and combining with (4.6), we have

$$|u^0|_{H^1(\Omega)} \le Ce^{C|\mathrm{Im}\,\lambda|}\Big[|f^0|_{H^1(\Omega)} + |f^1|_{L^2(\Omega)} + |u^0|_{L^2(\Gamma_0)}\Big]. \tag{4.7}$$

On the other hand, multiplying (4.2) by $\overline{u}^0$ and integrating it on Ω, it follows that

$$\begin{aligned} &\int_\Omega \Big(-\sum_{j,k=1}^n (a^{jk}u_j^0)_k + \lambda^2 u^0\Big)\cdot \overline{u}^0 dx \\ &= \lambda^2|u^0|^2_{L^2(\Omega)} + \sum_{j,k=1}^n \int_\Omega a^{jk}u_j^0\overline{u}_k^0 dx - \sum_{j,k=1}^n \int_{\partial\Omega} a^{jk}u_j^0\nu_k\overline{u}^0 dx \\ &= \lambda^2|u^0|^2_{L^2(\Omega)} + \sum_{j,k=1}^n \int_\Omega a^{jk}u_j^0\overline{u}_k^0 dx + \int_{\partial\Omega}(a\lambda u^0 + af^0)\overline{u}^0 dx. \end{aligned} \tag{4.8}$$

By taking the imaginary part on the both sides of (4.8), we find,

$$\begin{aligned} &|\mathrm{Im}\,\lambda|\int_{\partial\Omega} a|u^0|^2 dx \\ &\le \Big|-\sum_{j,k=1}^n (a^{jk}u_j^0)_k + \lambda^2 u^0\Big|_{L^2(\Omega)}|u^0|_{L^2(\Omega)} \\ &\qquad +2|\mathrm{Im}\,\lambda||\mathrm{Re}\,\lambda||u^0|^2_{L^2(\Omega)} + C|f^0|_{L^2(\partial\Omega)}|\sqrt{a}u^0|_{L^2(\partial\Omega)} \\ &\le C\Big[|(\lambda f^0 + f^1)|_{L^2(\Omega)}|u^0|_{L^2(\Omega)} \\ &\qquad +|\mathrm{Im}\,\lambda||\mathrm{Re}\,\lambda||u^0|^2_{L^2(\Omega)} + |f^0|_{H^1(\Omega)}|u^0|_{H^1(\Omega)}\Big]. \end{aligned} \tag{4.9}$$

Hence, combining (4.7) and (4.9), we have

$$\begin{aligned} &|u^0|_{H^1(\Omega)} \\ &\le Ce^{C|\mathrm{Im}\,\lambda|}\Big[|f^0|_{H^1(\Omega)} + |f^1|_{L^2(\Omega)} + |\mathrm{Im}\,\lambda||\mathrm{Re}\,\lambda||u^0|_{H^1(\Omega)}\Big]. \end{aligned} \tag{4.10}$$

We now take

$$Ce^{C|\mathrm{Im}\,\lambda|}|\mathrm{Im}\,\lambda||\mathrm{Re}\,\lambda| \le \frac{1}{2},$$

which holds, whenever $|\mathrm{Re}\,\lambda| \le -e^{C|\mathrm{Im}\,\lambda|}/C$ for some sufficiently large $C > 0$. Then, by (4.10), we have

$$|u^0|_{H^1(\Omega)} \le Ce^{C|\mathrm{Im}\,\lambda|}(|f^0|_{H^1(\Omega)} + |f^1|_{L^2(\Omega)}). \tag{4.11}$$

Recalling that $u^1 = f^0 + \lambda u^0$, it follows

$$\begin{aligned} |u^1|_{L^2(\Omega)} &\le |f^0|_{L^2(\Omega)} + |\lambda||u^0|_{L^2(\Omega)} \\ &\le Ce^{C|\mathrm{Im}\,\lambda|}(|f^0|_{H^1(\Omega)} + |f^1|_{L^2(\Omega)}). \end{aligned} \tag{4.12}$$

By (4.11)–(4.12), we know that $\mathcal{A} - \lambda I$ is injective. Therefore $\mathcal{A} - \lambda I$ is bi-injective from $D(\mathcal{A})$ to H. Moreover,

$$||(\mathcal{A} - \lambda I)^{-1}||_{\mathcal{L}(H,H)} \le Ce^{C|\mathrm{Im}\,\lambda|}, \quad \mathrm{Re}\,\lambda \in (-e^{C|\mathrm{Im}\,\lambda|}/C, 0), \quad |\lambda| \ge 1.$$

This completes the proof of Theorem 1.2. □

5 Proof of Theorem 1.1

This section is addressed to giving a proof of Theorem 1.1. It is now clear that once suitable resolvent estimates are established, the existing abstract semi-group results can be adopted to yield the desired energy decay rate. In this respect, we refer to [3, 5] for the energy decay for the wave equation on non-compact manifolds, [15] for the energy decay for the linear evolution equations on Hilbert spaces, and [2] for more general problems governed by bounded semi-groups on Banach spaces.

Proof of Theorem 1.1. Recalling the resolvent estimate established in Theorem 1.2, for any positive integer k, proceeding exactly as in [3, Théorème 3] and [15, Theorem 2.1], we conclude that

$$\left|\left|e^{t\mathcal{A}}u\frac{1}{(I-\mathcal{A})^k}\right|\right|_H \le \left(\frac{C}{\ln(2+t)}\right)^k ||u||_H, \qquad \forall\, t \ge 0, \tag{5.1}$$

i.e.,

$$||e^{t\mathcal{A}}u||_H \le \left(\frac{C}{\ln(2+t)}\right)^k ||u||_{D(\mathcal{A}^k)}, \qquad \forall\, t \ge 0. \tag{5.2}$$

Taking $k = 2$, then $D(\mathcal{A})$ is the interpolate space between $D(\mathcal{A}^0) = H$ and $D(\mathcal{A}^2)$. Note however that

$$||e^{t\mathcal{A}}u||_H \le C||u||_H. \tag{5.3}$$

Hence, combining (5.2) and(5.3), and using the standard interpolation technique, the desired decay rate result (1.7) in Theorem 1.1 follows. □

References

[1] C. Bardos, G. Lebeau and J. Rauch, *Sharp sufficient conditions for the observation, control, and stabilization of waves from the boundary, SIAM I. Control Optim.*, 30 (1992), 1024–1065.

[2] C. J. K. Batty and T. Duyckaerts, *Non-uniform stability for bounded semi-groups on Banach spaces, J. Evol. Equ.*, 8 (2008), 765–780.

[3] N. Burq, *Décroissance de l'énergie locale de l'équation des ondes pour le probléme extérieur et absence de résonance au voisinagage du réel,* Acta *Math.*, 180 (1998), 1–29.

[4] N. Burq and M. Hitrik, *Energy decay for damped wave equations on partially rectangular domains, Math. Res. Lett.*, 14 (2007), 35–47.

[5] H. Christianson, *Applications of cutoff resolvent estimates to the wave equations,* Preprint (see http://arxiv.org/pdf/0709.0555).

[6] T. Duyckaerts, *Optimal decay rates of the energy of a hyperbolic-parabolic system coupled by an interface, Asymptot. Anal.*, 51 (2007), 17–45.

[7] T. Duyckaerts, X. Zhang and E. Zuazua, *On the optimality of the observability inequalities for parabolic and hyperbolic systems with potentials,* Ann. Inst. H. Poincaré Anal. Non Linéaire, 25 (2008), 1–41.

[8] X. Fu, *A weighted identity for partial differential operators of second order and its applications, C. R. Math. Acad. Sci. Paris,* 342 (2006), 579–584.

[9] X. Fu, *Null controllability for the parabolic equation with a complex principal part,* Preprint (see http://arxiv.org/abs/0805.3808).

[10] X. Fu, J. Yong, X. Zhang, *Exact controllability for the multidimensional semilinear hyperbolic equations, SIAM J. Control Optim.*, 46 (2007), 1578–1614.

[11] A. V. Fursikov and O. Yu. Imanuvilov, *Controllability of Evolution Equations,* Lecture Notes Series 34, Research Institute of Mathematics, Seoul National University, Seoul, Korea, 1994.

[12] A. Haraux, *Systèmes Dynamiques Dissipatifs et Applications,* Masson, Paris, 1991.

[13] M. M. Lavrentév, V. G. Romanov and S. P. Shishat·skii, *Ill-posed Problems of Mathematical Physics and Analysis*, Translated from the Russian by J. R. Schulenberger, Translations of Mathematical Monographs, 64, American Mathematical Society, Providence, RI, 1986.

[14] G. Lebeau and L. Robbiano, *Stabilisation de l'équation des ondes par le bord, Duke Math. J.*, 86 (1997), 465–491.

[15] Z. Liu and B. Rao, *Characterization of polynomial decay rate for the solution of linear evolution equation, Z.* angew. *Math. Phys.*, 56 (2005), 630–644.

[16] K.-D. Phung, *Polynomial decay rate for the dissipative wave equation, J. Differential Equations,* 240 (2007), 92–124.

[17] K.-D. Phung, *Boundary stabilization for the wave equation in a bounded cylindrical domain, Discrete Contin. Dynam. Systems,* 20 (2008), 1057–1093.

[18] K.-D. Phung and X. Zhang, *Time reversal focusing of the initial state for Kirchoff plate, SIAM J. Appl. Math.*, 68 (2008), 1535–1556.

[19] J. Rauch, X. Zhang and E. Zuazua, *Polynomial decay of a hyperbolic-parabolic coupled system, J. Math. Pures Appl.*, 84 (2005), 407–470.

[20] G. Wang and L. Wang, *The Carleman inequality and its application to periodic optimal control governed by semilinear parabolic differential equations, J. Optim.* Theory *Appl.*, 118 (2003), 249–461.

[21] X. Zhang, *Explicit observability estimate for the wave equation with potential and its application,* R. Soc. Lond. Proc. Ser. A Math. Phys. Eng. *Sci.*, 456 (2000), 1101–1115.

[22] X. Zhang and E. Zuazua, *Long time behavior of a coupled heat-wave system arising in fluid-structure interaction,* Arch. Rat. Mech. Anal., 184 (2007), 49–120.

A Remark on the Controllability of a System of Conservation Laws in the Context of Entropy Solutions*

Olivier Glass
Université Pierre et Marie Curie-Paris6
UMR 7598 Laboratoire Jacques-Louis Lions
Paris, F-75005 France
Email: glass@ann.jussieu.fr

Abstract

In this paper, we consider a system of conservation laws introduced by DiPerna [12], from the point of view of boundary controllability, in the context of weak entropy solutions. Bressan and Coclite [5] have shown that this system is not controllable when the solutions are of small total variation. We study the use of a large shock wave for the control.

1 Introduction

1.1 Basic question and previous results

The problems of controllability for one-dimensional systems of conservation laws and more generally quasilinear hyperbolic systems have known much progress since the pioneering work of Cirinà [7], particularly in the framework of classical solutions of class C^1, see particularly Li and Rao [19] for an important work on this problem.

A general quasilinear hyperbolic system in one-dimension reads as follows:

$$u_t + A(u)u_x = 0 \text{ for } (t,x) \in \mathbb{R}_+ \times \mathbb{R}, \tag{1.1}$$

where $u : \mathbb{R}_+ \times \mathbb{R} \to \mathbb{R}^n$ is the unknown and the matrix $A(u) \in \mathcal{M}_n(\mathbb{R})$ satisfies the strict hyperbolic condition, that is, for any u in the state domain $\Omega \subset \mathbb{R}^n$, one has

$$A(u) \text{ has } n \text{ real distinct eigenvalues } \lambda_1 < \cdots < \lambda_n. \tag{1.2}$$

*The author is partly supported by the Agence Nationale de la Recherche, Project ContrôleFlux.

These eigenvalues are the characteristic speeds at which the system propagates; we associate the eigenvectors r_i with them. A very important particular case of hyperbolic systems is given by the systems of conservation laws:

$$u_t + (f(u))_x = 0 \text{ for } (t,x) \in \mathbb{R}_+ \times \mathbb{R}, \tag{1.3}$$

where the flux function f is regular from Ω to $\mathbb{R}^n$. Typically, t is the time and x is the position.

The general problem of controllability is the following. Consider the problem posed in the interval $[0,1]$ rather than in $\mathbb{R}$. In such a case one needs of course to prescribe boundary conditions on $[0,T] \times \{0,1\}$: here boundary conditions will be considered as a *control*, that is, a way to influence the system to make it behave in a prescribed way. Let us call $u(t,\cdot)$ the *state* of the system at time t. The question is: having given two possible states of the system, say u_0 and u_1, can we choose the control suitably, in order that the solution of the system starting from u_0, reaches u_1 at time T?

Let us underline that the boundary conditions for such hyperbolic systems of conservation laws are in general quite involved (particularly when the characteristic speeds can change sign). A way to overcome this difficulty is to reformulate the controllability problem in an *underdetermined* form: given u_0, u_1 and T, can we find a solution of (1.1) (*without boundary conditions*) satisfying

$$u_{|t=0} = u_0 \text{ and } u_{|t=T} = u_1?$$

A very general solution to this problem has been obtained by Li and Rao [19] in the case of solutions of class C^1 with small C^1 norm, when the characteristic speeds are strictly separated from zero.

Theorem 1.1 (Li-Rao, [19], 2002). *Consider the system* (1.1) *with the condition* $\lambda_1(u) < \cdots < \lambda_k(u) \leq -c < 0$ *and* $0 \leq c < \lambda_{k+1}(u) < \cdots < \lambda_n(u)$. *Then for all* $\phi, \psi \in C^1([0,1])$ *such that* $\|\phi\|_{C^1} + \|\psi\|_{C^1} < \varepsilon$, *there exists a solution* $u \in C^1([0,T] \times [0,1])$ *such that*

$$u_{|t=0} = \phi \textit{ and } u_{|t=T} = \psi.$$

In the same functional framework, a result has also been obtained in certain cases admitting vanishing characteristic speeds, see [11].

But the situation is far less well understood in the context of *entropy solutions* of systems of conservation laws (1.3). The origin of this theory stems from the fact that in general the solutions of these equations develop singularity in finite time. It is hence natural to consider discontinuous (weak) solutions. As is well known, such weak solutions are no longer unique, and it is natural to consider weak solutions which satisfy *entropy conditions* aiming at singling out the physically relevant solution. These entropy conditions are the following:

Definition 1.2. We define an entropy/entropy flux couple as a couple of functions (η, q) such that

$$\forall u \in \mathbb{R}_+^* \times \mathbb{R}, \quad D\eta(u).Df(u) = Dq(u).$$

Then entropy solutions are defined as weak solutions of the system,

$$u_t + (f(u))_x = 0,$$

which moreover satisfy that for all (η, q) entropy couple with η convex, stands, in the sense of distributions:

$$\eta(u)_t + q(u)_x \leq 0.$$

An important difference between the theory of entropy solutions and the one of classical solutions is that in the context of entropy solutions, the system is no longer reversible. This is, of course, of great significance for the study of these equations, and particularly for what concerns controllability problems. Of course, the C^1 solutions of the system are entropy solutions in particular.

To be more precise, in this paper, we will consider solutions *à la* Glimm [15], that is, entropy solutions in the sense above, of small total variation in x for all times. Note that the meaning of the boundary value in this context is intricate, especially when the characteristic speeds are not separated from zero, see the reference of Dubois and LeFloch [13] in particular. Hence the underdetermined version of the problem is particularly well suited here.

There are very few studies concerning the controllability problem for hyperbolic systems of conservation laws in the context of entropy solutions. Ancona and Marson [2] described the attainable set on a half line for convex scalar ($n = 1$) conservation laws. In the case of the Burgers equation, Horsin [16] considered the case of a bounded interval, when the initial data are not necessarily zero. His method relies on J.-M. Coron's so-called *return method*, to which we shall come back later. For what concerns systems of conservation laws ($n \geq 2$), Ancona and Coclite described the attainable set for the particular case of Temple systems [1]. Bressan and Coclite [5] showed that for a hyperbolic system of conservation laws with fields either linearly degenerate or genuinely nonlinear in the sense of Lax [17], with characteristic speeds strictly separated from zero, one can *asymptotically* converge toward any constant state. But for what concerns the finite time controllability, Bressan and Coclite showed the following very surprising result.

Theorem 1.3 (Bressan-Coclite, [5], 2002). *For a class of systems containing DiPerna's system [12]:*

$$\begin{cases} \partial_t \rho + \partial_x(\rho v) = 0, \\ \partial_t v + \partial_x \left(\frac{v^2}{2} + \frac{K^2}{\gamma - 1} \rho^{\gamma-1} \right) = 0, \end{cases} \tag{1.4}$$

there are initial conditions $\varphi \in BV([0,1])$ of arbitrary small total variation such that any entropy solution u remaining of small total variation for all times satisfies:

$$\text{for any } t,\ u(t,\cdot) := (\rho, v) \text{ is not constant.}$$

We see that the situation is strikingly different from the case of C^1 solutions. As we will see, DiPerna's system is strictly hyperbolic, has genuinely nonlinear characteristic fields, and there are large zones in which the two characteristic speeds are away from zero. Hence the Li-Rao theorem applies, and in the context of C^1 solutions, one can reach constant state in finite time, at least if one stays away from the critical state when one of the characteristic speeds vanishes. Hence Bressan and Coclite's result describes a particular phenomenon due to discontinuities. To describe very roughly their counterexample, the initial state they consider is constituted with a dense distribution of shock waves in $[0,1]$; a particular feature of DiPerna's system is that when two shocks of the same characteristic family interact, they merge into a larger shock and create an additional shock in the other characteristic family. This involves a permanent creation of shocks in the domain, hence the solution cannot be driven to a constant state.

Now the introduction of system (1.4) was motivated by isentropic fluid dynamics, which is described by a system very close to (1.4):

$$\begin{cases} \partial_t \rho + \partial_x m = 0, \\ \partial_t m + \partial_x (\frac{m^2}{\rho} + \kappa \rho^\gamma) = 0. \end{cases} \tag{1.5}$$

In the above equation, $\rho = \rho(t,x) \geq 0$ is the density of the fluid, $m(t,x)$ is the momentum ($v(t,x) = \frac{m(t,x)}{\rho(t,x)}$ is the velocity of the fluid), the pressure law is $p(\rho) = \kappa \rho^\gamma$, $\gamma \in (1,3]$. Equation (1.5) is formulated in Eulerian coordinates. The problem of one-dimensional isentropic gas dynamics is also frequently studied in Lagrangian coordinates:

$$\begin{cases} \partial_t \tau - \partial_x v = 0, \\ \partial_t v + \partial_x (\kappa \tau^{-\gamma}) = 0, \end{cases} \tag{1.6}$$

in which the system is referred to as the p-system; here $\tau = 1/\rho$ is the specific volume. What we have shown in [14] is that the particular behavior of system (1.4) does not occur in the case of equations (1.5) and (1.6):

Theorem 1.4 (G., [14], 2007). *Consider two constant states* $\overline{u}_0 := (\rho_0, m_0)$ *and* $\overline{u}_1 := (\rho_1, m_1)$ *in* $\mathbb{R}_+^* \times \mathbb{R}$*. There exist* $\varepsilon > 0$ *and* $T > 0$*, such that, for any* $u_0 \in BV([0,1])$ *satisfying:*

$$\|u_0 - \overline{u}_0\|_{L^1} \leq \varepsilon \text{ and } TV(u_0) \leq \varepsilon,$$

there is an entropy solution u *of* (1.5) *in* $[0,T] \times [0,1]$ *such that*

$$u_{|t=0} = u_0 \quad \text{and} \quad u_{|t=T} = \overline{u}_1.$$

The same result applies for equation (1.6).

Remark 1.5. Actually, the result of [14] describes a broader set of final states that can be reached via suitable boundary controls. Typically, this set contains all small C^1 states, and also states containing shocks, which fulfill a so-called Oleinik-type inequality. Also, one can see that in the case (1.5), no condition of separation of the characteristics speeds from zero is imposed, despite the fact that these speed can actually vanish.

The proof of Theorem 1.4 given in [14] relies in fact on two different methods for the case (1.5) and the case (1.6) and gives in fact slightly different results. Actually, the method we give for (1.6) applies also to equation (1.5) (see [14] for more details), and allows one to get the following property:

$$\text{if } u_0 - \overline{u}_1 \text{ is small in total variation, then the solution of the control problem can be chosen to be small as well.} \tag{1.7}$$

This does not mean that necessarily we will have for all times that $u(t,\cdot)$ is of total variation of order $TV(u_0 - \overline{u}_1)$; actually this is more likely $[TV(u_0 - \overline{u}_1)]^{1/3}$, but this is typically a behavior which is excluded for system (1.4). One of the main points is that for systems (1.5) and (1.6), when two shocks of the same characteristic family interact, they merge into a larger shock and create a rarefaction wave in the other characteristic family. The other method we present in [14] for (1.5) does not yield property (1.7), and does not apply to system (1.6). But what we are going to see in this paper is that it applies to system (1.4).

1.2 The result

What we show is the following.

Theorem 1.6. *Given* $\overline{u}_0 := (\rho_0, v_0), \overline{u}_1 := (\rho_1, v_1)$ *in* $\mathbb{R}_+^* \times \mathbb{R}$*, there exist* $\varepsilon > 0$ *and* $T > 0$*, such that, for any* $u_0 \in BV([0,1]; \mathbb{R}_+^* \times \mathbb{R})$ *satisfying:*

$$\|u_0 - \overline{u}_0\|_{L^1} \leq \varepsilon \text{ and } TV(u_0) \leq \varepsilon,$$

there is an entropy solution u *of* (1.4) *in* $[0,T] \times [0,1]$ *such that*

$$u_{|t=0} = u_0, \text{ and } u_{|t=T} = \overline{u}_1.$$

But of course, the solution we obtain does not remain of small total variation for all times!

1.3 Structure of the proof

As in [14], the proof of Theorem 1.6 consists in proving these two consecutive propositions.

Proposition 1.7. *Let* $u_0 \in BV([0,1]; \mathbb{R}_+^* \times \mathbb{R})$ *as in Theorem 1.6. Then there exist* $T_1 > 0$*, a constant state* $\omega_1 \in \Omega$*, and an entropy solution* $u : [0,T_1] \times [0,1] \to \Omega$ *of (1.4) such that*

$$u_{|t=0} = u_0 \tag{1.8}$$
$$u_{|t=T_1} = \omega_1. \tag{1.9}$$

This part is to show that the fact that the solution remains of small total variation is central in Theorem 1.3. This is connected to Coron's return method, which was introduced in [9]; see [10] for more details on it. Basically, this method advocates that in many situations, one has better controllability properties when the system goes far from the base point and returns to it. In the context here the Bressan-Coclite theorem shows that it is more or less necessary.

The second proposition, which can be seen as finite-dimensional control result, is the following.

Proposition 1.8. *For any* $(\omega, \omega') \in (\mathbb{R}_+^* \times \mathbb{R})^2$*, there is some* $T_2 > 0$ *and an entropy solution* u *of* (1.4) *in* $[0,T] \times [0,1]$*, such that:*

$$u_{|t=0} = \omega \tag{1.10}$$
$$u_{|t=T_2} = \omega'. \tag{1.11}$$

We show this two propositions in Sections 3 and 4, which establishes Theorem 1.6.

2 Characteristics of DiPerna's system

Let us briefly describe the main characteristics of system (1.4). The Jacobian matrix $A = df$ associated with (1.4) is the following

$$A(\rho, v) = \begin{pmatrix} v & \rho \\ K^2 \rho^{\gamma-2} & v \end{pmatrix}. \tag{2.1}$$

Hence it is easily seen that this system is strictly hyperbolic for $(\rho, v) \in \Omega := \mathbb{R}_+^* \times \mathbb{R}$, with eigenvalues

$$\lambda_1 = v - K\rho^\beta \text{ and } \lambda_2 = v + K\rho^\beta, \tag{2.2}$$

and eigenvectors

$$r_1 = \begin{pmatrix} -\rho \\ K\rho^\beta \end{pmatrix} \text{ and } r_2 = \begin{pmatrix} \rho \\ K\rho^\beta \end{pmatrix} \tag{2.3}$$

where

$$\beta := \frac{\gamma - 1}{2} \in (0, 1).$$

It is straightforward to check that the system is genuinely nonlinear in the sense of Lax [17]

$$r_i . \nabla \lambda_i > 0 \text{ in } \Omega.$$

Let us finally describe the wave curves associated with this system. The wave curves, that is, shock curves and rarefaction curves, are the set of states in Ω which can be connected to a given fixed state on the left $u_l := (\rho_l, v_l)$ via a shock wave or a rarefaction wave. Shock waves (associated with each characteristic family) are discontinuities satisfying Rankine-Hugoniot (in order to become a weak solution of the equation) relations

$$f(u_r) - f(u_l) = s[u_r - u_l], \tag{2.4}$$

and Lax's inequalities (in order to become entropic): for the i-th family of shocks,

$$\lambda_i(u_r) < s < \lambda_i(u_l) \tag{2.5}$$

$$\lambda_{i-1}(u_l) < s < \lambda_{i+1}(u_r). \tag{2.6}$$

where s is the speed of the shock, which gives the particular solution

$$u(t, x) = \begin{cases} u_l \text{ for } x/t < s, \\ u_r \text{ for } x/t > s. \end{cases}$$

Rarefaction waves are defined by introducing integral curves of r_i, and are discontinuity-free solutions:

$$\begin{cases} \frac{d}{d\sigma} W_i(\sigma) = r_i(W_i(\sigma)), \\ W_i(0) = u_l, \\ \sigma \geq 0. \end{cases}$$

The standard (right) shock curves at the point $(\rho_0, v_0) \in \Omega$ are given by the following

$$\begin{cases} v = v_0 - \frac{K}{\sqrt{\beta}} \sqrt{(\rho^{2\beta} - \rho_0^{2\beta}) \frac{\rho - \rho_0}{\rho + \rho_0}} \text{ with } \rho \geq \rho_0, \quad \text{along } R_1(\rho_0, v_0), \\ v = v_0 + \frac{K}{\sqrt{\beta}} \sqrt{(\rho^{2\beta} - \rho_0^{2\beta}) \frac{\rho - \rho_0}{\rho + \rho_0}} \text{ with } 0 < \rho \leq \rho_0, \quad \text{along } R_2(\rho_0, v_0). \end{cases} \tag{2.7}$$

The left shock curves at the point $(\rho_0, v_0) \in \Omega$ (when we fix the right state and look for the right one) are given by the following

$$\begin{cases} v = v_0 - \frac{K}{\sqrt{\beta}}\sqrt{(\rho^{2\beta} + \rho_0^{2\beta})\frac{\rho-\rho_0}{\rho+\rho_0}} \text{ with } 0 < \rho \leq \rho_0, \quad \text{along } L_1(\rho_0, v_0), \\ v = v_0 + \frac{K}{\sqrt{\beta}}\sqrt{(\rho^{2\beta} - \rho_0^{2\beta})\frac{\rho-\rho_0}{\rho+\rho_0}} \text{ with } \rho \geq \rho_0, \quad \text{along } L_2(\rho_0, v_0). \end{cases} \tag{2.8}$$

Instead of describing the rarefaction curves in the plane (ρ, v), we introduce the Riemann invariants associated with system (1.4). Precisely define

$$z = v - \frac{K}{\beta}\rho^\beta \text{ and } w = v + \frac{K}{\beta}\rho^\beta, \tag{2.9}$$

so that

$$r_1.\nabla w = r_2.\nabla z = 0 \text{ and } r_1.\nabla z > 0, \; r_2.\nabla w > 0.$$

In the (w, z)-plane, rarefaction curves are horizontal and vertical half-lines.

3 Proof of Proposition 1.7

The proof of Proposition 1.7 relies on large shocks for system (1.4). We will be able to treat them thanks to the next lemma.

Lemma 3.1. *All shocks (u_-, u_+) are Majda-stable in the sense that*

i. *s is not an eigenvalue of $A(u^\pm)$,*
ii. $\{r_j(u^+) \; / \; \lambda_j(u^+) > s\} \cup \{u^+ - u^-\} \cup \{r_j(u^-) \; / \; \lambda_j(u^-) < s\}$ *is a basis of $\mathbb{R}^2$ (for a j-shock).*

(3.1)

Proof. Taking account of the fact that Lax's inequalities are globally satisfied along the shock curves (see [12]) which means that 1-shocks (resp. 2-shocks) (u_l, u_r) satisfy

$$(r_1(u^-), u_+ - u_-) \text{ (resp. } (u_+ - u_-, r_2(u^+))) \text{ is a basis of } \mathbb{R}^2. \tag{3.2}$$

One of the properties of the system (1.4) as shown by DiPerna [12] is that its shock curves have special behavior in the plane given by the Riemann invariants. One can express all the wave curves in terms of $\eta := (2K/\beta)\rho^\beta$ and check that

$$\frac{\partial R_1}{\partial \eta}, \; \frac{\partial L_1}{\partial \eta} \leq 0 \text{ and } \frac{\partial R_2}{\partial \eta}, \; \frac{\partial L_2}{\partial \eta} \geq 0,$$

which involves that expressed in terms of w, the curves satisfy

$$-\infty \leq \frac{\partial R_1}{\partial w}, \; \frac{\partial L_1}{\partial w} \leq -1 \text{ and } -1 \leq \frac{\partial R_2}{\partial w}, \; \frac{\partial L_2}{\partial w} \leq 0.$$

This is referred to as property A_2 in [12]. Hence the shock curves are confined in cones which involve that (3.2) is satisfied since in the (w, z) plane, r_1 and r_2 are vertical and horizontal respectively. □

The other ingredient which appeared in [14] was the following.

Lemma 3.2. *For any $(\overline{\rho}_0, \overline{v}_0) \in \Omega$, there exists $(\rho, v) \in L_2(\omega)$, such that*

$$\lambda_2((\rho, v)) > \lambda_1((\rho, v)) \geq 3, \tag{3.3}$$

$$s((\overline{\rho}_0, \overline{v}_0), (\rho, v)) \geq 3. \tag{3.4}$$

Here s is the shock speed given by the Rankine-Hugoniot relation (2.4).

Proof. Consider $(\rho, v) \in L_2(\overline{\rho}_0, \overline{v}_0)$, with $\rho \to +\infty$. From the Rankine-Hugoniot relations, one easily computes

$$s((\overline{\rho}_0, \overline{v}_0), (\rho, v)) = \overline{v}_0 + \frac{K\rho}{\sqrt{\beta}\sqrt{\rho + \overline{\rho}_0}} \sqrt{\frac{\rho^{2\beta} - \overline{\rho}_0^{2\beta}}{\rho - \overline{\rho}_0}}$$

Hence clearly $s((\overline{\rho}_0, \overline{v}_0), (\rho, v)) \to +\infty$ as $\rho \to +\infty$. Next one sees that

$$\lambda_1((\rho, v)) = \overline{v}_0 + \frac{K}{\sqrt{\beta}} \sqrt{\frac{(\rho^{2\beta} - \overline{\rho}_0^{2\beta})(\rho - \overline{\rho}_0)}{\rho + \overline{\rho}_0}} - K\rho^{\beta}.$$

But since $\beta \in (0, 1)$, one has

$$\frac{K}{\sqrt{\beta}} > K.$$

Hence one deduces that as well $\lambda_1((\rho, v)) \to +\infty$ as $\rho \to +\infty$. With the global strict hyperbolicity this concludes the proof. □

Now given $\overline{u}_0 := (\overline{\rho}_0, \overline{v}_0)$ and u_0 as in Theorem 1.6, we introduce $\overline{u} = (\rho, v)$ as in Lemma 3.2. We introduce the following function $U_0 \in BV_{loc}(\mathbb{R}; \mathbb{R}_+^* \times \mathbb{R})$:

$$U_0(x) = \begin{cases} \overline{u} & \text{for } x < 0, \\ u_0(x) & \text{for } 0 \leq x \leq 1, \\ \overline{u}_0 & \text{for } x > 1. \end{cases} \tag{3.5}$$

Exactly as in [14], we can prove the following proposition.

Proposition 3.3. *If u_0 is small enough total variation, there is a global-in-time entropy solution U of (1.4) in $[0, +\infty) \times \mathbb{R}$ satisfying*

$$U(0, \cdot) = U_0 \text{ in } \mathbb{R}. \tag{3.6}$$

Moreover it satisfies:

$$U_{|\{1\} \times [0,1]} \text{ is constant.} \tag{3.7}$$

By simply taking the restriction of U to $[0,1] \times [0,1]$, we obtain a solution of the problem consisting in driving u_0 to a constant. The proof of Proposition 3.3 is exactly the same as in [14] (see also [6, 8, 18, 20, 21] for related problems). It relies only on the Majda-stability of the large shock and on the positivity of the propagation speeds on its left. Basically we show that the above initial condition is propagated for all times as a large shock plus small waves on both sides of it. However, due to condition (3.3), all these waves travel at positive speed and eventually leave the domain. The basic ingredient to prove this is the use of a front-tracking algorithm (see [4] for more details on this particular construction of solutions of systems of conservation laws). We refer to the above articles for a complete proof.

4 Proof of Proposition 1.8

This is almost exactly the same as in [14]. There are three zones in Ω with respect to the signs of the characteristic speeds: the zone $\Omega_- := \{(\rho, u) \ / \ u < -K\rho^\beta\}$ where both characteristic speeds are negative, the zone $\Omega_+ := \{(\rho, u) \ / \ u > K\rho^\beta\}$ where both characteristic speeds are positive and the zone $\Omega_\pm := \{(\rho, u) \ / \ -K\rho^\beta < u < K\rho^\beta\}$ where λ_1 is negative and λ_2 is positive. These three zones are separated by the two critical curves $\mathcal{C}_- := \{(\rho, u) \ / \ u = -K\rho^\beta\}$ and $\mathcal{C}_+ := \{(\rho, u) \ / \ u = K\rho^\beta\}$.

Now to prove Proposition 1.8, it suffices to prove that

1. Given ω and ω' in the same zone (Ω_-, Ω_+ or $\Omega_\pm$), one can find a solution from ω to ω'.
2. One can always find a solution from a given zone to another.
3. One can always go out to a critical curve or reach it from one of the above zones.

1. To prove the first point, let us limit ourselves to the case where ω and ω' are sufficiently close to each other. Then since the zones are path-connected and a path is compact, one easily deduces the general case.

Now one has to clarify which one does depend on the zone where the states occur. Given ω and ω' in Ω_+ and sufficiently close one to another, we solve the Riemann problem (ω', ω) (see [4, 17]). If the two states are sufficiently close to one another, the intermediate state is in Ω_+ as well, hence the two waves obtained in this Riemann problem are of fixed sign speed. Hence the Riemann solution of this problem indicates that if one waits long enough, the solution of this problem with ω' on $\mathbb{R}_-$ and ω on $\mathbb{R}_+$, will reach ω' in $[0,1]$.

If both states are in Ω_-, the idea is the same, but one has to let the waves enter from the right boundary, that is, one solves the Riemann problem (ω, ω'), where the separation between the states occurs at $x = 1$. Wait long enough, and ω' enters $[0, 1]$.

If both states are in $\Omega_\pm$, again, we manage in other that the intermediate state ω_m in the resolution of the Riemann problem (ω', ω) is in $\Omega_\pm$. Now we use the solution of the Riemann problem (ω, ω_m) with the states separated at $x = 1$ and the solution of the Riemann problem (ω', ω_m) with the states separated at $x = 0$ to join ω' from ω.

2. To prove the second point, we use roughly the same remark as for Lemma 3.2. If the state you consider is in Ω_- or $\Omega_\pm$, then by a large 2-shock on the left of the domain, you can reach Ω_+. In the same way, if the state you consider is Ω_-, then you can reach $\Omega_\pm$: reach the point on the second left shock curve for which $v = 0$ and observe that its speed is necessarily positive. The same (with 1-shocks on the right) can be done to go from Ω_+ to Ω_- or $\Omega_\pm$. Also, by the same method, one can leave a critical curve.

3. It remains to explain how to reach a critical curve. It is not difficult to see that one can arrive to the critical curve $\mathcal{C}_-$ by a small 1-shock that one lets enter by $x = 1$ (with the critical state on $x > 1$, and a non-critical state for $x < 1$), and that one can reach the critical curve $\mathcal{C}_+$ by a small 2-shock that one lets enter by $x = 0$ (with the critical state on $x < 0$, and a non-critical state for $x > 0$). It suffices to check that r_1 is transverse to $\mathcal{C}_-$ and r_2 is transverse to $\mathcal{C}_+$. This is easily established by noticing that $\beta < 1$.

Acknowledgements. The author would like to thank Professors Ta-Tsien Li, Bo-Peng Rao and Yue-Jun Peng for their warm greeting in Shanghai, and useful discussions.

References

[1] Ancona F., Coclite G.M., On the attainable set for Temple class systems with boundary controls, *SIAM J. Control Optim.*, **43**(6): 2166–2190, 2005.

[2] Ancona F., Marson A., On the attainable set for scalar nonlinear conservation laws with boundary control, *SIAM J. Control Optim.*, **36**(1): 290–312, 1998.

[3] Ancona F., Marson A., Asymptotic stabilization of systems of conservation laws by controls acting at a single boundary point. Control methods in PDE-dynamical systems, *Contemp. Math.*, **426**: 1–43, 2007.

[4] Bressan A., *Hyperbolic systems of conservation laws, the one-dimensional problem*, Oxford Lecture Series in Mathematics and its Applications 20, 2000.

[5] Bressan A., Coclite G.M., On the boundary control of systems of conservation laws, *SIAM J. Control Optim.*, **41**(2): 607–622, 2002.

[6] Bressan A., Colombo R. M., Unique solutions of 2×2 conservation laws with large data, *Indiana Univ. Math. J.*, **44**(3): 677–725, 1995.

[7] Cirinà M., Boundary controllability of nonlinear hyperbolic systems, *SIAM J. Control*, **7**: 198–212, 1969.

[8] Corli A., Sablé-Tougeron M., Perturbations of bounded variation of a strong shock wave, *J. Differential Equations*, **138**(2): 195–228, 1997.

[9] Coron J.-M., Global Asymptotic Stabilization for controllable systems without drift, *Math. Control Signal Systems*, **5**: 295–312, 1992.

[10] Coron J.-M., *Control and Nonlinearity*, Mathematical Surveys and Monographs 136, American Mathematical Society, Providence, RI, 2007.

[11] Coron J.-M., Glass O., Wang Z.Q., Exact boundary controllability for 1-D quasilinear hyperbolic systems with a vanishing characteristic speed, preprint 2009.

[12] DiPerna R. J., Global solutions to a class of nonlinear hyperbolic systems of equations, *Comm. Pure Appl. Math.*, **26**: 1–28, 1973.

[13] Dubois F., LeFloch P.G., Boundary conditions for nonlinear hyperbolic systems of conservation laws, *J. Differential Equations*, **71**(1): 93–122, 1988.

[14] Glass O., On the controllability of the 1-D isentropic Euler equation, *J. European Math. Soc.*, **9**(3): 427–486, 2007.

[15] Glimm J., Solutions in the large for nonlinear hyperbolic systems of equations, *Comm. Pure Appl. Math.*, **18**: 697–715, 1965.

[16] Horsin T., On the controllability of the Burgers equation, *ESAIM: Control Opt. Calc. Var.*, **3**: 83–95, 1998.

[17] Lax P. D., Hyperbolic Systems of Conservation Laws II, *Comm. Pure Appl. Math.*, **10**: 537–566, 1957.

[18] Lewicka, M., Trivisa K., On the L^1 well posedness of systems of conservation laws near solutions containing two large shocks, *J. Differential Equations*, **179**(1): 133–177, 2002.

[19] Li T.-T., Rao B.-P., Exact boundary controllability for quasi-linear hyperbolic systems, *SIAM J. Control Optim.*, **41**(6): 1748–1755, 2003.

[20] Sablé-Tougeron M., Stabilité de la structure d'une solution de Riemann à deux grands chocs, *Ann. Univ. Ferrara Sez. VII (N.S.)*, **44**: 129–172, 1998.

[21] Schochet S., Sufficient conditions for local existence via Glimm's scheme for large BV data, *J. Differential Equations*, **89**(2): 317–354, 1991.

Introduction to the Control of PDE's*

Vilmos Komornik
Département de Mathématique, Université de Strasbourg
7 rue René Descartes, 67084 Strasbourg Cedex, France
Email: komornik@math.u-strasbg.fr

Abstract

The purpose of this tutorial is to give a quick introduction to various basic results of observability, controllability and stabilization of linear partial differential equations with a time-reversible dynamics.

1 Introduction

The purpose of this tutorial is to give a quick introduction to various basic results of observability, controllability and stabilization of linear partial differential equations with a time-reversible dynamics.

In Section 2 we present the multiplier method which, in the hands of J.L. Lions, became a very powerful method in control theory. His landmark papers [26], [27] and his subsequent monography [28] stimulated a huge research activity resulting in hundreds of papers during the past twenty years.

In Section 3 we show a very efficient combination of the multiplier method and of harmonic analysis in order to obtain observability and controllability theorems in minimal time. Many other applications of this kind are given in the textbook [18].

In Section 4 we give an application of the Fourier series method in control theory. Numerous other results and examples are treated in our book [22] in collaboration with P. Loreti.

Finally, Section 5 contains a general linear stabilization method, originally exposed in [20].

---

*The author thanks the organizers of the Workshop of the French-Chinese Summer Institute on Applied Mathematics, Fudan University, Shanghai, September 2008, for their kind invitation.

2 The multiplier method and the wave equation

Throughout this section we fix a bounded domain $\Omega \subset \mathbb{R}^n$ of class C^2 with boundary Γ. We refer to the original works [27], [28] of Lions or to the textbook [18] for more results and for the proofs of some technical lemmas which are omitted here for the sake of brevity.

2.1 Observability

We investigate here the following system:

$$\begin{cases} u'' - \Delta u = 0 & \text{in} \quad \mathbb{R} \times \Omega, \\ u = 0 & \text{on} \quad \mathbb{R} \times \Gamma, \\ u(0) = u_0 \quad \text{and} \quad u'(0) = u_1 & \text{in} \quad \Omega. \end{cases} \tag{2.1}$$

The well posedness of this system is classical and well known (see [29]):

Proposition 2.1.

(a) If $(u_0, u_1) \in H_0^1(\Omega) \times L^2(\Omega)$, then the problem (2.1) *has a unique (weak) solution belonging to the space*

$$C(\mathbb{R}; H_0^1(\Omega)) \cap C^1(\mathbb{R}; L^2(\Omega)).$$

(b) The energy

$$E = \frac{1}{2} \int_\Omega |u'|^2 + |\nabla u|^2 \, dx$$

of the solutions is conserved (does not depend on $t \in \mathbb{R}$).

(c) If $(u_0, u_1) \in \big(H^2(\Omega) \cap H_0^1(\Omega)\big) \times H_0^1(\Omega)$, then the corresponding (strong) solution belongs to

$$C(\mathbb{R}; H^2(\Omega) \cap H_0^1(\Omega)) \cap C^1(\mathbb{R}; H_0^1(\Omega)).$$

The main result of this subsection is the following:

Theorem 2.2. *(J.L. Lions [27]) If Ω is contained in a ball of radius R, then for every bounded interval I of length $|I| > 2R$ there exist two positive constants c_1, c_2 such that*

$$c_1 E \le \int_I \int_\Gamma |\partial_\nu u|^2 \, d\Gamma \, dt \le c_2 E$$

for all solutions.

Remarks.

- The second "direct" inequality was discovered by Lasiecka and Triggiani [24], and a simpler proof was given by Lions [25] using the *multiplier method.*
- The first "inverse" inequality was discovered by Ho [9] and then improved by Lions [26].
- Later sharper results were obtained by Bardos, Lebeau and Rauch by employing deeper tools [3].

Let us outline the proof. We may assume that $\Omega \subset B(0,R)$. The proof is based on the following crucial identity where we write $m(x) = x$ and $Mu = 2m \cdot u + (n-1)u$ for brevity.

Lemma 2.3. *Every strong solution satisfies the identity*

$$\int_S^T \int_\Gamma (\partial_\nu u) Mu + (m \cdot \nu)((u')^2 - |\nabla u|^2) \, d\Gamma \, dt$$
$$= \Big[\int_\Omega u' Mu \, dx \Big]_S^T + \int_S^T \int_\Omega (u')^2 + |\nabla u|^2 \, dx \, dt \quad (2.2)$$

for all $-\infty < S < T < \infty$.

This identity is independent of the boundary and initial conditions in (2.1).

Proof. Integrating by parts we get

$$0 = \int_S^T \int_\Omega (u'' - \Delta u) Mu \, dx \, dt$$
$$= \Big[\int_\Omega u' Mu \, \mathrm{dx} \Big]_S^T - \int_S^T \int_\Gamma (\partial_\nu u) Mu \, d\Gamma \, dt$$
$$- \int_S^T \int_\Omega u' Mu' \, dx \, dt + \int_S^T \int_\Omega \nabla u \cdot \nabla (Mu) \, dx \, dt.$$

It remains to transform the last two integrals.

Using the equalities $Mu = 2m \cdot u + (n-1)u$ and $\operatorname{div} m = n$ we have

$$- \int_S^T \int_\Omega u' Mu' \, dx \, dt$$
$$= - \int_S^T \int_\Omega m \cdot \nabla (u')^2 + (n-1)(u')^2 \, dx \, dt$$
$$= - \int_S^T \int_\Gamma (m \cdot \nu)(u')^2 \, d\Gamma \, dt + \int_S^T \int_\Omega (u')^2 \, dx \, dt$$

and

$$\int_S^T \int_\Omega \nabla u \cdot \nabla(Mu)\ dx\ dt$$
$$= \int_S^T \int_\Omega m \cdot \nabla\big(|\nabla u|^2\big) + (n+1)|\nabla u|^2\ dx\ dt$$
$$= \int_S^T \int_\Gamma (m \cdot \nu)|\nabla u|^2\ d\Gamma\ dt + \int_S^T \int_\Omega |\nabla u|^2\ dx\ dt.$$

Substituting these expressions into the first equality we obtain (2.2). □

Using the boundary conditions the identity (2.2) is reduced to

$$\int_S^T \int_\Gamma (m \cdot \nu)(\partial_\nu u)^2\ d\Gamma\ dt = \Big[\int_\Omega u' Mu\ dx\Big]_S^T + 2(T-S)E. \quad (2.3)$$

In order to get a sharp estimate of the integral on the right side of (2.3) we establish (following [13]) a second identity:

Lemma 2.4. *We have*

$$\int_\Omega (Mu)^2\ dx = \int_\Omega |2m \cdot \nabla u|^2 + (1-n^2)u^2\ dx + (2n-2)\int_\Gamma (m \cdot \nu)u^2\ d\Gamma. \quad (2.4)$$

This identity is also independent of the boundary and initial conditions in (2.1).

Using the Dirichlet boundary conditions we conclude from the identity (2.4) that

$$\int_\Omega (Mu)^2\ dx \le \int_\Omega |2m \cdot \nabla u|^2\ dx,$$

and therefore

$$\Big|\int_\Omega u' Mu\ dx\Big| \le 2\|m\|_{L^\infty(\Omega)}\|u'\|_{L^2(\Omega)}\|\nabla u\|_{L^2(\Omega)} \le 2\|m\|_{L^\infty(\Omega)}E. \quad (2.5)$$

If $\Omega \subset B(0,R)$, then $\|m\|_{L^\infty(\Omega)} \le R$. Using this and the inequality (2.5) we deduce from the identity (2.3) that

$$\Big|\int_S^T \int_\Gamma (m \cdot \nu)(\partial_\nu u)^2\ d\Gamma\ dt - 2(T-S)E\Big| = \Big|\Big[\int_\Omega u' Mu\ dx\Big]_S^T\Big| \le 4RE.$$

We conclude for every bounded interval $I = (S,T)$ the identity

$$2(|I| - 2R)E \le \int_I \int_\Gamma (m \cdot \nu)(\partial_\nu u)^2\ d\Gamma\ dt \le 2(|I| + 2R)E.$$

This proves the theorem if Ω is star-shaped: $m \cdot \nu > 0$ on Γ. The general case requires minor modifications.

2.2 Controllability

In this subsection we investigate the nonhomogeneous system

$$\begin{cases} y'' - \Delta y = 0 \quad \text{in} \quad [0,T] \times \Omega, \\ y = v \quad \text{on} \quad [0,T] \times \Gamma, \\ y(0) = y_0 \quad \text{and} \quad y'(0) = y_1 \quad \text{in} \quad \Omega, \end{cases} \tag{2.6}$$

where $T > 0$ is a given number and v is considered a *control function.* Applying the transposition or duality method (see [29]) we deduce from Theorem 2.2 the following well posedness result:

Proposition 2.5. *If* $(y_0, y_1) \in L^2(\Omega) \times H^{-1}(\Omega)$ *and* $v \in L^2(0,T; L^2(\Gamma)$, *then the problem* (2.5) *has a unique (weak) solution belonging to the space*

$$C([0,T]; L^2(\Omega)) \cap C^1([0,T]; H^{-1}(\Omega)).$$

Moreover, the linear map

$$(y_0, y_1, v) \mapsto (y, y')$$

is continuous with respect to these topologies.

The main result of this subsection is the following:

Theorem 2.6. *(J.L. Lions [27]). If* $\Omega \subset B(x_0, R)$, $T > 2R$ *and* $(y^0, y^1) \in L^2(\Omega) \times H^{-1}(\Omega)$, *then there exists* $v \in L^2(0,T; L^2(\Gamma))$ *such that the solution of the problem satisfies*

$$y(T) = y'(T) = 0 \quad \textit{in} \quad \Omega.$$

Sketch of the proof by the Hilbert Uniqueness Method (HUM). The main idea is to seek a suitable control function in the form $v = \partial_\nu u$ where u is the solution of

$$\begin{cases} u'' - \Delta u = 0 & \text{in} \quad \mathbb{R} \times \Omega, \\ u = 0 & \text{on} \quad \mathbb{R} \times \Gamma, \\ u(0) = u_0 \quad \text{and} \quad u'(0) = u_1 & \text{in} \quad \Omega \end{cases}$$

for suitable initial data.

This will be a suitable control provided the solution of the homogeneous system

$$\begin{cases} y'' - \Delta y = 0 \quad \text{in} \quad [0,T] \times \Omega, \\ y = \partial_\nu u \quad \text{on} \quad [0,T] \times \Gamma, \\ y(T) = y'(T) = 0 \quad \text{in} \quad \Omega \end{cases}$$

satisfies $y(0) = y_0$ and $y'(0) = y_1$.

Thanks to the direct inequality

$$\|\partial_\nu u\|_{L^2(\Gamma)} \le c\|(u_0,u_1)\|_{H_0^1(\Omega)\times L^2(\Omega)}$$

of Theorem 2.2 and the well posedness of the nonhomogeneous problem, the formula

$$\Lambda(u_0,u_1) := (y'(0),-y(0))$$

defines a continuous linear map

$$\Lambda : H_0^1(\Omega)\times L^2(\Omega) \to H^{-1}(\Omega)\times L^2(\Omega).$$

It remains to prove that the operator Λ is surjective. For this first we establish the identity

$$\langle \Lambda(u_0,u_1),(u_0,u_1)\rangle_{H^{-1}(\Omega)\times L^2(\Omega),H_0^1(\Omega)\times L^2(\Omega)} = \int_0^T \int_\Gamma |\partial_\nu u|^2 \, d\Gamma \, dt$$

by a direct computation. Then, applying the Lax–Milgram theorem we conclude that Λ is even an *isomorphism*. □

2.3 Linear stabilization by natural feedbacks

Instead of ensuring $E(T)=0$ for some finite $T>0$ for the solutions of (2.6), it is more realistic and more practical to seek *automatic* or *feedback* controls yielding $E(\infty)=0$. (Watt's regulator device for the steam engine was one of the first such successful feedback controls; see [30].)

More precisely, we seek a function F such that the problem

$$\begin{cases} u''-\Delta u = 0 \quad \text{in} \quad \mathbb{R}_+\times\Omega, \\ F(u,u',\partial_\nu u) = 0 \quad \text{on} \quad \mathbb{R}_+\times\Gamma, \\ u(0)=u_0 \quad \text{and} \quad u'(0)=u_1 \quad \text{in} \quad \Omega \end{cases}$$

is well posed and its solutions satisfy $E(t)\to 0$ as $t\to\infty$.

One of the simplest expressions making the energy nondecreasing is the following. Fix a partititon $\{\Gamma_0,\Gamma_1\}$ of Γ, two nonnegative continuous functions a,b on Γ_1 and consider the problem

$$\begin{cases} u''-\Delta u = 0 \quad \text{in} \quad \mathbb{R}_+\times\Omega, \\ u=0 \quad \text{on} \quad \mathbb{R}_+\times\Gamma_0, \\ \partial_\nu u + au + bu' = 0 \quad \text{on} \quad \mathbb{R}_+\times\Gamma_1, \\ u(0)=u_0 \quad \text{and} \quad u'(0)=u_1 \quad \text{in} \quad \Omega. \end{cases} \tag{2.7}$$

The problem is well posed:

Proposition 2.7.

(a) If $(u_0, u_1) \in H_0^1(\Omega) \times L^2(\Omega)$*, then the problem* (2.7) *has a unique (weak) solution belonging to the space*

$$C(\mathbb{R}_+; H_0^1(\Omega)) \cap C^1(\mathbb{R}_+; L^2(\Omega)).$$

(b) The modified energy (or Liapunov function)

$$E(t) = \frac{1}{2}\int_\Omega |u'(t)|^2 + |\nabla u(t)|^2 \, dx + \frac{1}{2}\int_{\Gamma_1} a|u(t)|^2 \, d\Gamma$$

of the solutions is nonincreasing.

(c) If $(u_0, u_1) \in \big(H^2(\Omega) \cap H_0^1(\Omega)\big) \times H_0^1(\Omega)$*, then the corresponding (strong) solution belongs to*

$$C(\mathbb{R}_+; H^2(\Omega) \cap H_0^1(\Omega)) \cap C^1(\mathbb{R}_+; H_0^1(\Omega)).$$

Proof. We indicate only the proof of (b) for strong solutions and we refer to [15] or [18] for the full proof. Differentiating the modified energy we get

$$\begin{aligned} E' &= \int_\Omega u'u'' + \nabla u \cdot \nabla u' \, dx + \int_{\Gamma_1} auu' \, d\Gamma \\ &= \int_\Omega u'\Delta u + \nabla u' \cdot \nabla u \, dx + \int_{\Gamma_1} auu' \, d\Gamma \\ &= \int_{\Gamma_0} u'\partial_\nu \, d\Gamma + \int_{\Gamma_1} u'(\partial_\nu u + au) \, d\Gamma \\ &= -\int_{\Gamma_1} b(u')^2 \, d\Gamma \\ &\le 0. \end{aligned}$$

Since $E' \le 0$, E is nonincreasing. □

Under some additional assumptions the energy tends to zero. The first result of this kind is due to G. Chen [6], [7]:

Theorem 2.8. *Assume that* a, b *are positive on* Γ_1 *and that there is a point* $x_0 \in \mathbb{R}^n$ *such that setting* $m(x) = x - x_0$ *we have*

$$m \cdot \nu \le 0 \quad \text{on} \quad \Gamma_0 \quad \text{and} \quad m \cdot \nu \ge 0 \quad \text{on} \quad \Gamma_1. \tag{2.8}$$

Then there exist two constants $C, \omega > 0$ *such that*

$$E(t) \le CE(0)e^{-\omega t}, \quad t \ge 0$$

for all solutions of (2.7).

Let us consider the special case of the system (2.7) where Ω is the unit ball of $\mathbb{R}^3$, $x_0 = 0$, $\Gamma_0 = \varnothing$ and $a = b = 1$. Then (2.8) and the modified energy take the form

$$\begin{cases} u'' - \Delta u = 0 \quad \text{in} \quad \mathbb{R}_+ \times \Omega, \\ \partial_\nu u + u + u' = 0 \quad \text{on} \quad \mathbb{R}_+ \times \Gamma, \\ u(0) = u_0 \quad \text{and} \quad u'(0) = u_1 \quad \text{on} \quad \Omega \end{cases} \tag{2.9}$$

and

$$E = \frac{1}{2}\int_\Omega (u')^2 + |\nabla u|^2 \, dx + \frac{1}{2}\int_\Gamma u^2 \, d\Gamma.$$

The following theorem is a special case of a result obtained in [15].

Theorem 2.9. *The solutions of* (2.9) *satisfy the decay estimate*

$$E(t) \leq E(0)e^{1-\frac{t}{2}}$$

for all $t \geq 0$.

Proof. We recall the multiplier identity of Lemma 2.3:

$$\int_S^T \int_\Gamma (\partial_\nu u)Mu + (m\cdot\nu)((u')^2 - |\nabla u|^2) \, d\Gamma \, dt$$
$$= \Big[\int_\Omega u'Mu \, dx\Big]_S^T + \int_S^T \int_\Omega (u')^2 + |\nabla u|^2 \, dx \, dt$$

where $m(x) = x$ and $Mu = 2m\cdot\nabla u + 2u$.

Using the boundary condition it is simplified to

$$\int_S^T \int_\Gamma -2(u+u')(u + m\cdot\nabla u) + (u')^2 - |\nabla u|^2 + u^2 \, d\Gamma \, dt$$
$$= \Big[\int_\Omega u'Mu \, dx\Big]_S^T + 2\int_S^T E \, dt.$$

Since $|m| \leq 1$, we have

$$-2(u+u')(m\cdot\nabla u) \leq (u+u')^2 + |m\cdot\nabla u|^2 \leq (u+u')^2 + |\nabla u|^2,$$

and therefore

$$\begin{aligned}
&\Big[\int_\Omega u'Mu\ dx\Big]_S^T + 2\int_S^T E\ dt \\
&= \int_S^T \int_\Gamma -2(u+u')(u+m\cdot\nabla u) + (u')^2 - |\nabla u|^2 + u^2\ d\Gamma\ dt \\
&\le \int_S^T \int_\Gamma -2(u+u')u + (u+u')^2 + (u')^2 + u^2\ d\Gamma\ dt \\
&= 2\int_S^T \int_\Gamma (u')^2\ d\Gamma\ dt \\
&= 2E(S) - 2E(T).
\end{aligned}$$

Next we observe that (since $n = 3$ and $|m| \le 1$)

$$\begin{aligned}
\int_\Omega (Mu)^2\ dx &= \int_\Omega |2m\cdot\nabla u|^2 + (1-n^2)u^2\ dx \\
&\quad + (2n-2)\int_\Gamma (m\cdot\nu)u^2\ d\Gamma \\
&\le 4\int_\Omega |\nabla u|^2\ dx + 4\int_\Gamma u^2\ d\Gamma
\end{aligned}$$

and therefore

$$\begin{aligned}
\Big|\int_\Omega u'Mu\ dx\Big| &\le \int_\Omega (u')^2 + (1/4)(Mu)^2\ dx \\
&\le \int_\Omega (u')^2 + |\nabla u|^2\ dx + \int_\Gamma u^2\ d\Gamma \\
&= 2E.
\end{aligned}$$

Therefore we have

$$\begin{aligned}
-2E(S) - 2E(T) + 2\int_S^T E\ dt &\le \Big[\int_\Omega u'Mu\ dx\Big]_S^T + 2\int_S^T E\ dt \\
&\le 2E(S) - 2E(T).
\end{aligned}$$

Hence

$$2\int_S^T E\ dt \le 4E(S)$$

for all $0 \le S \le T < \infty$ and therefore

$$\frac{1}{2}\int_S^\infty E\ dt \le E(S)$$

for all $S \ge 0$.

The following "anti"-Gronwall lemma completes the proof:

Lemma 2.10. *If a nonincreasing function $E : \mathbb{R}_+ \to \mathbb{R}_+$ satisfies for some $\alpha > 0$ the condition*

$$\alpha \int_t^\infty E(s)\, ds \le E(t)$$

for all $t \ge 0$, then

$$E(t) \le E(0)e^{1-\alpha t}$$

for all $t \ge 0$.

Proof of the lemma. The function

$$f(x) := e^{\alpha x} \int_x^\infty E(s)\, ds, \quad x \ge 0$$

is nonincreasing because

$$f'(x) = e^{\alpha x}\Big(\alpha \int_x^\infty E(s)\, ds - E(x)\Big) \le 0$$

almost everywhere. Hence

$$\alpha e^{\alpha x} \int_x^\infty E(s)\, ds = \alpha f(x) \le \alpha f(0) = \alpha \int_0^\infty E(s)\, ds \le E(0)$$

for all $x \ge 0$. Since E is nonnegative and nonincreasing, we conclude that

$$E(0)e^{-\alpha x} \ge \alpha \int_x^\infty E(s)\, ds \ge \alpha \int_x^{x+\alpha^{-1}} E(s)\, ds \ge E(x+\alpha^{-1}).$$

Putting $t = x + \alpha^{-1}$ the inequality

$$E(x+\alpha^{-1}) \le E(0)e^{-\alpha x}, \quad x \ge 0$$

takes the form

$$E(t) \le E(0)e^{1-\alpha t}, \quad t \ge \alpha^{-1}.$$

The last inequality also holds for $0 \le t \le \alpha^{-1}$ because $E(t) \le E(0)$. □

Remark. The above approach can be adapted to more complex hyperbolic systems (see [17] on Maxwell's system, [19] on coupled systems, [1] on linear elastodynamic systems) and to various plate models (see, e.g., Lagnese [23]). However, many important models remain outside its applicability. Moreover, we cannot construct feedbacks of this type leading to arbitrarily large decay rates. We shall present in Section 5 another more general and more satisfactory approach.

2.4 Nonlinear stabilization by natural feedbacks

The method of the preceding subsection can be adapted for certain nonlinear feedbacks.

Consider the problem

$$\begin{cases} u'' - \Delta u = 0 \quad \text{in} \quad \mathbb{R}_+ \times \Omega, \\ u = 0 \quad \text{on} \quad \mathbb{R}_+ \times \Gamma_0, \\ \partial_\nu u + (m \cdot \nu) g(u') = 0 \quad \text{on} \quad \mathbb{R}_+ \times \Gamma_1, \\ u(0) = u_0 \quad \text{and} \quad u'(0) = u_1 \quad \text{in} \quad \Omega \end{cases} \tag{2.10}$$

where $\{\Gamma_0, \Gamma_1\}$ is a partition of Γ, $m(x) = x - x_0$ for some given point $x_0 \in \mathbb{R}^n$, and $g : \mathbb{R} \to \mathbb{R}$ is a given nondecreasing continuous function satisfying $g(0) = 0$.

Proposition 2.11.

(a) If $(u_0, u_1) \in H_0^1(\Omega) \times L^2(\Omega)$, *then the problem* (2.10) *has a unique (weak) solution belonging to the space*

$$C(\mathbb{R}_+; H_0^1(\Omega)) \cap C^1(\mathbb{R}_+; L^2(\Omega)).$$

(b) The energy

$$E(t) = \frac{1}{2} \int_\Omega u'(t)^2 + |\nabla u(t)|^2 \, dx$$

of the solutions is nonincreasing.

(c) If $(u_0, u_1) \in \big(H^2(\Omega) \cap H_0^1(\Omega)\big) \times H_0^1(\Omega)$, *then the corresponding (strong) solution belongs to*

$$C(\mathbb{R}_+; H^2(\Omega) \cap H_0^1(\Omega)) \cap C^1(\mathbb{R}_+; H_0^1(\Omega)).$$

Proof. As in the case of Proposition 2.7, we only indicate the proof of (b) for strong solutions and refer to [18] for the full proof. Differentiating

$$E = \frac{1}{2} \int_\Omega (u')^2 + |\nabla u|^2 \, dx$$

we get

$$\begin{aligned} E' &= \int_\Omega u' u'' + \nabla u \cdot \nabla u' \, dx \\ &= \int_\Omega u' \Delta u + \nabla u' \cdot \nabla u \, dx \\ &= \int_{\Gamma_0} u' \partial_\nu u \, d\Gamma + \int_{\Gamma_1} u' \partial_\nu u \, d\Gamma \\ &= -\int_{\Gamma_1} (m \cdot \nu) u' g(u') \, d\Gamma. \end{aligned}$$

Since the right side of the identity

$$E' = -\int_{\Gamma_1} (m \cdot \nu) u' g(u') \, d\Gamma$$

is ≤ 0, the assertion follows. □

Now we have the following theorem establishing the polynomial decay of the solutions:

Theorem 2.12. *Assume that*

- $n \geq 3$;
- $\Gamma_0 \neq \varnothing$;
- $m \cdot \nu \leq 0$ *on* Γ_0 *and* $m \cdot \nu \geq 0$ *on* Γ_1;
- $c_1|s|^p \leq |g(s)| \leq c_2|s|^{1/p}$ *if* $|s| \leq 1$;
- $c_3|s| \leq |g(s)| \leq c_4|s|$ *if* $|s| \geq 1$

with some real number $p > 1$ *and positive constants* c_1, c_2, c_3, c_4.

Then the solutions satisfy the energy estimate

$$E(t) \leq Ct^{\frac{2}{1-p}}$$

with a constant depending on the initial energy $E(0)$.

In the rest of this subsection we sketch the proof and refer to [18] for the details. Using the multiplier method we obtain the following identity which is reduced to that of the preceding subsection if $p = 1$. Setting $Mu := 2m \cdot \nabla u + (n-1)u$ and $d\Gamma_m := (m \cdot \nu) d\Gamma$ we have

$$\begin{aligned} 2\int_S^T E^{\frac{p+1}{2}} \, dt = &\int_S^T E^{\frac{p-1}{2}} \int_{\Gamma_0} |\partial_\nu u|^2 \, d\Gamma_m \, dt \\ &- \left[E^{\frac{p-1}{2}} \int_\Omega u' Mu \, dx \right]_S^T \\ &+ \frac{p-1}{2} \int_S^T E^{\frac{p-3}{2}} E' \int_\Omega u' Mu \, dx \, dt \\ &+ \int_S^T E^{\frac{p-1}{2}} \int_{\Gamma_1} (u')^2 - |\nabla u|^2 - g(u') Mu \, d\Gamma_m \, dt. \end{aligned}$$

The first term on the right side is ≤ 0. We show that the the next two terms are $\leq cE$.

Step 1. Estimate of $\int_\Omega u'Mu\ dx$. We have

$$\begin{aligned}\left|\int_\Omega u'Mu\ dx\right| &\le \int_\Omega (u')^2 + (Mu)^2\ dx\\ &= \int_\Omega (u')^2 + |2m\cdot\nabla u + (n-1)u|^2\ dx\\ &\le c\int_\Omega (u')^2 + |\nabla u|^2 + u^2\ dx\\ &\le cE.\end{aligned}$$

Hence

$$\left|E^{\frac{p-1}{2}}\int_\Omega u'Mu\ dx\right| \le cE^{\frac{p+1}{2}}$$

and

$$\left|E^{\frac{p-3}{2}}E'\int_\Omega u'Mu\ dx\right| \le -cE^{\frac{p-1}{2}}E' \le -c\left(E^{\frac{p+1}{2}}\right)'.$$

Step 2. Estimate of the right side. Since $E(t)$ is nonincreasing, it follows that

$$\begin{aligned}&2\int_S^T E^{\frac{p+1}{2}}\ dt\\ &\quad= \int_S^T E^{\frac{p-1}{2}}\int_{\Gamma_0}|\partial_\nu u|^2\ d\Gamma_m\ dt - \left[E^{\frac{p-1}{2}}\int_\Omega u'Mu\ dx\right]_S^T\\ &\qquad+\frac{p-1}{2}\int_S^T E^{\frac{p-3}{2}}E'\int_\Omega u'Mu\ dx\ dt\\ &\qquad+\int_S^T E^{\frac{p-1}{2}}\int_{\Gamma_1}(u')^2 - |\nabla u|^2 - g(u')Mu\ d\Gamma_m\ dt\\ &\quad\le cE^{\frac{p+1}{2}}(S) + \int_S^T E^{\frac{p-1}{2}}\int_{\Gamma_1}(u')^2 - |\nabla u|^2 - g(u')Mu\ d\Gamma_m\ dt\\ &\quad\le cE(S) + \int_S^T E^{\frac{p-1}{2}}\int_{\Gamma_1}(u')^2 - |\nabla u|^2 - g(u')Mu\ d\Gamma_m\ dt.\end{aligned}$$

Step 3. Estimate of the boundary integral. We have

$$\begin{aligned}
&\int_{\Gamma_1} (u')^2 - |\nabla u|^2 - g(u')Mu \; d\Gamma_m \\
&\quad = \int_{\Gamma_1} (u')^2 - |\nabla u|^2 - g(u')(2m \cdot \nabla u + (n-1)u) \; d\Gamma_m \\
&\quad \le \int_{\Gamma_1} (u')^2 - |\nabla u|^2 + \varepsilon |\nabla u|^2 + \varepsilon u^2 + c\varepsilon^{-1} g(u')^2 \; d\Gamma_m \\
&\quad \le c\varepsilon \int_\Omega |\nabla u|^2 \; dx \\
&\qquad + \int_{\Gamma_1} (u')^2 - |\nabla u|^2 + \varepsilon |\nabla u|^2 + c\varepsilon^{-1} g(u')^2 \; d\Gamma_m .
\end{aligned}$$

Choosing a small $\varepsilon > 0$ we get

$$\int_{\Gamma_1} (u')^2 - |\nabla u|^2 - g(u')Mu \; d\Gamma_m \le E + c \int_{\Gamma_1} (u')^2 + g(u')^2 \; d\Gamma_m .$$

Step 4. Simplified energy inequality. Using this estimate we obtain that

$$\begin{aligned}
&2 \int_S^T E^{\frac{p+1}{2}} \; dt \\
&\quad \le cE(S) + \int_S^T E^{\frac{p-1}{2}} \int_{\Gamma_1} (u')^2 - |\nabla u|^2 - g(u')Mu \; d\Gamma_m \; dt \\
&\quad \le cE(S) + \int_S^T E^{\frac{p+1}{2}} \; dt \\
&\qquad + c \int_S^T E^{\frac{p-1}{2}} \int_{\Gamma_1} (u')^2 + g(u')^2 \; d\Gamma_m \; dt
\end{aligned}$$

whence

$$\int_S^T E^{\frac{p+1}{2}} \; dt \le cE(S) + c \int_S^T E^{\frac{p-1}{2}} \int_{\Gamma_1} (u')^2 + g(u')^2 \; d\Gamma_m \; dt.$$

Step 5. Estimate of the last term. We will show that the last term is $\le cE$.

On $\Gamma_2 := \{x \in \Gamma_1 : |u'(x)| > 1\}$ we have $g(u'(x)) \asymp u'(x)$, so that

$$\begin{aligned}\int_S^T E^{\frac{p-1}{2}} \int_{\Gamma_2} (u')^2 + g(u')^2 \, d\Gamma_m \, dt \\ &\le c \int_S^T E^{\frac{p-1}{2}} \int_{\Gamma_2} u' g(u') \, d\Gamma_m \, dt \\ &\le c \int_S^T E^{\frac{p-1}{2}} \int_{\Gamma_1} u' g(u') \, d\Gamma_m \, dt \\ &= c \int_S^T E^{\frac{p-1}{2}} E' \, dt \\ &\le c \left(E^{\frac{p+1}{2}}(S) - E^{\frac{p+1}{2}}(T) \right) \\ &\le cE(S).\end{aligned}$$

Since for $|s| \le 1$ we have

$$|s|^{p+1} \le csg(s) \quad \text{and} \quad |g(s)|^{p+1} \le csg(s),$$

on $\Gamma_3 := \{x \in \Gamma_1 : |u'(x)| \le 1\}$ we have

$$\begin{aligned}\int_{\Gamma_3} (u')^2 + g(u')^2 \, d\Gamma_m &\le c \int_{\Gamma_3} (u' g(u'))^{\frac{2}{p+1}} \, d\Gamma_m \\ &\le c \left(\int_{\Gamma_3} u' g(u') \, d\Gamma_m \right)^{\frac{2}{p+1}} \\ &\le c \left(\int_{\Gamma_1} u' g(u') \, d\Gamma_m \right)^{\frac{2}{p+1}} \\ &= c(-E')^{\frac{2}{p+1}}.\end{aligned}$$

It follows that

$$\begin{aligned}\int_S^T E^{\frac{p-1}{2}} \int_{\Gamma_1} (u')^2 + g(u')^2 \, d\Gamma_m \, dt \\ &\le c \int_S^T E^{\frac{p-1}{2}} (-E')^{\frac{2}{p+1}} \, dt \\ &\le \int_S^T \varepsilon E^{\frac{p+1}{2}} - c(\varepsilon) E' \, dt \\ &\le \varepsilon \int_S^T E^{\frac{p+1}{2}} \, dt + c(\varepsilon) E(S)\end{aligned}$$

for any $\varepsilon > 0$.

Step 6. An integral inequality. Using the last estimates, we obtain that

$$\int_S^T E^{\frac{p+1}{2}}\,dt \le cE + c\int_S^T E^{\frac{p-1}{2}} \int_{\Gamma_1} (u')^2 + g(u')^2\,d\Gamma_m\,dt$$
$$\le c(\varepsilon)E(S) + c\varepsilon \int_S^T E^{\frac{p+1}{2}}\,dt.$$

Choosing $\varepsilon > 0$ such that $c\varepsilon < 1$, we get

$$\int_S^T E^{\frac{p+1}{2}}\,dt \le cE(S)$$

for all $T > S \ge 0$.

Letting $T \to \infty$ we conclude that

$$\int_S^\infty E^{\frac{p+1}{2}}\,dt \le cE(S)$$

for all $S \ge 0$.

The following lemma, which is reduced to Lemma 2.10 for $\alpha \to 0$, completes the proof:

Lemma 2.13. *If a nonincreasing function $E : \mathbb{R}_+ \to \mathbb{R}_+$ satisfies for some $\alpha > 0$ and $T > 0$ the condition*

$$\int_t^\infty E^{\alpha+1}(s)\,ds \le TE(0)^\alpha E(t)$$

for all $t \ge 0$, then

$$E(t) \le E(0)\left(\frac{T+\alpha t}{T+\alpha T}\right)^{\frac{-1}{\alpha}}$$

for all $t \ge 0$.

3 A Petrovsky system

The multiplier method can be adapted to some nonhyperbolic systems as various plate models. Usually a new difficulty arises when we try to determine the critical observability or controllability time. This can be overcome by applying a harmonic analysis argument. We present this approach on the example of the Petrovsky system

$$\begin{cases} u'' + \Delta^2 u = 0 & \text{in } \mathbb{R}\times\Omega,\\ u = \partial_\nu u = 0 & \text{on } \mathbb{R}\times\Gamma,\\ u(0) = u_0 \quad\text{and}\quad u'(0) = u_1 & \text{in } \Omega \end{cases} \tag{3.1}$$

where $\Omega \subset \mathbb{R}^N$ is a bounded domain of class C^4 with boundary Γ. We skip some details and refer to [18] for a full treatement.

The following results are classical and well known (see [29]):

Proposition 3.1.
(a) If $(u_0, u_1) \in H_0^2(\Omega) \times L^2(\Omega)$, then the problem (3.1) has a unique (weak) solution belonging to the space

$$C(\mathbb{R}; H_0^2(\Omega)) \cap C^1(\mathbb{R}; L^2(\Omega)).$$

(b) The energy

$$E = \frac{1}{2} \int_\Omega |u'|^2 + (\Delta u)^2 \, dx$$

of the solutions is conserved (does not depend on $t \in \mathbb{R}$).

We have the following observability result:

Theorem 3.2. *(J.L. Lions [27], E. Zuazua [31]) For every bounded interval I of length $|I| > 0$ there exist two positive constants c_1, c_2 such that*

$$c_1 E \leq \int_I \int_\Gamma (\Delta u)^2 \, d\Gamma \, dt \leq c_2 E \tag{3.2}$$

for all solutions.

Lions first proved the theorem for sufficiently large intervals I. Then Zuazua established the general case by an indirect compactness-uniqueness argument. We present here a constructive proof given in [14].

The proof of this theorem consists of four steps. We denote by $\lambda_1 < \lambda_2 < \cdots$ the eigenvalues of Δ^2 in $H_0^2(\Omega)$ and by Z_1, Z_2, ... the corresponding eigenspaces. We recall that $\lambda_k \to \infty$. Then the solutions of (3.1) may be written in the form

$$u(t) = \sum_{k \in K} u_k e^{i\omega_k t} \tag{3.3}$$

where $K = \mathbb{Z}^*$, $\omega_k = \pm\sqrt{\lambda_{|k|}}$ and the coefficients $u_k \in Z_{|k|}$ depend on the initial data. It suffices to consider finite sums: once the estimates (3.2) are established for finite sums, they remain valid by a density argument for all solutions.

Step 1. Using the same multiplier as for the wave equation (see Lemma 2.3 above), we obtain the estimates

$$c_1 E \leq \int_I \int_\Gamma (\Delta u)^2 \, d\Gamma \, dt \leq c_2 E$$

provided $|I| > 2R\lambda_1^{-1/4}$ where $\Omega \subset B(0,R)$ and λ_1 denotes the first eigenvalue of Δ^2 in $H_0^2(\Omega)$.

Applying this result to the the functions $u(t) = u_k e^{i\omega_k t}$ we obtain that

$$\Delta u_k = 0 \text{ on } \Gamma \Longrightarrow u_k = 0 \tag{3.4}$$

for all eigenfuctions $u_k \in Z_{|k|}$, $k = 1, 2, \ldots$.

Step 2. If we consider only solutions whose initial data are orthogonal to the first k eigenspaces, the proof of Step 1 shows that the estimates (3.2) hold under the weaker condition $|I| > 2R\lambda_{k+1}^{-1/4}$.

Step 3. We show that the condition $|I| > 2R\lambda_{k+1}^{-1/4}$ is in fact sufficient for *all* initial data. This is shown by adapting a method of Haraux [8] to trigonometric sums with *vector* coefficients. Introducing the seminorm

$$p(u) := \left(\int_\Gamma (\Delta u)^2 \, d\Gamma \right)^{1/2}$$

on the linear subspace spanned by the finite sums of the form (3.3), we need to show that

$$\int_I p(u)^2 \, dt \asymp \sum_{k \in K} \|u_k\|^2$$

where the notation $A \asymp B$ means that $c_1 A \leq B \leq c_2 A$ with suitable positive constants c_1 and c_2.

By Step 2 these estimates hold for all sums (3.3) satisfying $u_j = 0$ for all integers j satisfying $1 \leq |j| \leq k$, and by Step 1 the restriction of the seminorm p to each eigenspace Z_k is a *norm*. It remains to relax the assumptions $u_j = 0$ whenever $1 \leq |j| \leq k$. This will follow from the next lemma (see [21] and [22] for more general results):

Lemma 3.3. *Assume that for some interval I_0 and for some finite subset $K_0 \subset K$ we have*

$$\int_{I_0} p(u)^2 \, dt \asymp \sum_{k \in K} \|u_k\|^2 \tag{3.5}$$

for all sums (3.3) *satisfying*

$$u_k = 0 \quad \textit{for all} \quad k \in K_0.$$

Then for every interval I of length $|I| > |I_0|$ we have

$$\int_I p(u)^2 \, dt \asymp \sum_{k \in K} \|u_k\|^2 \tag{3.6}$$

for all *sums* (3.3).

Proof of the lemma. We start with two elementary observations. First, if one of the inequalities in (3.5) or (3.6) holds, then it also holds, with the same constant, for any translate of I_0, and I respectively. Secondly, by an induction argument we may assume that K_0 has only one element, say $K_0 = \{k_0\}$.

Proof of the direct inequality. Putting $z(t) := u_{k_0} e^{i\omega_{k_0} t}$ we have

$$\begin{aligned}\int_{I_0} p(u)^2 \, dt &\le 2 \int_{I_0} p(u-z)^2 \, dt + 2 \int_{I_0} p(z)^2 \, dt \\ &\le c \left(\sum_{k \ne k_0} \|u_k\|^2 \right) + 2|I_0| \cdot p(u_{k_0})^2 \\ &\le c \sum_{k \in K} \|u_k\|^2\end{aligned}$$

because p and $\|\cdot\|$ are equivalent norma on Z_{k_0}.

Covering I by finitely many translates of I_0 similar estimate is obtained on I with another constant c.

Preliminary inverse inequality. We first prove that

$$\sum_{k \ne k_0} \|u_k\|^2 \le \int_I p(u)^2 \, dt$$

for *all* sums (3.3).

We may assume by translation that $I_0 = (a, b)$ and $I = (a - \delta, b + \delta)$ with $\delta > 0$.

Turning to the proof, if

$$u(t) = \sum_{k \in K} u_k e^{i\omega_k t},$$

then setting

$$v(t) := u(t) - \frac{1}{2\delta} \int_{-\delta}^{\delta} e^{-i\omega_{k_0} s} u(t+s) \, ds$$

we obtain a function

$$v(t) = \sum_{k \in K} \left(1 - \frac{1}{2\delta} \int_{-\delta}^{\delta} e^{i(\omega_k - \omega_{k_0})s} \, ds \right) u_k e^{i\omega_k t} =: \sum_{k \in K} v_k e^{i\omega_k t}.$$

Observe that $v_{k_0} = 0$ and $\|v_k\| \asymp \|u_k\|$ for the remaining coefficients. Next we estimate $p(v)$. Observe that

$$\int_I p(v)^2 \, dt \le 4 \int_{I_0} p(u)^2 \, dt.$$

Indeed, for every fixed $t \in \mathbb{R}$ we have

$$\begin{aligned} p(v(t))^2 &\le 2p(u(t))^2 + 2p\Big(\frac{1}{2\delta}\int_{-\delta}^{\delta} e^{-i\omega_{k_0}s}u(t+s)\,ds\Big)^2 \\ &\le 2p(u(t))^2 + \frac{1}{2\delta^2}\Big|\int_{-\delta}^{\delta} p(u(t+s))\,ds\Big|^2 \\ &\le 2p(u(t))^2 + \frac{1}{\delta}\int_{-\delta}^{\delta} p(u(t+s))^2\,ds \\ &= 2p(u(t))^2 + \frac{1}{\delta}\int_{t-\delta}^{t+\delta} p(u(s))^2\,ds. \end{aligned}$$

It follows that

$$\begin{aligned} &\int_a^b p(v(t))^2\,dt \\ &\quad\le 2\int_a^b p(u(t))^2\,dt + \frac{1}{\delta}\int_a^b\int_{t-\delta}^{t+\delta} p(u(s))^2\,ds\,dt \\ &\quad= 2\int_a^b p(u(t))^2\,dt + \frac{1}{\delta}\int_{a-\delta}^{b+\delta}\int_{\max\{a,s-\delta\}}^{\min\{b,s+\delta\}} p(u(s))^2\,dt\,ds \\ &\quad\le 2\int_a^b p(u(t))^2\,dt + 2\int_{a-\delta}^{b+\delta} p(u(s))^2\,dt \\ &\quad\le 4\int_{a-\delta}^{b+\delta} p(u(s))^2\,dt. \end{aligned}$$

We have

$$\begin{aligned} \sum_{k\ne k_0}\|u_k\|^2 &\le c\sum_{k\ne k_0}\|v_k\|^2 \\ &\le c\int_{I_0} p(v)^2\,dt \\ &\le 4c\int_I p(u)^2\,dt. \end{aligned}$$

Proof of the inverse inequality. It remains to prove that

$$\|u_{k_0}\|^2 \le c\int_{I_0} p(u)^2\,dt.$$

Since the restriction of the seminorm p to Z_{k_0} is a norm, setting

$z(t) := u_{k_0}e^{i\omega_{k_0}t}$ we have

$$\begin{aligned}\|u_{k_0}\|^2 &\le cp(u_{k_0})^2\\ &= c\int_I p(z)^2\ dt\\ &\le 2c\int_I p(u)^2\ dt + 2c\int_I p(u-z)^2\ dt\\ &\le c\int_I p(u)^2\ dt + c\sum_{k\ne k_0}\|u_k\|^2\\ &\le c\int_I p(u)^2\ dt.\end{aligned}$$

The proof is completed. □

Step 4. Given an arbitrary bounded interval of positive length, we may choose a sufficiently large integer k satisfying this condition. Then by Step 3 the estimates (3.2) hold for all sums of the form (3.3).

The same approach as in Subsection 2.2 above yields now the following theorem:

Theorem 3.4. *If Ω is of class C^4, $T > 0$ and $(y^0, y^1) \in L^2(\Omega)\times H^{-2}(\Omega)$, then there exists a control function $v \in L^2(0,T;L^2(\Gamma))$ such that the solution of*

$$\begin{cases} y'' + \Delta^2 y = 0 \quad \textit{in} \quad [0,T]\times\Omega,\\ y = v \quad \textit{on} \quad [0,T]\times\Gamma,\\ y(0) = y_0 \quad \textit{and} \quad y'(0) = y_1 \quad \textit{in} \quad \Omega \end{cases}$$

satisfies

$$y(T) = y'(T) = 0 \quad \textit{in} \quad \Omega.$$

4 Observability by using Fourier series

Fourier series did a good service in the preceding section. Here we give another rather different application of Fourier series for the study of rectangular plates (and analogous higher-dimensional models). Many other examples and theorems are given in [22].

Let $\Omega \subset \mathbb{R}^N$ be an N-dimensional open interval ("brick") with boundary Γ. The problem

$$\begin{cases} u'' + \Delta^2 u = 0 & \text{in} \quad \mathbb{R}\times\Omega,\\ u = \Delta u = 0 & \text{on} \quad \mathbb{R}\times\Gamma,\\ u(0) = u_0 \quad \text{and} \quad u'(0) = u_1 & \text{in} \quad \Omega, \end{cases}$$

is well posed in $\left(H^2(\Omega)\cap H_0^1(\Omega)\right)\times L^2(\Omega)$.

Let Ω_0 be a nonempty open subset of Ω and I a nondegenerate bounded interval. A natural question is whether *it is possible to determine the initial data by observing the solution only in* $\Omega_0\times I$. The answer is yes:

Theorem 4.1. *(Haraux [8], Jaffard [11], [12], Komornik [16]) Given Ω_0 and I arbitrarily, there exist two constants $a_1, a_2 > 0$ such that*

$$a_1E(0)\le\int_I\int_{\Omega_0}|u'|^2\,dx\,dt\le a_2E(0)$$

for all solutions where the energy is defined by

$$E(t):=\frac{1}{2}\int_\Omega|u(t,x)|^2+|\Delta u(t,x)|^2+|u'(t,x)|^2\,dx.$$

Assume for simplicity that $\Omega=(0,\pi)^N$. Expanding the solution into Fourier series according to an orthonormal basis formed by eigenfunctions of the the infinitesimal generator of the underlying semi-group and setting

$$\{\omega_k : k\in K\}=\left\{\pm(|n|^2,n) : n\in\mathbb{Z}^N\right\}$$

the theorem is equivalent to the following estimates:

Proposition 4.2. *There exist positive constants a_1, a_2 such that*

$$a_1\sum_{k\in K}|c_k|^2\le\int_I\int_{\Omega_0}\left|\sum_{k\in K}c_ke^{i\omega_k\cdot(t,x)}\right|^2\,dt\,dx\le a_2\sum_{k\in K}|c_k|^2$$

for all square summable families $(c_k)_{k\in K}$ of complex numbers.

The proposition follows from Propositions 4.3–4.5 below. The first one is a vectorial generalization of a classical theorem of Ingham [10]:

Proposition 4.3. *(Baiocchi, K., Loreti [2]) If a family $(\omega_k)_{k\in K}$ of vectors in $\mathbb{R}^N$ satisfies the gap condition*

$$\gamma=\gamma(K):=\inf_{k\neq n}|\omega_k-\omega_n|>0,\tag{4.1}$$

then we have

$$\int_{B_R}\left|\sum_{k=-\infty}^{\infty}x_ke^{i\omega_k\cdot t}\right|^2\,dt\asymp\sum_{k=-\infty}^{\infty}|x_k|^2\tag{4.2}$$

for every $R>\frac{2\sqrt{\mu}}{\gamma}$ where B_R denotes the open ball of radius R in $\mathbb{R}^N$ and μ denotes the first eigenvalue of $-\Delta$ in the Sobolev space $H_0^1(B_1)$.

The next one is a strenghtening of the preceding result by weakening the assumptions on the radius R of the ball B_R; this generalizes a one-dimensional result of Beurling [4]:

Proposition 4.4. *Given a family* $(\omega_k)_{k\in K} \subset \mathbb{R}^N$ *satisfying* (4.1), *the estimates* (4.2) *still hold if*

$$R > R_0 := \frac{2\sqrt{\mu}}{\gamma(K_1)} + \cdots + \frac{2\sqrt{\mu}}{\gamma(K_m)}$$

for a suitable finite partition $K = K_1 \cup \cdots \cup K_m$ *of* K.

The final step is a combinatorial result:

Proposition 4.5. *If*

$$\{\omega_k : k \in K\} = \left\{(\pm|n|^2, n) : n \in \mathbb{Z}^N\right\},$$

then for every $\varepsilon > 0$ *there exists a finite partition* $K = K_1 \cup \cdots \cup K_m$ *of* K *such that*

$$\frac{1}{\gamma(K_1)} + \cdots + \frac{1}{\gamma(K_m)} < \varepsilon.$$

Example. For $N = 1$ the exponents ω_k lie on two parabolas. If m is a positive integer, then setting

$$\{\omega_k : k \in K_j\} = \left\{(\pm n^2, n) : n \in \mathbb{Z}, \quad n \equiv j \mod m\right\}$$

we have $\gamma(K_j) \geq m^2$ for every $j = 1, \ldots, m$.

5 A general method of stabilization

At the end of Subsection 2.3 we mentioned some drawbacks of the linear stabilization by natural feedbacks. In this section we present another more general and powerful method. For a full exposition we refer to [20].

The following two theorems are due to J.L. Lions [25], [27].

Let Ω be a bounded nonempty open set of class C^2 in $\mathbb{R}^N$. We denote by ν the unit outward normal vector to its boundary Γ. Consider the following controllability problem:

$$\begin{cases} y'' - \Delta y = 0 & \text{in } \Omega \times (0, \infty), \\ y(0) = y_0 \quad \text{and} \quad y'(0) = y_1 & \text{in } \Omega, \\ y = u & \text{on } \Gamma \times (0, \infty). \end{cases}$$

First we reformulate Theorem 2.6.

Theorem 5.1.

(a) For $u \in L^2_{loc}(0,\infty; L^2(\Gamma))$ the problem is well posed in

$$\mathcal{H} := L^2(\Omega) \times H^{-1}(\Omega).$$

(b) For $T > 0$ sufficiently large, the problem is exactly controllable in time T: for every $(y_0, y_1) \in \mathcal{H}$ there exists u such that the solution satisfies

$$y(T) = y'(T) = 0 \quad in \quad \Omega.$$

Remarks. This theorem was proved by a general method (HUM). The controls u were constructed explicitly, and the overall proof presented few technical difficulties.

Concerning the stabilization the following result holds true:

Theorem 5.2. *There exist two bounded linear maps*

$$P : H^{-1}(\Omega) \to H^1_0(\Omega), \; Q : L^2(\Omega) \to H^1_0(\Omega)$$

and two constants $M, \omega > 0$ such that, putting

$$u = \partial_\nu (Py' + Qy)$$

the problem is well posed in $\mathcal{H} := L^2(\Omega) \times H^{-1}(\Omega)$, and its solutions satisfy the estimate

$$\|(y(t), y'(t))\|_{\mathcal{H}} \leq M \|(y_0, y_1)\|_{\mathcal{H}} e^{-\omega t}$$

for all $(y_0, y_1) \in \mathcal{H}$ and $t \geq 0$.

Remarks. This theorem was also proved by a general method. However, there was no construction of the feedbacks P and Q and, due to some indirect arguments, no explicit decay rate was provided by the proof. Finally, the proof of this theorem was technically much more involved than that of Theorem 5.1: infinite-dimensional Riccati equations had to be solved.

The following similar theorem was obtained in [20]:

Theorem 5.3. *Fix $\omega > 0$ arbitrarily. There exist two bounded linear maps*

$$P : H^{-1}(\Omega) \to H^1_0(\Omega), \; Q : L^2(\Omega) \to H^1_0(\Omega)$$

and a constant M such that, putting

$$u = \partial_\nu (Py' + Qy)$$

the problem is well posed in $\mathcal{H} := L^2(\Omega) \times H^{-1}(\Omega)$, *and its solutions satisfy the estimate*

$$\|(y(t), y'(t))\|_{\mathcal{H}} \leq M\|(y_0, y_1)\|_{\mathcal{H}} e^{-\omega t}$$

for all $(y_0, y_1) \in \mathcal{H}$ *and* $t \geq 0$.

Remarks. Compared with Theorem 5.2, this result was also proved by a general method. Moreover, the feedbacks P and Q were constructed and explicit decay rates were obtained by a constructive proof, which was substantially simpler than that of Theorem 5.2. Finally, arbitrarily high decay rates may be ensured by suitable constructions of the feedbacks.

In order to compare these two theorems and their proofs, first we review Lions's approach:

1. *Observability of the dual problem.* Taking $u = 0$ in and using multipliers we get

$$\|(y_0, y_1)\|_{H_0^1(\Omega)\times L^2(\Omega)} \asymp \|\partial_\nu y\|_{L^2(\Gamma\times(0,T))}.$$

2. *Controllability of the primal problem.* By duality, for every given $(y_0, y_1) \in L^2(\Omega) \times H^{-1}(\Omega)$ there exists $u \in L^2(\Gamma \times (0,T))$ such that

$$y(T) = y'(T) = 0 \quad \text{in} \quad \Omega.$$

3. *Stabilization.* We minimize the cost function

$$J(y, u) := \int_0^\infty \int_\Omega y^2 \, dx \, dt + \int_0^\infty \int_\Gamma u^2 \, d\Gamma \, dt$$

where (y, u) solves our problem. Then (Riccati) y and u are related by the feedback equation. The decay estimate follows from a theorem of Datko.

The Hilbert Uniqueness Method is based on the implication

$$\text{observability} \Longrightarrow \text{controllability}.$$

The Riccati equation approach is based on the implication chain

$$\text{observability} \Longrightarrow \text{controllability} \Longrightarrow \text{stabilizability}.$$

By contrast, Theorem 5.3 is obtained by establishing directly the implication

$$\text{observability} \Longrightarrow \text{stabilizability}.$$

Let us explain the new approach in an abstract setting.

Primal-dual problems. Consider the abstract problem

$$x' = Ax + Bu, \qquad x(0) = x_0 \tag{5.1}$$

and its dual

$$\varphi' = -A^*\varphi, \qquad \varphi(0) = \varphi_0, \qquad \psi = B^*\varphi, \tag{5.2}$$

where $A : H \to H$ and $B : G \to H$ are densely defined closed linear maps is some Hilbert spaces H and G. We assume the following:

(H1) *(reversability)* A^* generates a *group* e^{sA^*} in H';

(H2) *(weakened continuity of B)* We have $D(A^*) \subset D(B^*)$ and there exist $\lambda \in \mathbb{C}$ and $C > 0$ such that

$$\|B^*\varphi\| \le C\|(A - \lambda I)^*\varphi\|$$

for all $\varphi_0 \in D(A^*)$;

(H3) *(direct inequality)* There exist $T' > 0$ and $c' > 0$ such that

$$\|\psi\|_{L^2(0,T';G')} \le c'\|\varphi_0\|_{H'}$$

for all $\varphi_0 \in D(A^*)$;

(H4) *(inverse inequality)* There exist $T > 0$ and $c > 0$ such that

$$\|\psi\|_{L^2(0,T';G')} \ge c\|\varphi_0\|_{H'}$$

for all $\varphi_0 \in D(A^*)$.

An equivalent norm. Fix $\omega > 0$ and set

$$T_\omega := T + \frac{1}{2\omega}, \quad e_\omega(s) = \begin{cases} e^{-2\omega s} & \text{if } 0 \le s \le T, \\ 2\omega e^{-2\omega T}(T_\omega - s) & \text{if } T \le s \le T_\omega. \end{cases}$$

Identify G' with G. Then

$$\Lambda_\omega := \int_0^{T_\omega} e_\omega(s) e^{-sA} BB^* e^{-sA^*} \, ds$$

defines an isomorphism Λ_ω of H' onto H, and the formula

$$\|x\|_\omega := \langle \Lambda_\omega^{-1} x, x \rangle_{H',H}^{1/2}$$

an equivalent norm on H.

Now we have the following abstract stabilization theorem:

Theorem 5.4. *Assume (H1) to (H4) and fix $\omega > 0$ arbitrarily large. Then the problem*

$$x' = (A - BB^*\Lambda_\omega^{-1})x, \qquad x(0) = x_0$$

is well posed in H, and its solutions satisfy the estimate

$$\|x(t)\|_\omega \le \|x_0\|_\omega e^{-\omega t}$$

for all $x_0 \in H$ and $t \ge 0$.

Remark. For $\omega = 0$ this is reduced to the observability operator Λ in Subsection 2.2. Note that $\omega > 0$ in the above theorem.

Formal proof. The solutions satisfy the following identity:

$$\frac{d}{dt}\langle \Lambda_\omega^{-1}x, x\rangle_{H',H} = \langle \Lambda_\omega^{-1}x, (A\Lambda_\omega + \Lambda_\omega A^* - 2BB^*)\Lambda_\omega^{-1}x\rangle_{H',H}.$$

Two different evaluations of

$$\int_0^{T_\omega} \frac{d}{ds}\left(e_\omega(s)e^{-sA}BB^*e^{-sA^*}\right)\, ds$$

show that

$$A\Lambda_\omega + \Lambda_\omega A^* - 2BB^* \le -2\omega\Lambda_\omega.$$

Hence

$$\frac{d}{dt}\langle \Lambda_\omega^{-1}x, x\rangle_{H',H} \le -2\omega\langle \Lambda_\omega^{-1}x, x\rangle_{H',H}$$

and

$$\|x(t)\|_\omega \le \|x_0\|_\omega e^{-\omega t}.$$

The two evaluations are as follows. First, using Leibniz's rule we have

$$\begin{aligned}
&\int_0^{T_\omega} \frac{d}{ds}\left(e_\omega(s)e^{-sA}BB^*e^{-sA^*}\right)\, ds \\
&\qquad = \int_0^{T_\omega} e_\omega'(s)e^{-sA}BB^*e^{-sA^*}\, ds - A\Lambda_\omega - \Lambda_\omega A^* \\
&\qquad \le -2\omega\Lambda_\omega - A\Lambda_\omega - \Lambda_\omega A^*.
\end{aligned}$$

On the other hand, applying the Newton–Leibniz formula we obtain that

$$\begin{aligned}
\int_0^{T_\omega} \frac{d}{ds}\left(e_\omega(s)e^{-sA}BB^*e^{-sA^*}\right)\, ds &= e_\omega(T)e^{-TA}BB^*e^{-TA^*} - BB^* \\
&\ge -BB^* \\
&\ge -2BB^*.
\end{aligned}$$

Hence

$$A\Lambda_\omega + \Lambda_\omega A^* + 2\omega\Lambda_\omega \le 2BB^*. \qquad \square$$

Proof of Theorem 5.3. Let us rewrite the systems (2.1) and (2.6) in the form (5.2) and (5.1) as follows.

First, putting $\varphi = (u, u')$, $\varphi_0 = (u_0, u_1)$ and introducing the linear operators A^* and B^* by the formulas

$$\begin{aligned} D(A^*) = D(B^*) = (H^2(\Omega) \cap H_0^1(\Omega)) \times H_0^1(\Omega),\\ A^*(z_0, z_1) = -(z_1, \Delta z_0),\\ B^*(z_0, z_1) = \partial_\nu z_0, \end{aligned}$$

we may rewrite (2.1) with the observation of $\partial_\nu u$ in the abstract form (5.2).

We claim that choosing $H' = H_0^1(\Omega) \times L^2(\Omega)$ and $G' = L^2(\Gamma)$ the assumptions (H1)–(H4) are satisfied. Indeed, (H1) is well known and is related to the energy conservation, see [29]. Property (H2) follows (with $\lambda = 0$) from the definition of A^*, B^* and from the elliptic regularity theory for $z_0 \in H^2(\Omega) \cap H_0^1(\Omega)$:

$$\begin{aligned} \|B^*(z_0, z_1)\|_{L^2(\Gamma)} &= \|\partial_\nu z_0\|_{L^2(\Gamma)} \le c\|z_0\|_{H^2(\Omega)}\\ &\le c\|\Delta z_0\|_{L^2(\Omega)} \le c\|A^*(z_0, z_1)\|_{H_0^1(\Omega)\times L^2(\Omega)}. \end{aligned}$$

Finally, (H3) and (H4) are equivalent to the inequalities proved in theorem 2.2.

Now a standard computation of duality shows that the dual problem (5.1) is just another form of the problem (2.6) if we introduce the notations $x = (-y', y)$, $x_0 = (-y_1, y_0)$ and if we put $G := G'' = L^2(\Gamma)$ and $H := H'' = H^{-1}(\Omega) \times L^2(\Omega)$.

Therefore Theorem 5.3 follows from Theorem 5.4. $\square$

Remark. Many numerical simulations and physical experiments were made by F. Bourquin and his collaborators: see [5].

References

[1] F. Alabau, V. Komornik, *Boundary observability, controllability and stabilization of linear elastodynamic systems,* SIAM J. Control Optim. 37 (1998), 521–542.

[2] C. Baiocchi, V. Komornik, and P. Loreti, *Ingham type theorems and applications to control theory,* Bol. Un. Mat. Ital. B (8) 2 (1999), no. 1, 33–63.

[3] C. Bardos, G. Lebeau, and J. Rauch, *Sharp sufficient conditions for the observation, control and stabilization of waves from the boundary,* SIAM J. Control Optim. 30 (1992), 1024–1065.

[4] J.N.J.W.L. Carleson and P. Malliavin, editors, *The Collected Works of Arne Beurling*, Volume 2, Birkhäuser, 1989.

[5] F. Bourquin, M. Collet, M. Joly et L. Ratier, *An efficient feedback control algorithm for beams: experimental investigations*, J. Sound Vibr. 278 (2004), 1–2, 181–206.

[6] G. Chen, *Energy decay estimates and exact boundary value controllability for the wave equation in a bounded domain*, J. Math. Pures Appl. 58 (1979), 249–274.

[7] G. Chen, *A note on the boundary stabilization of the wave equation*, SIAM J. Control Opt. 19 (1981), 106–113.

[8] A. Haraux, *Séries lacunaires et contrôle semi-interne des vibrations d'une plaque rectangulaire*, J. Math. Pures Appl. 68 (1989), 457–465.

[9] L.F. Ho, *Observabilité frontière de l'équation des ondes*, C. R. Acad. Sci. Paris Sér. I Math. 302 (1986), 443–446.

[10] A. E. Ingham, *Some trigonometrical inequalities with applications in the theory of series*, Math. Z. 41 (1936), 367–379.

[11] S. Jaffard, *Contrôle interne exact des vibrations d'une plaque carrée*, C. R. Acad. Sci. Paris Sér. I Math. 307 (1988), 759–762.

[12] S. Jaffard, *Contrôle interne exact des vibrations d'une plaque rectangulaire*, Portugalia Math. 47 (1990), 423–429.

[13] V. Komornik, *Contrôlabilité exacte en un temps minimal*, C. R. Acad. Sci. Paris Sér. I Math. 304 (1987), 223–225.

[14] V. Komornik, *A new method of exact controllability in short time and applications*, Ann. Fac. Sci. Toulouse 10 (1989), 415–464.

[15] V. Komornik, *Rapid boundary stabilization of the wave equation*, SIAM J. Control Optim. 29 (1991), 197–208.

[16] V. Komornik, *On the exact internal controllability of a Petrowsky system*, J. Math. Pures Appl. (9) 71 (1992), 331–342.

[17] V. Komornik, *Boundary stabilization, observation and control of Maxwell's equations*, Panamer. Math. J. 4 (1994), Number 4, 47–61.

[18] V. Komornik, *Exact Controllability and Stabilization. The Multiplier Method*, Masson, Paris, and John Wiley & Sons, Chicester, 1994.

[19] V. Komornik, B. Rao, *Boundary stabilization of compactly coupled wave equations*, Asymptotic Anal. 14 (1997), 339–359.

[20] V. Komornik, *Rapid boundary stabilization of linear distributed systems*, SIAM J. Control Optim. 35 (1997), 1591–1613.

[21] V. Komornik and P. Loreti, *Observability of compactly perturbed systems*, J. Math. Anal. Appl. 243 (2000), 409–428.

[22] V. Komornik and P. Loreti, *Fourier Series in Control Theory*, Springer-Verlag, New York, 2005.

[23] J. E. Lagnese, *Boundary Stabilization of Thin Plates*, SIAM Studies in Appl. Math., Philadelphia, 1989.

[24] I. Lasiecka and R. Triggiani, *Regularity of hyperbolic equations under $L_2(0,T;L_2(\Gamma))$ boundary terms*, Appl. Math. and Optimiz. 10 (1983), 275–286.

[25] J.-L. Lions, *Contrôle des systèmes distribués singuliers*, Gauthier-Villars, Paris, 1983.

[26] J.-L. Lions, *Contrôlabilité exacte des systèmes distribués*, C. R. Acad. Sci. Paris Sér. I Math. 302 (1986), 471–475.

[27] J.-L. Lions, *Exact controllability, stabilizability, and perturbations for distributed systems*, Siam Rev. 30 (1988), 1–68.

[28] J.-L. Lions, *Contrôlabilité exacte et stabilisation de systèmes distribués I-II*, Masson, Paris, 1988.

[29] J.-L. Lions and E. Magenes, *Non-homogeneous Boundary Value Problems and Applications I-III*, Die Grundlehren der mathematischen Wissenschaften, Bände 181–183. Springer-Verlag, New York-Heidelberg, 1972–73.

[30] L. S. Pontriaguine, *Équations différentielles ordinaires*, Mir, Moscou, 1975.

[31] E. Zuazua, *Contrôlabilité exacte en un temps arbitrairement petit de quelques modèles de plaques*, Appendice 1 in [28].

Exact Controllability and Exact Observability for Quasilinear Hyperbolic Systems: Known Results and Open Problems*

Tatsien Li
School of Mathematical Sciences, Fudan University
Shanghai 200433, China
Email: dqli@fudan.edu.cn

Bopeng Rao
Institut de Recherche Mathématique Avancée
Université de Strasbourg, 67084 Strasbourg, France
Email: rao@math.u-strasbg.fr

Abstract

In this paper we give known results and open problems on the exact controllability and the exact observability for quasilinear hyperbolic systems.

1 Introduction and known results

Consider the following first order quasilinear hyperbolic systems

$$\frac{\partial u}{\partial t} + A(u)\frac{\partial u}{\partial x} = F(u), \tag{1.1}$$

where $u = (u_1, \cdots, u_n)^T$ is the unknown vector function of (t, x), $A(u)$ is an $n \times n$ matrix with suitably smooth entries $a_{ij}(u)$ $(i, j = 1, \cdots, n)$ and $F(u) = (f_1(u), \cdots, f_n(u))^T$ is a suitably smooth vector function with

$$F(0) = 0. \tag{1.2}$$

By (1.2), $u = 0$ is an equilibrium of system (1.1).

*Supported by the Basic Research Program of China (No. 2007CB814800).

By hyperbolicity, for any given u on the domain under consideration, the matrix $A(u)$ possesses n real eigenvalues

$$\lambda_1(u), \lambda_2(u), \cdots, \lambda_n(u) \tag{1.3}$$

and a complete set of left eigenvectors $l_i(u) = (l_{i1}(u), \cdots, l_{in}(u))$ $(i = 1, \cdots, n)$:

$$l_i(u)A(u) = \lambda_i(u)l_i(u). \tag{1.4}$$

Suppose that all $\lambda_i(u)$ and $l_i(u)$ $(i = 1, \cdots, n)$ have the same regularity as $A(u) = (a_{ij}(u))$, and there are no zero eigenvalues:

$$\lambda_r(u) < 0 < \lambda_s(u) \quad (r = 1, \cdots, m;\ s = m+1, \cdots, n). \tag{1.5}$$

Let

$$v_i = l_i(u)u \quad (i = 1, \cdots, n). \tag{1.6}$$

where v_i is called the diagonal variable corresponding to $\lambda_i(u)$ $(i = 1, \cdots, n)$. In a neighbourhood of $u = 0$, $v = (v_1, \cdots, v_n)^T$ is a diffeomorphism of $u = (u_1, \cdots, u_n)^T$. When (1.1) is a system of diagonal form, i.e., $A(u)$ is a diagonal matrix:

$$A(u) = \mathrm{diag}\{\lambda_1(u), \cdots, \lambda_n(u)\}, \tag{1.7}$$

v is just u.

Under the previous assumptions, on the domain $\{(t, x) |\, t \geq 0,\ 0 \leq x \leq L\}$ the most general boundary conditions which guarantee the well-posedness of the forward problem can be written as

$$x = 0: \ v_s = G_s(t, v_1, \cdots, v_m) + H_s(t) \quad (s = m+1, \cdots, n), \tag{1.8}$$

$$x = L: \ v_r = G_r(t, v_{m+1}, \cdots, v_n) + H_r(t) \quad (r = 1, \cdots, m), \tag{1.9}$$

where G_i and H_i $(i = 1, \cdots, n)$ are all smooth and, without loss of generality, we may suppose that

$$G_i(t, 0, \cdots, 0) \equiv 0 \quad (i = 1 \cdots, n). \tag{1.10}$$

By (1.5), on the boundary $x = 0$ the characteristics $\frac{dx}{dt} = \lambda_s(u)$ $(s = m+1, \cdots, n)$ corresponding to all the positive eigenvalues are called the coming characteristics since they reach the boundary $x = 0$ from the interior of the domain. Similarly, the characteristics $\frac{dx}{dt} = \lambda_r(u)$ $(r = 1, \cdots, m)$ corresponding to all the negative eigenvalues are the coming characteristics on the boundary $x = L$.

Thus, the characters of boundary conditions (1.8)–(1.9) are as follows (cf. [1]):

1. on $x = 0$, the number of the boundary conditions = the number of the coming characteristics = the number of positive eigenvalues = $n - m$,

while, on $x = L$, the number of the boundary conditions = the number of the coming characteristics = the number of negative eigenvalues = m.

2. The boundary conditions (1.8) on $x = 0$ are written in the form that all the diagonal variables v_s $(s = m+1, \cdots, n)$ corresponding to the coming characteristics on $x = 0$ are explicitly expressed by the diagonal variables v_r $(r = 1, \cdots, m)$ corresponding to other characteristics. Similarly, the boundary conditions (1.9) on $x = L$ are written in the form that all the diagonal variables v_r $(r = 1, \cdots, m)$ corresponding to the coming characteristics on $x = L$ are explicitly expressed by the diagonal variables v_s $(s = m+1, \cdots, n)$ corresponding to other characteristics.

In what follows we give the known results on controllability and observability for quasilinear hyperbolic systems [2–3], which are obtained by the theory of semi-global C^1 solution [4].

1.1 Exact boundary controllability

1.1.1 Two-sided exact boundary controllability [5]

Let

$$T > T_0 \overset{\text{def.}}{=} L \max_{\substack{r=1,\cdots,m \\ s=m+1,\cdots,n}} \left(\frac{1}{|\lambda_r(0)|}, \frac{1}{\lambda_s(0)} \right). \tag{1.11}$$

For any given initial data $\phi(x)$ and final data $\Phi(x)$ with small $C^1[0, L]$ norm, there exist boundary controls $H_i(t)$ $(i = 1, \cdots, n)$ with small $C^1[0, T]$ norm, such that the corresponding mixed initial-boundary value problems (1.1), (1.8)–(1.9) and

$$t = 0: \quad u = \phi(x), \quad 0 \le x \le L \tag{1.12}$$

admit a unique semi-global C^1 solution $u = u(t, x)$ with small C^1 norm on the domain $R(T) = \{(t, x) \,\big|\, 0 \le t \le T, \ 0 \le x \le L\}$, which exactly satisfies the final condition

$$t = L: \quad u = \Phi(x), \quad 0 \le x \le L. \tag{1.13}$$

1.1.2 One-sided exact boundary controllability [6]

Suppose that the number of positive eigenvalues is not larger than that of negative ones:

$$\bar{m} \overset{\text{def.}}{=} n - m \le m, \quad \text{i.e.,} \quad n \le 2m. \tag{1.14}$$

Suppose furthermore that in a neighbourhood of $u = 0$, the boundary condition (1.8) on $x = 0$ (the side with less coming characteristics) can be equivalently rewritten as

$$x = 0: \ v_{\bar{r}} = \bar{G}_{\bar{r}}(t, v_{\bar{m}+1}, \cdots, v_m, v_{m+1}, \cdots, v_n) + \bar{H}_{\bar{r}}(t) \quad (\bar{r} = 1, \cdots, \bar{m}) \tag{1.15}$$

with

$$\bar{G}_{\bar{r}}(t, 0, \cdots, 0) \equiv 0 \quad (\bar{r} = 1, \cdots, \bar{m}). \tag{1.16}$$

Let

$$T > T_0 \stackrel{\text{def.}}{=} L\Big(\max_{r=1,\cdots,m} \frac{1}{|\lambda_r(0)|} + \max_{s=m+1,\cdots,n} \frac{1}{\lambda_s(0)}\Big). \tag{1.17}$$

For any given initial data $\phi(x)$ and final data $\Phi(x)$ with small $C^1[0, L]$ norm and any given $H_s(t)$ $(s = m+1, \cdots, n)$ with small $C^1[0, T]$ norm, such that the conditions of C^1 compatibility are satisfied at the points $(t, x) = (0, 0)$ and $(T, 0)$, respectively, there exist boundary controls $H_r(t)$ $(r = 1, \cdots, m)$ with small $C^1[0, T]$ norm on $x = L$ (the side with more coming characteristics), such that the corresponding mixed initial-boundary value problems (1.1), (1.8)–(1.9) and (1.12) admit a unique semi-global C^1 solution $u = u(t, x)$ with small C^1 norm on the domain $R(T) = \{(t, x) \,\big|\, 0 \le t \le T,\ 0 \le x \le L\}$, which satisfies exactly the final condition (1.13).

1.2 Exact boundary observability

1.2.1 Two-sided exact boundary observability [7]

Suppose that $T > 0$ satisfies (1.11). Suppose furthermore that $\phi(x)$ with small $C^1[0, L]$ norm and $H_i(t)$ $(i = 1, \cdots, n)$ with small $C^1[0, T]$ norm are given and the conditions of C^1 compatibility are satisfied at the points $(t, x) = (0, 0)$ and $(T, 0)$, respectively. Then, for the mixed initial-boundary value problems (1.1), (1.8)–(1.9) and (1.12), the boundary observations $v_r = \bar{v}_r(t)$ $(r = 1, \cdots, m)$ at $x = 0$ and the boundary observations $v_s = \bar{\bar{v}}_s(t)$ $(s = m+1, \cdots, n)$ at $x = L$ in the interval $[0, T]$ can uniquely determine the initial data $\phi(x)$, and the following observability inequality holds:

$$\|\varphi\|_{C^1[0,L]} \le C\Big(\sum_{r=1}^{m} \|\bar{v}_r\|_{C^1[0,T]} + \sum_{s=m+1}^{n} \|\bar{\bar{v}}_s\|_{C^1[0,T]} + \|H\|_{C^1[0,T]}\Big), \tag{1.18}$$

where C^1 is a positive constant.

1.2.2 One-sided exact boundary observability [7]

Suppose that (1.14) holds. Suppose furthermore that in neighbourhood of $u = 0$, boundary condition (1.9) on $x = L$ (the side with more coming characteristics) implies

$$x = L: \quad v_s = \bar{G}_s(t, v_1, \cdots, v_{\bar{m}}, v_{\bar{m}+1}, \cdots, v_m) \quad (s = m+1, \cdots, n) \tag{1.19}$$

with

$$\bar{G}_s(t, 0, \cdots, 0) \equiv 0 \quad (s = m+1, \cdots, n). \tag{1.20}$$

Let $T > 0$ satisfies (1.17). For the same mixed initial-boundary value problems (1.1), (1.8)–(1.9) and (1.12), the boundary observations $v_r = \bar{v}_r(t)$ $(r = 1, \cdots, m)$ at $x = 0$ (the side with less coming characteristics) in the interval $[0, T]$ can uniquely determine the initial data $\phi(x)$, and the following observability inequality holds:

$$\|\varphi\|_{C^1[0,L]} \le C\Big(\sum_{r=1}^{m} \|\bar{v}_r\|_{C^1[0,T]} + \|H\|_{C^1[0,T]}\Big), \tag{1.21}$$

where C^1 is a positive constant.

2 Remarks and open problems

2.1. In the previous results of controllability and observability, the estimates on the controllability time coincide with the estimates on the observability time, and all these estimates are sharp. On the other hand, generally speaking, the number of boundary controls or boundary observations on the interval $[0, T]$ can not be reduced.

2.2. In practice, it is natural to ask if it is possible to reduce the number of boundary controls or boundary observations for a problem under consideration. Moreover, in some control problems, according to the physical meaning, certain boundary conditions do not contain terms which can be used as boundary controls, namely, some of $H_i(t)$ $(i = 1, \cdots, n)$ are identically equal to zero. This situation also leads to the study on the lack of boundary controls.

Problem 1: Upon what additional hypotheses about system (1.1) and boundary conditions (1.8)–(1.9), is it possible (resp. not possible) to get the previous results of controllability and observability by means of less boundary controls or boundary observations in the interval $[0, T]$, including the case that certain boundary controls or boundary observations only partially play their role, i.e., only act in an interval whose length is less than T_0 defined by (1.11) or (1.17)?

Particularly, is it possible to realize the two-sided exact boundary controllability or observability by means of part (not whole!) boundary controls or boundary observations on each side?

Problem 2: When the requirement of Problem 1 can not be realized in the interval $[0, T]$, on what additional hypotheses about (1.1) and (1.8)–(1.9), is it possible to realize it in an enlarged interval $[0, \tilde{T}]$ with $\tilde{T} > T$?

2.3. In the case of one-sided exact boundary controllability, there are two related problems as follows:

Problem 3: Suppose that the number of positive eigenvalues is less than that of negative ones:

$$\bar{m} \overset{\text{def.}}{=} n - m < m, \quad i.e., \ n < 2m. \tag{2.1}$$

Upon what additional hypotheses about (1.1) and (1.8)–(1.9), is it possible(resp. not possible) to get the one-sided exact boundary controllability by means of boundary controls on $x = 0$ (the side with less coming characteristics) instead of on $x = L$ (the side with more coming characteristics) in the interval $[0, T]$?

Problem 4: When the requirement of Problem 3 can not be realized in the interval $[0, T]$, is it possible to realize it at an enlarged interval $[0, \tilde{T}]$ with $\tilde{T} > T$?

Moreover, in the previous result of the one-sided exact boundary controllability, the boundary conditions on $x = 0$ (the side with less coming characteristics) are required to satisfy hypothesis (1.15), which guarantees the well-posedness of the corresponding backward problem and is a generalization of the Group Condition [8] in the linear case with the assumption $n = 2m$, i.e., the number of positive eigenvalues is equal to that of negative ones.

Problem 5: Upon what additional hypotheses about (1.1) and (1.8)–(1.9), is it possible to get the one-sided exact boundary controllability without hypothesis (1.15)?

Partial results of Problem 5 can be found in [9–10].

Here, we point out that an affirmative answer can be given to Problem 3 and Problem 5 on the following hypotheses:

a. The final data are specially taken as

$$t = T: \quad u = 0 \quad (0 \le x \le L), \tag{2.2}$$

namely, the so-called zero controllability is considered.

b. For Problem 3, the boundary conditions on $x = L$ (the side with more coming characteristics) are specially prescribed as

$$x = L: \ v_r = G_r(t, v_{m+1}, \cdots, v_n) \quad (r = 1, \cdots, m), \tag{2.3}$$

namely, $H_r(t) \equiv 0$ $(r = 1, \cdots, m)$ and then $u = 0$ is an equilibrium of systems (1.1) and (2.3). While, for Problem 5, the boundary conditions on $x = 0$ (the side with less coming characteristics) are specially prescribed as

$$x = 0: \ v_s = G_s(t, v_1, \cdots, v_m) \quad (s = m + 1, \cdots, n), \tag{2.4}$$

namely, $H_s(t) \equiv 0$ $(s = m + 1, \cdots, n)$ and then $u = 0$ is an equilibrium of systems (1.1) and (2.4).

Upon these additional hypotheses, by means of boundary controls $H_s(t)$ $(s = m+1, \cdots, n)$ (for Problem 3) or boundary controls $H_r(t)$ $(r = 1, \cdots, m)$ (for Problem 5), the one-sided exact boundary controllability can be realized in the interval $[0, T]$, where $T > 0$ satisfies (1.17).

In fact, even though the corresponding backward mixed initial-boundary value problem is not well posed in this case, we can simply take $u \equiv 0$ as its solution, and then the constructive method given in [6] can be still applied effectively. A similar idea was introduced by D. Russell [8] in the linear case with the assumption $n = 2m$, however, our constructive method gives a more clear way to get the result.

We call the zero controllability as a special kind of controllability the weak controllability, and the usual controllability the strong controllability (cf. [11]). Obviously, the strong controllability implies the weak controllability, however, the weak controllability can not imply the strong controllability generically. For the weak controllability, we can also turn to the corresponding Problems 1 and 2.

2.4. In the case of one-sided exact boundary observability, there are two similar problems as follows:

Problem 6: Suppose that (2.1) holds. Upon what additional hypotheses about (1.1) and (1.8)–(1.9), is it possible to get the one-sided exact boundary observability by means of boundary observations on $x = L$ (the side with more coming characteristics) instead of on $x = 0$ (the side with less coming characteristics) at the interval $[0, T]$?

Problem 7: When the requirement of Problem 6 can not be realized at the interval $[0, T]$, is it possible to realize it at an enlarged interval $[0, \tilde{T}]$ with $\tilde{T} > T$?

Moreover, in the previous result of the one-sided exact boundary observability, the boundary conditions on $x = L$ (the side with more coming characteristics) are required to satisfy hypothesis (1.19), which guarantees the well-posedness of the corresponding backward problem and is a generalization of the Group Condition [8] in the linear case with the assumption $n = 2m$.

Problem 8: Upon what additional hypotheses about (1.1) and (1.8)–(1.9), is it possible to get the one-sided exact boundary observability without hypothesis (1.19)?

Partial results of Problem 8 can be found in [12].

D. Russell has introduced a kind of observability in [8]. In his definition, the boundary observations are required to uniquely determine the initial data for the backward problem, namely, to uniquely determine the final data for the forward problem. In the special case that $n = 2m$ (the number of positive eigenvalues is equal to that of negative ones) and the boundary conditions (1.8)–(1.9) can be equivalently rewritten as

$$x = 0: \ v_r = \bar{G}_r(t, v_{m+1}, \cdots, v_n) + \bar{H}_r(t) \quad (r = 1, \cdots, m) \qquad (2.5)$$

and

$$x = L: \ v_s = \bar{G}_s(t, v_1, \cdots, v_m) + \bar{H}_s(t) \quad (s = m+1, \cdots, n) \tag{2.6}$$

with

$$\bar{G}_i(t, 0, \cdots, 0) \equiv 0 \quad (i = 1, \cdots, n), \tag{2.7}$$

the backward problems (1.1), (1.8)–(1.9) and (1.13) are still well posed, then, to uniquely determining the final data $\Phi(x)$ is equivalent to uniquely determining the initial data $\phi(x)$ and, consequently, the definition given by D. Russell coincides with the previous definition of observability. However, in the general situation, even if the forward problem is well posed, the corresponding backward problem with the same boundary conditions might not be well posed, then the usual definition of observability is stronger than the definition given by D. Russell. From this point of view, we call the observability previously defined the strong observability, and the observability defined by D. Russell the weak observability [11].

For the weak observability we can still go to the corresponding Problems 1 and 2. Moreover, since, in order to get the weak observability, it is not necessary to solve the corresponding backward problem, we can give an affirmative solution to Problem 6 and Problem 8 without any additional hypotheses about (1.1) and (1.8)–(1.9). Correspondingly, we have

Problem 9: Upon what additional hypotheses about (1.1) and (1.8)–(1.9), is it possible to get the one-sided weak exact boundary observability by means of less boundary observations on $x = L$ in the interval $[0, T]$?

Problem 10: When the requirement of Problem 9 can not be realized at the interval $[0, T]$, is it possible to realize it in an enlarged interval $[0, \tilde{T}]$ with $\tilde{T} > T$?

2.5. In the previous discussion we always suppose that all the observed values are accurate, i.e., there is no measuring error in the observation. However, one can not eliminate errors in practical observation, leading to the following problem on the stability of observation.

Problem 11: Is it possible to control the error of the initial data (for the strong observability) or the final data (for the weak observability) by means of the error of observed values?

An affirmative solution to Problem 11 in the case of strong observability can be found in [13].

In the case of controllability, a related problem is

Problem 12: Is it possible to estimate the error of the determined final data by means of the error of both boundary controls and the initial data?

2.6. Since all the problems are considered in the framework of classical solutions in a neighbourhood of the equilibrium $u = 0$, they are all related to the local exact boundary controllability and the local exact boundary observability. However, in some special cases, the global exact boundary controllability with arbitrarily large distance between the initial data and the final data can be obtained [14–19].

Problem 13: Upon what additional hypotheses about (1.1) and (1.8)–(1.9), is it possible to get the global exact boundary controllability?

Problem 14: Upon what additional hypotheses about (1.1) ad (1.8)–(1.9), is it possible to get the global exact boundary observability?

2.7. When there are zero eigenvalues:

$$\lambda_p(u) < \lambda_q(u) \equiv 0 < \lambda_r(u) \tag{2.8}$$

$$(p = 1, \cdots, l;\ q = l+1, \cdots, m;\ r = m+1, \cdots, n),$$

the situation is quite different from the case without zero eigenvalues.

In the case of controllability, suitable boundary controls corresponding to non-zero eigenvalues and suitable internal controls corresponding to zero eigenvalues can be used to realize the exact controllability [20]. However, according to the physical meaning, certain equations corresponding to zero eigenvalues do not contain terms which can be used as internal controls, then we should face the difficulty of the lack of internal controls.

Problem 15: Upon what additional hypotheses about (1.1) and (1.8)–(1.9), is it possible to realize the exact controllability without internal controls or by less internal controls in the case that (2.5) holds?

A very preliminary consideration on Problem 15 can bc found in [21].

In the case of observation, by a constructive method together with the theory as to the semi-global C^1 solution for a special kind of boundary value problem [22], suitable boundary observations corresponding to non-zero eigenvalues and suitable internal observations corresponding to zero eigenvalues can be used to realize the exact observability [23].

Problem 16: Upon what additional hypotheses about (1.1) and (1.8)–(1.9), is it possible to realize the exact observability without internal observations or by means of less internal observations in the case that (2.5) holds?

2.8. Quite different from the autonomous case, for the nonautonomous first order quasilinear hyperbolic system

$$\frac{\partial u}{\partial t} + A(t, x, u)\frac{\partial u}{\partial x} = F(t, x, u), \tag{2.9}$$

the exact controllability and the exact observability possess various possibilities and should be studied in a more delicate way (cf. [24–25]).

Problem 17: Study the previous problems for the corresponding mixed initial-boundary value problem related to the nonautonomous system (2.9).

2.9. Higher order hyperbolic equations (systems) are also of great importance in practice. The related study on controllability and observability can be found in [8], [26] and the references therein for the linear case and in [27–31] and [10] for the quasilinear case. Since the corresponding study in the quasilinear case is just at the beginning, we have

Problem 18: Establish a complete theory about the controllability and observability for higher order quasilinear hyperbolic systems.

2.10. For higher dimensional hyperbolic equations (systems), the study on controllability and observability can be found in [8] and [26] for the linear case and in [33–37] for the semilinear case, however, for the quasilinear case, the only result obtained up to now is that by means of a boundary control of Dirichlet type on the whole boundary, the local exact boundary controllability can be realized for quasilinear wave equations [38–39]. Compared with the corresponding results in the linear case and in the 1-D quasilinear case, this result is far from what we want to get. As to the observability, it seems to be still open in higher dimensional quasilinear case.

Problem 19: Establish a complete theory as to the exact controllability in higher dimensional quasilinear hyperbolic case.

Problem 20: Establish the exact observability in higher dimensional quasilinear hyperbolic case.

References

[1] Tatsien Li, Wenci Yu, Boundary Value Problems for Quasilinear Hyperbolic Systems, Duke University Mathematics Series V, 1985.

[2] Tatsien Li, *Controllability and observability: From ODEs to quasilinear hyperbolic systems*, 6th International Congress on Industrial and Applied Mathematics (ICIAM 07), Zürich, Switzerland, 16–20 July 2007, Invited Lectures (Editors: Rolf Jeltsch, Gerhard Wanner), European Mathematical Society, 2009, 251–278.

[3] Tatsien Li, *Controllability and Observability for Quasilinear Hyperbolic Systems*, AIMS Series on Applied Mathematics, Volume 3, American Institute of Mathematical Sciences, 2009.

[4] Tatsien Li, Yi Jin, Semi-global C^1 solution to the mixed initial-boundary value problem for quasilinear hyperbolic systems, Chin. Ann. Math., 22B(2001), 325–336.

[5] Tatsien Li, Bopeng Rao, *Exact boundary controllability for quasilinear hyperbolic systems*, SIAM J. Control Optim., 41(2003), 1748–1755.

[6] Tatsien Li, Bopeng Rao, *Local exact boundary Controllability for a class of quasilinear hyperbolic systems*, Chin. Ann. Math., 23B(2002), 209–218.

[7] Tatsien Li, *Exact boundary observability for quasilinear hyperbolic systems*, ESAIM: Control, Optimization and Calculus of Variations, 14(2008), 759–766.

[8] D. L. Russell, *Controllability and stabilizability theory for linear partial differential equations, Recent progress and open questions*, SIAM Rev., 20(1978), 639–739.

[9] Tatsien Li, Bopeng Rao, Zhiqiang Wang, *A note on the one-side exact boundary controllability for quasilinear hyperbolic systems*, Commun. Pure Appl. Anal., 8(2009), 405–418.

[10] Lixin Yu, *Exact boundary controllability for a kind of second order quasilinear hyperbolic systems and its applications*, to appear in Math. Meth. Appl. Sci.

[11] Tatsien Li, Bopeng Rao, *Strong (weak) exact controllability and strong (weak) exact observability for quasilinear hyperbolic systems*, to appear.

[12] Tatsien Li, Bopeng Rao, Zhiqiang Wang, *A note on the one-side exact boundary observability for quasilinear hyperbolic systems*, Georgian Mathematical Journal, 15(2008), 571–580.

[13] Zhiqiang Wagn, *Stability of boundary observations for first order quasilinear hyperbolic systems*, to appear.

[14] M. Gugat, G. Leugering, *Global boundary controllability of the de St. Venant equations between steady states*, Ann. Inst. H. Poincaré, Anal. Non Linéaire, 20(2003), 1–11.

[15] M. Gugat, G. Leugering, *Global boundary controllability of the Saint-Venant system for sloped canals with firction*, Ann. Inst. H. Poincaré, Anal. Non Linéaire, 26(2009), 257–270.

[16] Tatsien Li, Bingyu Zhang, *Global exact boundary controllability of a class of quasilinear hyperbolic systems*, J. Math. Anal. Appl., 225(1998), 289–311.

[17] Tatsien Li, Zhiqiang Wang, *Global exact boundary controllability for first order quasilinear hyperbolic systems of diagonal form*, Int. J. Dynamical Systems and Differential Equations, 1(2007), 12–19.

[18] Zhiqiang Wang, *Global exact controllability for quasilinear hyperbolic systems of diagonal form with linearly degenerate characteristics*, Nonlinear Analysis, 69(2008), 510–522.

[19] Tatsien Li, *Global exact boundary controllability for first order quasilinear hyperbolic systems*, to appear in DCDS-B.

[20] Tatsien Li, Lixin Yu, *Exact controllability for first order quasilinear hyperbolic systems with zero eigenvalues*, Chin. Ann. Math., 24B(2003), 415–422.

[21] Tatsien Li, Bopeng Rao, *Exact controllability for first order quasilinear hyperbolic systems with vertical characteristics*, Acta Mathematica Scientia, 29B(2009), 980–990.

[22] Lina Guo, *A kind of boundary value problems for first order quasilinear hyperbolic systems with zero eigenvalues*, to appear.

[23] Lina Guo, Tatsien Li, *Exact observability for first order quasilinear hyperbolic systems with zero eigenvalues*, to appear.

[24] Tatsien Li, Zhiqiang Wang, *A note on the exact controllability for nonautonomous hyperbolic systems*, Commun. Pure Appl. Anal., 6(2007), 229–235.

[25] Lina Guo, Zhiqiang Wang, *Exact boundary observability for nonautonomous first order quasilinear hyperbolic systems*, Math. Meth. Appl. Sci., 31(2008), 1956–1971.

[26] J. -L. Lions, *Contrôlabilité Exacte, Perturbations et Stabilisation de Systèmes Distribués*, Vol. I, II, Masson, 1988.

[27] Tatsien Li, Lixin Yu, *Exact boundary controllability for 1-D quasilinear wave equations*, SIAM J. Control Optim., 45(2006), 1074–1083.

[28] Tatsien Li, *Exact boundary observability for 1-D quasilinear wave equations*, Math. Meth. Appl. Sci., 29(2006), 1543–1553.

[29] Lixin Yu, *Exact boundary controllability for higher order quasilinear hyperbolic equations*, Appl. Math. J. Chinese Univ., 20B(2005), 127–141.

[30] Lixin Yu, *Exact boundary controllability for second order quasilinear hyperbolic systems (in Chinese)*, Chinese J. Eng. Math., 22(2005), 199–211.

[31] Zhiqiang Wang, *Exact boundary controllability for nonautonomous quasilinear wave equations*, Math. Meth. Appl. Sci., 30(2007), 1311–1327.

[32] Lina Guo, Zhiqiang Wang, *Exact boundary observability for nonautonomous quasilinear wave equations*, to appear.

[33] O. Yu. Emanuilov, *Boundary control by semilinear evolution equations*, Russian Math. Surveys, 44(3)(1989), 183–184.

[34] A. V. Fursikov, O. Yu Imanuvilov, *Controllability of Evolution Equations*, Lecture Notes, Ser. 34, Seoul National University, Seoul, 1996.

[35] I. Lasiecka, R. Triggiani, *Exact controllability of semilinear abstract systems with applications to waves and plates boundary control problems*, Appl. Math. Optim., 23(1991), 109–154.

[36] E. Zuazua, *Exact controllability for the semilinear wave equation*, J. Math. Pures et Appl., 69(1990), 1–31.

[37] E. Zuazua, *Exact controllability for semilinear wave equations*, Ann. Inst. H. Poincaré, Anal. Non Linéaire, 10(1993), 109–129.

[38] Yi Zhou, Zhen Lei, *Local exact boundary controllability for nonlinear wave equations*, SIAM J. Control Optim., 46(2007), 1022–1051.

[39] Pengfei Yao, *Boundary controllability for the quasilinear wave equation*, to appear.

Waves, Damped Wave and Observation*

Kim Dang Phung
Yangtze Center of Mathematics, Sichuan University
Chengdu 610064, China
Email: kim_dang_phung@yahoo.fr

Abstract

This article describes some applications of two kinds of observation estimates for the wave equation and for the damped wave equation in a bounded domain where the geometric control condition of C. Bardos, G. Lebeau and J. Rauch may fail.

1 The wave equation and observation

We consider the wave equation in the solution $u = u(x,t)$

$$\left\{ \begin{array}{rl} \partial_t^2 u - \Delta u = 0 & \text{in } \Omega \times \mathbb{R} \ , \\ u = 0 & \text{on } \partial\Omega \times \mathbb{R} \ , \\ (u, \partial_t u)(\cdot, 0) = (u_0, u_1) & , \end{array} \right. \tag{1.1}$$

consisting in a bounded open set Ω in $\mathbb{R}^n$, $n \geq 1$, either convex or C^2, to be connected with boundary $\partial\Omega$. It is well known that for any initial data $(u_0, u_1) \in H^2(\Omega) \cap H_0^1(\Omega) \times H_0^1(\Omega)$, the above problem is well posed and has a unique strong solution.

Linked to exact controllability and strong stabilization for the wave equation (see [Li]), it appears the following observability problem which consists in proving the following estimate

$$\|(u_0, u_1)\|^2_{H_0^1(\Omega) \times L^2(\Omega)} \leq C \int_0^T \int_\omega |\partial_t u(x,t)|^2 \, dxdt$$

for some constant $C > 0$ independent of the initial data. Here, $T > 0$ and ω is a non-empty open subset in Ω. Due to finite speed of propagation,

*This work is supported by the NSF of China under grants 10525105 and 10771149. Part of this talk was done when the author visited Fudan University with a financial support from the "French-Chinese Summer Institute on Applied Mathematics" (September 1–21, 2008).

the time T has to be chosen to be large enough. Dealing with high frequency waves, i.e., waves which propagate according to the law of geometrical optics, the choice of ω can not be arbitrary. In other words, the existence of trapped rays (e.g, constructed with gaussian beams (see [Ra]) implies the requirement of some kinds of geometric conditions on (ω, T) (see [BLR]) in order that the above observability estimate may hold.

Now, we want to know what kind of estimate we may expect in a geometry with trapped rays. Let us introduce the quantity

$$\Lambda = \frac{\|(u_0, u_1)\|_{H^2 \cap H_0^1(\Omega) \times H_0^1(\Omega)}}{\|(u_0, u_1)\|_{H_0^1(\Omega) \times L^2(\Omega)}},$$

which can be seen as a measure of the frequency of the wave. In this paper, we present the two following inequalities

$$\|(u_0, u_1)\|^2_{H_0^1(\Omega) \times L^2(\Omega)} \leq e^{C\Lambda^{1/\beta}} \int_0^T \int_\omega |\partial_t u(x,t)|^2 \, dx dt \tag{1.2}$$

and

$$\|(u_0, u_1)\|^2_{H_0^1(\Omega) \times L^2(\Omega)} \leq C \int_0^{C\Lambda^{1/\gamma}} \int_\omega |\partial_t u(x,t)|^2 \, dx dt \tag{1.3}$$

where $\beta \in (0, 1)$, $\gamma > 0$. We will also give their applications to control theory.

The strategy to get estimate (1.2) is now well known (see [Ro2],[LR]) and a sketch of the proof will be given in Appendix for completeness. More precisely, we have the following results.

Theorem 1.1. *For any ω non-empty open subset in Ω, for any $\beta \in (0, 1)$, there exist $C > 0$ and $T > 0$ such that for any solution u of (1.1) with non-identically zero initial data $(u_0, u_1) \in H^2(\Omega) \cap H_0^1(\Omega) \times H_0^1(\Omega)$, the inequality (1.2) holds.*

Now, we can ask whether it is possible to get another weight function of Λ other than the exponential one, a polynomial weight function with a geometry (Ω, ω) with trapped rays in particular. Here we present the following results.

Theorem 1.2. *There exists a geometry (Ω, ω) with trapped rays such that for any solution u of (1.1) with non-identically zero initial data $(u_0, u_1) \in H^2(\Omega) \cap H_0^1(\Omega) \times H_0^1(\Omega)$, the inequality (1.3) holds for some $C > 0$ and $\gamma > 0$.*

The proof of Theorem 1.2 is given in [Ph1]. With the help of Theorem 2.1 below, it can also be deduced from [LiR], [BuH].

2 The damped wave equation and our motivation

We consider the following damped wave equation in the solution $w = w(x,t)$

$$\begin{cases} \partial_t^2 w - \Delta w + 1_\omega \partial_t w = 0 & \text{in } \Omega \times (0,+\infty) \ , \\ w = 0 & \text{on } \partial\Omega \times (0,+\infty) \ , \end{cases} \tag{2.1}$$

consisting in a bounded open set Ω in $\mathbb{R}^n$, $n \geq 1$, either convex or C^2, to be connected with boundary $\partial\Omega$. Here ω is a non-empty open subset in Ω with trapped rays and 1_ω denotes the characteristic function on ω. Further, for any $(w, \partial_t w)(\cdot, 0) \in H^2(\Omega) \cap H_0^1(\Omega) \times H_0^1(\Omega)$, the above problem is well posed for any $t \geq 0$ and has a unique strong solution.

Denote for any $g \in C\left([0,+\infty); H_0^1(\Omega)\right) \cap C^1\left([0,+\infty); L^2(\Omega)\right)$,

$$E(g,t) = \frac{1}{2}\int_\Omega \left(\left|\nabla g(x,t)\right|^2 + \left|\partial_t g(x,t)\right|^2 \right) dx \ .$$

Then for any $0 \leq t_0 < t_1$, the strong solution w satisfies the following formula

$$E(w,t_1) - E(w,t_0) + \int_{t_0}^{t_1} \int_\omega \left|\partial_t w(x,t)\right|^2 dxdt = 0 \ . \tag{2.2}$$

2.1 The polynomial decay rate

Our motivation for establishing estimate (1.3) comes from the following result.

Theorem 2.1. *The following two assertions are equivalent. Let* $\delta > 0$.

(i) *There exists* $C > 0$ *such that for any solution* w *of (2.1) with the non-null initial data* $(w, \partial_t w)(\cdot, 0) = (w_0, w_1) \in H^2(\Omega) \cap H_0^1(\Omega) \times H_0^1(\Omega)$, *we have*

$$\left\|(w_0,w_1)\right\|^2_{H_0^1(\Omega)\times L^2(\Omega)} \leq C \int_0^{C\left(\frac{E(\partial_t w,0)}{E(w,0)}\right)^{1/\delta}} \int_\omega \left|\partial_t w(x,t)\right|^2 dxdt \ .$$

(ii) *There exists* $C > 0$ *such that the solution* w *of (2.1) with the initial data* $(w, \partial_t w)(\cdot, 0) = (w_0, w_1) \in H^2(\Omega) \cap H_0^1(\Omega) \times H_0^1(\Omega)$ *satisfies*

$$E(w,t) \leq \frac{C}{t^\delta} \left\|(w_0,w_1)\right\|^2_{H^2 \cap H_0^1(\Omega)\times H_0^1(\Omega)} \quad \forall t > 0 \ .$$

Remark. It is not difficult to see (e.g., [Ph2]) by a classical decomposition method, a translation in time and (2.2), that the inequality (1.3) with the exponent γ for the wave equation implies the inequality of (i) in Theorem 2.1 with the exponent $\delta = 2\gamma/3$ for the damped wave equation. And conversely, the inequality of (i) in Theorem 2.1 with the exponent δ for the damped wave equation implies the inequality (1.3) with the exponent $\gamma = \delta/2$ for the wave equation.

Proof of Theorem 2.1.

$(ii) \Rightarrow (i)$. Suppose that

$$E\left(w,T\right) \leq \frac{C}{T^{\delta}} \left\|(w_0,w_1)\right\|^2_{H^2\cap H^1_0(\Omega)\times H^1_0(\Omega)} \qquad \forall T>0\ .$$

Therefore from (2.2)

$$E\left(w,0\right) \leq \frac{C}{T^{\delta}} \left\|(w_0,w_1)\right\|^2_{H^2\cap H^1_0(\Omega)\times H^1_0(\Omega)} + \int_0^T \int_\omega \left|\partial_t w\left(x,t\right)\right|^2 dxdt\ .$$

By choosing

$$T = \left(2C \frac{\left\|(w_0,w_1)\right\|^2_{H^2\cap H^1_0(\Omega)\times H^1_0(\Omega)}}{E\left(w,0\right)}\right)^{1/\delta} ,$$

we get the desired estimate

$$E\left(w,0\right) \leq 2 \int_0^{\left[2C \frac{\left\|(w_0,w_1)\right\|^2_{H^2\cap H^1_0(\Omega)\times H^1_0(\Omega)}}{E(w,0)}\right]^{1/\delta}} \int_\omega \left|\partial_t w\left(x,t\right)\right|^2 dxdt\ .$$

$(i) \Rightarrow (ii)$. Conversely, suppose the existence of a constant $c>1$ such that the solution w of (2.1) with the non-null initial data $(w,\partial_t w)\left(\cdot,0\right) = (w_0,w_1) \in H^2\left(\Omega\right) \cap H^1_0\left(\Omega\right) \times H^1_0\left(\Omega\right)$ satisfies

$$E\left(w,0\right) \leq c \int_0^{c\left(\frac{E(w,0)+E(\partial_t w,0)}{E(w,0)}\right)^{1/\delta}} \int_\omega \left|\partial_t w\left(x,t\right)\right|^2 dxdt\ .$$

We obtain the following inequalities by a translation on the time variable and by using (2.2). $\forall s \geq 0$

$$\begin{aligned} \frac{E(w,s)}{E(w,0)+E(\partial_t w,0))} &\leq c \int_s^{s+c\left(\frac{(E(w,0)+E(\partial_t w,0))}{E(w,s)}\right)^{1/\delta}} \int_\omega \frac{|\partial_t w(x,t)|^2}{E(w,0)+E(\partial_t w,0)} dxdt \\ &\leq c\left(\frac{E(w,s)}{E(w,0)+E(\partial_t w,0)} - \frac{E\left(w,s+c\left(\frac{E(w,0)+E(\partial_t w,0)}{E(w,s)}\right)^{1/\delta}\right)}{E(w,0)+E(\partial_t w,0)}\right)\ . \end{aligned}$$

Denoting $G(s)=\frac{E(w,s)}{E(w,0)+E(\partial_t w,0)}$, we deduce using the decreasing of G that

$$G\left(s+c\left(\frac{1}{G(s)}\right)^{1/\delta}\right)\leq G(s)\leq c\left[G(s)-G\left(s+c\left(\frac{1}{G(s)}\right)^{1/\delta}\right)\right]$$

which gives

$$G\left(s+c\left(\frac{1}{G(s)}\right)^{1/\delta}\right)\leq\frac{c}{1+c}G(s)\ .$$

Let $c_1=\left(\frac{1+c}{c}\right)^{1/\delta}-1>0$ and denote $d(s)=\left(\frac{c}{c_1}\frac{1}{s}\right)^{\delta}$. We distinguish two cases.

If $c_1 s\leq c\left(\frac{1}{G(s)}\right)^{1/\delta}$, then $G(s)\leq\left(\frac{c}{c_1}\frac{1}{s}\right)^{\delta}$ and

$$G((1+c_1)s)\leq d(s)\ .$$

If $c_1 s>c\left(\frac{1}{G(s)}\right)^{1/\delta}$, then $s+c\left(\frac{1}{G(s)}\right)^{1/\delta}<(1+c_1)s$ and the decreasing of G gives $G((1+c_1)s)\leq G\left(s+c\left(\frac{1}{G(s)}\right)^{1/\delta}\right)$ and then

$$G((1+c_1)s)\leq\frac{c}{1+c}G(s)\ .$$

Consequently, we have that $\forall s>0$, $\forall n\in\mathbb{N}$, $n\geq 1$,

$$\begin{aligned}G((1+c_1)s)\leq\max\Big[d(s),\tfrac{c}{1+c}d\left(\tfrac{s}{(1+c_1)}\right),\cdots,\\ \left(\tfrac{c}{1+c}\right)^{n}d\left(\tfrac{s}{(1+c_1)^n}\right),\left(\tfrac{c}{1+c}\right)^{n+1}G\left(\tfrac{s}{(1+c_1)^n}\right)\Big]\ .\end{aligned}$$

Now, remark that with our choice of c_1, we get

$$\frac{c}{1+c}d\left(\frac{s}{(1+c_1)}\right)=d(s)\quad\forall s>0\ .$$

Thus, we deduce that $\forall n\geq 1$

$$\begin{aligned}G((1+c_1)s)&\leq\max\left(d(s),\left(\tfrac{c}{1+c}\right)^{n+1}G\left(\tfrac{s}{(1+c_1)^n}\right)\right)\\&\leq\max\left(d(s),\left(\tfrac{c}{1+c}\right)^{n+1}\right)\qquad\text{because } G\leq 1\ ,\end{aligned}$$

and conclude that $\forall s>0$

$$\frac{E(w,s)}{E(w,0)+E(\partial_t w,0)}=G(s)\leq d\left(\frac{s}{1+c_1}\right)=\left(\frac{c(1+c_1)}{c_1}\right)^{\delta}\frac{1}{s^{\delta}}\ .$$

This completes the proof.

2.2 The approximate controllability

The goal of this section consists in giving an application of estimate (1.2).

For any ω non-empty open subset in Ω, for any $\beta \in (0,1)$, let $T>0$ be given in Theorem 1.1.

Let $(v_0, v_1, v_{0d}, v_{1d}) \in \left(H^2(\Omega) \cap H_0^1(\Omega) \times H_0^1(\Omega)\right)^2$ and u be the solution of (1.1) with initial data $(u, \partial_t u)(\cdot, 0) = (v_0, v_1)$.

For any integer $N>0$, let us introduce

$$f_N(x,t) = -1_\omega \sum_{\ell=0}^{N} \left[\partial_t w^{(2\ell+1)}(x,t) + \partial_t w^{(2\ell)}(x, T-t)\right] , \tag{2.3}$$

where $w^{(0)} \in C\left([0,T]; H^2(\Omega) \cap H_0^1(\Omega)\right)$ is the solution of the damped wave equation (2.1) with initial data

$$\left(w^{(0)}, \partial_t w^{(0)}\right)(\cdot, 0) = (v_{0d}, -v_{1d}) - (u, -\partial_t u)(\cdot, T) \text{ in } \Omega ,$$

and for $j \geq 0$, $w^{(j+1)} \in C\left([0,T]; H^2(\Omega) \cap H_0^1(\Omega)\right)$ is the solution of the damped wave equation (2.1) with initial data

$$\left(w^{(j+1)}, \partial_t w^{(j+1)}\right)(\cdot, 0) = \left(-w^{(j)}, \partial_t w^{(j)}\right)(\cdot, T) \text{ in } \Omega .$$

Introduce

$$M = \sup_{j \geq 0} \left\| w^{(j)}(\cdot, 0), \partial_t w^{(j)}(\cdot, 0)\right\|^2_{H^2(\Omega) \times H_0^1(\Omega)} .$$

Our main result is as follows.

Theorem 2.2 . *Suppose that $M < +\infty$. Then there exists $C>0$ such that for all $N>0$, the control function f_N given by (2.3) drives the system*

$$\begin{cases} \partial_t^2 v - \Delta v = 1_{\omega \times (0,T)} f_N & \text{in } \Omega \times (0,T) , \\ v = 0 & \text{on } \partial\Omega \times (0,T) , \\ (v, \partial_t v)(\cdot, 0) = (v_0, v_1) & \text{in } \Omega , \end{cases}$$

to the desired data (v_{0d}, v_{1d}) approximately at time T, i.e.,

$$\left\| v(\cdot, T) - v_{0d}, \partial_t v(\cdot, T) - v_{1d}\right\|^2_{H_0^1(\Omega) \times L^2(\Omega)} \leq \frac{C}{\left[\ln(1+2N)\right]^{2\beta}} M ,$$

and satisfies

$$\|f_N\|_{L^\infty(0,T; L^2(\Omega))} \leq C(N+1) \left\|(v_0, v_1, v_{0d}, v_{1d})\right\|_{\left(H_0^1(\Omega) \times L^2(\Omega)\right)^2} .$$

Remark. For any $\varepsilon > 0$, we can choose N such that

$$\frac{C}{\left[\ln\left(1+2N\right)\right]^{2\beta}} M \simeq \varepsilon^2 \text{ and } (2N+1) \simeq e^{\left(\frac{\sqrt{CM}}{\varepsilon}\right)^{1/\beta}} ,$$

in order that

$$\left\| v\left(\cdot, T\right) - v_{0d}, \partial_t v\left(\cdot, T\right) - v_{1d} \right\|_{H_0^1(\Omega)\times L^2(\Omega)} \leq \varepsilon ,$$

and

$$\|f\|_{L^\infty(0,T;L^2(\Omega))} \leq e^{\left[\left(\frac{C}{\varepsilon}\sqrt{M}\right)^{1/\beta}\right]} \left\|(v_0, v_1, v_{0d}, v_{1d})\right\|_{\left(H_0^1(\Omega)\times L^2(\Omega)\right)^2} .$$

In [Zu], a method was proposed to construct an approximate control. It consists of minimizing a functional depending on the parameter ε. However, no estimate of the cost is given. On the other hand, estimate of the form (1.2) was originally established by [Ro2] to give the cost (see [Le]). Here, we present a new way to construct an approximate control by superposing different waves. Given a cost to be not overcome, we construct a solution which will be closed in the above sense to the desired state. It takes ideas from [Ru] and [BF] like an iterative time reversal construction.

2.2.1 Proof

Consider the solution

$$V\left(\cdot, t\right) = \sum_{\ell=0}^{N} \left[w^{(2\ell+1)}\left(\cdot, t\right) + w^{(2\ell)}\left(\cdot, T-t\right)\right] .$$

We deduce that for $t \in (0, T)$

$$\begin{cases} \partial_t^2 V\left(\cdot, t\right) - \Delta V\left(\cdot, t\right) = -1_\omega \sum_{\ell=0}^{N} \left[\partial_t w^{(2\ell+1)}\left(\cdot, t\right) + \partial_t w^{(2\ell)}\left(\cdot, T-t\right)\right] , \\ V = 0 \quad \text{on } \partial\Omega \times (0, T) , \\ (V, \partial_t V)\left(\cdot, 0\right) = 0 \quad \text{in } \Omega . \end{cases}$$

Now, from the definition of $w^{(0)}$, the property of $\left(w^{(j+1)}, \partial_t w^{(j+1)}\right)(\cdot, 0)$ and a change of variable, we obtain that

$$\begin{aligned} (V, \partial_t V)\left(\cdot, T\right) &= \left(w^{(0)}, -\partial_t w^{(0)}\right)(\cdot, 0) + \left(w^{(2N+1)}, \partial_t w^{(2N+1)}\right)(\cdot, T) \\ &= (v_{0d}, v_{1d}) - (u, \partial_t u)\left(\cdot, T\right) + \left(w^{(2N+1)}, \partial_t w^{(2N+1)}\right)(\cdot, T) \end{aligned}$$

Finally, the solution $v = V + u$ satisfies

$$\begin{cases} \partial_t^2 v - \Delta v = 1_{\omega\times(0,T)} f_N \quad \text{in } \Omega \times (0, T) , \\ v = 0 \quad \text{on } \partial\Omega \times (0, T) , \\ (v, \partial_t v)\left(\cdot, 0\right) = (v_0, v_1) \quad \text{in } \Omega , \\ (v, \partial_t v)\left(\cdot, T\right) = (v_{0d}, v_{1d}) + \left(w^{(2N+1)}, \partial_t w^{(2N+1)}\right)(\cdot, T) \quad \text{in } \Omega . \end{cases}$$

Clearly,

$$\|v(\cdot,T) - v_{0d}, \partial_t v(\cdot,T) - v_{1d}\|^2_{H_0^1(\Omega)\times L^2(\Omega)} = 2E\left(w^{(2N+1)},T\right) .$$

It remains to estimate $E\left(w^{(2N+1)},T\right)$. We claim that

$$\exists C>0 \qquad \forall N\geq 1 \qquad E\left(w^{(2N+1)},T\right) \leq \frac{C}{\left[\ln(1+2N)\right]^{2\beta}} M .$$

Indeed, from Theorem 1.1, we can easily see by a classical decomposition method that there exist $C>0$ and $T>0$ such that for any $j\geq 0$,

$$\begin{aligned} &\left\|w^{(j+1)}(\cdot,0), \partial_t w^{(j+1)}(\cdot,0)\right\|^2_{H_0^1(\Omega)\times L^2(\Omega)} \\ &\leq C\exp\left(C\frac{\left\|w^{(j+1)}(\cdot,0),\partial_t w^{(j+1)}(\cdot,0)\right\|_{H^2(\Omega)\times H^1(\Omega)}}{\left\|w^{(j+1)}(\cdot,0),\partial_t w^{(j+1)}(\cdot,0)\right\|_{H_0^1(\Omega)\times L^2(\Omega)}}\right)^{1/\beta} \\ &\int_0^T\int_\omega \left|\partial_t w^{(j+1)}(x,t)\right|^2 dxdt . \end{aligned}$$

Since

$$E\left(w^{(j+1)},0\right) = E\left(w^{(j)},T\right) \qquad \forall j\geq 0 ,$$

we deduce from (2.2) that for any $j\geq 0$

$$\begin{aligned} E\left(w^{(j+1)},0\right) \leq C\exp\left(C\frac{M}{\left\|w^{(j+1)}(\cdot,0),\partial_t w^{(j+1)}(\cdot,0)\right\|^2_{H_0^1(\Omega)\times L^2(\Omega)}}\right)^{1/(2\beta)} \\ \left[E\left(w^{(j)},T\right) - E\left(w^{(j+1)},T\right)\right] . \end{aligned}$$

Let

$$d_j = E\left(w^{(j+1)},T\right) .$$

By using the decreasing property of the sequence d_j, that is $d_j \leq d_{j-1}$, we obtain that for any integer $0\leq j\leq 2N$

$$d_j \leq Ce^{\left(C\frac{M}{d_{2N}}\right)^{1/(2\beta)}}\left[d_{j-1}-d_j\right] .$$

By summing over $[0,2N]$, we deduce that

$$(2N+1)\,d_{2N} \leq Ce^{\left(C\frac{M}{d_{2N}}\right)^{1/(2\beta)}}\left[d_{-1}-d_{2N}\right] .$$

Finally, using the fact that $d_{-1}\leq M$, it follows that

$$d_{2N} \leq \frac{C}{\left[\ln(1+2N)\right]^{2\beta}} M .$$

This completes the proof of our claim.

On the other hand, the computation of the bound of f_N is immediate. Therefore, we check that for some $C > 0$ and $T > 0$,

$$\|f_N\|_{L^\infty(0,T;L^2(\Omega))} \leq C(N+1)\|(v_0, v_1, v_{0d}, v_{1d})\|_{\left(H_0^1(\Omega)\times L^2(\Omega)\right)^2},$$

$$\|v(\cdot,T) - v_{0d}, \partial_t v(\cdot,T) - v_{1d}\|^2_{H_0^1(\Omega)\times L^2(\Omega)} \leq \frac{C}{[\ln(1+2N)]^{2\beta}} M,$$

for any $\beta \in (0,1)$ and any integer $N > 0$. This completes the proof of our Theorem.

2.2.2 Numerical experiments

Here, we perform numerical experiments to investigate the practical applicability of the approach proposed to construct an approximate control. For simplicity, we consider a square domain $\Omega = (0,1) \times (0,1)$, $\omega = (0,1/5) \times (0,1)$. The time of controllability is given by $T = 4$.

For convenience we recall some well-known formulas. Denote by $\{e_j\}_{j\geq 1}$ the Hilbert basis in $L^2(\Omega)$ formed by the eigenfunctions of the operator $-\Delta$ with eigenvalues $\{\lambda_j\}_{j\geq 1}$, such that $\|e_j\|_{L^2(\Omega)} = 1$ and $0 < \lambda_1 < \lambda_2 \leq \lambda_3 \leq \cdots$, i.e.,

$$\begin{cases} \lambda_j = \pi^2\left(k_j^2 + \ell_j^2\right), & k_j, \ell_j \in \mathbb{N}^*, \\ e_j(x_1,x_2) = 2\sin(\pi k_j x_1)\sin(\pi \ell_j x_2) . \end{cases}$$

The solution of

$$\begin{cases} \partial_t^2 v - \Delta v = f & \text{in } \Omega \times (0,T), \\ v = 0 & \text{on } \partial\Omega \times (0,T), \\ (v, \partial_t v)(\cdot, 0) = (v_0, v_1) & \text{in } \Omega, \end{cases}$$

where f is in the form

$$f(x_1,x_2) = -1_\omega \sum_{j\geq 1} f_j(t)\, e_j(x_1,x_2),$$

is given by the formula

$$\begin{aligned} v(x_1,x_2,t) = \lim_{G\to+\infty} \sum_{j=1}^{G} \Big\{ & a_j^0 \cos\left(t\sqrt{\lambda_j}\right) + a_j^1 \tfrac{1}{\sqrt{\lambda_j}} \sin\left(t\sqrt{\lambda_j}\right) \\ & + \tfrac{1}{\sqrt{\lambda_j}} \textstyle\int_0^t \sin\left((t-s)\sqrt{\lambda_j}\right) R_j(s)\, ds \Big\}\, e_j(x_1,x_2), \end{aligned}$$

where

$$\begin{cases} v_0(x_1,x_2) = \lim_{G\to+\infty} \sum_{j=1}^{G} a_j^0 e_j(x_1,x_2), & \sum_{j\geq1} \lambda_j \left|a_j^0\right|^2 < +\infty , \\ v_1(x_1,x_2) = \lim_{G\to+\infty} \sum_{j=1}^{G} a_j^1 e_j(x_1,x_2), & \sum_{j\geq1} \left|a_j^1\right|^2 < +\infty , \\ R_j(t) = -\lim_{G\to+\infty} \sum_{i=1}^{G} \left(\int_\omega e_i e_j dx_1 dx_2\right) f_i(t) . \end{cases}$$

Here, G will be the number of Galerkin mode. The numerical results are shown below. The approximate solution of the damped wave equation is established via a system of ODE solved by MATLAB.

Example 1 : low frequency The initial condition and desired target are specifically as follows: $(v_0, v_1) = (0, 0)$ and $(v_{0d}, v_{1d}) = (e_1 + e_2, e_1)$. We take the number of Galerkin mode $G = 100$ and the number of iterations in the time reversal construction $N = 30$.

Below, we plot the graph of the desired initial data v_{0d} and the controlled solution $v(\cdot, t = T = 4)$.

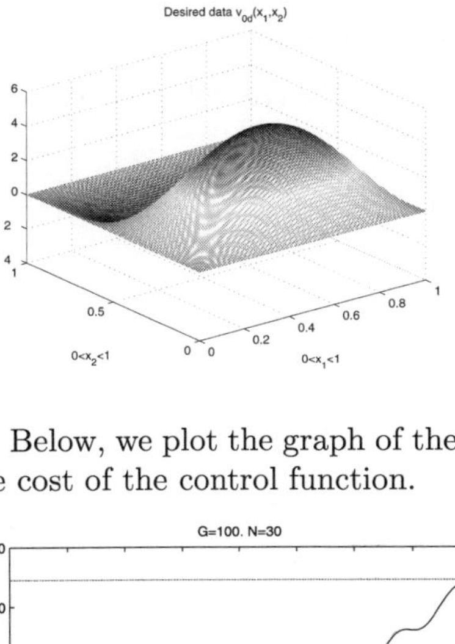

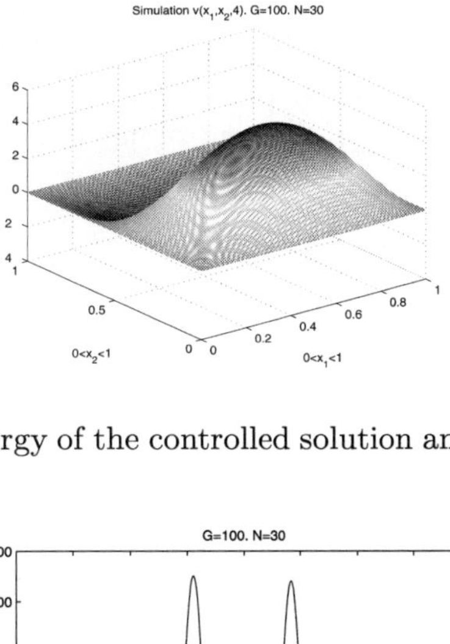

Below, we plot the graph of the energy of the controlled solution and the cost of the control function.

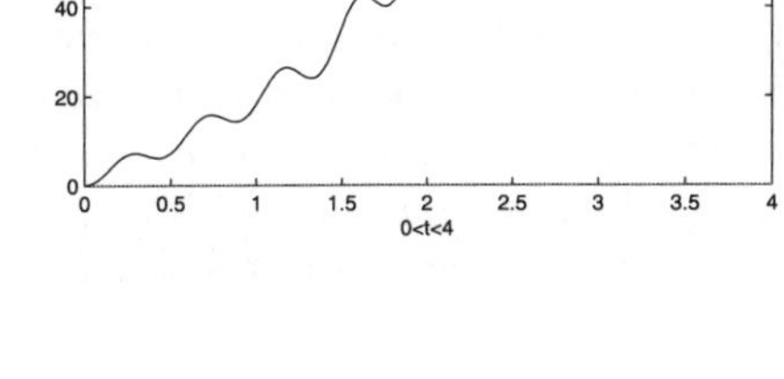

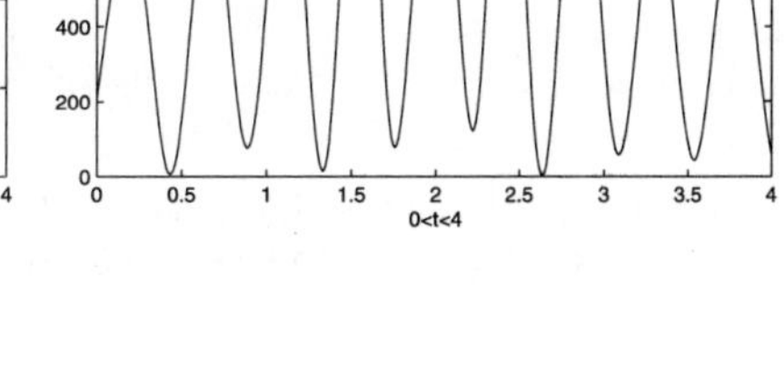

Example 2 : high frequency The initial condition and desired target are specifically as follows: $(v_{0d}, v_{1d}) = (0,0)$ and with $(k_o, a_o, b_o,) = (200, 1/2, 10000)$, for $(x_1, x_2) \in (0,1) \times (0,1)$,

$$\begin{cases} v_0(x_1,x_2) = \sum_{j=1}^{G} \left(\int_0^1 \int_0^1 g_0(x_1,x_2)\, e_j(x_1,x_2)\, dx_1 dx_2 \right) e_j(x_1,x_2) \ , \\ v_1(x_1,x_2) = \sum_{j=1}^{G} \left(\int_0^1 \int_0^1 g_1(x_1,x_2)\, e_j(x_1,x_2)\, dx_1 dx_2 \right) e_j(x_1,x_2) \ , \\ g_0(x_1,x_2) = e^{-\frac{k_o a_o}{2}(x_1 - x_{o1})^2} e^{-\frac{k_o b_o}{2}(x_2 - x_{o2})^2} \cos\left(k_o (x_2 - x_{o2})/2\right) \ , \\ g_1(x_1,x_2) = e^{-\frac{k_o a_o}{2}(x_1 - x_{o1})^2} e^{-\frac{k_o b_o}{2}(x_2 - x_{o2})^2} \\ \qquad [k_o b_o (x_2 - x_{o2}) \cos\left(k_o (x_2 - x_{o2})/2\right) \\ \qquad + (k_o/2 + a_o) \sin\left(k_o (x_2 - x_{o2})/2\right) \\ \qquad - k_o a_o^2 (x_1 - x_{o1})^2 \sin\left(k_o (x_2 - x_{o2})/2\right)] \ . \end{cases}$$

Notice that we have chosen as initial data the G-first projections on the basis $\{e_j\}_{j \geq 1}$ of a gaussian beam $g(x_1, x_2, t)$ such that $g(\cdot, t=0) = g_0$, $\partial_t g(\cdot, t=0) = g_1$, which propagate in the direction $(0,1)$.

We take the number of Galerkin mode $G = 1000$ and the number of iterations in the time reversal construction $N = 100$.

Below, we plot the graph of the energy of the controlled solution and the cost of the control function.

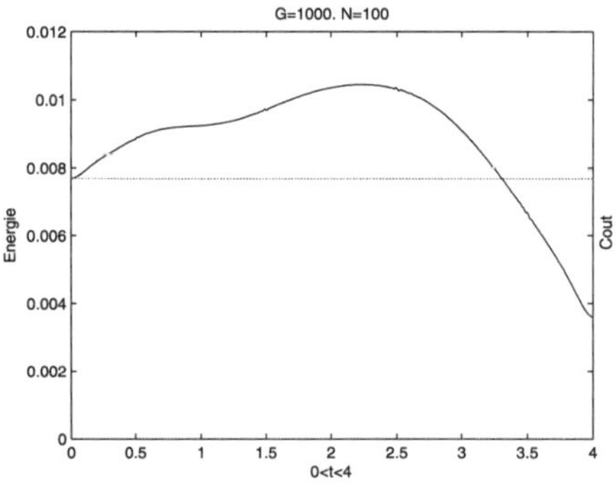

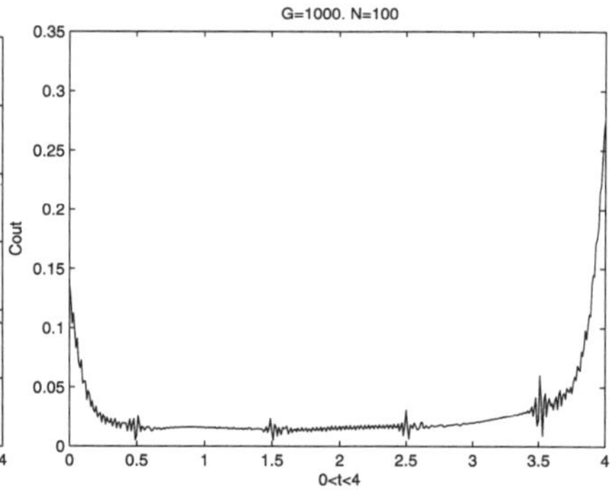

3 Conclusion

In this paper, we have considered the wave equation in a bounded domain (eventually convex). Two kinds of inequalities are described when there occur trapped rays. Applications to control theory are given. First, we link such kind of estimate with the damped wave equation and its decay rate. Next, we describe the design of an approximate control function by an iterative time reversal method. We also provide a numerical simulation in a square domain. I'm grateful to Prof. Jean-Pierre Puel, the "French-Chinese Summer Institute on Applied Mathematics" and Fudan University for the kind invitation and the support to my visit.

4 Appendix

In this appendix, we recall most of the materials from the works by I. Kukavica [Ku2] and L. Escauriaza [E] for the elliptic equation and from the works by G. Lebeau and L. Robbiano [LR] for the wave equation.

In the original paper dealing with doubling property and frequency function, N. Garofalo and F.H. Lin [GaL] studied the monotonicity property of the following quantity

$$\frac{r\int_{B_{0,r}} \left|\nabla v\left(y\right)\right|^2 dy}{\int_{\partial B_{0,r}} \left|v\left(y\right)\right|^2 d\sigma\left(y\right)} .$$

However, it seems more natural in our context to consider the monotonicity properties of the frequency function (see [Ze]) defined by

$$\frac{\int_{B_{0,r}} \left|\nabla v\left(y\right)\right|^2 \left(r^2-\left|y\right|^2\right) dy}{\int_{B_{0,r}} \left|v\left(y\right)\right|^2 dy} .$$

4.1 Monotonicity formula

Following the ideas of I. Kukavica ([Ku2], [Ku], [KN], see also [E], [AE]), one obtains the following three lemmas. Detailed proofs are given in [Ph3].

Lemma A. *Let $D\subset\mathbb{R}^{N+1}$, $N\geq 1$, be a connected bounded open set such that $\overline{B_{y_o,R_o}}\subset D$ with $y_o\in D$ and $R_o>0$. If $v=v\left(y\right)\in H^2\left(D\right)$ is a solution of $\Delta_y v=0$ in D, then*

$$\Phi\left(r\right)=\frac{\int_{B_{y_o,r}} \left|\nabla v\left(y\right)\right|^2 \left(r^2-\left|y-y_o\right|^2\right) dy}{\int_{B_{y_o,r}} \left|v\left(y\right)\right|^2 dy}$$

is non-decreasing on $0<r<R_o$, and

$$\frac{d}{dr}\ln\int_{B_{y_o,r}} \left|v\left(y\right)\right|^2 dy=\frac{1}{r}\left(N+1+\Phi\left(r\right)\right) .$$

Lemma B. *Let $D\subset\mathbb{R}^{N+1}$, $N\geq 1$, be a connected bounded open set such that $\overline{B_{y_o,R_o}}\subset D$ with $y_o\in D$ and $R_o>0$. Let r_1, r_2, r_3 be three* *real numbers such that* $0 < r_1 < r_2 < r_3 < R_o$. *If* $v = v(y) \in H^2(D)$ *is a solution of* $\Delta_y v = 0$ *in* D, *then*

$$\int_{B_{y_o,r_2}} |v(y)|^2 dy \leq \left(\int_{B_{y_o,r_1}} |v(y)|^2 dy \right)^{\alpha} \left(\int_{B_{y_o,r_3}} |v(y)|^2 dy \right)^{1-\alpha},$$

where $\alpha = \frac{1}{\ln \frac{r_2}{r_1}} \left(\frac{1}{\ln \frac{r_2}{r_1}} + \frac{1}{\ln \frac{r_3}{r_2}} \right)^{-1} \in (0,1)$.

The above two results are still available when we are closed to a part Γ of the boundary $\partial\Omega$ under the homogeneous Dirichlet boundary condition on Γ, as follows.

Lemma C. *Let* $D \subset \mathbb{R}^{N+1}$, $N \geq 1$, *be a connected bounded open set with boundary* ∂D. *Let* Γ *be a non-empty Lipschitz open subset of* ∂D. *Let* r_o, r_1, r_2, r_3, R_o *be five real numbers such that* $0 < r_1 < r_o < r_2 < r_3 < R_o$. *Suppose that* $y_o \in D$ *satisfies the following three conditions:*

i). $B_{y_o,r} \cap D$ *is star-shaped with respect to* y_o $\forall r \in (0, R_o)$,
ii). $B_{y_o,r} \subset D$ $\forall r \in (0, r_o)$,
iii). $B_{y_o,r} \cap \partial D \subset \Gamma$ $\forall r \in [r_o, R_o)$.

If $v = v(y) \in H^2(D)$ *is a solution of* $\Delta_y v = 0$ *in* D *and* $v = 0$ *on* Γ, *then*

$$\int_{B_{y_o,r_2} \cap D} |v(y)|^2 dy \leq \left(\int_{B_{y_o,r_1}} |v(y)|^2 dy \right)^{\alpha} \left(\int_{B_{y_o,r_3} \cap D} |v(y)|^2 dy \right)^{1-\alpha}$$

where $\alpha = \frac{1}{\ln \frac{r_2}{r_1}} \left(\frac{1}{\ln \frac{r_2}{r_1}} + \frac{1}{\ln \frac{r_3}{r_2}} \right)^{-1} \in (0,1)$.

4.1.1 Proof of Lemma B

Let

$$H(r) = \int_{B_{y_o,r}} |v(y)|^2 dy .$$

By applying Lemma A, we know that

$$\frac{d}{dr} \ln H(r) = \frac{1}{r} (N + 1 + \Phi(r)) .$$

Next, from the monotonicity property of Φ, one deduces the following two inequalities

$$\begin{aligned} \ln \left(\frac{H(r_2)}{H(r_1)} \right) &= \int_{r_1}^{r_2} \frac{N+1+\Phi(r)}{r} dr \\ &\leq (N + 1 + \Phi(r_2)) \ln \frac{r_2}{r_1} , \end{aligned}$$

$$\begin{aligned}\ln\left(\frac{H(r_3)}{H(r_2)}\right) &= \int_{r_2}^{r_3}\frac{N+1+\Phi(r)}{r}dr \\ &\geq (N+1+\Phi(r_2))\ln\frac{r_3}{r_2} \ .\end{aligned}$$

Consequently,

$$\frac{\ln\left(\frac{H(r_2)}{H(r_1)}\right)}{\ln\frac{r_2}{r_1}} \leq (N+1)+\Phi(r_2) \leq \frac{\ln\left(\frac{H(r_3)}{H(r_2)}\right)}{\ln\frac{r_3}{r_2}} \ ,$$

and therefore the desired estimate holds

$$H(r_2) \leq (H(r_1))^{\alpha}(H(r_3))^{1-\alpha} \ ,$$

where $\alpha = \frac{1}{\ln\frac{r_2}{r_1}}\left(\frac{1}{\ln\frac{r_2}{r_1}}+\frac{1}{\ln\frac{r_3}{r_2}}\right)^{-1}$.

4.1.2 Proof of Lemma A

We introduce the following two functions H and D for $0 < r < R_o$:

$$\begin{aligned}H(r) &= \int_{B_{y_o,r}}|v(y)|^2 dy \ , \\ D(r) &= \int_{B_{y_o,r}}|\nabla v(y)|^2\left(r^2-|y-y_o|^2\right)dy \ .\end{aligned}$$

First, the derivative of $H(r) = \int_0^r\int_{S^N}|v(\rho s+y_o)|^2\rho^N d\rho d\sigma(s)$ is given by $H'(r) = \int_{\partial B_{y_o,r}}|v(y)|^2 d\sigma(y)$. Next, recall the Green formula

$$\begin{aligned}&\int_{\partial B_{y_o,r}}|v|^2\partial_\nu G d\sigma(y) - \int_{\partial B_{y_o,r}}\partial_\nu\left(|v|^2\right)G d\sigma(y) \\ &= \int_{B_{y_o,r}}|v|^2\Delta G dy - \int_{B_{y_o,r}}\Delta\left(|v|^2\right)G dy \ .\end{aligned}$$

We apply it with $G(y) = r^2-|y-y_o|^2$ where $G_{|\partial B_{y_o,r}} = 0$, $\partial_\nu G_{|\partial B_{y_o,r}} = -2r$, and $\Delta G = -2(N+1)$. It gives

$$\begin{aligned}H'(r) &= \frac{1}{r}\int_{B_{y_o,r}}(N+1)|v|^2 dy + \frac{1}{2r}\int_{B_{y_o,r}}\Delta\left(|v|^2\right)\left(r^2-|y-y_o|^2\right)dy \\ &= \frac{N+1}{r}H(r) + \frac{1}{r}\int_{B_{y_o,r}}div(v\nabla v)\left(r^2-|y-y_o|^2\right)dy \\ &= \frac{N+1}{r}H(r) + \frac{1}{r}\int_{B_{y_o,r}}\left(|\nabla v|^2+v\Delta v\right)\left(r^2-|y-y_o|^2\right)dy \ .\end{aligned}$$

Consequently, when $\Delta_y v = 0$,

$$H'(r) = \frac{N+1}{r}H(r) + \frac{1}{r}D(r) \ , \qquad \text{(A.1)}$$

that is $\frac{H'(r)}{H(r)} = \frac{N+1}{r} + \frac{1}{r}\frac{D(r)}{H(r)}$ the second equality in Lemma A.

Now, we compute the derivative of $D(r)$.

$$\begin{aligned} D'(r) &= \tfrac{d}{dr}\left(r^2 \int_0^r \int_{S^N} \left|(\nabla v)_{|\rho s+y_o}\right|^2 \rho^N d\rho d\sigma(s)\right) \\ &\quad - \int_{S^N} r^2 \left|(\nabla v)_{|rs+y_o}\right|^2 r^N d\sigma(s) \\ &= 2r \int_0^r \int_{S^N} \left|(\nabla v)_{|\rho s+y_o}\right|^2 \rho^N d\rho d\sigma(s) \\ &= 2r \int_{B_{y_o,r}} |\nabla v|^2 dy \ . \end{aligned} \tag{A.2}$$

On the other hand, we have by integrations by parts that

$$\begin{aligned} 2r \int_{B_{y_o,r}} |\nabla v|^2 dy &= \tfrac{N+1}{r} D(r) + \tfrac{4}{r} \int_{B_{y_o,r}} |(y-y_o)\cdot\nabla v|^2 dy \\ &\quad - \tfrac{1}{r} \int_{B_{y_o,r}} \nabla v \cdot (y-y_o)\, \Delta v \left(r^2 - |y-y_o|^2\right) dy \ . \end{aligned} \tag{A.3}$$

Therefore,

$$\begin{aligned} &(N+1) \int_{B_{y_o,r}} |\nabla v|^2 \left(r^2 - |y-y_o|^2\right) dy \\ &= 2r^2 \int_{B_{y_o,r}} |\nabla v|^2 dy - 4 \int_{B_{y_o,r}} |(y-y_o)\cdot\nabla v|^2 dy \\ &\quad +2 \int_{B_{y_o,r}} (y-y_o)\cdot\nabla v \Delta v \left(r^2 - |y-y_o|^2\right) dy \ , \end{aligned}$$

and this is the desired estimate (A.3).

Consequently, from (A.2) and (A.3), we obtain, when $\Delta_y v = 0$, the following formula

$$D'(r) = \frac{N+1}{r} D(r) + \frac{4}{r} \int_{B_{y_o,r}} |(y-y_o)\cdot\nabla v|^2 dy \ . \tag{A.4}$$

The computation of the derivative of $\Phi(r) = \frac{D(r)}{H(r)}$ gives

$$\Phi'(r) = \frac{1}{H^2(r)} \left[D'(r) H(r) - D(r) H'(r)\right] \ ,$$

which implies using (A.1) and (A.4) that

$$H^2(r)\,\Phi'(r) = \frac{1}{r}\left(4 \int_{B_{y_o,r}} |(y-y_o)\cdot\nabla v|^2 dy H(r) - D^2(r)\right) \geq 0 \ ,$$

indeed, thanks to an integration by parts and using Cauchy-Schwarz inequality, we have

$$\begin{aligned} D^2(r) &= 4 \left(\int_{B_{y_o,r}} v\nabla v \cdot (y-y_o)\, dy\right)^2 \\ &\leq 4 \left(\int_{B_{y_o,r}} |(y-y_o)\cdot\nabla v|^2 dy\right) \left(\int_{B_{y_o,r}} |v|^2 dy\right) \\ &\leq 4 \left(\int_{B_{y_o,r}} |(y-y_o)\cdot\nabla v|^2 dy\right) H(r) \ . \end{aligned}$$

Therefore, we have proved the desired monotonicity for Φ and this completes the proof of Lemma A.

4.1.3 Proof of Lemma C

Under the assumption $B_{y_o,r} \cap \partial D \subset \Gamma$ for any $r \in [r_o, R_o)$, we extend v by zero in $\overline{B_{y_o,R_o} \backslash D}$ and denote by $\overline{v}$ its extension. Since $v = 0$ on Γ, we have

$$\begin{cases} \overline{v} = v 1_D & \text{in } \overline{B_{y_o,R_o}} \ , \\ \overline{v} = 0 & \text{on } B_{y_o,R_o} \cap \partial D \ , \\ \nabla \overline{v} = \nabla v 1_D & \text{in } B_{y_o,R_o} \ . \end{cases}$$

Now, we denote $\Omega_r = B_{y_o,r} \cap D$, when $0 < r < R_o$. Particularly, $\Omega_r = B_{y_o,r}$, when $0 < r < r_o$. We introduce the following three functions:

$$\begin{aligned} H(r) &= \int_{\Omega_r} |v(y)|^2 \, dy \ , \\ D(r) &= \int_{\Omega_r} |\nabla v(y)|^2 \left(r^2 - |y - y_o|^2 \right) dy \ , \end{aligned}$$

and

$$\Phi(r) = \frac{D(r)}{H(r)} \geq 0 \ .$$

Our goal is to show that Φ is a non-decreasing function. Indeed, we will prove that the following equality holds

$$\frac{d}{dr} \ln H(r) = (N+1) \frac{d}{dr} \ln r + \frac{1}{r} \Phi(r) \ . \tag{C.1}$$

Therefore, from the monotonicity of Φ, we will deduce (in a similar way to that in the proof of Lemma A) that

$$\frac{\ln \left(\frac{H(r_2)}{H(r_1)} \right)}{\ln \frac{r_2}{r_1}} \leq (N+1) + \Phi(r_2) \leq \frac{\ln \left(\frac{H(r_3)}{H(r_2)} \right)}{\ln \frac{r_3}{r_2}} \ ,$$

and this will imply the desired estimate

$$\int_{\Omega_{r_2}} |v(y)|^2 \, dy \leq \left(\int_{B_{y_o,r_1}} |v(y)|^2 \, dy \right)^{\alpha} \left(\int_{\Omega_{r_3}} |v(y)|^2 \, dy \right)^{1-\alpha} ,$$

where $\alpha = \frac{1}{\ln\frac{r_2}{r_1}}\left(\frac{1}{\ln\frac{r_2}{r_1}} + \frac{1}{\ln\frac{r_3}{r_2}}\right)^{-1}$.

First, we compute the derivative of $H(r) = \int_{B_{y_o,r}} |\overline{v}(y)|^2 \, dy$.

$$\begin{aligned} H'(r) &= \int_{S^N} |\overline{v}(rs+y_o)|^2 r^N d\sigma(s) \\ &= \tfrac{1}{r}\int_{S^N} |\overline{v}(rs+y_o)|^2 rs \cdot sr^N d\sigma(s) \\ &= \tfrac{1}{r}\int_{B_{y_o,r}} div\left(|\overline{v}(y)|^2 (y-y_o)\right) dy \\ &= \tfrac{1}{r}\int_{B_{y_o,r}} \left((N+1)|\overline{v}(y)|^2 + \nabla|\overline{v}(y)|^2 \cdot (y-y_o)\right) dy \\ &= \tfrac{N+1}{r} H(r) + \tfrac{2}{r}\int_{\Omega_r} v(y)\nabla v(y)\cdot(y-y_o)\,dy \,. \end{aligned} \tag{C.2}$$

Next, when $\Delta_y v = 0$ in D and $v_{|\Gamma} = 0$, we remark that

$$D(r) = 2\int_{\Omega_r} v(y)\nabla v(y)\cdot(y-y_o)\,dy \;, \tag{C.3}$$

indeed,

$$\begin{aligned} &\int_{\Omega_r} |\nabla v|^2 \left(r^2 - |y-y_o|^2\right) dy \\ &= \int_{\Omega_r} div\left[v\nabla v\left(r^2 - |y-y_o|^2\right)\right] dy - \int_{\Omega_r} v div\left[\nabla v\left(r^2-|y-y_o|^2\right)\right] dy \\ &= -\int_{\Omega_r} v\Delta v\left(r^2-|y-y_o|^2\right) dy - \int_{\Omega_r} v\nabla v\cdot\nabla\left(r^2-|y-y_o|^2\right) dy \\ &\quad \text{because on } \partial B_{y_o,r},\ r=|y-y_o| \ \text{ and } v_{|\Gamma}=0 \\ &= 2\int_{\Omega_r} v\nabla v\cdot(y-y_o)\,dy \quad \text{because } \Delta_y v = 0 \text{ in } D \,. \end{aligned}$$

Consequently, from (C.2) and (C.3), we obtain

$$H'(r) = \frac{N+1}{r}H(r) + \frac{1}{r}D(r) \;, \tag{C.4}$$

and this is (C.1).

On the other hand, the derivative of $D(r)$ is

$$\begin{aligned} D'(r) &= 2r\int_0^r \int_{S^N} \left|(\nabla\overline{v})_{|\rho s+y_o}\right|^2 \rho^N d\rho d\sigma(s) \\ &= 2r\int_{\Omega_r} |\nabla v(y)|^2 \, dy \,. \end{aligned} \tag{C.5}$$

Here, when $\Delta_y v = 0$ in D and $v_{|\Gamma} = 0$, we will remark that

$$\begin{aligned} 2r\int_{\Omega_r} |\nabla v(y)|^2 dy &= \tfrac{N+1}{r} D(r) + \tfrac{4}{r}\int_{B_{y_o,r}} |(y-y_o)\cdot\nabla v(y)|^2 dy \\ &\quad + \tfrac{1}{r}\int_{\Gamma\cap B_{y_o,r}} |\partial_\nu v|^2 \left(r^2 - |y-y_o|^2\right)(y-y_o)\cdot\nu d\sigma(y) \end{aligned} \tag{C.6}$$

indeed,

$$\begin{aligned}
&(N+1)\int_{\Omega_r}|\nabla v|^2\left(r^2-|y-y_o|^2\right)dy\\
&=\int_{\Omega_r}div\left(|\nabla v|^2\left(r^2-|y-y_o|^2\right)(y-y_o)\right)dy\\
&\quad-\int_{\Omega_r}\nabla\left(|\nabla v|^2\left(r^2-|y-y_o|^2\right)\right)\cdot(y-y_o)\,dy\\
&=\int_{\Gamma\cap B_{y_o,r}}|\nabla v|^2\left(r^2-|y-y_o|^2\right)(y-y_o)\cdot\nu d\sigma(y)\\
&\quad-\int_{\Omega_r}\partial_{y_i}\left(|\nabla v|^2\left(r^2-|y-y_o|^2\right)\right)(y_i-y_{oi})\,dy\\
&=\int_{\Gamma\cap B_{y_o,r}}|\nabla v|^2\left(r^2-|y-y_o|^2\right)(y-y_o)\cdot\nu d\sigma(y)\\
&\quad-\int_{\Omega_r}2\nabla v\partial_{y_i}\nabla v\left(r^2-|y-y_o|^2\right)(y_i-y_{oi})\,dy\\
&\quad+2\int_{\Omega_r}|\nabla v|^2|y-y_o|^2\,dy\ ,
\end{aligned}$$

and

$$\begin{aligned}
&-\int_{\Omega_r}\partial_{y_j}v\partial^2_{y_iy_j}v\left(r^2-|y-y_o|^2\right)(y_i-y_{oi})\,dy\\
&=-\int_{\Omega_r}\partial_{y_j}\left((y_i-y_{oi})\,\partial_{y_j}v\partial_{y_i}v\left(r^2-|y-y_o|^2\right)\right)dy\\
&\quad+\int_{\Omega_r}\partial_{y_j}(y_i-y_{oi})\,\partial_{y_j}v\partial_{y_i}v\left(r^2-|y-y_o|^2\right)dy\\
&\quad+\int_{\Omega_r}(y_i-y_{oi})\,\partial^2_{y_j}v\partial_{y_i}v\left(r^2-|y-y_o|^2\right)dy\\
&\quad+\int_{\Omega_r}(y_i-y_{oi})\,\partial_{y_j}v\partial_{y_i}v\partial_{y_j}\left(r^2-|y-y_o|^2\right)dy\\
&=-\int_{\Gamma\cap B_{y_o,r}}\nu_j\left((y_i-y_{oi})\,\partial_{y_j}v\partial_{y_i}v\left(r^2-|y-y_o|^2\right)\right)d\sigma(y)\\
&\quad+\int_{\Omega_r}|\nabla v|^2\left(r^2-|y-y_o|^2\right)dy\\
&\quad+0\quad\text{because } \Delta_y v=0 \text{ in } D\\
&\quad-\int_{\Omega_r}2\,|(y-y_o)\cdot\nabla v|^2\,dy\ .
\end{aligned}$$

Therefore, when $\Delta_y v = 0$ in D, we have

$$\begin{aligned}
&(N+1)\int_{\Omega_r}|\nabla v|^2\left(r^2-|y-y_o|^2\right)dy\\
&=\int_{\Gamma\cap B_{y_o,r}}|\nabla v|^2\left(r^2-|y-y_o|^2\right)(y-y_o)\cdot\nu d\sigma(y)\\
&\quad-2\int_{\Gamma\cap B_{y_o,r}}\partial_{y_j}v\nu_j\left((y_i-y_{oi})\,\partial_{y_i}v\right)\left(r^2-|y-y_o|^2\right)d\sigma(y)\\
&\quad+2r^2\int_{\Omega_r}|\nabla u|^2\,dy-4\int_{\Omega_r}|(y-y_o)\cdot\nabla v|^2\,dy\ .
\end{aligned}$$

By the fact that $v_{|\Gamma} = 0$, we get $\nabla v = (\nabla v\cdot\nu)\,\nu$ on Γ and deduce that

$$\begin{aligned}
&(N+1)\int_{\Omega_r}|\nabla v|^2\left(r^2-|y-y_o|^2\right)dy\\
&=-\int_{\Gamma\cap B_{y_o,r}}|\partial_\nu v|^2\left(r^2-|y-y_o|^2\right)(y-y_o)\cdot\nu d\sigma(y)\\
&\quad+2r^2\int_{\Omega_r}|\nabla v|^2\,dy-4\int_{\Omega_r}|(y-y_o)\cdot\nabla v|^2\,dy\ ,
\end{aligned}$$

and this is (C.6).

Consequently, from (C.5) and (C.6), when $\Delta_y v = 0$ in D and $v_{|\Gamma} = 0$, we have

$$\begin{array}{rl} D'(r) = & \frac{N+1}{r} D(r) + \frac{4}{r} \int_{\Omega_r} \left|(y-y_o)\cdot\nabla v(y)\right|^2 dy \\ & + \frac{1}{r} \int_{\Gamma\cap B_{y_o,r}} \left|\partial_\nu v\right|^2 \left(r^2 - \left|y-y_o\right|^2\right)(y-y_o)\cdot\nu d\sigma(y) \ . \end{array} \quad (C.7)$$

The computation of the derivative of $\Phi(r) = \frac{D(r)}{H(r)}$ gives

$$\Phi'(r) = \frac{1}{H^2(r)}\left[D'(r)H(r) - D(r)H'(r)\right] \ ,$$

which implies from (C.4) and (C.7) that

$$\begin{array}{rl} H^2(r)\Phi'(r) = & \frac{1}{r}\left(4\int_{\Omega_r} \left|(y-y_o)\cdot\nabla v(y)\right|^2 dy \ H(r) - D^2(r)\right) \\ & + \frac{H(r)}{r}\int_{\Gamma\cap B_{y_o,r}} \left|\partial_\nu v\right|^2 \left(r^2 - \left|y-y_o\right|^2\right)(y-y_o)\cdot\nu d\sigma(y) \end{array}$$

Thanks to (C.3) and Cauchy-Schwarz inequality, we obtain that

$$0 \leq 4\int_{\Omega_r} \left|(y-y_o)\cdot\nabla v(y)\right|^2 dy \ H(r) - D^2(r) \ .$$

The inequality $0 \leq (y-y_o)\cdot\nu$ on Γ holds when $B_{y_o,r}\cap D$ is star-shaped with respect to y_o for any $r \in (0, R_o)$. Therefore, we get the desired monotonicity for Φ which completes the proof of Lemma C.

4.2 Quantitative unique continuation property for the Laplacian

Let $D \subset \mathbb{R}^{N+1}$, $N \geq 1$, be a connected bounded open set with boundary ∂D. Let Γ be a non-empty Lipschitz open part of ∂D. We consider the Laplacian in D, with a homogeneous Dirichlet boundary condition on $\Gamma \subset \partial\Omega$:

$$\left\{\begin{array}{rll} \Delta_y v = 0 & \text{in } D \ , \\ v = 0 & \text{on } \Gamma \ , \\ v = v(y) \in H^2(D) & . \end{array}\right. \quad (D.1)$$

The goal of this section is to describe interpolation inequalities associated with solutions v of (D.1).

Theorem D. *Let ω be a non-empty open subset of D. Then, for any $D_1 \subset D$, $\partial D_1 \cap \partial D \Subset \Gamma$ and $\overline{D_1}\backslash(\Gamma\cap\partial D_1) \subset D$, there exist $C > 0$ and $\mu \in (0,1)$ such that for any v solution of* (D.1), *we have*

$$\int_{D_1} \left|v(y)\right|^2 dy \leq C\left(\int_\omega \left|v(y)\right|^2 dy\right)^\mu \left(\int_D \left|v(y)\right|^2 dy\right)^{1-\mu} \ .$$

Or in an equivalent way and by a minimization technique, there occur the following results:

Theorem D'. *Let ω be a non-empty open subset of D. Then, for any $D_1 \subset D$, $\partial D_1 \cap \partial D \Subset \Gamma$ and $\overline{D_1} \backslash (\Gamma \cap \partial D_1) \subset D$, there exist $C > 0$ and $\mu \in (0,1)$ such that for any v solution of* (D.1), *we have*

$$\int_{D_1} |v(y)|^2 \, dy \leq C \left(\frac{1}{\varepsilon}\right)^{\frac{1-\mu}{\mu}} \int_\omega |v(y)|^2 \, dy + \varepsilon \int_D |v(y)|^2 \, dy \quad \forall \varepsilon > 0 \, .$$

Proof of Theorem D. We divide the proof into two steps.

Step 1. We apply Lemma B, and use a standard argument (see e.g., [Ro]) which consists of constructing a sequence of balls chained along a curve. More precisely, we claim that for any non-empty compact sets in D, K_1 and K_2, meas $(K_1) > 0$, there exists $\mu \in (0,1)$ such that for any $v = v(y) \in H^2(D)$, solution of $\Delta_y v = 0$ in D, we have

$$\int_{K_2} |v(y)|^2 \, dy \leq \left(\int_{K_1} |v(y)|^2 \, dy\right)^\mu \left(\int_D |v(y)|^2 \, dy\right)^{1-\mu} . \tag{D.2}$$

Step 2. We apply Lemma C, and choose y_o in a neighborhood of the part Γ such that the conditions *i, ii, iii* hold. Next, by an adequate partition of D, we deduce from (D.2) that for any $D_1 \subset D$, $\partial D_1 \cap \partial D \Subset \Gamma$ and $\overline{D_1} \backslash (\Gamma \cap \partial D_1) \subset D$, there exist $C > 0$ and $\mu \in (0,1)$ such that for any $v = v(y) \in H^2(D)$, $\Delta_y v = 0$ on D and $v = 0$ on Γ, we have

$$\int_{D_1} |v(y)|^2 \, dy \leq C \left(\int_\omega |v(y)|^2 \, dy\right)^\mu \left(\int_D |v(y)|^2 \, dy\right)^{1-\mu} .$$

This completes the proof.

4.3 Quantitative unique continuation property for the elliptic operator $\partial_t^2 + \Delta$

In this section, we present the following result.

Theorem E. *Let Ω be a bounded open set in $\mathbb{R}^n$, $n \geq 1$, either convex or C^2 connected. We choose $T_2 > T_1$ and $\delta \in (0, (T_2 - T_1)/2)$. Let $f \in L^2(\Omega \times (T_1, T_2))$. We consider the elliptic operator of the second order in $\Omega \times (T_1, T_2)$ with a homogeneous Dirichlet boundary condition on $\partial\Omega \times (T_1, T_2)$,*

$$\begin{cases} \partial_t^2 w + \Delta w = f & \text{in } \Omega \times (T_1, T_2) \, , \\ w = 0 & \text{on } \partial\Omega \times (T_1, T_2) \, , \\ w = w(x,t) \in H^2(\Omega \times (T_1, T_2)) \, . \end{cases} \tag{E.1}$$

Then, for any $\varphi \in C_0^\infty(\Omega \times (T_1,T_2))$, $\varphi \neq 0$, *there exist* $C > 0$ *and* $\mu \in (0,1)$ *such that for any* w *solution of* (E.1), *we have*

$$\begin{aligned} &\int_{T_1+\delta}^{T_2-\delta}\int_\Omega |w(x,t)|^2\,dxdt \\ &\leq C\left(\int_{T_1}^{T_2}\int_\Omega |w(x,t)|^2\,dxdt\right)^{1-\mu} \\ &\quad\left(\int_{T_1}^{T_2}\int_\Omega |\varphi w(x,t)|^2\,dxdt + \int_{T_1}^{T_2}\int_\Omega |f(x,t)|^2\,dxdt\right)^{\mu} . \end{aligned}$$

Proof. First, by a difference quotient technique and a standard extension at $\Omega \times \{T_1,T_2\}$, we check the existence of a solution $u \in H^2(\Omega \times (T_1,T_2))$ solving

$$\begin{cases} \partial_t^2 u + \Delta u = f & \text{in } \Omega \times (T_1,T_2) \ , \\ u = 0 & \text{on } \partial\Omega \times (T_1,T_2) \cup \Omega \times \{T_1,T_2\} \ , \end{cases}$$

such that

$$\|u\|_{H^2(\Omega\times(T_1,T_2))} \leq c\,\|f\|_{L^2(\Omega\times(T_1,T_2))} \ ,$$

for some $c > 0$ only depending on (Ω, T_1, T_2). Next, we apply Theorem D with $D = \Omega\times(T_1,T_2)$, $\Omega\times(T_1+\delta, T_2-\delta) \subset D_1$, $y = (x,t)$, $\Delta_y = \partial_t^2 + \Delta$, and $v = w - u$.

4.4 Application to the wave equation

From the idea of L. Robbiano [Ro2] which consists of using an interpolation inequality of Hölder type for the elliptic operator $\partial_t^2 + \Delta$ and the Fourier-Bros-Iagolnitzer transform introduced by G. Lebeau and L. Robbiano [LR], we obtain the following estimate of logarithmic type.

Theorem F. *Let* Ω *be a bounded open set in* $\mathbb{R}^n$, $n \geq 1$, *either convex or* C^2 *connected. Let* ω *be a non-empty open subset in* Ω. *Then, for any* $\beta \in (0,1)$ *and* $k \in \mathbb{N}^*$, *there exist* $C > 0$ *and* $T > 0$ *such that for any solution* u *of*

$$\begin{cases} \partial_t^2 u - \Delta u = 0 & \text{in } \Omega \times (0,T) \ , \\ u = 0 & \text{on } \partial\Omega \times (0,T) \ , \\ (u, \partial_t u)(\cdot,0) = (u_0,u_1) \ , \end{cases}$$

with non-identically zero initial data $(u_0,u_1) \in D\left(A^{k-1}\right)$, *we have*

$$\|(u_0,u_1)\|_{D(A^{k-1})} \leq C e^{\left(C\frac{\|(u_0,u_1)\|_{D(A^{k-1})}}{\|(u_0,u_1)\|_{L^2(\Omega)\times H^{-1}(\Omega)}}\right)^{1/(\beta k)}} \|u\|_{L^2(\omega\times(0,T))} \ .$$

Proof. First, recall that with a standard energy method, we have that

$$\forall t \in \mathbb{R} \qquad \|(u_0,u_1)\|^2_{H_0^1(\Omega)\times L^2(\Omega)} = \int_\Omega \left(|\partial_t u(x,t)|^2 + |\nabla u(x,t)|^2\right) dx \ , \tag{F.1}$$

and there exists a constant $c > 0$ such that for all $T \geq 1$,

$$T \left\|(u_0, u_1)\right\|^2_{L^2(\Omega)\times H^{-1}(\Omega)} \leq c \int_0^T \int_\Omega \left|u(x,t)\right|^2 dx \ . \qquad \text{(F.2)}$$

Next, let $\beta \in (0,1)$, $k \in \mathbb{N}^*$, and choose $N \in \mathbb{N}^*$ such that $0 < \beta + \frac{1}{2N} < 1$ and $2N > k$. Put $\gamma = 1 - \frac{1}{2N}$. For any $\lambda \geq 1$, the function $F_\lambda(z) = \frac{1}{2\pi}\int_{\mathbb{R}} e^{iz\tau} e^{-\left(\frac{\tau}{\lambda^\gamma}\right)^{2N}} d\tau$ is holomorphic in $\mathbb{C}$, and there exists four positive constants C_o, c_0, c_1 and c_2 (independent of λ) such that

$$\begin{cases} \forall z \in \mathbb{C} \quad \left|F_\lambda(z)\right| \leq C_o \lambda^\gamma e^{c_0 \lambda \left|\mathrm{Im} z\right|^{1/\gamma}} , \\ \left|\mathrm{Im} z\right| \leq c_2 \left|\mathrm{Re} z\right| \Rightarrow \left|F_\lambda(z)\right| \leq C_o \lambda^\gamma e^{-c_1 \lambda \left|\mathrm{Re} z\right|^{1/\gamma}} , \end{cases} \qquad \text{(F.3)}$$

(see [LR]).

Now, let $s, \ell_o \in \mathbb{R}$, we introduce the following Fourier-Bros-Iagolnitzer transformation in [LR]:

$$W_{\ell_o,\lambda}(x,s) = \int_{\mathbb{R}} F_\lambda(\ell_o + is - \ell)\Phi(\ell)u(x,\ell)d\ell \ , \qquad \text{(F.4)}$$

where $\Phi \in C_0^\infty(\mathbb{R})$. As u is solution of the wave equation, $W_{\ell_o,\lambda}$ satisfies:

$$\begin{cases} \partial_s^2 W_{\ell_o,\lambda}(x,s) + \Delta W_{\ell_o,\lambda}(x,s) \\ \quad = \int_{\mathbb{R}} -F_\lambda(\ell_o + is - \ell)\left[\Phi''(\ell)u(x,\ell) + 2\Phi'(\ell)\partial_t u(x,\ell)\right] d\ell \ , \\ W_{\ell_o,\lambda}(x,s) = 0 \quad \text{for } x \in \partial\Omega \ , \\ W_{\ell_o,\lambda}(x,0) = \left(F_\lambda * \Phi u(x,\cdot)\right)(\ell_o) \quad \text{for } x \in \Omega \ . \end{cases} \qquad \text{(F.5)}$$

On the other hand, we also have for any $T > 0$,

$$\begin{aligned} \left\|\Phi u(x,\cdot)\right\|_{L^2\left(\left(\frac{T}{2}-1,\frac{T}{2}+1\right)\right)} &\leq \left\|\Phi u(x,\cdot) - F_\lambda * \Phi u(x,\cdot)\right\|_{L^2\left(\left(\frac{T}{2}-1,\frac{T}{2}+1\right)\right)} \\ &\quad + \left\|F_\lambda * \Phi u(x,\cdot)\right\|_{L^2\left(\left(\frac{T}{2}-1,\frac{T}{2}+1\right)\right)} \\ &\leq \left\|\Phi u(x,\cdot) - F_\lambda * \Phi u(x,\cdot)\right\|_{L^2(\mathbb{R})} \\ &\quad + \left(\int_{t\in\left(\frac{T}{2}-1,\frac{T}{2}+1\right)} \left|W_{t,\lambda}(x,0)\right|^2 dt\right)^{1/2} . \end{aligned} \qquad \text{(F.6)}$$

Denoting $\mathcal{F}(f)$ the Fourier transform of f, by using Parseval equality and $\mathcal{F}(F_\lambda)(\tau) = e^{-\left(\frac{\tau}{\lambda^\gamma}\right)^{2N}}$, one obtains

$$\begin{aligned} &\left\|\Phi u(x,\cdot) - F_\lambda * \Phi u(x,\cdot)\right\|_{L^2(\mathbb{R})} \\ &= \frac{1}{\sqrt{2\pi}} \left\|\mathcal{F}\left(\Phi u(x,\cdot) - F_\lambda * \Phi u(x,\cdot)\right)\right\|_{L^2(\mathbb{R})} \\ &= \frac{1}{\sqrt{2\pi}} \left(\int_{\mathbb{R}} \left|\left(1 - e^{-\left(\frac{\tau}{\lambda^\gamma}\right)^{2N}}\right) \mathcal{F}\left(\Phi u(x,\cdot)\right)(\tau)\right|^2 d\tau\right)^{1/2} \\ &\leq C \left(\int_{\mathbb{R}} \left|\left(\frac{\tau}{\lambda^\gamma}\right)^k \mathcal{F}\left(\Phi u(x,\cdot)\right)(\tau)\right|^2 d\tau\right)^{1/2} \quad \text{because } k < 2N \\ &\leq C \frac{1}{\lambda^{\beta k}} \left(\int_{\mathbb{R}} \left|\mathcal{F}\left(\partial_t^k\left(\Phi u(x,\cdot)\right)\right)(\tau)\right|^2 d\tau\right)^{1/2} \quad \text{because } \beta < \gamma \\ &\leq C \frac{1}{\lambda^{\beta k}} \left\|\partial_t^k\left(\Phi u(x,\cdot)\right)\right\|_{L^2(\mathbb{R})} . \end{aligned} \qquad \text{(F.7)}$$

Therefore, from (F.6) and (F.7), one gets

$$\begin{array}{l} \left\| \Phi u\left(x,\cdot\right)\right\|_{L^2\left(\left(\frac{T}{2}-1,\frac{T}{2}+1\right)\right)} \\ \leq C\frac{1}{\lambda^{\beta k}}\left\|\partial_t^k\left(\Phi u(x,\cdot)\right)\right\|_{L^2(\mathbb{R})} + \left(\int_{t\in\left(\frac{T}{2}-1,\frac{T}{2}+1\right)}\left|W_{t,\lambda}(x,0)\right|^2 dt\right)^{1/2} . \end{array} \tag{F.8}$$

Now, recall that from the Cauchy's theorem we have:

Proposition 1. *Let f be a holomorphic function in a domain $D\subset\mathbb{C}$. Let $a,b>0$, $z\in\mathbb{C}$. We suppose that*

$$D_o = \left\{(x,y)\in\mathbb{R}^2 \simeq \mathbb{C}\setminus\left|x-\mathrm{Re}z\right|\leq a,\ \ \left|y-\mathrm{Im}z\right|\leq b\right\}\subset D ,$$

then

$$f\left(z\right) = \frac{1}{\pi ab}\int\int_{\left|\frac{x-\mathrm{Re}z}{a}\right|^2+\left|\frac{y-\mathrm{Im}z}{b}\right|^2\leq 1} f\left(x+iy\right)dxdy .$$

Choosing $z=t\in\left(\frac{T}{2}-1,\frac{T}{2}+1\right)\subset\mathbb{R}$ and $x+iy=\ell_o+is$, we deduce that

$$\begin{array}{rl} \left|W_{t,\lambda}(x,0)\right| & \leq \frac{1}{\pi ab}\int_{\left|\ell_o-t\right|\leq a}\int_{|s|\leq b}\left|W_{\ell_o+is,\lambda}(x,0)\right|d\ell_o ds \\ & \leq \frac{1}{\pi ab}\int_{\left|\ell_o-t\right|\leq a}\int_{|s|\leq b}\left|W_{\ell_o,\lambda}(x,s)\right|dsd\ell_o \\ & \leq \frac{2}{\pi\sqrt{ab}}\left(\int_{\left|\ell_o-t\right|\leq a}\int_{|s|\leq b}\left|W_{\ell_o,\lambda}(x,s)\right|^2 dsd\ell_o\right)^{1/2} , \end{array} \tag{F.9}$$

and with $a=2b=1$,

$$\begin{array}{l} \int_{t\in\left(\frac{T}{2}-1,\frac{T}{2}+1\right)}\left|W_{t,\lambda}(x,0)\right|^2 dt \\ \leq \int_{t\in\left(\frac{T}{2}-1,\frac{T}{2}+1\right)}\left(\int_{\left|\ell_o-t\right|\leq 1}\int_{|s|\leq 1/2}\left|W_{\ell_o,\lambda}(x,s)\right|^2 dsd\ell_o\right)dt \\ \leq \int_{t\in\left(\frac{T}{2}-1,\frac{T}{2}+1\right)}\int_{\ell_o\in\left(\frac{T}{2}-2,\frac{T}{2}+2\right)}\int_{|s|\leq 1/2}\left|W_{\ell_o,\lambda}(x,s)\right|^2 dsd\ell_o dt \\ \leq 2\int_{\ell_o\in\left(\frac{T}{2}-2,\frac{T}{2}+2\right)}\int_{|s|\leq 1/2}\left|W_{\ell_o,\lambda}(x,s)\right|^2 dsd\ell_o . \end{array} \tag{F.10}$$

Consequently, from (F.8), (F.10) and integrating over Ω, we get the existence of $C>0$ such that

$$\begin{array}{l} \int_\Omega\int_{t\in\left(\frac{T}{2}-1,\frac{T}{2}+1\right)}\left|\Phi(t)u(x,t)\right|^2 dtdx \\ \leq C\frac{1}{\lambda^{2\beta k}}\int_\Omega\int_{\mathbb{R}}\left|\partial_t^k\left(\Phi\left(t\right)u(x,t)\right)\right|^2 dtdx \\ \quad +4\int_{\ell_o\in\left(\frac{T}{2}-2,\frac{T}{2}+2\right)}\left(\int_\Omega\int_{|s|\leq 1/2}\left|W_{\ell_o,\lambda}(x,s)\right|^2 dsdx\right)d\ell_o . \end{array} \tag{F.11}$$

Now recall the following quantification result for unique continuation of elliptic equation with Dirichlet boundary condition (Theorem E applied to $T_1=-1$, $T_2=1$, $\delta=1/2$, $\varphi\in C_0^\infty\left(\omega\times(-1,1)\right)$):

Proposition 2. *Let Ω be a bounded open set in $\mathbb{R}^n$, $n\geq 1$, either convex or C^2 connected. Let ω be a non-empty open subset in Ω. Let*

$f = f(x,s) \in L^2(\Omega \times (-1,1))$. *Then there exists* $\widetilde{c} > 0$ *such that for all* $w = w(x,s) \in H^2(\Omega \times (-1,1))$ *solution of*

$$\begin{cases} \partial_s^2 w + \Delta w = f & \text{in } \Omega \times (-1,1)\ , \\ w = 0 & \text{on } \partial\Omega \times (-1,1)\ , \end{cases}$$

for all $\varepsilon > 0$*, we have :*

$$\begin{array}{l} \int_{|s|\leq 1/2} \int_\Omega |w(x,s)|^2\, dxds \\ \leq \widetilde{c} e^{\widetilde{c}/\varepsilon} \left(\int_{|s|\leq 1} \int_\omega |w(x,s)|^2\, dxds + \int_{|s|\leq 1} \int_\Omega |f(x,s)|^2\, dxds \right) \\ \quad + e^{-4c_0/\varepsilon} \int_{|s|\leq 1} \int_\Omega |w(x,s)|^2\, dxds\ . \end{array}$$

Applying to $W_{\ell_0,\lambda}$, from (F.5) we deduce that for all $\varepsilon > 0$,

$$\begin{array}{l} \int_{|s|\leq 1/2} \int_\Omega |W_{\ell_0,\lambda}(x,s)|^2\, dxds \\ \leq e^{-4c_0/\varepsilon} \int_{|s|\leq 1} \int_\Omega |W_{\ell_0,\lambda}(x,s)|^2\, dxds \\ \quad + \widetilde{c} e^{\widetilde{c}/\varepsilon} \int_{|s|\leq 1} \int_\omega |W_{\ell_0,\lambda}(x,s)|^2\, dxds \\ \quad + \widetilde{c} e^{\widetilde{c}/\varepsilon} \int_{|s|\leq 1} \int_\Omega \Big| \int_{\mathbb{R}} -F_\lambda(\ell_0 + is - \ell) \\ \qquad\qquad [\Phi''(\ell) u(x,\ell) + 2\Phi'(\ell) \partial_t u(x,\ell)]\, d\ell|^2\, dxds\Big|^2\, dxds\ . \end{array} \tag{F.12}$$

Consequently, from (F.11) and (F.12), there exists a constant $C > 0$, such that for all $\varepsilon > 0$,

$$\begin{array}{l} \int_\Omega \int_{t\in\left(\frac{T}{2}-1,\frac{T}{2}+1\right)} |\Phi(t)u(x,t)|^2\, dtdx \\ \leq C \frac{1}{\lambda^{2\beta k}} \int_\Omega \int_{\mathbb{R}} \left|\partial_t^k (\Phi(t)\, u(x,t)))\right|^2 dtdx \\ \quad + 4e^{-4c_0/\varepsilon} \int_{\ell_o\in\left(\frac{T}{2}-2,\frac{T}{2}+2\right)} \left(\int_{|s|\leq 1} \int_\Omega |W_{\ell_0,\lambda}(x,s)|^2\, dxds \right) d\ell_o \\ \quad + 4Ce^{C/\varepsilon} \int_{\ell_o\in\left(\frac{T}{2}-2,\frac{T}{2}+2\right)} \left(\int_{|s|\leq 1} \int_\omega |W_{\ell_0,\lambda}(x,s)|^2\, dxds \right) d\ell_o \\ \quad + 4Ce^{\widetilde{c}/\varepsilon} \int_{\ell_o\in\left(\frac{T}{2}-2,\frac{T}{2}+2\right)} \Big(\int_{|s|\leq 1} \int_\Omega \Big| \int_{\mathbb{R}} -F_\lambda(\ell_0 + is - \ell) \\ \qquad\qquad [\Phi''(\ell)u(x,\ell) + 2\Phi'(\ell)\partial_t u(x,\ell)]\, d\ell|^2\, dxds \Big)\, d\ell_o\ . \end{array} \tag{F.13}$$

Let us define $\Phi \in C_0^\infty(\mathbb{R})$ more precisely now: we choose $\Phi \in C_0^\infty((0,T))$, $0 \leq \Phi \leq 1$, $\Phi \equiv 1$ on $\left(\frac{T}{4}, \frac{3T}{4}\right)$. Furthermore, let $K = \left[0, \frac{T}{4}\right] \cup \left[\frac{3T}{4}, T\right]$ such that $\mathrm{supp}(\Phi') = K$ and $\mathrm{supp}(\Phi'') \subset K$.

Let $K_0 = \left[\frac{3T}{8}, \frac{5T}{8}\right]$. Particularly, dist $(K, K_o) = \frac{T}{8}$. Let us define $T > 0$ more precisely now: we choose $T > 16\max(1, 1/c_2)$ in order that $\left(\frac{T}{2} - 2, \frac{T}{2} + 2\right) \subset K_0$ and $\mathrm{dist}(K, K_o) \geq \frac{2}{c_2}$.

Now, we will choose $\ell_0 \in \left(\frac{T}{2} - 2, \frac{T}{2} + 2\right) \subset K_0$ and $s \in [-1,1]$. Consequently, for any $\ell \in K$, $|\ell_0 - \ell| \geq \frac{2}{c_2} \geq \frac{1}{c_2}|s|$ and it will imply from the second line of (F.3) that

$$\forall \ell \in K \quad |F_\lambda(\ell_o + is - \ell)| \leq A\lambda^\gamma e^{-c_1\lambda\left(\frac{T}{8}\right)^{1/\gamma}}\ . \tag{F.14}$$

Till the end of the proof, C and C_T will denote a generic positive constant independent of ε and λ but dependent on Ω and respectively (Ω, T), whose value may change all along the line.

The first term on the right hand side of (F.13) becomes, using (F.1),

$$\frac{1}{\lambda^{2\beta k}} \int_\Omega \int_\mathbb{R} \left|\partial_t^k \left(\Phi\left(t\right) u(x,t)\right)\right|^2 dtdx \leq C_T \frac{1}{\lambda^{2\beta k}} \left\|(u_0,u_1)\right\|^2_{D(A^{k-1})} \ . \tag{F.15}$$

The second term on the right hand side of (F.13) becomes, using the first line of (F.3),

$$\begin{array}{l} e^{-4/\varepsilon} \int_{\ell_o \in \left(\frac{T}{2}-2,\frac{T}{2}+2\right)} \left(\int_{|s|\leq 1} \int_\Omega \left|W_{\ell_0,\lambda}(x,s)\right|^2 dxds\right) d\ell_o \\ \leq \left(C_o \lambda^\gamma e^{\lambda c_0}\right)^2 e^{-4c_0/\varepsilon} \int_{\ell_o \in \left(\frac{T}{2}-2,\frac{T}{2}+2\right)} \left[\int_{|s|\leq 1} \int_\Omega \left|\int_0^T \left|u(x,\ell)\right| d\ell\right|^2 dxds\right] \\ \leq C_T \lambda^{2\gamma} e^{2\lambda c_0} e^{-4c_0/\varepsilon} \left\|(u_0,u_1)\right\|^2_{H^1_0(\Omega)\times L^2(\Omega)} \ . \end{array} \tag{F.16}$$

The third term on the right hand side of (F.13) becomes, using the first line of (F.3),

$$\begin{array}{l} e^{C/\varepsilon} \int_{\ell_o \in \left(\frac{T}{2}-2,\frac{T}{2}+2\right)} \left(\int_{|s|\leq 1} \int_\omega \left|W_{\ell_0,\lambda}(x,s)\right|^2 dxds\right) d\ell_o \\ \leq \left(C_o \lambda^\gamma e^{\lambda c_0}\right)^2 e^{C/\varepsilon} \int_{\ell_o \in \left(\frac{T}{2}-2,\frac{T}{2}+2\right)} \left[\int_{|s|\leq 1} \int_\omega \left|\int_0^T \left|u(x,\ell)\right| d\ell\right|^2 dxds\right] d\ell_o \\ \leq C \lambda^{2\gamma} e^{C\lambda} e^{C/\varepsilon} \int_\omega \int_0^T \left|u(x,t)\right|^2 dtdx \ . \end{array} \tag{F.17}$$

The fourth term on the right hand side of (F.13) becomes, using (F.14) and the choice of Φ,

$$\begin{array}{l} e^{\widetilde{c}/\varepsilon} \int_{\ell_o \in \left(\frac{T}{2}-2,\frac{T}{2}+2\right)} \left(\int_{|s|\leq 1} \int_\Omega \left|\int_\mathbb{R} -F_\lambda(\ell_0 + is - \ell)\right.\right. \\ \qquad \left.\left.\left[\Phi''(\ell) u(x,\ell) + 2\Phi'(\ell)\partial_t u(x,\ell)\right] d\ell\right|^2 dxds\right) d\ell_o \\ \leq C \left(A\lambda^\gamma e^{-c_1 \lambda \left(\frac{T}{8}\right)^{1/\gamma}}\right)^2 e^{\widetilde{c}/\varepsilon} \int_\Omega \left|\int_K \left(\left|u(x,\ell)\right| + \left|\partial_t u(x,\ell)\right|\right) d\ell\right|^2 dx \\ \leq C \lambda^{2\gamma} e^{-2c_1 \lambda \left(\frac{T}{8}\right)^{1/\gamma}} e^{\widetilde{c}/\varepsilon} \left\|(u_0,u_1)\right\|^2_{H^1_0(\Omega)\times L^2(\Omega)} \ . \end{array} \tag{F.18}$$

We finally obtain from (F.15), (F.16), (F.17), (F.18) and (F.13) that

$$\begin{array}{l} \int_\Omega \int_{t \in \left(\frac{T}{2}-1,\frac{T}{2}+1\right)} \left|\Phi(t) u(x,t)\right|^2 dtdx \\ \leq C_T \frac{1}{\lambda^{2\beta k}} \left\|(u_0,u_1)\right\|^2_{D(A^{k-1})} \\ \quad + C_T \lambda^{2\gamma} e^{2\lambda c_0} e^{-4c_0/\varepsilon} \left\|(u_0,u_1)\right\|^2_{H^1_0(\Omega)\times L^2(\Omega)} \\ \quad + C \lambda^{2\gamma} e^{C\lambda} e^{C/\varepsilon} \int_\omega \int_0^T \left|u(x,t)\right|^2 dtdx \\ \quad + C \lambda^{2\gamma} e^{-2c_1 \lambda \left(\frac{T}{8}\right)^{1/\gamma}} e^{\widetilde{c}/\varepsilon} \left\|(u_0,u_1)\right\|^2_{H^1_0(\Omega)\times L^2(\Omega)} \ . \end{array} \tag{F.19}$$

We begin to choose $\lambda = \frac{1}{\varepsilon}$ in order that

$$\begin{aligned}
&\int_\Omega \int_{t\in(\frac{T}{2}-1,\frac{T}{2}+1)} |\Phi(t)u(x,t)|^2 \, dtdx \\
&\le \varepsilon^{2\beta k} C_T \|(u_0,u_1)\|^2_{D(A^{k-1})} \\
&\quad + e^{-2c_0/\varepsilon} \frac{1}{\varepsilon^{2\gamma}} C_T \|(u_0,u_1)\|^2_{H_0^1(\Omega)\times L^2(\Omega)} \\
&\quad + e^{C/\varepsilon} C \int_\omega \int_0^T |u(x,t)|^2 \, dtdx \\
&\quad + C\frac{1}{\varepsilon^{2\gamma}} \exp\left(\left(-2c_1\left(\frac{T}{8}\right)^{1/\gamma} + \widetilde{c}\right)\frac{1}{\varepsilon}\right) \|(u_0,u_1)\|^2_{H_0^1(\Omega)\times L^2(\Omega)} \ .
\end{aligned} \tag{F.20}$$

We finally need to choose $T > 16\max(1, 1/c_2)$ large enough such that $\left(-2c_1\left(\frac{T}{8}\right)^{1/\gamma} + \widetilde{c}\right) \le -1$ that is $8\left(\frac{1+\widetilde{c}}{2c_1}\right)^\gamma \le T$, to deduce the existence of $C > 0$ such that for any $\varepsilon \in (0,1)$,

$$\begin{aligned}
\int_\Omega \int_{t\in(\frac{T}{2}-1,\frac{T}{2}+1)} |u(x,t)|^2 \, dtdx &\le \int_\Omega \int_{t\in(\frac{T}{2}-1,\frac{T}{2}+1)} |\Phi(t)u(x,t)|^2 \, dtdx \\
&\le C\varepsilon^{2\beta k} \|(u_0,u_1)\|^2_{D(A^{k-1})} \\
&\quad + Ce^{C/\varepsilon} \int_\omega \int_0^T |u(x,t)|^2 \, dtdx \ .
\end{aligned} \tag{F.21}$$

Now we conclude from (F.2) that there exist a constant $c > 0$ and a time $T > 0$ large enough such that for all $\varepsilon > 0$ we have

$$\begin{aligned}
&\|(u_0,u_1)\|^2_{L^2(\Omega)\times H^{-1}(\Omega)} \\
&\le e^{c/\varepsilon} \int_\omega \int_0^T |u(x,t)|^2 \, dtdx + \varepsilon^{2\beta k} \|(u_0,u_1)\|^2_{D(A^{k-1})} \ .
\end{aligned} \tag{F.22}$$

Finally, we choose

$$\varepsilon = \left(\frac{\|(u_0,u_1)\|_{L^2(\Omega)\times H^{-1}(\Omega)}}{\|(u_0,u_1)\|_{D(A^{k-1})}}\right)^{1/(\beta k)} .$$

Theorem 1.1 is deduced by applying Theorem F to $\partial_t u$.

References

[AE] V. Adolfsson and L. Escauriaza, $C^{1,\alpha}$ domains and unique continuation at the boundary, Comm. Pure Appl. Math., 50 (1997), 935–969.

[BF] C. Bardos and M. Fink, Mathematical foundations of the time reversal mirror, Asymptotic Anal. 29 (2002), 157–182.

[BLR] C. Bardos, G. Lebeau and J. Rauch, Sharp sufficient conditions for the observation, control and stabilization of waves from the boundary, SIAM J. Control Optim. 30 (1992), 1024–1065.

[BuH] N. Burq and M. Hitrik, Energy decay for damped wave equations on partially rectangular domains, Math. Res. Lett. 14 (2007), 35–47.

[E] L. Escauriaza, Doubling property and linear combinations of eigenfunctions, manuscript.

[GaL] N. Garofalo and F H. Lin, Monotonicity properties of variational integrals, A_p-weights and unique continuation, Indiana Univ. Math. J. 35 (1986), 245–268.

[Ku] I. Kukavica, Level sets for the stationary Ginzburg-Landau equation, Calc. Var. 5 (1997), 511–521.

[Ku2] I. Kukavica, Quantitative uniqueness for second order elliptic operators, Duke Math. J. 91 (1998), 225–240.

[KN] I. Kukavica and K. Nyström, Unique continuation on the boundary for Dini domains, Proc. of Amer. Math. Soc. 126 (1998), 441–446.

[Le] G. Lebeau, Contrôle analytique I: estimations a priori, Duke Math. J. 68 (1992), 1–30.

[Li] J.-L. Lions, Contrôlabilité Exacte, Stabilisation et Perturbation des Systèmes Distribués, 1, collection R.M.A., vol. 8. Editions Masson, Paris, 1988.

[LiR] Z. Liu and B. Rao, Characterization of polynomial decay rate for the solution of linear evolution equation, Z. Angew. Math. Phys. 56 (2005), 630–644.

[LR] G. Lebeau and L. Robbiano, Stabilisation de l'équation des ondes par le bord, Duke Math. J. 86 (1997), 465–491.

[Ph1] K.-D. Phung, Polynomial decay rate for the dissipative wave equation, Journal of Differential Equations 240 (2007), 92–124.

[Ph2] K.-D. Phung, Boundary stabilization for the wave equation in a bounded cylindrical domain, Discrete and Continuous Dynamical Systems Series A, 20 (2008), 1057–1093.

[Ph3] K.-D. Phung, Observation et stabilisation d'ondes: géométrie et coût du contrôle, Habilitation à diriger des recherches, (2007).

[Ra] J. Ralston, Gaussian beams and propagation of singularities, in Studies in Partial Differential Equations, Littman éd., MAA studies in Mathematics, 23 (1982), 206–248.

[Ro] L. Robbiano, Théorème d'unicité adapté au contrôle des solutions des problèmes hyperboliques, Comm. Part. Diff. Eq. 16 (1991), 789–800.

[Ro2] L. Robbiano, Fonction de coût et contrôle des solutions des équations hyperboliques, Asymptotic Anal. 10 (1995), 95–115.

[Ru] D. Russel, Exact boundary value controllability theorems for wave and heat processes in star-complemented regions, in Differential Games and Control Theory, Roxin, Liu, Sternberg, eds., Marcel Dekker, New York, 1974.

[Ze] S. Zelditch, Local and global analysis of eigenfunctions on Riemannian manifolds, arXiv:09033420.

[Zu] E. Zuazua, Controllability and Observability of Partial Differential Equations: Some results and open problems, in Handbook of Differential Equations: Evolutionary Differential Equations, vol. 3, C. M. Dafermos and E. Feireisl eds., Elsevier science, Amsterdam, 2006, 527–621.

Control Problems for Fluid Equations

Jean-Pierre Puel
Laboratoire de Mathématiques de Versailles
Université de Versailles Saint-Quentin
45 avenue des Etats Unis, 78035 Versailles, France
Email: jppuel@math.uvsq.fr

Abstract

Controllability problems for fluid equations are an important topic and have been the object of a lot of studies in recent years. We present here, in a first part, the resonable objectives which can be attempted, namely, the exact controllability to trajectories, then the results and methods which have been used in the recent work of the incompressible fluid systems. In the last sections, we discuss the possibility of global controllability and we pose some open problems for compressible fluid systems as well as for lagrangian control.

1 Introduction

In the late 1980's, J.-L. Lions gave a systematic study of controllability problems in [19] and, among others, raised the issue of "controllability" of fluid flows.

The relevant notion of "controllability" was not very precise at that time.

For Euler equations, J.-L. Lions showed that the linearized problem around zero was not controllable in any sense as the curl of the solution had to be constant in time.

For Navier-Stokes equations there is no hope for obtaining "exact controllability" because of dissipativity and irreversability of the problem. J.-L. Lions was clearly expecting "approximate controllability". But this notion is not really relevant here : even if we have approximate controllability at time T, then what should we do after time T to preserve the neighborhood of a target? Anyway this could be an important step in understanding how we can act on the Navier-Stokes system. Anyway, for classical boundary conditions, this is still an open problem.

In the early 1990's, A. Fursikov and O. Imanuvilov showed that the viscous Burger's equation was not approximately controllable (see [7] for a general monograph and the references therein). One essential reason why Burger's equation is not controllable is that it is too much stable, whereas Navier-Stokes equations are much less stable.

A basic philosophy claimed by J.-L. Lions was the following: the more unstable a system is, the more controllable it will be.

So what could be the situation for incompressible fluids? This was an exciting and important challenge $\cdots$ and a few years after came some very interesting positive answers.

In 1996, J.-M. Coron, in [3], proved exact controllability for Euler's equations in 2-d. This result was extended to the 3-d case by O. Glass in [8]. J.-M. Coron used successfully his return method which consists in linearizing the system around a special nonzero trajectory going from zero to zero (around which the linearized problem is controllable), then using an implict function theorem to obtain a local result, and finishing by a scaling argument to obtain a global result.

He then used this result in [4] to prove approximate controllability for Navier-Stokes equations (again in 2-d) with Navier boundary conditions (which include a term on the curl on the boundary). With these boundary conditons, one can prove convergence of the Navier-Stokes sytem toward the Euler system when the viscosity tends to zero and J.-M. Coron used this fact plus a scaling argument to obtain his result.

About the same time, L. Robbiano and G. Lebeau ([18]), A. Fursikov and O. Imanuvilov ([7]), by different methods, proved "null controllability" for the heat equation.

A. Fursikov and O.Imanuvilov extended their method to the Navier-Stokes system with boundary conditions on the *curl* of the solution, proving that one can reach in finite time any stationnary solution (for example), even an unstable one, if the initial condition is not too far from this stationnary solution.

This was a real breakthrough with a number of extensions later on. It contained the idea of exact controllability to trajectories: even if you cannot reach any point in the state space, you can reach (in finite time) any point on the trajectories of the same operator.

In 1998, O. Imanuvilov proved, in a very important article [13], a result of local exact controllability to trajectories for Navier-Stokes equations with classical Dirichlet boundary conditions, under rather strong regularity assumptions. This paper was improved in [14].

A lot of work has been done since then to improve and extend his result. Recently, in articles written in collaboration with O. Imanuvilov and J.-P. Puel ([15]) and E. Fernandez-Cara, S. Guerrero, O. Imanuvilov and J.-P. Puel ([6]), we have given a rather strong extension of Imanuvilov's first result with systematic proofs of each step of the argument, with some of the results being now optimal.

We will present here the notion of exact controllability to trajectories, the (local) results obtained for Navier-Stokes equations (and Boussinesq equations) with Dirichlet boundary conditions and some ideas of the methods used for proving these results. Then we will give some global controllability results of a simplfied model and we will discuss some important pen problems.

2 Exact controllability to trajectories

2.1 Abstract setting

We consider a nonlinear evolution system with a control variable v

$$\begin{cases} \frac{\partial Y}{\partial t} + LY + N(Y) = F + Bv \text{ in } (0,T), \\ Y(0) = Y_0. \end{cases} \tag{2.1}$$

L is for example an elliptic operator and N is a nonlinear perturbation.

We can think of a nonlinear convection-diffusion equation or Navier-Stokes equations or other related examples.

On the other hand, we have an uncontrolled trajectory of the same operator, which we call the "ideal" trajectory we want to reach

$$\begin{cases} \frac{\partial \bar{Y}}{\partial t} + L\bar{Y} + N(\bar{Y}) = F \text{ in } (0,T), \\ \bar{Y}(0) = \bar{Y}_0. \end{cases} \tag{2.2}$$

Exact controllability to trajectories is the following question: can we find a control v such that

$$Y(T) = \bar{Y}(T)?$$

(In the linear case, taking the difference between the two systems, we can speak of null controllability: we look for v such that $Y(T) = 0$.)

A local version of this question is the following: provided $(Y_0 - \bar{Y}_0)$ is "small" in a suitable norm, can we find a control v such that

$$Y(T) = \bar{Y}(T)?$$

Remark: If the answer is positive and if our evolution system is well posed, after time T we can switch off the control and the system will follow the "ideal" trajectory.

An important case is the case where $\bar{Y}$ is a stationnary solution (with F independent of time t), namely,

$$L\bar{Y} + N(\bar{Y}) = F. \tag{2.3}$$

Many important nonlinear stationnary systems of this type may have several solutions, unstable solutions in particular. In this case, if $\bar{Y}$ is such an unstable solution and if the problem of exact controllability to trajectories has a positive answer, it corresponds to stabilizing (and exactly reaching) an unstable solution.

2.2 Case of Navier-Stokes equations

What follows is also valid for Boussinesq equations and related coupled systems.

Let $(\bar{\mathbf{y}}, \bar{p})$ be a fixed "ideal" solution of Navier-Stokes equations, for example, a stationnary solution.

$$\begin{cases} \frac{\partial \bar{\mathbf{y}}}{\partial t} - \nu\Delta\bar{\mathbf{y}} + \bar{\mathbf{y}}.\nabla\bar{\mathbf{y}} + \nabla\bar{p} = \mathbf{f} \text{ in } \Omega \times (0,T), \\ \operatorname{div}\bar{\mathbf{y}} = 0 \text{ in } \Omega \times (0,T), \\ \bar{\mathbf{y}} = 0 \text{ on } \Gamma \times (0,T) \\ \bar{\mathbf{y}}(0) = \bar{\mathbf{y}}_0 \text{ in } \Omega. \end{cases} \tag{2.4}$$

Let us consider a solution of the controlled system, starting from a different initial value

$$\begin{cases} \frac{\partial \mathbf{y}}{\partial t} - \nu\Delta\mathbf{y} + \mathbf{y}.\nabla\mathbf{y} + \nabla p = \mathbf{f} + \mathbf{v}.1\!\!1_\omega \text{ in } \Omega \times (0,T), \\ \operatorname{div}\mathbf{y} = 0 \text{ in } \Omega \times (0,T), \\ \mathbf{y} = 0 \quad \text{on} \quad \Gamma \times (0,T) \\ \mathbf{y}(0) = \mathbf{y}_0 \text{ in } \Omega, \end{cases} \tag{2.5}$$

where $1\!\!1_\omega$ is the characteristic function of a (little) subset ω of Ω.

Exact Controllability to Trajectories for this sytem is the following question: Can we find a control $\mathbf{v}$ such that

$$\mathbf{y}(T) = \bar{\mathbf{y}}(T)?$$

i.e. can we reach exactly in finite time the "ideal" trajectory $\bar{\mathbf{y}}$?

The local version is the same question, provided $||\mathbf{y}_0 - \bar{\mathbf{y}}_0||$ is small enough.

Remark 2.1. If there exists such a control $\mathbf{v}$, then, after time T, just switch off the control ($\mathbf{v} = 0$) and the system can stay (or will stay depending on the uniqueness problem) on the "ideal" trajectory.

2.3 Linearization

In the abstract setting, if we substract the equation for $\bar{Y}$ from the equation for Y and then linearize the control problem around the trajectory $\bar{Y}$, we obtain (still writing Y for the solution and Y_0 for the difference $Y_0 - \bar{Y}_0$) a linear system of the form

$$\begin{cases} \frac{\partial Y}{\partial t} + LY = Bv \text{ in } (0,T), \\ Y(0) = Y_0. \end{cases} \tag{2.6}$$

Now we look for v such that $Y(T) = 0$.

First step: let us consider a parameter $\epsilon > 0$ and the following functional

$$J_\epsilon(v) = \frac{1}{2\epsilon}||Y(T)||^2 + \frac{1}{2}\int_0^T ||v||^2 dt.$$

From standard optimal control arguments, we know that there exists a unique v_ϵ such that

$$J_\epsilon(v_\epsilon) = \min_v J_\epsilon(v)$$

and v_ϵ is characterized by the following optimality system ($Y_\epsilon = Y(v_\epsilon)$)

$$\begin{cases} \frac{\partial Y_\epsilon}{\partial t} + LY_\epsilon = Bv_\epsilon \ \text{ in } (0,T), \\ Y_\epsilon(0) = Y_0, \\ \\ -\frac{\partial \Phi}{\partial t} + L^*\Phi = 0 \ \text{ in } (0,T), \\ \Phi(T) = \frac{1}{\epsilon} Y_\epsilon(T), \\ \\ B^*\Phi + v_\epsilon = 0. \end{cases} \tag{2.7}$$

Multiplying the first equation by Φ we obtain immediately

$$\frac{1}{\epsilon}||Y_\epsilon(T)||^2 + \int_0^T ||B^*\Phi(t)||^2 dt = (Y_0, \Phi(0)).$$

Second step: estimates.

Let us assume that we know an Observability Inequality for solutions of adjoint equation of the following type:

$$||\Phi(0)||^2 \leq C \int_0^T ||B^*\Phi(t)||^2 dt.$$

Then we obtain an estimate on the control

$$\int_0^T ||v_\epsilon||^2 dt = \int_0^T ||B^*\Phi^\epsilon||^2 dt \leq C||Y_0||^2.$$

We also obtain

$$\frac{1}{\epsilon}||Y_\epsilon(T)||^2 \leq C||Y_0||^2.$$

Third step: Passage to the limit when $\epsilon \to 0$.

After extraction of or a subsequence v_ϵ converges weakly to v, and therefore

$Y(v_\epsilon)(T) \to Y(v)(T)$ and $||Y(v_\epsilon)(T)|| \leq \sqrt{\epsilon C}||Y_0||$, which implies

$$Y(v)(T) = 0.$$

We have then solved the null controllability problem and obtained some additional properties $\cdots$

The problem is now to know how to obtain the Observability Inequality for adjoint system!!

This is the most difficult part of the argument and has to be done for each specific sytem we are considering.

Let us then consider the linearized controlled Navier-Stokes equations around the trajectory $\bar{y}$ which can be written as follows:

$$\begin{cases} \frac{\partial \mathbf{y}}{\partial t} - \nu\Delta\mathbf{y} + \nabla\cdot(\bar{\mathbf{y}}\otimes\mathbf{y} + \mathbf{y}\otimes\bar{\mathbf{y}}) + \nabla p = \mathbf{v}.\mathbb{1}_\omega \\ \qquad\qquad\qquad\qquad\qquad\qquad \text{in } \Omega\times(0,T), \\ \operatorname{div}\mathbf{y} = 0 \quad \text{in} \quad \Omega\times(0,T), \\ \mathbf{y} = 0 \quad \text{on} \quad \Gamma\times(0,T) \\ \mathbf{y}(0) = \mathbf{y}_0 \quad \text{in} \quad \Omega. \end{cases} \tag{2.8}$$

The real adjoint system can be replaced by the following pseudo-adjoint system (which is a backward equation) by changing the pressure term:

$$\begin{cases} -\frac{\partial\varphi}{\partial t} - \nu\Delta\varphi - \bar{\mathbf{y}}\cdot D(\varphi) + \nabla\pi = 0 \text{ in } \Omega\times(0,T), \\ \operatorname{div}\varphi = 0 \quad \text{in} \quad \Omega\times(0,T), \\ \varphi = 0 \quad \text{on} \quad \Gamma\times(0,T) \\ \varphi(T) = \varphi_0 \quad \text{in} \quad \Omega, \end{cases} \tag{2.9}$$

where $D\varphi = \nabla\varphi + \nabla\varphi^T$.

We want to show the Observability Inequality

$$|\varphi(0)|^2_{L^2(\Omega)} \le C\int_0^T\int_\omega |\varphi|^2 dxdt \tag{2.10}$$

(there is no reference to the "initial" value φ_0 in this inequality).

It turns out that to prove this observability inequality requires the proof of a Global Carleman estimate (which is the difficult part) plus standard energy estimates.

Let us describe how we can obtain the desired Carleman estimate. This requires several difficult steps. First of all we have to define some weights involved in the Carleman estimate.

For later technical reasons, let us consider two non-empty open sets $\omega_2 \subset\subset \omega_1 \subset\subset \omega$. We know that there exists $\eta^0 \in C^2(\overline{\Omega})$ such that

$$\eta^0 > 0 \text{ in } \Omega,\ \eta^0 = 0 \text{ on } \partial\Omega,\ |\nabla\eta^0| > 0 \text{ in } \overline{\Omega}\setminus\omega_2.$$

From this function η^0 and for $s,\lambda \ge 1$ and $m > 4$, we construct the following weight functions:

$$\begin{cases} \alpha(x,t) = \dfrac{e^{(5/4)\lambda m\|\eta^0\|_\infty} - e^{\lambda(m\|\eta^0\|_\infty + \eta^0(x))}}{t^4(T-t)^4}, \\ \xi(x,t) = \dfrac{e^{\lambda(m\|\eta^0\|_\infty + \eta^0(x))}}{t^4(T-t)^4}. \end{cases} \tag{2.11}$$

These precise weights were considered in [6].

Let us introduce a notation corresponding to weighted Sobolev norms

$$I(s,\lambda;\varphi) = s^{-1}\iint_Q e^{-2s\alpha}\xi^{-1}|\varphi_t|^2\,dx\,dt + s^{-1}\iint_Q e^{-2s\alpha}\xi^{-1}|\Delta\varphi|^2\,dx\,dt$$
$$+ s\lambda^2\iint_Q e^{-2s\alpha}\xi|\nabla\varphi|^2\,dx\,dt + s^3\lambda^4\iint_Q e^{-2s\alpha}\xi^3|\varphi|^2\,dx\,dt.$$

Using the results of Fursikov-Imanuvilov [7] of the heat equation, the result of Imanuvilov-Puel [15] of general non homogeneous elliptic equation (for the pressure here) and regularity results of Stokes system ([20]) we can obtain the following Carleman estimates.

Proposition 2.2. *There exist three positive constants* C_0*,* s_0 *and* λ_0 *depending on* Ω *and* ω_1 *such that, for every* $\varphi^0 \in H$*, the solution* (φ,π) *satisfies,*

$$\left\{\begin{array}{l}\displaystyle\iint_Q e^{-2s\alpha}|\nabla\pi|^2\,dx\,dt + s^2\lambda^2\iint_Q e^{-2s\alpha}\xi^2|\pi|^2\,dx\,dt + I(s,\lambda;\varphi)\\ \displaystyle\qquad\le C_0\left(s^3\lambda^4\iint_{\omega_1\times(0,T)} e^{-2s\alpha}\xi^3\,|\varphi|^2\,dx\,dt\right.\\ \displaystyle\qquad\qquad\left.+\, s^2\lambda^2\iint_{\omega_1\times(0,T)} e^{-2s\alpha}\xi^2\,|\pi|^2\,dx\,dt\right)\end{array}\right. \tag{2.12}$$

This gives us particularly

$$s^3\lambda^4\iint_Q e^{-2s\alpha}\xi^3|\varphi|^2\,dx\,dt \le C_0\left(s^3\lambda^4\iint_{\omega_1\times(0,T)} e^{-2s\alpha}\xi^3\,|\varphi|^2\,dx\,dt\right.$$
$$\left.+\, s^2\lambda^2\iint_{\omega_1\times(0,T)} e^{-2s\alpha}\xi^2\,|\pi|^2\,dx\,dt\right).$$

We have to notice that on the right hand side we have two local terms: one on the velocity (which is a good term) and the other in the pressure.

The method used in [6] (dimension 3) consists in getting rid of the local term on the pressure and this gives rise to a long and difficult argument. Moreover, if we consider another related system (like Boussinesq system or other coupling with diffusion-convectin equations), we need to write this argument again for each separate system.

An alternative method has been given by M. Gonzales-Burgos, S. Guerrero, J.-P. Puel in [10]. This method turns out to be quite interesting if we have a coupling with other equations (like diffusion convection equations, for example, for Boussinesq system, oceans models, etc.).

This is also valid for boundary control on a part Γ_0 of the boundary. In this case, we extend the domain Ω by $\tilde{\Omega}$ such that $\Gamma - \Gamma_0 \in \partial\tilde{\Omega}$ and we take a subset $\tilde{\omega} \in \tilde{\Omega} - \Omega$. We are then in the context of a distributed control.

The method relies on the consideration of another control which is fictitious and should be removed in a further step. We will take two controls in ω: one on the right hand side and the other on the divergence!!.

Let us take a nonempty open set ω_1 with $\overline{\omega}_1 \subset \omega$ and let us consider $\zeta \in C_0^\infty(\mathbb{R}^N)$ such that

$$0 \leq \zeta(x) \leq 1,\ \forall x \in \mathbb{R}^N,\ \zeta(x) = 1,\ \forall x \in \omega_1,\ \operatorname{Supp}\zeta \subset \omega.$$

For linearized Navier-Stokes equations around $\bar{y}$ we consider the following control problem

$$\begin{cases} \frac{\partial \mathbf{y}}{\partial t} - \nu\Delta\mathbf{y} + \nabla\cdot(\bar{\mathbf{y}}\otimes\mathbf{y} + \mathbf{y}\otimes\bar{\mathbf{y}}) + \nabla p = \mathbf{v}.1\!\!1_\omega \\ \qquad\qquad\qquad\qquad\qquad\qquad \text{in } \Omega\times(0,T), \\ \operatorname{div}\mathbf{y} = h\,\zeta \quad \text{in} \quad \Omega\times(0,T), \\ \mathbf{y} = 0 \quad \text{on} \quad \Gamma\times(0,T) \\ \mathbf{y}(0) = \mathbf{y}_0 \quad \text{in} \quad \Omega. \end{cases} \tag{2.13}$$

Using similar arguments as the ones presented above, it is immediately shown that the controllability problem for this sytem corresponds to the following Observability Inequality on the adjoint system

$$|\varphi(0)|^2_{L^2(\Omega)} \leq C\Big(\int_0^T\int_{\omega_1}|\varphi|^2dxdt + \int_0^T\int_{\omega_1}|\pi|^2dxdt\Big). \tag{2.14}$$

We have to notice that we allow here the presence of a local term in the pressure on the right hand side.

From (2.13) together with standard energy estimate we immediately obtain our observability inequality (2.14) and therefore our null controllability property with two controls.

Moreover, Carleman estimate (2.12) tells us that we can choose a control which is more regular than expected:

The control v can be taken bounded with additional integrability properties.

The fictitious control $h.\zeta$ can be shown to be regular in space and time.

Then we can lift this control $h.\zeta$ by a regular function (in space and time) Z such that

$$\operatorname{div} Z = h.\zeta$$

and Z has support contained in ω. By a simple translation, adding some function of Z to the control v (which stays supported in ω), we can get rid of the fictitious control h.

This shows the null controllability for our linearized Navier-Stokes system (with only the v control).

The next step is then to apply a fixed point theorem and we can use here Kakutani fixed point theorem to obtain a local result of exact controllability to trajectories which is analogous to the one obtained in [6].

Theorem 2.3. *If $\bar{y} \in L^\infty$ and $\bar{y}_t \in L^2(0,T;L^q(\Omega))$ with $q > N/3$ (and solution to the Navier-Stokes system) with $\bar{y}_0 \in W^{s,p}(\Omega)$ with s,p suitably chosen, then there exists $\delta > 0$ such that if $||y_0 - \bar{y}_0||_{W^{s,p}} \leq \delta$, there exists a control $v \in L^2(Q)^N$ such that $y(T) = \bar{y}(T)$.*

3 Global controllability

The previous result is a local result of exact controllability to trajectories. It is then natural to ask whether this result could be extended to a global result. For the Navier-Stokes system with Dirichlet boundary conditions (at least on a part of the boundary) the problem is completely open and there is no strong conjecture on it. When the control acts on the whole boundary, the type of boundary conditions is of course at our choice and combining the results of Coron (approximate controllability) and Fursikov-Imanuvilov (on local controllability) one can obtain a global controllability result ([5]), but this is no longer valid if the control does not act on the whole boundary.

It is then interesting to consider simpler models to try to understand the situation.

For 1-d viscous Burgers equation in an interval, O. Imanuvilov and S. Guerrero have given in [11] a counter-example, even in the case where the control acts on both extremities of the interval.

For the 2-d Burgers equation, the global (boundary) controllability question has been studied by O. Imanuvilov and J.-P. Puel in [16]. We have obtained positive results and counter examples depending on the geometry of the region where the control acts. More precisely, let us consider the following 2-d Burgers equation

$$\frac{\partial u}{\partial t} - \Delta u + \frac{\partial u^2}{\partial x_1} + \frac{\partial u^2}{\partial x_2} = f \quad \text{in } Q = (0,T) \times \Omega, \tag{3.1}$$

$$u|_{\Gamma_0} = 0, \quad u|_{\Gamma_1} = h, \tag{3.2}$$

$$u(0,\cdot) = u_0, \tag{3.3}$$

$$u(T,\cdot) = 0. \tag{3.4}$$

The control h acts on the part Γ_1 of the boundary and we take the homogeneous Dirichlet boundary conditions on the complementary part of the boundary. Without loss of generality we may assume that Ω is included in the rectangle $0 \leq x_1 - x_2 \leq A, \quad -B \leq x_1 + x_2 \leq B$ with A and B being two positive constants.

We obtain the following positive result of global controllability to zero

Theorem 3.1. *Let us assume that*

$$\Gamma_0 \subset \{x \in \Gamma \mid x_1 - x_2 = 0\} \tag{3.5}$$

(or Γ_0 is empty which is allowed). Suppose that $f \in L^2(0,T; L^2(\Omega))$ and that there exists $T_0 \in (0,T)$ such that $f(t,x) = 0, \forall t \geq T_0$.

Then for every $u_0 \in L^2(\Omega)$ there exists a solution $u \in L^2(0,T; H_0^1(\Omega)) \cap C([0,T]; L^2(\Omega))$ such that $t^2.u \in H^{1,2}(Q) = H^1(0,T; L^2(\Omega)) \cap L^2(0,T; H^2(\Omega) \cap H_0^1(\Omega))$ to problems (3.1)–(3.4) (and a corresponding control h).

The proof is related to Coron's return method but different from it. We make use of a special solution U of Burgers equation such that NU is again a solution of the Burgers equation and we can drive this solution to zero by an action on Γ_1 in arbitrary small time whenever we want. We use the boundary value of NU on Γ_1 as control in a short interval of time and we prove that for N sufficiently large, the solution is driven in an ϵ-neighborhood of NU with ϵ as small as we wish. Then we use a control which drives NU to zero and the difference between the solution and NU is kept constant so that the solution is now close to zero. In the last step, we use a result of local controllability to achieve exactly zero.

4 Open problems

4.1 Compressible fluids

What is the situation for compressible viscous fluids, even for local controllability? This is a completely open question. The system to be considered is, for example,

$$\begin{cases} \frac{\partial \rho}{\partial t} + \operatorname{div}(\rho.\mathbf{y}) = 0, \\ \rho(\frac{\partial \mathbf{y}}{\partial t} + \mathbf{y}.\nabla\mathbf{y}) - \nu\Delta\mathbf{y} + \nabla p = \mathbf{f} + \mathbf{v}.1\!\mathrm{I}_\omega, \\ \mathbf{y} = 0 \quad \text{on} \quad \Gamma \times (0,T) \\ \mathbf{y}(0) = \mathbf{y}_0, \\ p = C\rho^\gamma \end{cases} \tag{4.1}$$

where ρ is the density of the fluid, $\mathbf{v}$ the velocity and p the pressure and γ gets as close as possible to 1.4. To our knowledge, nothing has been done for the controllability of such a system.

One possibility of approach could be the use of S. Klainerman and A. Majda's result ([17]): when Mach number tends to $+\infty$, at first order a compressible fluid behaves like an incompressible fluid on which you have propagation of acoustic waves. We could try to control this coupled system, but the first question would be: do we have to put the control on both equations or only on the incompressible motion?

4.2 Nonlinear control. Lagrangian control

This is a wide area with very interesting applications where nonlinear control problems arise.

The first question is the following:

Given two densities of mass ρ_0 and ρ_1 (for example, characteristic functions) does there exist a vector field u such that we have

$$\frac{\partial \rho}{\partial t} + u.\nabla\rho = 0, \text{ in } (0,T)\times\Omega \tag{4.2}$$

$$\rho(0) = \rho_0, \tag{4.3}$$

$$\rho(T) = \rho_1? \tag{4.4}$$

This can be formulated as a problem for transport of mass and it should be related to optimal transportation problems (Monge Kantorovitch problem) where we try to find a "mapping" transporting ρ_0 to ρ_1 (Brenier [2], Benamou-Brenier [1], Villani [21]$\cdots$) which has to be optimal for some "cost" function which behaves like a distance function. Here we don't look for any optimality in a first step but we would like to transport the masses by a flow or a generalized flow.

The second question could be: can u be a divergence free vector field? (of course with compatibility conditions on ρ_0 and ρ_1)

The third question is: can u be solution of Euler equations or Navier-Stokes equations with a control acting on a subdomain?

The fourth question is: can u be a motion of a compressible fluid?

Simpler problems in 1-d and 2-d have been considered by T.Horsin with the heat equation and Burgers equation with positive results (see, for example, [12]).

Very recent interesting results have been given by Glass and Horsin in [9] on approximate controllability for the transport of Jordan curves in 2-d by a vector field which is a velocity satisfying the Euler equations.

References

[1] J.-D. Benamou, Y. Brenier: A Computational Fluid Mechanics solution to the Monge-Kantorovich mass transfer problem, *Numer. Math.* **84**: 375–393, 2000.

[2] Y. Brenier: Extended Monge-Kantorovich theory, in *Optimal transportation and Applications*, Lecture Notes in Mathematics 1813, 91–121, Springer, 2003.

[3] J.-M. Coron: On the controllability of 2-D incompressible perfect fluids, *J. Math. Pures & Appliquées*, **75**: 155–188, 1996.

[4] J.-M. Coron: On the controllability of the 2-D incompressible Navier-Stokes equations with the Navier slip boundary conditions, *ESAIM: Control, Optimisation and Calculus of Variations, www.emath.fr/cocv/*, **1**: 35–75, 1996.

[5] J.-M. Coron, A. Fursikov: On the controllability of the 2-D incompressible Navier-Stokes equations on a manifold without boundary, *Russian J. of Math. Physics*, **4**(3): 1–20, 1996.

[6] E. Fernandez-Cara, S. Guerrero, O. Yu. Imanuvilov, J.-P. Puel: Local exact controllability of the Navier-Stokes system, *Journal de Math. Pures et Appl.* , **83**(12): 1501–1542, 2004.

[7] A. Fursikov, O. Imanuvilov: *Controllability of Evolution Equations,* Lecture Notes Series 34, RIM-GARC, Seoul National University, 1996.

[8] O. Glass: Exact boundary controllability of 3-D Euler equation, *ESAIM: Control, Optimisation and Calculus of Variations, www.emath.fr/cocv/*, **5**: 1–44, 2000.

[9] O.Glass, T.Horsin: Approximate Lagrangian controllability for the 2-D Euler equation. Application to the control of the shape of vortex patches, to appear in *J. Math. Pures et Appliquées.*

[10] M. Gonzàlez-Burgos, S. Guerrero, J.-P. Puel: Local Exact Controllability to the Trajectories of the Boussinesq System via a Fictitious Control on the Divergence Equation, *Communications in Pure and Applied Analysis*, **8**: 311–333, 2009.

[11] S. Guerrero , O. Yu. Imanuvilov: Remarks on global controllability for the Burgers equation with two control forces, Annales de l'Institut Henri Poincaré, *Analyse Non Linéaire*, **24**: 897–906, 2007.

[12] T. Horsin: Local exact Lagrangian controllability of the Burgers viscous equation, Annales de l'Institut Henri Poincaré, *Analyse non linaire*, **25**(2): 219–230, 2008.

[13] O. Imanuvilov: On exact controllability for the Navier-Stokes equations, *ESAIM: COCV*, www.esaim-cocv.org, **3**: 97–131, 1998.

[14] O. Imanuvilov: Remarks on exact controllability for Navier-Stokes equations, *ESAIM: COCV*, www.esaim-cocv.org, **6**: 39–72, 2001.

[15] O.Yu. Imanuvilov, J.-P. Puel: Global Carleman estimates for weak elliptic non homogeneous Dirichlet problem, *Int. Math. Research Notices*, **16**: 883–913, 2003.

[16] O.Yu. Imanuvilov, J.-P. Puel: On Global Controllability of the 2*d* Burgers Equation, Discete and Continuous Dynamical Systems, **23**(1/2): 299–313, 2009.

[17] S.Klainerman, A.Majda: Singular limits of quasilinear hyperbolic systems with large parameters and the incompressible limit of compressible fluids, *Comm. Pure Appl. Math.*, **34**(4): 481–524, 1981.

[18] G. Lebeau, L. Robbiano: Contrôle exact de l'équation de la chaleur, *Comm. Part. Diff. Eq.*, **20**: 335–356, 1995.

[19] J.-L. Lions: *Contrôlabilité exacte et stabilisation de systèmes distribués*, Vol. 1, Masson, Paris, 1988.

[20] R. Temam: *Navier-Stokes equations. Theory and numerical analysis. Studies in Mathematics and its applications,* **2**; *North Holland Publishing Co., Amsterdam-New York-Oxford, 1977.*

[21] C. Villani: Topics in Optimal Transportation. Graduate Studies in Mathematics 58, American Mathematical Society, Providence, 2003

Global Existence of a Degenerate Kirchhoff System with Boundary Dissipation*

Qiong Zhang
Department of Mathematics
Beijing Institute of Technology
Beijing, 100081, China
Email: qiongzhg@gmail.com, zhangqiong@bit.edu.cn

Abstract

A degenerate nonlinear system of Kirchhoff type with boundary damping is studied. We prove that this system has a unique global solution if the domain and initial data satisfy some assumptions. We also obtain the polynomial decay of the global solution of the system.

1 Introduction

In 1880's, G. Kirchhoff introduced a model describing the transversal vibration of a nonlinear elastic string ([9]), which is later called Kirchhoff string.

$$\begin{cases} u''(x,t) - (a + b\|u_x(x,t)\|^2)u_{xx}(x,t) = 0 & \text{in } (0,L)\times R^+, \\ u(0,t) = u(L,t) = 0, & \\ u(x,0) = u_0(x),\ u'(x,0) = u_1(x) & \text{in } (0,L), \end{cases} \tag{1.1}$$

where u is the transverse displacement of the string, a and b are positive constants, L is the length of the string, $\|\cdot\|$ represents the norm on $L^2(0,L)$, and u_0, u_1 are initial data. Both ends of the string are fixed.

Later, more general models were considered by Carrier ([4]) and Lions ([12]) such as

$$\begin{cases} u''(x,t) - M(\|A^{\frac{1}{2}}u(x,t)\|_H^2)Au(x,t) = 0 & \text{in } \Omega\times R^+, \\ u(x,0) = u_0(x),\ u'(x,0) = u_1(x) & \text{in } \Omega, \end{cases} \tag{1.2}$$

*Supported by the National Natural Science Foundation of China (60504001).

where A is a selfadjoint linear nonnegative operator on Hilbert space H with dense domain $D(A)$, $M : [0, \infty) \to [0, \infty)$ is a differential function and $\Omega \subset \mathbb{R}^+$ is a bounded open domain. System (1.2) is called a non-degenerate equation when $M(\xi) \geq \tilde{M} > 0$ for all $\xi \geq 0$, and a degenerate one when $M(\xi) \geq 0$ for all $\xi \geq 0$. In the case of $M(\xi) \equiv \hat{M} > 0$, system (1.2) is a usual linear elastic equation.

Since then, the local and global solvability of nonlinear systems of Kirchhoff type has been studied under various assumptions concerned with function M, the dissipative term, and the regularity of the initial data (see [1, 6, 8]). Among the literature, the latest results about local solvability of system (1.2) can be found in [8]. They proved the existence, uniqueness and regularity of the local solution for non-dissipative model (1.2) when M is a degenerate differential function and $M(\xi) \neq 0$ for ξ belonging to the neighborhood of $\|A^{\frac{1}{2}}u_0\|^{2\gamma}$.

It is clear that an extra dissipative term is necessary to obtain the existence, uniqueness and regularity of the global solution of system (1.2). In the non-degenerate case, system (1.2) describes a pre-stressed structure. When initial data are small enough, Brito [3], Nishihara [13] and Cavalcanti [5] proved that there exists a unique global solution with exponential decay property for non-degenerate system if a damping is applied in Ω and M is a C^1 function. The same result was obtained by Lasiecka and Ong [11] if the dissipative term is on the boundary.

Degenerate nonlinear systems of Kirchhoff type were considered by Nishihara and Yamada ([14]), Ono ([15]), Ghisi ([7]) and references therein. They proved that when $M(\xi)$ behaves like ξ^γ ($\gamma > 0$), there exists a unique global solution of the system and the solution decays polynomially if a velocity dissipative term is applied in Ω and initial data are small and regular enough. It is more difficult to handle a degenerate case than a non-degenerate case. The difficulty increases when the damping is applied on the boundary. The method in references can not be applied directly to the degenerate Kirchhoff system with boundary damping.

In this paper we study the global existence and decay properties of the solution of a degenerate Kirchhoff wave equation system with natural boundary damping. The nonlinear coefficient function is assumed to be $M(\xi) = \xi^\gamma$ ($\gamma > 1$). First, we prove the existence, uniqueness and regularity of the global solution of the system. We introduce several estimates of higher order energy divided by $\|A^{\frac{1}{2}}u\|^s$ ($s > 0$) to overcome difficulties due to the degenerateness of $M(\xi)$. Then by the idea of iterativeness of the time and assumptions on initial data, we show that $M(\|A^{\frac{1}{2}}u(t)\|) > 0$ for all $t > 0$, i.e., the degenerateness situation never occurs because of the dissipation. Thus, the local solution of the system can be continued globally in time. We also obtain the polynomial decay of the solution of the system by the classical Gronwall lemma.

This paper is organized as follows. In Section 2 we present the system and main results. In Section 3 we prove the global existence of the solution of the system.

2 Main results

In this section we state main results of this paper. Let Ω be an open bounded set in $\mathbb{R}^n$ with smooth boundary $\Gamma = \Gamma_0 \cup \Gamma_1$. Assume that Γ_0 has positive boundary measure and $\overline{\Gamma}_0 \bigcap \overline{\Gamma}_1 = \emptyset$ (this assumption excludes simply connected regions). In what follows, $H^r(\Omega)$ denotes a usual Sobolev space for any $r \in \mathbb{R}$. $H^1_{\Gamma_0}(\Omega)$ denotes space $\{w \in H^1(\Omega) : w = 0 \text{ on } \Gamma_0\}$. Let X be a Banach space. We denote by $C^m([0,T];\, X)$ the space of all m times continuously differentiable functions defined on $[0,T]$ with values in X, and write $C([0,T];\, X)$ for $C^0([0,T];\, X)$. The scalar product and norm in $L^2(\Omega)$ and $L^2(\Gamma)$ are represented by $\|\cdot\|$, $(\cdot\,,\,\cdot)$ and $|\cdot|_\Gamma$, $\langle\cdot\,,\,\cdot\rangle_\Gamma$, respectively.

We consider the following degenerate nonlinear wave equation of Kirchhoff type with boundary damping.

$$\begin{cases} u''(x,t) - \|\nabla u(x,t)\|^{2\gamma}\Delta u(x,t) = 0 & \text{in } \Omega \times \mathbb{R}^+, \\ u(x,t) = 0 & \text{on } \Gamma_0 \times \mathbb{R}^+, \\ \|\nabla u(x,t)\|^{2\gamma}\partial_\nu u(x,t) + ku'(x,t) = 0 & \text{on } \Gamma_1 \times \mathbb{R}^+, \\ u(x,0) = u_0(x),\ u'(x,0) = u_1(x) & \text{in } \Omega, \end{cases} \tag{2.1}$$

where u is the transverse displacement of the wave, $\gamma > 1$ is a real number, $\nu(x)$ denotes a unit outward normal vector at $x \in \Gamma$, u_0, u_1 are initial data, ku' $(k > 0)$ is boundary damping.

The natural energy of system (2.1) is defined by

$$E(u(t)) = \frac{1}{2}\|u'(t)\|^2 + \frac{1}{2\gamma+2}\|\nabla u(t)\|^{2\gamma+2}. \tag{2.2}$$

A direct computation gives that

$$\frac{d}{dt}E(u(t)) = -k|u'(t)|^2_{\Gamma_1}, \tag{2.3}$$

i.e., energy function (2.2) for system (2.1) decreases on $[0,\infty)$.

For completeness, we introduce the following local existence theorem, which can be proved by the contraction mapping theorem.

Theorem 2.1. *Let $\gamma > 0$, and initial data $u_0 \in H^2(\Omega) \bigcap H^1_{\Gamma_0}(\Omega)$ and $u_1 \in H^1_{\Gamma_0}(\Omega)$. Suppose $\|\nabla u_0\| > 0$. Then there exists $T = T(\|\nabla u_0\|) > 0$ such that system (2.1) has a unique local solution u with regularity*

$$u \in C\big([0,T];\, H^2(\Omega)\cap H^1_{\Gamma_0}(\Omega)\big)\cap C^1\big([0,T];\, H^1_{\Gamma_0}(\Omega)\big)\cap C^2\big([0,T];\, L^2(\Omega)\big). \tag{2.4}$$

We note that the dissipative term on the boundary does not play role in the local existence result, while it plays a crucial role in the global existence one.

Now, we present an assumption about geometry of the domain, called "geometric optic conditions". It guarantees that all rays of geometric optics meet Γ_1 to which the memory damping is applied. This assumption is useful to get estimations about higher order energy when we prove the global existence of the solution of system (2.1).

(B) $m(x) \cdot \nu(x) \le 0$ on Γ_0 and $m(x) \cdot \nu(x) \ge \delta > 0$ on Γ_1, where $m(x) \doteq x - x_0$, x_0 is an arbitrary fixed point in $\mathbb{R}^n$.

The main result of this paper is as follows. Its proof is in Section 3.

Theorem 2.2. *Assume (B) holds and $\gamma \ge 1$. Let initial data $u_0 \in H^2(\Omega) \bigcap H^1_{\Gamma_0}(\Omega)$, $u_1 \in H^1_{\Gamma_0}(\Omega)$ satisfy*

$$\|\nabla u_0\| > 0, \tag{2.5}$$

$$\frac{1}{2}\|u_1\|^2 + \frac{1}{2\gamma+2}\|\nabla u_0\|^{2\gamma+2} \le 1, \tag{2.6}$$

and there exists $\kappa > 0$, depending on Ω, k and γ, such that

$$\frac{\|\Delta u_0\|^2}{\|\nabla u_0\|^2} + \frac{\|\nabla u_1\|^2}{\|\nabla u_0\|^{2\gamma+2}} + \frac{\|\Delta u_0\|^2}{\|\nabla u_0\|^{2\gamma+2}} + \frac{\|\nabla u_1\|^2}{\|\nabla u_0\|^{4\gamma+2}} < \kappa. \tag{2.7}$$

Then system (2.1) has a unique global solution $u(x,t)$ with regularity that

$$u \in C\big([0,\infty); H^2(\Omega)\cap H^1_{\Gamma_0}(\Omega)\big)\cap C^1\big([0,\infty); H^1_{\Gamma_0}(\Omega)\big)\cap C^2\big([0,\infty); L^2_{\Gamma_0}(\Omega)\big).$$

Remark 2.1. *The key point is to prove that the degeneration situation never occurs. Firstly, we introduce several higher order energy functions containing $\dfrac{\|\Delta u\|^2}{\|\nabla u\|^s}$ and $\dfrac{\|\nabla u'\|^2}{\|\nabla u\|^s}$ $(s > 0)$. Then, we show that $\|\nabla u(t)\| > 0$ for all $t \ge 0$ by the assumption on initial data. Therefore, the local solution of system (2.1) can be continued globally in time.*

After getting the existence, uniqueness and regularity of the solution of system (2.1), we study the asymptotic behavior of the global solution of system (2.1). In fact, we have the following decay properties by using proper estimates and the classical Gronwall lemma.

Theorem 2.3. *Assume that all conditions of Theorem 2.2 are satisfied. Suppose that*

$$\|\Delta u_0\|,\ \|\nabla u_0\|,\ \|\nabla u_1\| < \kappa_1,$$

where $\kappa_1(\Omega, k, \gamma) > 0$ *is a constant. Let* $u(x,t)$ *be the global solution of system (2.1). Then there exists positive constant* C *depending on* Ω, k, γ *and initial data such that*

$$E(t) \le \frac{C}{(1+t)^{\frac{2\gamma+2}{\gamma}}}, \qquad \forall\, t \ge 0. \tag{2.8}$$

Remark 2.2. *The proof of Theorem 2.3 is regular and we just give a sketch here. The key is to introduce an auxiliary function*

$$\mathcal{L}(t) \doteq \eta\big(E(t)\big)^{p+\frac{\gamma+2}{2\gamma+2}} + \big(E(t)\big)^p \rho(t),$$

where $p > 0$, $\eta > 0$ *and*

$$\rho(t) \doteq 2\big(u'(t),\, m\cdot\nabla u(t)\big) + (n-1)\big(u'(t),\, u(t)\big).$$

Then we can choose a proper constant η *such that*

$$\big[\mathcal{L}(t)\big]^\alpha + C\frac{d}{dt}\,\mathcal{L}(t) \le 0, \quad \alpha \doteq \frac{p+1}{p+\frac{\gamma+2}{2\gamma+2}} > 1, \quad C > 0. \tag{2.9}$$

Therefore, the decay property is obtained by using the Gronwall lemma.

As a corollary, we have immediately the following polynomial decay results.

Theorem 2.4. *Assume that all conditions of Theorem 2.3 are satisfied. Let* $u(x,t)$ *be the global solution of system (2.1). Then the following estimates satisfy*

$$0 < \|\nabla u(t)\|^2 \le \frac{C}{(1+t)^{\frac{2}{\gamma}}},$$

$$\|u'(t)\|^2,\ \|\Delta u(t)\|^2 \le \frac{C}{(1+t)^{\frac{2\gamma+2}{\gamma}}},$$

$$\|\nabla u'(t)\|^2 \le \frac{C}{(1+t)^{\frac{4\gamma+2}{\gamma}}},$$

with a positive constant C *for all* $t \ge 0$.

3 Proof of Theorem 2.2

In this section we prove the global existence of the solution of system (2.1). First, we set

$$T_1 \doteq \sup\big\{t \in [0,\infty) \, : \, \|\nabla u(\tau)\| > 0, \quad \forall\, 0 \le \tau < t\big\}.$$

Then $T_1 > 0$, $\|\nabla u(t)\| > 0$ for $0 \le t < T_1$, and $\|\nabla u(T_1)\| = 0$. For any $0 \le t < T_1$, define

$$\begin{aligned} H_s(t) &= \frac{\|\Delta u(x,t)\|^2}{\|\nabla u(x,t)\|^{s-4\gamma}} + \frac{\|\nabla u'(x,t)\|^2}{\|\nabla u(x,t)\|^{s-2\gamma}}, \\ \rho_s(t) &= \frac{\big(\Delta u(x,t),\, 2m(x)\cdot \nabla u'(x,t)\big) + (n-1)\big(\Delta u(x,t),\, u'(x,t)\big)}{\|\nabla u(x,t)\|^{s-2\gamma}}, \end{aligned}$$

where $s \ge 4\gamma$ is a real number. It is clear that for any $0 \le t < T_1$,

$$|\rho_s(t)| \le H_{s+2\gamma}(t) + \big(2R+n-1\big)^2 H_s(t), \quad R \doteq \max_{x\in\overline{\Omega}} |m(x)|. \tag{3.1}$$

Set

$$\mathcal{E}_s(t) \doteq \rho_s(t) + \lambda_1 H_s(t) + \lambda_2 H_{s+2\gamma}(t),$$

where λ_1 and λ_2 are positive constants satisfying

$$\begin{aligned} \lambda_1 &\ge \max\Big\{(2R+n-1)^2 + 1,\ \frac{R}{k}\Big\}, \\ \lambda_2 &\ge \max\Big\{2,\ \frac{2R^2k}{\delta} + k(n-1)^2\Big\}. \end{aligned} \tag{3.2}$$

Suppose $s > 4\gamma$. We differentiate the first equation of (2.1) with respect to t, multiply the result by $2m\cdot\nabla u + (n-1)u$ and integrate it in Ω,

$$\begin{aligned} &H_s + \frac{d}{dt}\mathcal{E}_s \\ &\le (R-\lambda_1 k)\frac{|u''|^2_{\Gamma_1}}{\|\nabla u\|^s} + \Big[\frac{2R^2k^2}{\delta} + k^2(n-1)^2 - \lambda_2 k\Big]\frac{|u''|^2_{\Gamma_1}}{\|\nabla u\|^{s+2\gamma}} \\ &\quad + \Big(\frac{8R^2k^2\gamma^2}{\delta} + 4\lambda_2 k\gamma^2\Big)\frac{\big(\nabla u,\nabla u'\big)^2|u'|^2_{\Gamma_1}}{\|\nabla u\|^{s+2\gamma+4}} + \frac{4\lambda_1 k\gamma^2\,(\nabla u,\nabla u')^2\,|u'|^2_{\Gamma_1}}{\|\nabla u\|^{s+4}} \\ &\quad + \frac{2k\gamma(n-1)\big(\nabla u,\nabla u'\big)|u'|^2_{\Gamma_1}}{\|\nabla u\|^{s+2}} + \frac{|u'|^2_{\Gamma_1}}{4\|\nabla u\|^{s-2\gamma}} - \frac{(s-2\gamma)\big(\nabla u,\nabla u'\big)\rho_s(t)}{\|\nabla u\|^2} \\ &\quad - \frac{\lambda_1(s-2\gamma)\,(\nabla u,\nabla u')}{\|\nabla u\|^2}H_s - \frac{\lambda_2 s\,(\nabla u,\nabla u')}{\|\nabla u\|^{2\gamma+2}}H_s, \quad \forall\ 0 \le t < T_1. \end{aligned} \tag{3.3}$$

Here we use assumption (B). Combining (3.2) with (3.3) yields

$$
\begin{aligned}
&H_s + \frac{d}{dt}\mathcal{E}_s \\
&\le \Big(\frac{8R^2k^2\gamma^2}{\delta} + 4\lambda_2 k\gamma^2\Big)\frac{\big(\nabla u, \nabla u'\big)^2|u'|^2_{\Gamma_1}}{\|\nabla u\|^{s+2\gamma+4}} + \frac{4\lambda_1 k\gamma^2\,(\nabla u, \nabla u')^2\,|u'|^2_{\Gamma_1}}{\|\nabla u\|^{s+4}} \\
&\quad + \frac{2k\gamma(n-1)\big(\nabla u, \nabla u'\big)|u'|^2_{\Gamma_1}}{\|\nabla u\|^{s+2}} + \frac{|u'|^2_{\Gamma_1}}{4\|\nabla u\|^{s-2\gamma}} - \frac{(s-2\gamma)\big(\nabla u, \nabla u'\big)\rho_s(t)}{\|\nabla u\|^2} \\
&\quad - \frac{\lambda_1(s-2\gamma)\,(\nabla u, \nabla u')}{\|\nabla u\|^2}H_s - \frac{\lambda_2 s\,(\nabla u, \nabla u')}{\|\nabla u\|^{2\gamma+2}}H_s, \quad \forall\ 0 \le t < T_1.
\end{aligned}
\tag{3.4}
$$

Secondly, for a positive constant β specified later, we define

$$
T_2 \doteq \sup\left\{ t \in [0,\infty)\ :\ \frac{\big|(\nabla u(\tau), \nabla u'(\tau))\big|}{\|\nabla u(\tau)\|^{2\gamma+2}} < \beta, \quad \forall\ 0 \le \tau < t \right\}.
$$

It follows that $T_2 > 0$. Suppose

$$
T_2 < T_1. \tag{3.5}
$$

Then,

$$
\frac{\big|(\nabla u(T_2), \nabla u'(T_2))\big|}{\|\nabla u(T_2)\|^{2\gamma+2}} = \beta. \tag{3.6}
$$

Therefore, we have from (3.4) that for any $0 \le t \le T_2$,

$$
\begin{aligned}
H_s + \frac{d}{dt}\mathcal{E}_s \ &\le \beta^2\Big(\frac{8R^2k^2\gamma^2}{\delta} + 4\lambda_2 k\gamma^2\Big)\frac{|u'|^2_{\Gamma_1}}{\|\nabla u\|^{s-2\gamma}} + \frac{4\beta^2\lambda_1 k\gamma^2\,|u'|^2_{\Gamma_1}}{\|\nabla u\|^{s-4\gamma}} \\
&\quad + \frac{2\beta k\gamma(n-1)|u'|^2_{\Gamma_1}}{\|\nabla u\|^{s-2\gamma}} + \frac{|u'|^2_{\Gamma_1}}{4\|\nabla u\|^{s-2\gamma}} + \beta(s-2\gamma)\|\nabla u\|^{2\gamma}\rho_s \\
&\quad + \beta\lambda_1(s-2\gamma)\|\nabla u\|^{2\gamma}H_s + \beta\lambda_2 s H_s.
\end{aligned}
\tag{3.7}
$$

Let $s = 4\gamma + 2$. It follows from (3.1) and (3.7) that for any $0 \le t \le T_2$,

$$
H_{4\gamma+2} + \frac{d}{dt}\mathcal{E}_{4\gamma+2} \le \Big(\frac{1}{4} + \beta^2\tilde{\mu}_1(t) + \beta\mu_2\Big)\frac{|u'|^2_{\Gamma_1}}{\|\nabla u\|^{2\gamma+2}} + \beta\tilde{\mu}_3(t)H_{4\gamma+2}, \tag{3.8}
$$

where

$$
\begin{aligned}
&\tilde{\mu}_1(t) \doteq \frac{8R^2k^2\gamma^2}{\delta} + 4\lambda_2 k\gamma^2 + 4\lambda_1 k\gamma^2\|\nabla u(t)\|^{2\gamma}, \\
&\mu_2 \doteq 2k\gamma(n-1), \\
&\tilde{\mu}_3(t) \doteq 2\gamma + 2 + \lambda_2(4\gamma+2) + (2\gamma+2)[(2R+n-1)^2 + \lambda_1]\|\nabla u(t)\|^{2\gamma}.
\end{aligned}
$$

By using (2.3) and (2.6),

$$\tilde{\mu}_1(t) \le \mu_1 \doteq \frac{8R^2k^2\gamma^2}{\delta} + 4\lambda_2 k\gamma^2 + 4\lambda_1 k\gamma^2(2\gamma+2)^{\frac{2\gamma}{2\gamma+2}},$$

$$\tilde{\mu}_3(t) \le \mu_3 \doteq 2\gamma + 2 + \lambda_2(4\gamma+2) + (2\gamma+2)^{1+\frac{2\gamma}{2\gamma+2}}[(2R+n-1)^2 + \lambda_1].$$

Not that $|u'|^2_{\Gamma_1} \le C_\Omega \|\nabla u'\|^2$, $(C_\Omega > 0)$. Consequently, we have from (3.8) that

$$H_{4\gamma+2}(t) + \frac{d}{dt}\mathcal{E}_{4\gamma+2}(t) \le \Big[\frac{C_\Omega}{4} + \beta^2\mu_1 C_\Omega + \beta(\mu_2 C_\Omega + \mu_3)\Big] H_{4\gamma+2}(t). \quad (3.9)$$

We can choose β such that

$$\beta^2\mu_1 C_\Omega + \beta(\mu_2 C_\Omega + \mu_3) < \frac{C_\Omega}{4}. \quad (3.10)$$

Thus, it follows from (3.9)–(3.10) that

$$\frac{d}{dt}\mathcal{E}_{4\gamma+2}(t) \le 0, \quad \forall\ 0 \le t \le T_2. \quad (3.11)$$

By (3.1), (3.2) and (3.11), we get that for any $0 \le t \le T_2$,

$$H_{4\gamma+2}(t) + H_{6\gamma+2}(t) \le \lambda\big[H_{4\gamma+2}(0) + H_{6\gamma+2}(0)\big], \quad (3.12)$$

where

$$\lambda \doteq \frac{\max\big\{\lambda_1 + (2R+n-1)^2,\ \lambda_2 + 1\big\}}{\min\big\{\lambda_1 - (2R+n-1)^2,\ \lambda_2 - 1\big\}}.$$

On the other hand, it is clear that for any $0 \le t < T_1$,

$$\frac{\big|(\nabla u(t), \nabla u'(t))\big|}{\|\nabla u(t)\|^{2\gamma+2}} \le \frac{\|\nabla u'(t)\|}{\|\nabla u(t)\|^{2\gamma+1}} \le H^{\frac{1}{2}}_{6\gamma+2}(t). \quad (3.13)$$

Combining (3.5), (3.12) with (3.13) yields that for any $0 \le t \le T_2 < T_1$,

$$\frac{\big|(\nabla u(t), \nabla u'(t))\big|}{\|\nabla u(t)\|^{2\gamma+2}} \le \big[\lambda\big(H_{4\gamma+2}(0) + H_{6\gamma+2}(0)\big)\big]^{\frac{1}{2}}. \quad (3.14)$$

We set

$$\frac{\|\Delta u_0\|^2}{\|\nabla u_0\|^2} + \frac{\|\nabla u_1\|^2}{\|\nabla u_0\|^{2\gamma+2}} + \frac{\|\Delta u_0\|^2}{\|\nabla u_0\|^{2\gamma+2}} + \frac{\|\nabla u_1\|^2}{\|\nabla u_0\|^{4\gamma+2}} < \frac{\beta^2}{4\lambda}. \quad (3.15)$$

From (3.14) and (3.15),

$$\frac{\big|(\nabla u(t), \nabla u'(t))\big|}{\|\nabla u(t)\|^{2\gamma+2}} < \frac{\beta}{2}. \quad (3.16)$$

Then, we get a contradiction to (3.6). Therefore, $T_1 \le T_2$.

Finally, we show that $\|\nabla u(t)\| > 0$ for all $t > 0$. By the definition of T_1, we have that

$$\|\nabla u(T_1)\| = 0. \tag{3.17}$$

Then we have from (3.12) and (3.17) that

$$\|\Delta u(T_1)\| = \|\nabla u'(T_1)\| = 0. \tag{3.18}$$

Induce a variable $v(t) \doteq u(T_1 - t)$. Then $v(t)$ satisfies

$$\begin{cases} v'' - \|\nabla v\|^{2\gamma}\Delta v = 0 & \text{in } \ \Omega \times [0, T_1], \\ v = 0 & \text{on } \ \Gamma_0 \times [0, T_1], \\ \|\nabla v\|^{2\gamma}\partial_\nu v - kv' = 0 & \text{on } \ \Gamma_1 \times [0, T_1], \\ v(0) = 0, \ v'(0) = 0 & \text{in } \ \Omega. \end{cases} \tag{3.19}$$

It is clear that

$$\frac{d}{dt}E(v(t)) = k|v'(t)|^2_{\Gamma_1}. \tag{3.20}$$

By the trace theorem, there exists a positive constant $\tilde{C}_\Omega$ such that

$$|v'(t)|^2_{\Gamma_1} \le \tilde{C}_\Omega \|\nabla v'(t)\| \, \|v'(t)\|, \quad \forall \ 0 \le t \le T_1. \tag{3.21}$$

Moreover, we have from (3.12) that

$$\|\nabla u'(T_1 - t)\|^2 \le \lambda\big[H_{4\gamma+2}(0) + H_{6\gamma+2}(0)\big]\|\nabla u(T_1 - t)\|^{2\gamma+2}. \tag{3.22}$$

Combining (3.21) with (3.22) yields

$$\begin{aligned} |v'(t)|^2_{\Gamma_1} \le \tilde{C}_\Omega(2\gamma+2)\big[\lambda\big(H_{4\gamma+2}(0) + H_{6\gamma+2}(0)\big)\big]^{\frac{1}{2}} \\ E(v(t)), \quad \forall \ 0 \le t \le T_1. \end{aligned} \tag{3.23}$$

Hence, by (3.20) and (3.23), we get that

$$\begin{aligned} \frac{d}{dt}E(v(t)) \le k\tilde{C}_\Omega(2\gamma+2)\big[\lambda\big(H_{4\gamma+2}(0) + H_{6\gamma+2}(0)\big)\big]^{\frac{1}{2}} \\ E(v(t)), \quad \forall \ 0 \le t \le T_1. \end{aligned} \tag{3.24}$$

Notice that $E(v(0)) = 0$. Thus, it follows from (3.24) that $E(v(T_1)) = 0$. Consequently, $\|\nabla u(0)\| = 0$, which contradicts assumption (2.5). Therefore, $\|\nabla u(t)\| > 0$ for all $t > 0$, and (3.12) holds for all $t \ge 0$. The proof of Theorem 2.2 is completed.

References

[1] A. Arosio and S. Garavaldi, *On the mildly degenerate Kirchhoff string,* Mathematical Methods in Applied Science, **14** (1991), 177–195.

[2] V. Barbu, I. Lasiecka, and M. A. Rammaha, *On nonlinear wave equations with degenerate damping and source terms,* Transactions of the American Mathematical Society, **357** (2005), 2571–2611.

[3] E. H. de Brito, *Decay estimates for the generalized damped extensible string and beam equation,* Nonlinear Analysis, **8** (1984), 1489–1496.

[4] G.F. Carrier, *On the non-linear vibration problem of the elastic string,* Quarterly of Applied Mathematics, **3** (1945), 157–165.

[5] M. M. Cavalcanti, U V. N. Domingos Cavalcanti, J. S. Prates Filho, and J. A. Soriano, *Existence and exponential decay for a kirchhoff-carrier model with viscosity,* Journal of Mathematical Analysis and Application, **226** (1998), 40–60.

[6] P. D'Ancona and S. Spagnolo, *Nonlinear perturbations of the Kirchhoff equation,* Communications on Pure and Applied Mathematics, **47** (1994), 1005–1029.

[7] M. Ghisi, *Global solutions for dissipative Kirchhoff strings with non-Lipschitz nonlinear term,* Journal of Differential Equations, **230** (2006), 128–139.

[8] R. Izaguirrea, R. Fuentesb and M. M. Miranda *Existence of local solutions of the kirchhoffCcarrier equation in Banach spaces,* Nonlinear Analysis, **68** (2008), 3565–3580.

[9] G. Kirchhoff, *Vorlesungen über Mechanik,* Teubner, Leipzig, 1883.

[10] V. Komornik, *Exact Controllability and Stabilization: The Multiplier Method,* Masson-John Wiley, Paris, 1994.

[11] I. Lasiceka and J. Ong, *Global solvability and uniform decays of solutions to quasilinear equation with nonlinear boundary dissipation,* Communications in Partial Differential Equations, **24** (1999), 2069–2107.

[12] J. L. Lions, *On some questions in boundary value problem of mathematical physics,* in: G.M. de La Penha, L.A. Medeiros (Eds.), Contemporary development in continuum mechanics and partial differential equations, North-Holland, Amsterdam, 1978, 285–346.

[13] K. Nishihara, *Global existence and asymptotic behaviour of the solution of some quasilinear hyperbolic equation with linear damping,* Funkcialaj Ekvacioj, **32** (1989), 343–355.

[14] K. Nishihara and Y. Yamada, *On global solutions of some degenerate quasilinear hyperbolic equations with dissipative terms,* Funkcialaj Ekvacioj, **33** (1990), 151–159.

[15] K. Ono, *Global existence, decay, and blowup of solutions for some mildly degenerate nonlinear Kirchhoff strings,* Journal of Differential Equations **137** (1997), 273–301.

[16] M. L. Santos, J. Ferreira, D. C. Pereira and C. A. Raposo, *Global existence and stability for wave equation of Kirchhoff type with memory condition at the boundary,* Nonlinear Anaysis, **54** (2003), 959–976.

[17] R. Temam, *Infinite Dimensional Dynamical Systems in Mechanics and Physics,* Springer-Verlag, Applied Mathematical Sciences, Vol. 68, New York, 1988.

[18] Q. Zhang, *Global Existence and Decay of Solutions for a Mildly Degenerate Kirchhoff Wave with Boundary Damping,* submitted.

Remarks on the Controllability of Some Quasilinear Equations*

Xu Zhang

Academy of Mathematics and Systems Sciences
Chinese Academy of Sciences, Beijing 100190, China
&
Yangtze Center of Mathematics
Sichuan University, Chengdu 610064, China
Email: xuzhang@amss.ac.cn

Abstract

In this paper, we review the main existing results, methods, and some key open problems on the controllability of nonlinear hyperbolic and parabolic equations. Especially, we describe our recent universal approach to solve the local controllability problem of quasilinear time-reversible evolution equations, which is based on a new unbounded perturbation technique. It is also worthy to mention that the technique we developed can also be applied to other problems for quasilinear equations, say local existence, stabilization, etc.

1 Introduction

Consider the following controlled evolution equation:

$$\begin{cases} \dfrac{d}{dt}y(t) = A(y(t))y(t) + Bu(t), \quad t \in (0,T), \\ y(0) = y_0. \end{cases} \tag{1.1}$$

Here, the time $T > 0$ is given, $y(t) \in Y$ is the state variable, $u(t) \in U$ is the control variable, $y_0 (\in Y)$ is the initial state; Y and U are respectively

*This work was supported by the NSF of China under grants 10525105, 10831007 and 60821091, the Chunhui program (State Education Ministry of China) under grant Z007-1-61006, and the project MTM2008-03541 of the Spanish Ministry of Science and Innovation. Part of this work was done when the author visited Fudan University, with a financial support from the "French-Chinese Summer Institute on Applied Mathematics" (September 1–21, 2008).

the state space and control space, both of which are some Hilbert space; $A(\cdot)$ is a suitable (nonlinear and usually unbounded) operator on Y, while the control operator B maps U into Y. Many control problems for relevant nonlinear Partial Differential Equations (PDEs, for short) enter into this context, for instance, the quasilinear/semilinear parabolic equation, wave equation, plate equation, Schrödinger equation, Maxwell equations, and Lamé system, etc.

In this paper, we shall describe some existing methods, results and main open problems on the controllability of these systems, especially these for nonlinear hyperbolic and parabolic equations.

System (1.1) is said to be exactly controllable in Y at time T if for any $y_0, y_1 \in Y$, there is a control $u \in L^2(0, T; U)$ such that the solution of system (1.1) with this control satisfies

$$y(T) = y_1. \tag{1.2}$$

When $\dim Y = \infty$ (We shall focus on this case later unless otherwise stated), sometimes one has to relax the requirement (1.2), and this leads to various notions and degrees of controllability: approximate controllability, null controllability, etc. Note however that for time reversible system, the notion of exact controllability is equivalent to that of null controllability.

Roughly speaking, the controllability problem for an evolution equation consists in driving the state of the system (the solution of the controlled equation under consideration) to a prescribed final target state (exactly or in some approximate way) in finite time. Problems of this type are common in science and engineering and, particularly, they arise often in the context of flow control, in the control of flexible structures appearing in flexible robots and in large space structures, in quantum chemistry, etc.

The controllability theory about finite dimensional linear systems was introduced by R.E. Kalman [14] at the very beginning of the 1960s. Thereafter, many authors were devoted to developing it for more general systems including infinite dimensional ones, and its nonlinear and stochastic counterparts.

The controllability theory of PDEs depends heavily on its nature and, on its time-reversibility properties in particular. To some extent, the study of controllability for linear PDEs is well developed although many challenging problems are still unsolved. Classical references in this field are D.L. Russell [29] and J.L. Lions [21]. Updated progress can be found in a recent survey by E. Zuazua ([43]). Nevertheless, much less is known for nonlinear controllability problems for PDEs although several books on this topic have been available, say J.M. Coron [6], A.V. Fursikov & O.Yu. Imanuvilov [11], T.T. Li [16], and X. Zhang [36]. Therefore, in this paper, we concentrate on controllability problems for systems governed by nonlinear PDEs.

The main results in this paper can be described as follows: Assume that $(A(0), B)$ is exactly controllable in Y. Then, under some assumptions on the structure of $A(y)$ (for concrete problems, which needs more regularity on the state space, say $\mathcal{D}(A(0)^k)$ for sufficiently large k), system (1.1) is locally exactly controllable in $\mathcal{D}(A(0)^k)$.

The main approach we employed to show the above controllability result is a new perturbation technique. The point is that the perturbation is *unbounded* but *small*. Note however that this approach does NOT work for the null controllability problem of the time-irreversible systems, and therefore, one has to develop different methods to solve the local null controllability of quasilinear parabolic equations.

For simplicity, in what follows, we consider mainly the case of internal control, i.e. $B \in \mathcal{L}(U, Y)$. Also, we will focus on the local controllability of the quasilinear wave equation. However, our approach is universal, and therefore, it can be extended to other quasilinear PDEs, say, quasilinear plate equation, Schrödinger equation, Maxwell equations, and Lamé system, etc.

On the other hand, we mention that the technique developed in this paper can also be applied to other problems for quasilinear equations. For example, stabilization problem for system (1.1) (with small initial data) can be considered similarly. Indeed, although there does not exist the same equivalence between exact controllability and stabilization in the nonlinear setting, the approaches to treat them can be employed one other.

The rest of this paper is organized as follows. In Section 2, we review the robustness of the controllability in the setting of Ordinal Differential Equations (ODEs, for short). In Section 3, we recall some known perturbation result of the exact controllability of abstract evolution equations. Then, in Section 4, we show a new perturbation result of the exact controllability of general evolution equations. Sections 5 and 6 are addressed to present local controllability results of multidimensional quasilinear hyperbolic equations and parabolic equations, respectively. Finally, in Section 7, we collect some open problems, which seem to be important in the field of controllability of PDEs.

2 Starting point: the case of ODEs

Consider the following controlled ODE:

$$\begin{cases} \dfrac{d}{dt}y = Ay + Bu, & t \in (0, T), \\ y(0) = y_0, \end{cases} \tag{2.1}$$

where $A \in \mathbb{R}^{n\times n}$ and $B \in \mathbb{R}^{n\times m}$. It is well known ([14]) that system (2.1) is exactly controllable in $(0, T)$ if and only if

$$B^* e^{A^* t} x_0 = 0, \quad \forall\, t \in (0, T) \Rightarrow x_0 = 0.$$

Note that this condition is also equivalent to the following Kalman rank condition:

$$rank(B, AB, A^2B, \cdots, A^{n-1}B) = n. \tag{2.2}$$

From (2.2), it is clear that if (A, B) is exactly controllable, then there exists a small $\varepsilon = \varepsilon(A, B) > 0$ such that $(\widetilde{A}, \widetilde{B})$ is still exactly controllable provided that $||\widetilde{A} - A|| + ||\widetilde{B} - B|| < \varepsilon$. Therefore, the exact controllability of system (2.1) is robust under small perturbation.

Because of the above robustness, the local exact controllability of nonlinear OPEs is quite easy. Indeed, consider the following controlled system:

$$\begin{cases} \dfrac{d}{dt} y = Ay + f(y) + Bu, & t \in (0, T), \\ y(0) = y_0, \end{cases} \tag{2.3}$$

with $f(\cdot) \in C^1(\mathbb{R}^n)$ and $f(y) = O(|y|^{1+\delta})$ when y is small, for some $\delta > 0$. The local exact controllability of system (2.3) follows from a standard perturbation argument.

However, the corresponding problem in PDE setting is much more complicated, as we shall see below.

3 Known perturbation result of exact controllability

In this section, we recall some known perturbation results of the exact controllability of abstract evolution equations. These results are based on the following two tools:

• **Duality argument** (e.g. [20, 21, 36]): In the linear setting (i.e., $A(y) \equiv A$ is independent of y and linear, and further A generates a C_0-group $\{e^{At}\}_{t\in\mathbb{R}}$ on Y), the null controllability of system (1.1) is equivalent to the following observability estimate:

$$|e^{A^*T} z^*|_{Y^*}^2 \le C \int_0^T |B^* e^{A^* s} z^*|_{U^*}^2 ds, \qquad \forall\, z^* \in Y^*, \tag{3.1}$$

for some constant $C > 0$.

• **Variation of constants formula**: In the setting of semigroup, for a bounded perturbation $P \in \mathcal{L}(Y)$:

$$e^{(A+P)t} x = e^{At} x + \int_0^t e^{A(t-s)} Pxds, \qquad \forall\, x \in Y. \tag{3.2}$$

Combining (3.1) and (3.2), it is easy to establish the following well-known (bounded) perturbation result of the exact controllability:

Theorem 3.1. *Assume that A generates a C_0-group $\{e^{At}\}_{t\in\mathbb{R}}$ on Y and $B \in \mathcal{L}(U, Y)$. If (A, B) is exactly controllable, then so is $(A+P, B)$ provided that $||P||_{\mathcal{L}(Y)}$ is small enough.*

The above perturbation P can also be time-dependent. In this case, one needs the language of evolution system. In the sequel, for a simple presentation, we consider only the time-independent case.

As a consequence of Theorem 3.1 and the standard fixed point technique, one can easily deduce a local exact controllability result of some semilinear equations, say, the counterpart of system (2.3):

$$\begin{cases} \frac{d}{dt}z = Az + f(z) + Bv, \qquad t \in (0, T), \\ z(0) = z_0. \end{cases} \tag{3.3}$$

More precisely, we have

Corollary 3.1. *Assume that A generates a C_0-group $\{e^{At}\}_{t\in\mathbb{R}}$ on Y, $B \in \mathcal{L}(U, Y)$, and (A, B) is exactly controllable. If the nonlinearity $f(\cdot)$: $Y \to Y$ satisfies $f(\cdot) \in C^1(Y)$ and, for some $\delta > 0$, $|f(z)|_Y = O(|z|_Y^{1+\delta})$ as $|z|_Y \to 0$, then system (3.3) is locally exactly controllable in Y.*

Clearly, both the time reservability of the underlying system and the variation of constants formula (3.2) play a key role in the above perturbation-type results.

When the system is time-irreversible, the above perturbation technique does not work. The typical example is the controlled heat equation. In this case, one has to search for other robust methods to derive the desired controllability, say, Carleman estimate. We shall consider this case in Section 6.

When the perturbation operator P is unbounded, formula (3.2) may fail to work, and in this case things become much more delicate even for the semigroup theory itself. Nevertheless, there do exist some special cases, for which the perturbation P is unbounded but the above variation of constants formula still works (in the usual sense), say, when the semigroup $\{e^{At}\}_{t\geq 0}$ has some smooth effect. In this case, one can find some perturbation results of exact controllability in the articles by S. Boulite, A. Idrissi and L. Maniar [3], S. Hadd [12], and H. Leiva [15]. However, it does not seem that these perturbation results can be adapted to solve the nonlinear controllability problems, especially for quasilinear equations.

4 A new perturbation result of exact controllability

In this section, we present a new perturbation result of the exact controllability of general evolution equations. The idea is simple, and the key point is that the generation of a C_0-semigroup $\{e^{At}\}_{t\geq 0}$ is robust with respect to a small perturbation of the same "order" with respect to the generator A.

Stimulated by quasilinear problem, we consider the following small perturbation of the same "order":

$$P = P_0 A,$$

where $P_0 \in \mathcal{L}(Y)$ and $||P_0|| < 1$. That is, the perturbed operator reads: $(I + P_0)A$. It is easy to show that if A generated a contractive C_0-semigroup, then so is $(I + P_0)A$. Indeed, it is obvious that $(I + P_0)A$ is dissipative in Y with the new scalar product $((I + P_0)^{-1}\cdot, \cdot)$, which induces a norm, equivalent to the original one. Nevertheless, we remark that the variation of constants formula does not work for $e^{(I+P_0)At}$ for this general case.

Thanks to the above observation, a new perturbation result for exact controllability is shown in [38], which reads as follows:

Theorem 4.1. *Assume that A generates a unitary group $\{e^{At}\}_{t\in\mathbb{R}}$ on Y and $B \in \mathcal{L}(U, Y)$. If (A, B) is exactly controllable, then so is $(A+P, B) \equiv ((I + P_0)A, B)$ provided that $||P_0||_{\mathcal{L}(Y)}$ is small enough.*

Since the variation of constants formula does not work for $e^{(I+P_0)At}$, the above result can not be derived as Theorem 3.1. Instead, we need to use Laplace transform and some elementary tools from complex analysis to prove the desired result.

The above simple yet useful perturbation-type controllability result can be employed to treat the local controllability problems for quasilinear evolution-type PDEs with time-reversibility, as we shall see in the next section.

5 Local exact controllability for multidimensional quasilinear hyperbolic equations

This section is addressed to the local exact controllability of quasilinear hyperbolic equations in any space dimensions.

To begin with, let us recall the related known controllability results of controlled quasilinear hyperbolic equations. The problem is well understood in one space dimension. To the author's best knowledge, the first paper in this direction was written by M. Cirina [5]. Recent rich results are made available by T.T. Li & B.P. Rao [17], T.T. Li & B.Y. Zhang [23], T.T. Li & L.X. Yu [19], Z.Q. Wang [32], and especially the above mentioned book by T.T. Li [16]. As for the corresponding controllability results in multi-space dimensions, we refer to P. F. Yao [35] and Y. Zhou & Z. Lei [41].

Let Ω be a bounded domain in $\mathbb{R}^n$ with a sufficiently smooth boundary Γ. Put $Q = (0,T)\times\Omega$ and $\Sigma = (0,T)\times\Gamma$. Let ω be a nonempty open subset of Ω. We consider the following controlled quasilinear hyperbolic equations:

$$\begin{cases} z_{tt} - \displaystyle\sum_{i,j=1}^{n} \partial_{x_i}(a_{ij}(x)z_{x_j}) \\ \quad = G(t,x,z,\nabla_{t,x}z,\nabla^2_{t,x}z) + \phi_\omega(x)u, & \text{in } Q, \\ z = 0, & \text{in } \Sigma, \\ z(0) = z_0, z_t(0) = z_1, & \text{in } \Omega, \end{cases} \tag{5.1}$$

where the coefficients $a_{ij}(\cdot) \in C^2(\overline{\Omega})$ $(i,j = 1,\cdots,n)$ satisfy $a_{ij} = a_{ji}$, and for some constant $\rho > 0$,

$$\sum_{i,j=1}^{n} a_{ij}(x)\xi_i\xi_j \geq \rho|\xi|^2, \quad \forall\ (x,\xi) = (x,\xi_1,\cdots,\xi_n) \in \overline{\Omega}\times\mathbb{R}^n,$$

and following [41], the nonlinearity $G(\cdot)$ is taken to be of the form

$$\begin{aligned} &G(t,x,\nabla_{t,x}z,\nabla^2_{t,x}z) \\ &= \sum_{i=1}^{n}\sum_{\alpha=0}^{n} g_{i\alpha}(t,x,\nabla_{t,x}z)\partial^2_{x_ix_\alpha}z + O(|u|^2 + |\nabla_{t,x}z|^2), \end{aligned}$$

$g_{i\alpha}(t,x,0,0) = 0$ and $x_0 = t$; ϕ_ω is a nonnegative smooth function defined on $\overline{\Omega}$, satisfying $\min\limits_{x\in\omega}\phi_\omega(x) > 0$.

Denote by χ_ω the characteristic function of ω. We need to introduce the following

Assumption (H): *Assume the linear hyperbolic equation*

$$\begin{cases} y_{tt} - \displaystyle\sum_{i,j=1}^{n} \partial_{x_i}(a_{ij}(x)y_{x_j}) = \chi_\omega(x)u, & \textit{in } Q, \\ y = 0, & \textit{in } \Sigma, \\ y(0) = y_0, y_t(0) = y_1, & \textit{in } \Omega \end{cases} \tag{5.2}$$

is exactly controllable in $H^1_0(\Omega)\times L^2(\Omega)$.

The following controllability result for quasilinear hyperbolic equations is shown in [38]:

Theorem 5.1. *Let Assumption (H) hold. Then, for any* $s > \frac{n}{2} + 1$, *system (5.1) is locally exactly controllable in* $(H^{s+1}(\Omega) \cap H_0^1(\Omega)) \times H^s(\Omega)$, *provided that some compatible conditions are satisfied for the initial and final data.*

Clearly, Theorem 5.1 covers the main results in [35, 41]. The above result follows by combining our new perturbation result for exact controllability, i.e. Theorem 4.1 and the fixed point technique developed in [41].

Remark 5.1. *The boundary control problem can be considered similarly although the technique is a little more complicated.*

Remark 5.2. *The key point of our approach is to reduce the local exact controllability of quasilinear equations to the exact controllability of the linear equation. This method is general and simple. The disadvantage is that we can not construct the control explicitly. Therefore, this approach does not replace the value of [41], and the deep results of the corresponding* $1-d$ *problem, obtained by T. T. Li and his collaborators, as mentioned before. Especially, from the computational point of view, the later approach might be more useful.*

We now return to Assumption (H), and review the known results and unsolved problems for exact controllability of the linear hyperbolic equation but we concentrate on the case of boundary control although similar things can be said for the case of internal control.

Denote by $\mathcal{A}$ the elliptic operator appearing in the first equation of system (5.2). We consider the following controlled linear hyperbolic equation with a boundary controller:

$$\begin{cases} y_{tt} + \mathcal{A}y = 0, & \text{in } Q, \\ y = \chi_{\Sigma_0} u, & \text{in } \Sigma, \\ y(0) = y_0, y_t(0) = y_1, & \text{in } \Omega, \end{cases} \tag{5.3}$$

where $\emptyset \neq \Sigma_0 \subset \Sigma$ is the controller. It is easy to show that system (5.3) is exactly controllable in $L^2(\Omega) \times H^{-1}(\Omega)$ at time T by means of control $u \in L^2(\Sigma_0)$ if and only if there is a constant $C > 0$ such that solutions of its dual system

$$\begin{cases} w_{tt} + \mathcal{A}w = 0, & \text{in } Q \\ w = 0, & \text{in } \Sigma \\ w(0) = w_0, w_t(0) = w_1, & \text{in } \Omega \end{cases} \tag{5.4}$$

satisfy the following observability estimate:

$$|w_0|^2_{H_0^1(\Omega)} + |w_1|^2_{L^2(\Omega)} \leq C \int_{\Sigma_0} \left| \frac{\partial_{\mathcal{A}} w}{\partial \nu} \right|^2 d\Sigma_0, \qquad \forall\, (w_0, w_1) \in H_0^1(\Omega) \times L^2(\Omega). \tag{5.5}$$

When $\mathcal{A} = -\Delta$, $\Sigma_0 = (0,T) \times \Gamma_0$ with Γ_0 being a suitable subset of $\partial\Omega$, L.F. Ho [13] established (5.5) by means of the classical Rellich-type multiplier. Later, K. Liu [22] gave a nice improvement for the case of internal control. When $\mathcal{A}$ is a general elliptic operator of second order, and Σ_0 is a general (maybe non-cylinder) subset of Σ, J.L. Lions [21] posed an open problem on "under which condition, does inequality (5.5) hold?". When $\Sigma_0 = (0,T) \times \Gamma_0$ is a cylinder subset of Σ, Lions's problem is almost solved. In this case, typical results are as follows:

1) Geometric Optics Condition (GOC for short) introduced by C. Bardos, G. Lebeau & J. Rauch [1], which is a sufficient and (almost) necessary condition for inequality (5.5) to hold. GOC is perfect except the three disadvantage: one is that it needs considerably high regularity on both the coefficients and $\partial\Omega$ (N. Burq [4] gives some improvement in this respect); one is that this condition is not easy to verify; one is that the observability constant derived from GOC is not explicit because it involves the contradiction argument to absorb the undesired lower order terms appearing in the observability estimate.
2) Rellich-type multiplier conditions introduced by L.F. Ho [13], K. Liu [22], A. Osses [27] et al., which require less smooth conditions than GOC but they are not necessary conditions for inequality (5.5) to hold.
3) There exist some other sufficient conditions for inequality (5.5) to hold, say, the vector field condition by A. Wyler [33], and the curvature condition by P.F. Yao [34]. Later, it is shown by S.J. Feng & D.X. Feng [9] that these two conditions are equivalent although they are introduced through different tools.
4) Mixed tensor/vector field condition introduced by X. Zhang & E. Zuazua [40], which covers the conditions in 2) and 3).

Remark 5.3. *It is shown by L. Miller [26] that when the data are sufficiently smooth, the conditions in 2) and 3) are special cases of GOC. Nevertheless, as far as I know, it is an unsolved problem on the minimal assumption on data for GOC.*

When $\Sigma_0 \neq (0,T) \times \Gamma_0$, especially when it is NOT a cylinder subset of Σ, there exists almost no nontrivial progress on Lions's problem (which seems to be a challenging mathematical problem), even for the simplest $1-d$ wave equation! The only related results are as follows:

a) For $1-d$ wave equation and $\Sigma_0 = E \times \Gamma_0$ with $E \subset (0,T)$ to be a Lebesgue measurable set with positive measure, P. Martinez & J. Vancostenoble [24] show that (5.5) holds.

b) G. Wang [31] obtains an interesting internal observability estimate for the heat equation in multi-space dimensions, where the observer is $E \times \omega$ with E being the same as in the above case and ω to be any nonempty open subset of Ω.

6 Local null controllability for quasilinear parabolic equations

In this section, we consider the local exact controllability of quasilinear parabolic equations in any space dimensions.

As mentioned before, the perturbation technique does not apply to the time irreversible system, exactly the case of parabolic equations. Therefore, one has to search for other robust methods to derive the desired null controllability, say, Carleman estimate even if the perturbation to the null-controllable system is very small (even in the linear setting!).

We consider the following controlled quasilinear parabolic system

$$\begin{cases} y_t - \sum_{i,j=1}^{n} (a_{ij}(y) y_{x_i})_{x_j} = \chi_\omega u & \text{in } Q, \\ y = 0 & \text{on } \Sigma, \\ y(0) = y_0 & \text{in } \Omega, \end{cases} \tag{6.1}$$

where $a_{ij}(\cdot) : \mathbb{R} \to \mathbb{R}$ are twice the continuously differentiable functions satisfying similar conditions in the last section.

In the last decades, there were many papers devoted to the controllability of linear and semilinear parabolic equations (see [11, 43] and the rich references therein). However, as far as we know, nothing is known about the controllability of quasilinear parabolic equations except for the case of one space dimension. In [2], the author proves the local null controllability of a $1-d$ quasilinear diffusion equation by means of the Sobolev embedding relation $L^\infty(0,T; H_0^1(\Omega)) \subseteq L^\infty(Q)$, which is valid only for one space dimension.

The following local null controllability result for a class of considerably general multidimensional quasilinear parabolic equations, system (6.1), is shown in [23].

Theorem 6.1. *There is a constant $\gamma > 0$ such that for any initial value $y_0 \in C^{2+\frac{1}{2}}(\overline{\Omega})$ satisfying $|y_0|_{C^{2+\frac{1}{2}}(\overline{\Omega})} \le \gamma$ and the first order compatibility condition, one can find a control $u \in C^{\frac{1}{2},\frac{1}{4}}(\overline{Q})$ with $\operatorname{supp} u \subseteq \omega \times [0,T]$ so that the solution y of system (6.1) satisfies $y(T) = 0$ in Ω. Moreover,*

$$|u|_{C^{\frac{1}{2},\frac{1}{4}}(\overline{Q})} \le Ce^{e^{CA}} |y_0|_{L^2(\Omega)},$$

where $A = \sum\limits_{i,j=1}^{n} \left(1 + \sup\limits_{|s|\leq 1} |a_{ij}(s)|^2 + \sup\limits_{|s|\leq 1} |a'_{ij}(s)|^2\right)$, *and* C *depends only on* ρ, n, Ω *and* T.

The key point in the proof of Theorem 6.1 is to improve the regularity of the control function for smooth data, which is a consequence of a new observability inequality for linear parabolic equations with an explicit estimate on the observability constant in terms of the C^1-norm of the coefficients in the principle operator. The latter is based on a new global Carleman estimate for the parabolic operator.

7 Open problems

Although great progress has been made on the controllability theory of PDEs, the field is still full of open problems. In some sense, the linear theory is well understood and there exist extensive works on the controllability of linear PDEs. But, still, even for the linear setting, some fundamental problems remain to be solved, as we shall explain later. The controllability theory of nonlinear system originated in the middle of 1960s but the progress is very slow. Similar to other nonlinear problems, controllability of infinite dimensional nonlinear system is usually very little. Due to the underlying properties of the equation, the progress of the exact controllability theory for nonlinear hyperbolic equations is even slower. Nevertheless, nonlinear problems are not always difficult than linear ones. Indeed, as we have shown in Theorem 5.1, local exact controllability of quasilinear hyperbolic equations is a consequence of the exact controllability of linear hyperbolic equations. One may then ask such a question: "How to judge whether a nonlinear result is good or not?" To the author's opinion, except for some famous unsolved problems, the point is either "whether the result is optimal or not in some nontrivial sense", or "whether some new phenomenon is discovered or not".

From the above "criteria", our result of the local exact controllability of quasilinear hyperbolic equations is not good at all. Indeed, there is no evidence to show that the result is optimal. Therefore,

> *how to establish the "optimal" local exact controllability result of quasilinear equations*

is one of the most challenging problems in the field of control of PDEs. As we shall see below, this problem is also highly nontrivial even in the semilinear setting!

We now review the exact controllability for the following semilinear hyperbolic equations:

$$\begin{cases} z_{tt} + \mathcal{A}z = f(z) + \chi_\omega(x)u(t,x), & \text{in } Q, \\ z = 0, & \text{in } \Sigma, \\ z(0) = z_0, z_t(0) = z_1, & \text{in } \Omega. \end{cases} \tag{7.1}$$

For some very general nonlinearity $f(\cdot)$ and a suitable controller ω, E. Zuazua [42] obtains the local exact controllability for system (7.1). Recently, B. Dehman & G. Lebeau [7] made a significant improvement. However, as far as I know, no optimality on the controllability results is analyzed in these works, which seems also to be a challenging problem.

Remark 7.1. *The possible optimality on the local exact controllability for semilinear equations should be largely related to PDEs with lower regularity dada. This is a very rapid developing field in recent years.*

Remark 7.2. *There exists big difference between the controllability problems and pure PDEs problems. Indeed, the exact controllability problem for the system*

$$\begin{cases} z_{tt} + \mathcal{A}z = f(z_t) + \chi_\omega(x)u(t,x), & \textit{in } Q \\ z = 0, & \textit{in } \Sigma \\ z(0) = z_0, z_t(0) = z_1, & \textit{in } \Omega \end{cases} \tag{7.2}$$

in the natural energy space $H_0^1(\Omega) \times L^2(\Omega)$ is not clear even if $f(\cdot)$ is global Lipchtiz continuous. But, of course, the well-posedness of the corresponding pure PDE problem (i.e. the control $u \equiv 0$) is trivial.

Global exact controllability for semilinear equations is generally a very difficult problem. We refer to [36] for known global controllability results of the semilinear hyperbolic equation when the nonlinearity is global Lipschitz continuous. For system (7.1), if the nonlinearity $f(\cdot)$ grows too fast, say,

$$\lim_{|s|\to\infty} |f(s)||s|^{-1}\log^{-r}|s| = 0, \qquad r > 2, \tag{7.3}$$

the solution may blow up, and therefore, global exactly controllability is impossible in this case. Recently, based on X. Fu, J. Yong & X. Zhang [10] and V.Z. Meshkov [25], T. Duyckaerts, X. Zhang & E. Zuazua [8] showed that, if

$$\lim_{|s|\to\infty} |f(s)||s|^{-1}\log^{-r}|s| = 0, \qquad r < 3/2, \tag{7.4}$$

then system (7.1) is globally exactly controllable. Moreover, it is also shown that the above index "3/2" is optimal in some sense (i.e., whether the linearization argument works or not) when $n \geq 2$. But this number is not optimal in $1-d$.

Remark 7.3. *The same "3/2"-phenomenon happens also for parabolic equations when $n \geq 2$. Surprisingly, the $1-d$ problem is unsolved. That is, it is not clear whether the index "3/2" is optimal or not in $1-d$! This means, sometimes, the $1-d$ problem is more difficult than the multidimensional one.*

Remark 7.4. *Note that for the pure PDE problems, the same phenomenon described above does not happen. This indicates that the study of the controllability problem for nonlinear PDEs has some independent interest, which is far from a sub-PDE-problem.*

Remark 7.5. *Another strongly related longstanding unsolved problem is the exact controllability of the linear time- and space-dependent hyperbolic equation under the GOC. It seems that this needs to combine cleverly the tool from micro-local analysis and the technique of Carleman estimate. But nobody knows how to do it.*

To end this paper, we list the following further open problems.

- **Controllability of the coupled and/or higher order systems by using minimal number of controls**. As shown by X. Zhang & E. Zuazua [39], the study of the related controllability problem is surprisingly complicated and highly nontrivial even for the systems in one space dimension!
- **Constrained controllability**. As shown by K.D. Phung, G. Wang & X. Zhang [28], the problem is unexpected difficult even for the simplest $1-d$ wave equation and heat equation.
- **Controllability of parabolic PDEs with memory, or retard argument and/or other nonlocal terms**. Consider the following controlled heat equations with a memory term:

$$\begin{cases} z_t - \Delta z = \displaystyle\int_0^t a(s,x)z(s)ds + \chi_\omega(x)u, & \text{in } Q, \\ z = 0, & \text{in } \Sigma, \\ z(0) = z_0, & \text{in } \Omega. \end{cases}$$

  The PDE problem itself is not difficult. But, as far as I know, the controllability problem for the above equation is unsolved even if the memory kernel $a(\cdot,\cdot)$ is small!
- **Controllability/observability of stochastic PDEs**. There exist only very few nontrivial results, say, [30, 37] and the reference cited therein. I believe this is a very hopeful direction for the control of PDEs in the near future.
- **Controllability of PDEs in non-reflexive space**. There exists almost no nontrivial result in this direction!

- **Other types of controllability**. Different notions of controllability, say, periodic controllability, may lead to new and interesting problems for PDEs.

References

[1] C. Bardos, G. Lebeau and J. Rauch, *Sharp sufficient conditions for the observation, control and stabilization of waves from the boundary, SIAM J. Control Optim.*, 30 (1992), 1024–1065.

[2] M. Beceanu, *Local exact controllability of the diffusion equation in one dimension, Abstr. Appl. Anal.*, 14 (2003), 793–811.

[3] S. Boulite, A. Idrissi and L. Maniar, *Robustness of controllability under some unbounded perturbations, J. Math. Anal. Appl.*, 304 (2005), 409–421.

[4] N. Burq, *Contrôlabilité exacte de l'équation des ondes dans des ouverts peu réguliers, Asymptot. Anal.*, 14 (1997), 157–191.

[5] M. Cirina, *Boundary controllability of nonlinear hyperbolic systems, SIAM J. Control,* 7 (1969), 198–212.

[6] J.M. Coron, *Control and Nonlinearity*, Mathematical Surveys and Monographs, vol. 136, American Mathematical Society, Providence, RI, 2007.

[7] B. Dehman and G. Lebeau, *Analysis of the HUM control operator and exact controllability for semilinear waves in uniform time, SIAM J. Control Optim.*, 48 (2009), 521–550.

[8] T. Duyckaerts, X. Zhang and E. Zuazua, *On the optimality of the observability inequalities for parabolic and hyperbolic systems with potentials, Ann. Inst. H. Poincaré Anal. Non Linéaire,* 25 (2008), 1–41.

[9] S.J. Feng and D.X. Feng, *A note on geometric conditions for boundary control of wave equations with variable coefficients, J. Math. Anal. Appl.*, 271 (2002), 59–65.

[10] X. Fu, J. Yong and X. Zhang, *Exact controllability for the multidimensional semilinear hyperbolic equations, SIAM J. Control Optim.*, 46 (2007), 1578–1614.

[11] A.V. Fursikov and O.Yu. Imanuvilov, *Controllability of Evolution Equations,* Lecture Notes Series 34, Research Institute of Mathematics, Seoul National University, Seoul, Korea, 1994.

[12] S. Hadd, *Exact controllability of infinite dimensional systems persists under small perturbations, J. Evol. Equ.*, 5 (2005) 545–555.

[13] L.F. Ho, *Observabilité frontière de l'équation des ondes, C. R. Math. Acad. Sci. Paris,* 302 (1986), 443–446.

[14] R.E. Kalman, *On the general theory of control systems, Proc. 1st IFAC Congress,* Moscow, 1960, Butterworth, London, 1961, Vol: 1, 481–492.

[15] H. Leiva, *Unbounded perturbation of the controllability for evolution equations, J. Math. Anal. Appl.*, 280 (2003), 1–8.

[16] T.T. Li, *Controllability and Observability for Quasilinear Hyperbolic Systems,* Springer, 2009.

[17] T.T. Li and B.P. Rao, *Exact boundary controllability for quasi-linear hyperbolic systems, SIAM J. Control Optim.*, 41 (2003), 1748–1755.

[18] T.T. Li and B.Y. Zhang, *Global exact boundary controllability of a class of quasilinear hyperbolic systems, J. Math. Anal. Appl.*, 225 (1998), 289–311.

[19] T.T. Li and L.X. Yu, *Contrôlabilité exacte frontière pour les équations des ondes quasi linéaires unidimensionnelles, C. R. Math. Acad. Sci. Paris,* 337 (2003), 271–276.

[20] X. Li and J. Yong, *Optimal Control Theory for Infinite-Dimensional Systems,* Systems & Control: Foundations & Applications, *Birkhäuser Boston, Inc., Boston, MA,* 1995.

[21] J.L. Lions, *Exact controllability, stabilization and perturbations for distributed systems, SIAM Rev.*, 30 (1988), 1–68.

[22] K. Liu, *Locally distributed control and damping for the conservative systems, SIAM J. Control Optim.*, 35 (1997), 1574–1590.

[23] X. Liu and X. Zhang, *On the local controllability of a class of multidimensional quasilinear parabolic equations,* Preprint.

[24] P. Martinez, J. Vancostenoble, *Exact controllability in "arbitrarily short time" of the semilinear wave equation,* Discrete Contin. Dyn. Syst. B, 9 (2003), 901–924.

[25] V.Z. Meshkov, *On the possible rate of decrease at infinity of the solutions of second-order partial differential equations, (Russian),* Mat. Sb., **182** (1991), 364–383; *translation in* Math. USSR-Sb., **72** (1992), 343–361.

[26] L. Miller, *Escape function conditions for the observation, control and stabilization of the wave equation, SIAM* Journal on Control and Optimization, 41 (2003), 1554–1566.

[27] A. Osses, *A rotated multiplier applied to the controllability of waves, elasticity, and tangential Stokes control, SIAM J. Control Optim.*, 40 (2001), 777–800.

[28] K.D. Phung, G. Wang and X. Zhang, *On the existence of time optimal control of some linear evolution equations,* Discrete Contin. Dyn. Syst. B, 8 (2007), 925–941.

[29] D.L. Russell, *Controllability and stabilizability theory for linear partial differential equations: recent progress and open problems, SIAM* Rev., 20 (1978), 639–739.

[30] S. Tang and X. Zhang, *Null controllability for forward and backward stochastic parabolic equations,* in submission.

[31] G. Wang, *L^∞-null controllability for the heat equation and its consequences for the time optimal control problem, SIAM J. Control and Optim.*, 47 (2008), 1701–1720.

[32] Z.Q. Wang, *Global exact controllability for quasilinear hyperbolic systems of diagonal form with linearly degenerate characteristics, Nonlinear Anal.*, 69 (2008), 510–522.

[33] A. Wyler, *Stability of wave equations with dissipative boundary conditions in a bounded domain,* Differential Integral Equations, 7 (1994), 345–366.

[34] P.F. Yao, *On the observability inequalities for exact controllability of wave equations with variable coefficients, SIAM J. Control Optim.*, 37 (1999), 1568–1599.

[35] P.F. Yao, *Boundary controllability for the quasilinear wave equation,* Arxiv preprint math.AP/0603280, 2006.

[36] X. Zhang, *Exact Controllability of Semi-linear Distributed Parameter Systems,* Gaodeng Jiaoyu Chubanshe (Higher Education Press), Beijing, 2004 (in Chinese).

[37] X. Zhang, *Carleman and observability estimates for stochastic wave equations, SIAM J. Math. Anal.*, 40 (2008), 851–868.

[38] X. Zhang, *Controllability and stabilization of some quasilinear partial differential equations, an unbounded perturbation approach,* In preparation.

[39] X. Zhang and E. Zuazua, *Polynomial decay and control of a* $1-d$ *hyperbolic-parabolic coupled system, J. Differential Equations,* 204 (2004), 380–438.

[40] X. Zhang and E. Zuazua, Book in press.

[41] Y. Zhou and Z. Lei, *Local exact boundary controllability for nonlinear wave equations, SIAM J. Control Optim.*, 46 (2007), 1022–1051.

[42] E. Zuazua, *Exact controllability for the semilinear wave equation, J. Math. Pures Appl.*, 69 (1990), 1–31.

[43] E. Zuazua, *Controllability and observability of partial differential equations: some results and open problems,* in *Handbook of Differential Equations: Evolutionary Equations, vol. 3,* Elsevier Science, 2006, 527–621.